The Enzymes of Biological Membranes

SECOND EDITION

Volume 1

Membrane Structure and Dynamics

THE ENZYMES OF BIOLOGICAL MEMBRANES
Second Edition

Volume 1: Membrane Structure and Dynamics
Volume 2: Biosynthesis and Metabolism
Volume 3: Membrane Transport
Volume 4: Bioenergetics of Electron and Proton Transport

The Enzymes of Biological Membranes

SECOND EDITION

Volume 1

Membrane Structure and Dynamics

Edited by

Anthony N. Martonosi

State University of New York
Syracuse, New York

Plenum Press • New York and London

Library of Congress Cataloging in Publication Data

Main entry under title:

The Enzymes of biological membranes.

Bibliography: v. 1, p.
Includes index.
Contents: v. 1. Membrane structure and dynamics.
1. Membranes (Biology)—Collected works. 2. Enzymes—Collected works. I. Martonosi, Anthony, 1928–
QH601.E58 1984 574.87′5 84-8423

ISBN 978-1-4684-4600-5 ISBN 978-1-4684-4598-5 (eBook)
DOI 10.1007/978-1-4684-4598-5

A Division of Plenum Publishing Corporation
233 Spring Street, New York, N.Y. 10013
Softcover reprint of the hardcover 1st edition 1985

Contributors

Dennis Chapman, Department of Biochemistry and Chemistry, Royal Free Hospital School of Medicine, University of London, London NW3 2PF, England

Carl M. Cohen, Department of Biomedical Research, St. Elizabeth's Hospital, Boston, Massachusetts; and Departments of Medicine and of Molecular Biology and Microbiology, Tufts University School of Medicine, Boston, Massachusetts

P. R. Cullis, Department of Biochemistry, University of British Columbia, Vancouver, British Columbia, V6T 1W5, Canada

Philippe F. Devaux, Institut de Biologie Physico-Chimique, 75005 Paris, France

Peter B. Garland, Department of Biochemistry, University of Dundee, Dundee DD1 4HN, Scotland, United Kingdom

M. J. Hope, Department of Biochemistry, University of British Columbia, Vancouver, British Columbia, V6T 1W5, Canada

Frances Jay, Institute for Cell Biology, Federal Institute of Technology, ETH-Hönggerberg, CH-8093 Zürich, Switzerland

Pauline Johnson, Department of Biochemistry, University of Dundee, Dundee DD1 4HN, Scotland, United Kingdom

B. de Kruijff, Department of Molecular Biology, State University of Utrecht, 3584 CH Utrecht, The Netherlands

Jeff Leaver, Department of Biochemistry and Chemistry, Royal Free Hospital School of Medicine, University of London, London NW3 2PF, England

Jack A. Lucy, Department of Biochemistry and Chemistry, Royal Free Hospital School of Medicine, University of London, London NW3 2PF, England

Kurt Mühlethaler, Institute for Cell Biology, Federal Institute of Technology, ETH-Hönggerberg, CH-8093 Zürich, Switzerland

Camillo Peracchia, Department of Physiology, School of Medicine, University of Rochester, Rochester, New York

Hans Sigrist, Institute of Biochemistry, University of Berne, CH-3012 Berne, Switzerland

Deborah K. Smith, Department of Biomedical Research, St. Elizabeth's Hospital, Boston, Massachusetts. *Present address:* Department of Physiology, Duke University Medical Center, Durham, North Carolina

T. F. Taraschi, Hahnemann Medical College, Department of Pathology, Philadelphia, Pennsylvania

David D. Thomas, Department of Biochemistry, University of Minnesota, Medical School, Minneapolis, Minnesota

Dan W. Urry, Laboratory of Molecular Biophysics, University of Alabama in Birmingham, School of Medicine, Birmingham, Alabama

C. J. A. Van Echteld, Department of Biochemistry, State University of Utrecht, 3584 CH Utrecht, The Netherlands

A. J. Verkleij, Department of Molecular Biology, State University of Utrecht, 3584 CH Utrecht, The Netherlands

Alan S. Waggoner, Center for Fluorescence Research in Biomedical Sciences and Department of Biological Sciences, Carnegie-Mellon University, Pittsburgh, Pennsylvania

Evangelina E. Yguerabide, Department of Biology, University of California, San Diego, La Jolla, California

Juan Yguerabide, Department of Biology, University of California, San Diego, La Jolla, California

Peter Zahler, Institute of Biochemistry, University of Berne, CH-3012 Berne, Switzerland

Preface to the Second Edition

In the first edition of *The Enzymes of Biological Membranes,* published in four volumes in 1976, we collected the mass of widely scattered information on membrane-linked enzymes and metabolic processes up to about 1975. This was a period of transition from the romantic phase of membrane biochemistry, preoccupied with conceptual developments and the general properties of membranes, to an era of mounting interest in the specific properties of membrane-linked enzymes analyzed from the viewpoints of modern enzymology. The level of sophistication in various areas of membrane research varied widely; the structures of cytochrome *c* and cytochrome b_5 were known to atomic detail, while the majority of membrane-linked enzymes had not even been isolated.

In the intervening eight years our knowledge of membrane-linked enzymes expanded beyond the wildest expectations. The purpose of the second edition of *The Enzymes of Biological Membranes* is to record these developments. The first volume describes the physical and chemical techniques used in the analysis of the structure and dynamics of biological membranes. In the second volume the enzymes and metabolic systems that participate in the biosynthesis of cell and membrane components are discussed. The third and fourth volumes review recent developments in active transport, oxidative phosphorylation and photosynthesis.

The topics of each volume represent a coherent group in an effort to satisfy specialized interests, but this subdivision is to some extent arbitrary. Several subjects of the first edition were omitted either because they were extensively reviewed recently or because there was not sufficient new information to warrant review at this time. New chapters cover areas where major advances have taken place in recent years. As a result, the second edition is a fundamentally new treatise that faithfully and critically reflects the major transformation and progress of membrane biochemistry in the last eight years. For a deeper insight into membrane function, the coverage includes not only well-defined enzymes, but several membrane proteins with noncatalytic functions.

We hope that *The Enzymes of Biological Membranes* will catalyze the search for general principles that may lead to better understanding of the structure and function of membrane proteins. We ask for your comments and criticisms that may help us to achieve this aim.

My warmest thanks to all who contributed to this work.

Anthony N. Martonosi

Syracuse, New York

Contents of Volume 1

Contents of Volume 2

Contents of Volume 3

Contents of Volume 4

1

Electron Microscopy of Biological Membranes

Kurt Mühlethaler and Frances Jay

I. INTRODUCTION

The study of the molecular organization of cellular membranes has been considerably hampered by their extreme thinness. For about 100 years it was postulated that the outermost zone of the cell is covered by a special organized layer called the cell membrane (Nägeli and Cramer, 1855; Overton, 1899; Gorter and Grendel, 1925; Danielli and Davson, 1935). With the introduction of the electron microscope in 1933, it was hoped that the presence and structure of this delicate cell component could be visualized. In the first electron micrographs of cultured cells, a definite boundary at the periphery of the cytoplasm could be seen, but new information regarding its structure was not obtained (Porter *et al.*, 1945). A new substructure was observed by Robertson in 1958 using fixation with the potassium permanganate method as introduced by Luft (1956). The cell membrane appeared as a triple-layered structure, consisting of two dense strata, each about 2.5 nm thick, bordering a light central zone of about equal thickness. These dimensions were in good agreement with the model put forward by Danielli and Davson (1935). Since various authors observed similar tripartite membranes in a number of cells, the unit membrane concept became generally accepted. It was concluded that all biological membranes consisted of the same kind of fundamental structure pattern, i.e., a lipid bilayer arranged with the polar heads of the molecules pointing outwards and covered by a protein layer.

The view that all the membrane types in organisms ranging from bacteria to man should be similar in structure and to a great extent also in composition was not convincing to the biologist. Considering the variety of different physiological processes occurring in these structures, such a simple model seemed rather unrealistic. Again,

Kurt Mühlethaler and Frances Jay • Institute for Cell Biology, Federal Institute of Technology, ETH-Hönggerberg, CH-8093 Zürich, Switzerland.

the old question which had been raised at the beginning of specimen preparation for light microscopy in the last century came into discussion, namely, to what extent living structures are changed due to the use of fixative. It seemed reasonable to assume that the triple-layered structure seen after permanganate fixation could also be an artifact. The same year that Robertson postulated his "unit membrane concept," work was going on in our laboratory to develop alternative methods for specimen preparation in the frozen state (Müller, 1957). Subsequently, a special apparatus was constructed to splinter frozen objects following the concept proposed by Steere (1957). We soon recognized that this technique, which we called freeze-etching (Moor *et al.*, 1961), was well suited for membrane studies, since fracturing occurred preferentially along cellular membrane systems, thus permitting extended views into the membrane.

II. METHODS USED FOR STUDYING BIOLOGICAL MEMBRANES IN THE ELECTRON MICROSCOPE

A. Sectioning

With the introduction of plastic embedding, thermal specimen advancement, and glass knives by Newman *et al.* (1949) and Latta and Hartmann (1950), sectioning developed into a routine procedure. Today we still use the same basic technique as 30 years ago. Instead of using methacrylate as embedding medium, new polymers have been introduced such as Epon 812, Araldite, Durcupan, and Lowicryl resulting in an improvement in cell morphology.

Membranes can be visualized in tissue or cell sections by introducing heavy metals (electron-dense) into the membrane. This can be carried out by chemical fixation with osmium tetroxide (OsO_4) or by staining with uranyl acetate or lanthanum nitrate. Their use for structural studies on the molecular level is, however, limited, as the chemical reactions taking place during fixation with osmium tetroxide or potassium permanganate ($KMnO_4$) indicated that these reagents led to membrane destabilization.

Using erythrocyte ghosts as a model system, Luftig *et al.* (1977) showed that an increase in OsO_4 concentration resulted in an increased release of protein from the membrane. After exposure to 1% OsO_4 for 2 hr they noted that 5 mM EDTA solubilized over 80% of the membrane proteins. Based on optical rotatory dispersion (ORD) and circular dichroism (CD) measurements, the α-helical content of membrane proteins is least altered by glutaraldehyde (Lenard and Singer, 1968). Glutaraldehyde contains significant amounts of α,β-unsaturated aldehydes, oligomers, and polymers which, according to Richards and Knowles (1968), are thought to form stable Michael-type adducts with amino groups which result in the crosslinkage of proteins. Among the lipids, phosphatidyl ethanolamine and phosphatidyl serine are also fixed by glutaraldehyde (Gigg and Payne, 1969).

In 1956, Luft introduced $KMnO_4$ as an alternative fixative to OsO_4. This fixative became very popular since, by this means, the triple-layered structure as postulated by Robertson (1958) was easily achieved. However, this fixative, like OsO_4, causes extensive extraction of background cellular substances. Permanganates penetrate into the tissue faster than do the more commonly used fixatives, but in most cells major alterations, such as swelling of mitochondria and plastids, loss of ribosomes, lipid droplets, and microtubules, and other artifacts result. As shown by nuclear resonance studies, Mn^{2+} binds preferentially to the phosphate group in nucleotides (Kotowycz and Suzuki, 1973). Potassium permanganate causes drastic conformational changes in the primary structure of protein chains (Hake, 1965). Since $KMnO_4$ causes these severe artifacts, this fixative does not preserve membranes in their native form and its use should be restricted to simple studies regarding general membrane topology arrangements in cells.

B. Negative Staining

Membranes consist primarily of molecules containing carbon, hydrogen, oxygen, and nitrogen with low electron-scattering power which cannot be studied by simply mounting whole membranes on a grid. Due to this lack of contrast, the various components, such as proteins and lipids, cannot be sufficiently differentiated in the electron micrograph. In order to increase specimen contrast, Brenner and Horne (1959) introduced negative staining for virus particles. Due to its simplicity, this technique has become an important method in membrane research. The only restriction in its application is that membranes must be isolated prior to staining. Inorganic reagents such as phosphotungstic acid ($H_3PO_4 \cdot 12\ WO_3 \cdot 24H_2O$), uranyl acetate [$UO_2(CH_3COO)_2 \cdot 2H_2O$], and various other compounds are used as staining reagents. Using negative staining, Fernández-Morán (1962) observed stalked globules of 90 Å diameter on the inner mitochondrial membrane. Since they were not observed in sectioned material, a long-lasting controversy followed as to whether or not these particles were real or artifactual. Each stalk of particles was thought to contain one complete set of the electron transport chain (Lehninger *et al.*, 1958). Later, Racker and his collaborators were able to identify the so-called "submitochondrial particles" as ATPase complexes, termed "coupling factor F_1" (Racker *et al.*, 1965). Subsequently, similar globules were found in other membrane systems such as photosynthetic membranes (thylakoids) by Mühlethaler (1971a,b), Garber and Steponkus (1974), and Oleszko and Moudrianakis (1974). In recent years, negative staining has been widely applied to image reconstruction studies. The stain provides both high contrast and resistance to the electron beam. As the specimens become less radiation-sensitive, a tilt series can be carried out on a single specimen. The disadvantage of using negatively stained membranes is that the resolution is limited to about 2 nm. This means that only the overall molecular outlines can be observed.

C. Freeze-Etching and Freeze-Fracturing

As mentioned earlier, the sectioning work led to the generally accepted view that all membranes in the cell were uniform and triple-layered, whereas negative staining disclosed a particulate structure. In order to develop an alternative procedure for specimen preparation, we tried to freeze the specimens and to infiltrate the frozen objects with monomeric methacrylate. Polymerization was achieved with ultraviolet light in a cooled air stream. This procedure has again become frequently used in combination with low-temperature embedding media (Armbruster *et al.*, 1982). As shown by Müller (1957), our early results with this low-temperature embedding procedure were variable as most of the specimens were destroyed by ice crystals. Another problem was the interaction of the monomeric methacrylate with the specimen, where the former acted as a lipid solvent. As a result, the membrane system became blurred and even after osmic vapor staining did not show much contrast. The whole procedure was time consuming, due to the slow dehydration of the frozen specimen. A frozen piece of leaf tissue of about 200 mg, dried for 80 hr at $-60°C$, still contained 10% water. For comparison, the same water content was achieved after sublimation at $-20°C$ for only 8 hr. This experiment indicated clearly that lowering the temperature increased the drying time enormously. On the other hand, the tissue must be kept at a very low temperature in order to avoid local melting or recrystallization. In order to avoid the removal of lipidic material by monomeric methacrylate during embedding, the technique first introduced by Steere (1957) was modified and a new device constructed (Moor *et al.*, 1961). Replication of a frozen specimen has the advantage that the object is preserved in a hydrated chemically unaltered form and therefore potentially viable state. It was soon recognized that this method, which we called freeze-etching, was well suited for membrane studies because fracturing of the frozen cell occurred preferentially along the membrane system, thus permitting extended surface views of these structures. In contradiction to the then generally accepted "unit membrane image" in chemically fixed specimens, the freeze-fracture replica disclosed a considerable variation in the structural appearance of cytological membrane systems. In photosynthetic membranes we observed that the thylakoids were covered with a large number of particles in the size range of 3–12 nm in diameter (Figure 1). It became evident that some of the particles were only loosely attached to the membrane whereas others seemed to be embedded in the homogeneous basic layer (Mühlethaler *et al.*, 1965). The frequently observed cross sections through frozen membranes did not show the tripartite structure typical of separated proteins and lipid. This early concept of a distinct particle distribution in an amorphous basic layer was taken as proof of the mixed arrangement of the two main components of the membrane. As shown in Figure 2, the model put forward at this time has now been generally accepted. When this concept was proposed, the supporters of the trilamellar concept objected that during freezing a phase-separation effect would occur thus leading to structural changes. In order to determine whether cryofixation changes the location of membrane components, we have to relate the lateral diffusion time for membrane proteins with the actual freezing time. Lateral diffusion coefficients have been determined for a number of different membrane proteins (Cherry, 1979; Edidin, 1982). The values found for proteins range from 3×10^{-10} cm^2 sec^{-1} for peptide hormones to less than 3×10^{-12} cm^2 sec^{-1} for acetylcholine receptors in cultured myotubes. These very low diffusion coefficients would not interfere with the quenching process, which only lasts milliseconds. These results indicate that free lateral diffusion of proteins is restricted, probably due to the interaction of integral proteins with each other, influence of the surrounding membrane lipids, or by interactions of membrane proteins with cytoplasmic proteins or with those of the cytoskeleton. Lateral diffusion of membrane proteins

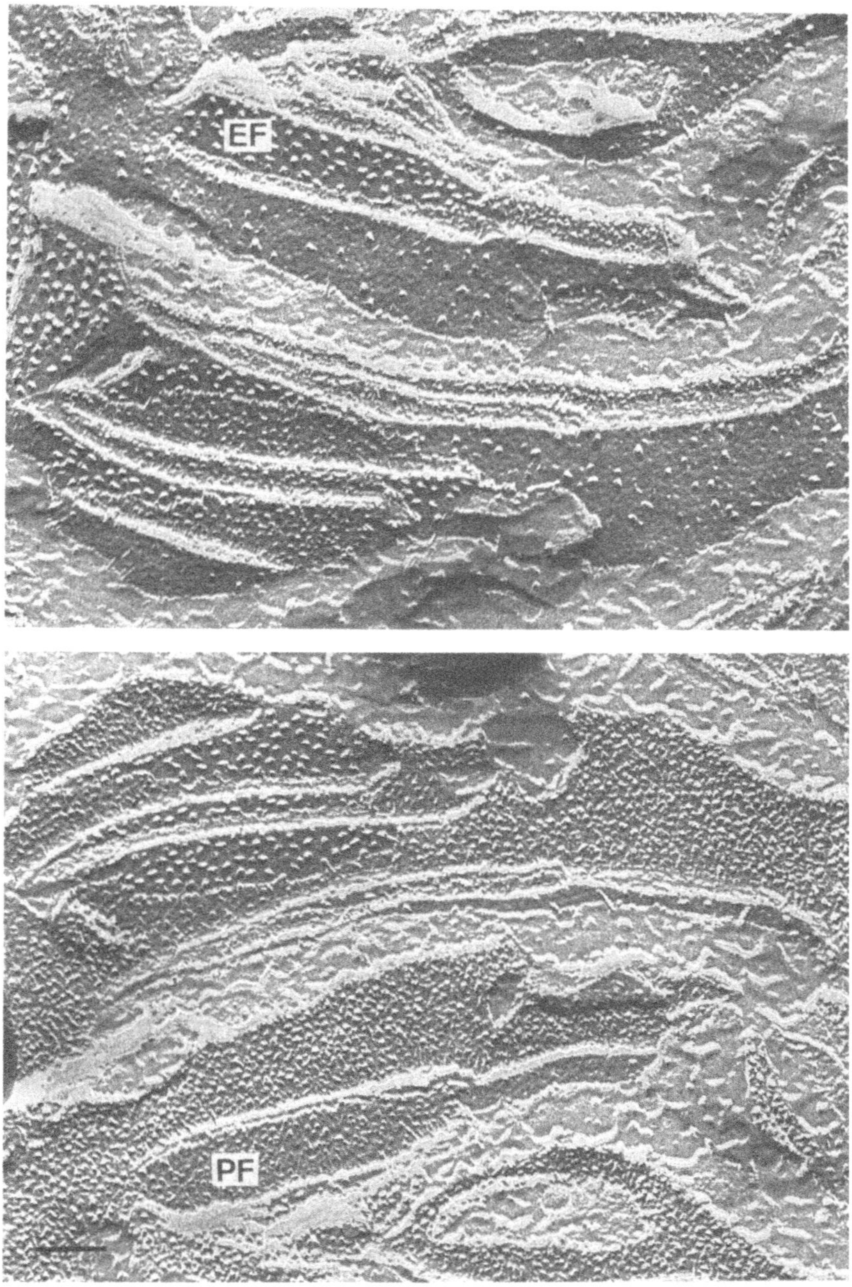

Figure 1. Two complementary replicas of a grana region, showing corresponding freeze-fractured thylakoids of *B. kortensis* chloroplasts. The cytoplasmic and exoplasmic fracture faces (PF and EF) are covered with particles containing polypeptides of the photosynthetic apparatus. Scale bar: 0.1 μm. Photograph from E. Wehrli.

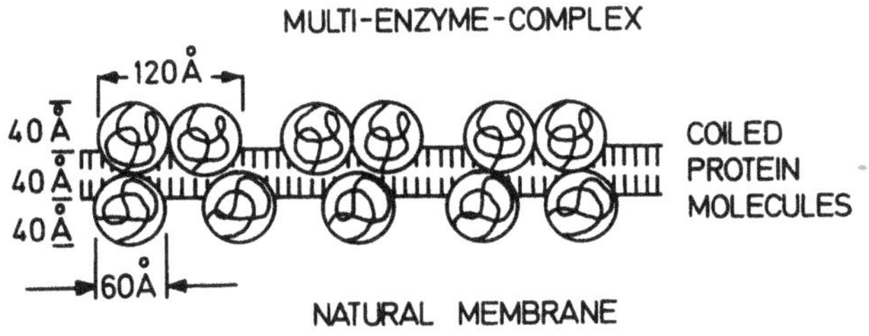

Figure 2. Early scheme of the arrangement of photosynthetic membrane particles based on freeze-fracture experiments. From Mühlethaler *et al.* (1965).

has been studied by Höchli and Hackenbrock (1976) on the mitochondrial inner membrane using the freeze-fracture technique. In freshly frozen membranes, the integral particles are randomly distributed. This arrangement changes drastically if the membranes are equilibrated at temperatures below the phase transition temperature of the lipids. Instead of the random distribution, a separation becomes visible between the smooth, particle-free regions and particle dense areas. Upon warming to physiological temperatures, this lateral phase separation is reversed and the former random distribution recovers. Of importance is the fact that the thermotrophic, reversible, lipid–protein separation does not destroy the functional state of the membrane such as electron transfer or oxidative phosphorylation (Höchli and Hackenbrock, 1976).

This mobility of membrane proteins can also be studied by labeling the membranes with immunoglobulins. If the coupling factor (CF_1) in photosynthetic membranes is labeled with its antibody, aggregates of these 14-nm complexes are formed by lateral movement of the particles at the membrane surface (Berzborn *et al.*, 1974). Since the mean distance of the particles in the control was larger than the maximum length which can be bridged by an antibody, we may assume that the antibody molecule first reacts by only one of its binding sites with a CF_1 complex. This particle then moves laterally within the thylakoid until it meets another complex where the antibody then reacts with its second binding site. Thus, the observed aggregates can be explained by a repetition of the proposed process. Apparently, the reaction with the antibodies does not by itself induce the capacity of lateral movement but it must be a property of the 14-nm particles *in vivo*. This observation in thylakoid membranes is in agreement with the work of Frye and Edidin (1970), who observed, after labeling with fluorescent antibody, lateral mobility of membrane antigens at the surface. In order to determine the lateral diffusion coefficient of intramembrane particles in the mitochondrial inner membrane, Sowers and Hackenbrock (1981) combined electrophoresis and freeze-fracture electron microscopy. An electric current was passed through a suspension of purified inner membrane vesicles, which caused an electrophoretic migration of intramembrane particles in the membrane plane into a single, crowded patch facing the positive electrode. The membrane vesicles were quenched at specific times after removing the electrophoretic force and while the particles were diffusing back into a random distribution. With this method, Sowers and Hackenbrock (1981) calculated the diffusion coefficient of the integral proteins in the mitochondrial inner membrane to be 8.3×10^{-10} cm^2 sec^{-1} at a temperature of 20°C. From this value, a root-mean-square displacement of 57 nm in 10 msec can be calculated. This diffusion coefficient for the intramembrane particles is consistent with reported values for a variety of proteins of various cellular membranes (Cherry, 1979). As shown by Schneider *et al.* (1980), lateral diffusion and collision of interacting membrane components such as proteins may be necessary for eliciting specific catalytic events in membranes. For this study, Schneider *et al.* (1980) developed an *in vitro* method for increasing the lipid bilayer surface area of the mitochondrial energy-transducing membrane by fusion of liposomes with the native membrane. In this way, the mitochondrial inner membranes became enriched with phospholipid. During the increase of the surface area, the lateral distribution of the integral proteins was measured by freeze-fracturing and related to the electron transfer rates in various segments of the electron transfer sequence. As a result, it was found that electron transfer is diffusion-limited between the dehydrogenases and cytochromes bc_1. Schneider *et al.* (1980) found, in addition, that the dehydrogenases and cytochromes bc_1 complex, as well as ubiquinone, are independent diffusible membrane oxidation–reduction components. Phase separation phenomena have also been observed in plant cells. As shown by Pike and Berry (1980), the phase separation temperatures of total leaf phospholipids from warm- and cool-climate plants are different and related to a species' thermal habitat. Winter active species in the Mojave Desert showed lower phase separation temperatures than the summer active species. A group of warm-climate annual grasses showed separation temperatures distinctly higher than those of a cool environment. Winter active ephemerals seem to be genetically programmed to synthesize a mixture of phospholipids which will not phase separate in the usual growth conditions. In desert plants, the lipids showed pronounced seasonal shifts in the phase separation temperature. According to the studies of Pike and Berry (1980), plants normally growing in a certain environment synthesize a mixture of phospholipids which will not phase separate in the usual thermal regime. The compositions of lipids for ephemeral plants appear to be largely genetically determined, whereas for perennial plants they are mainly adaptations to a changing environment.

In recent years, cell fusion of animal or plant cells has become an important area in research because it allowed the relationship between the nucleus and the cytoplasm to be determined. Also here, phase phenomenon may be of importance. Since most of these results just discussed are based on freeze-fractured or freeze-etched specimens, we must be sure that the specimens are not altered during preparation. An evaluation of freezing methods used in freeze-fracture electron microscopy using low-temperature X-ray diffraction was carried out by Gulik-Krzywicki and Costello (1978). X-ray studies of several lipid–water and protein–lipid–water phases have been conducted and the results correlated with freeze-fracture data in order to investigate some of the structural perturbations which can occur during preparation of freeze-fracture samples. It was found that the most important perturbations of these multilamellar model systems were the formation of ice crystals between bilayers and the rearrangement of lipid molecules in the planes of the bilayers. Structural perturbations are considerably reduced using thin preparations and rapid cooling rates. Fast quenching can be achieved by high-pressure freezing (Riehle, 1968; Höchli, 1974), spray-freezing (Bachmann and Schmitt, 1971), and more recently with the propane-jet freezer (Müller *et al.*, 1980), whereby tissues can be frozen without the addition of cryoprotectants. A summary of theoretical and experimental data from several authors on the cooling process of different objects is given in a paper by Winkelmann (1981). If fast cooling rates are used, the original structural arrangement can be preserved. Some precautions should also be taken during the fracturing and replication process. The molecular events taking place during splintering are still not fully understood. Freezing is thought to weaken hydrophobic bonds (Deamer and Branton, 1967). Based on model experiments, it is now generally accepted that cleavage occurs along nonpolar interaction areas (Branton, 1966). Utilizing this splintering mechanism, techniques have been developed for the production of replicas from complementary fracture faces of a membrane (Figure 1; Wehrli *et al.*, 1970; Steere and Moseley, 1970; Sleytr, 1970). A careful study of these fractured membrane halves indicated that the condensation of material onto the fracture face of the freshly cleaved objects can be one of the main sources of artifacts. Between the fracturing and replication process, gases may condense onto the exposed fracture phase which may mask, or enhance, structural details by specific decoration (Gross *et al.*, 1978b). Such a contamination on the freshly cleaved membrane face can be prevented if the specimen is enclosed in a liquid nitrogen- or helium-cooled cold trap or cold shroud (Steere, 1973; Sleytr and Umrath, 1976). The alternative method is the use of a higher vacuum whereby all condensable gases are trapped (Gross *et al.*, 1978a; Escaig and Nicolas, 1976). An improved vacuum also enables fracturing at lower temperatures whereby the probability of plastic deformation is reduced. If fracturing is carried out at $-196°C$, an ultrahigh vacuum (UHV) of the order of 10^{-9} torr is needed. This high experimental expenditure is rewarded by an improved structural presentation of the fracture face (Gross *et al.*, 1978a). This can be demonstrated with the well-known yeast cell plasmalemma structure as first portrayed by Moor and Mühlethaler in 1963. The paracrystalline plasmalemma structure (Figures 3a and b) prepared by UHV freeze-fracturing at $-196°C$ shows the improved topographic resolution very clearly. The improvement is explained by less structural distortion during the fracturing at lower temperatures. The optical diffraction patterns of these paracrystalline regions demonstrate the improvement of the structural preservation by showing well-defined third- and fourth-order spots of the reciprocal lattice in the optical transforms of both halves of the membrane. Conventional preparation produces diffuse diffraction spots limited to the second order. The existence of finer detailed micrographs is substantiated by image processing, i.e., removal of the aperiodic noise components by Fourier filtration and reconstruction of the filtered data (insets in Figures 3a and b). Under these controlled conditions in the vacuum chamber, contamination as a decoration phenomenon can also be studied. Contamination is mainly caused by water vapor which is the major residual gas in conventional high-vacuum systems. As shown by Staehelin and Bertaud (1971), condensation of residual water vapor may create small ice cubes on the fracture faces resembling intramembrane particles. If fracturing is carried out in a conventional apparatus, care should be taken that the object temperature doesn't fall below $-100°C$ at a vacuum of 10^{-6} torr, otherwise ice particles condense on the specimen. If statistical analysis of particle densities is being carried out, this effect may interfere with interpretation of the real membrane particles.

As shown by Gross *et al.* (1978b), condensation phenomena may be helpful in

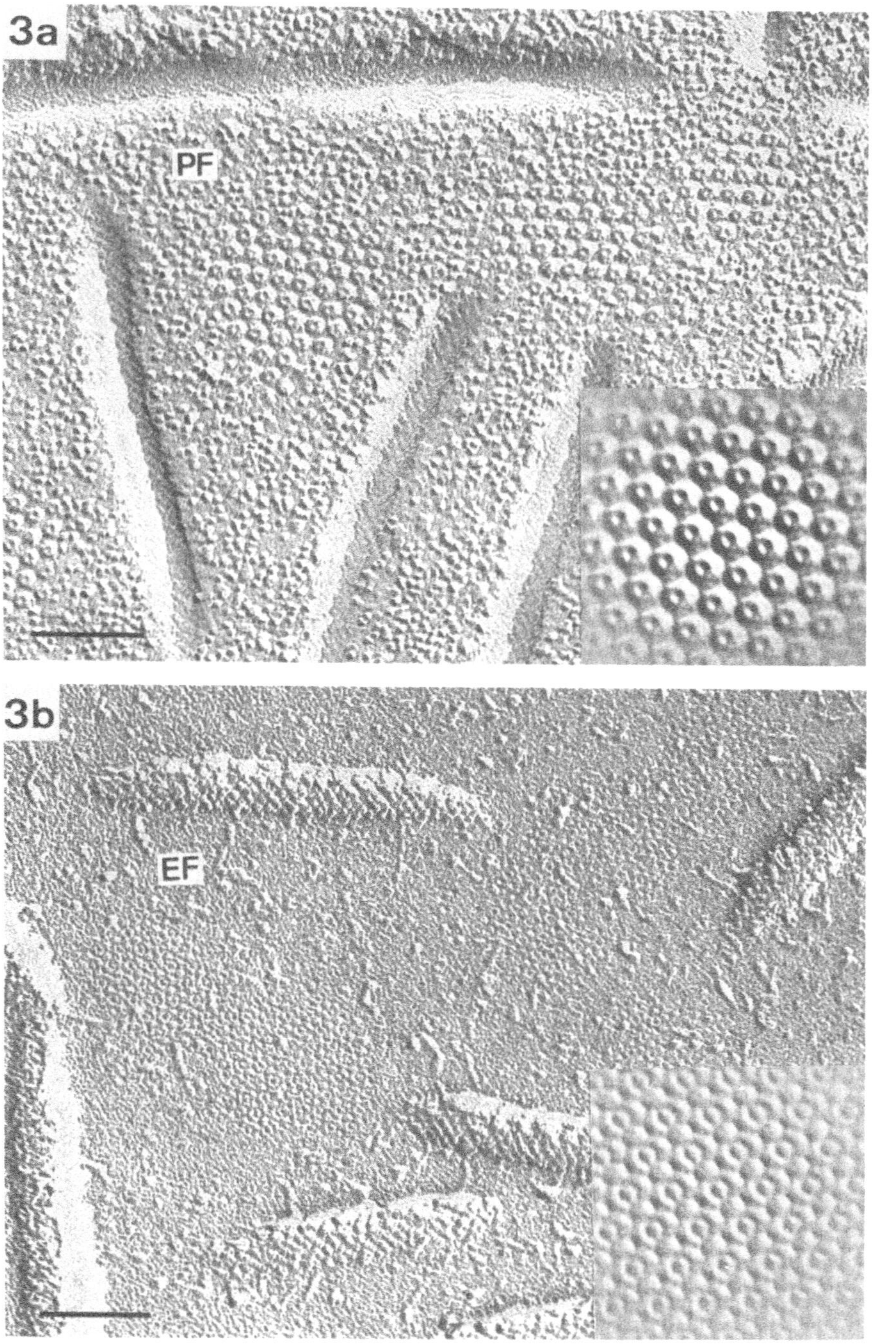

Figure 3. Yeast plasmalemma freeze-fractured at $-196°C$ and 1×10^{-9} torr. (a) The crater-like particles on the plasma fracture face (PF) are more easily visible than complementary depressions on the (b) extraplasmic fracture face (EF). The inset shows hexagonally ordered arrays after Fourier filtration. The complementarity of particles on PF (a) and depressions on EF (b) is clearly revealed. On the EF, an additional hexagonal substructure becomes visible which is not discernible before image filtration. Scale bar: 0.1 μm. From Gross *et al.* (1978a,b) and Kübler *et al.* (1978).

the determination of certain physicochemical properties of the membrane components. Ice crystals formed from condensed water vapor preferentially deposit on hydrophilic structural elements (Walzthöny *et al.*, 1981). This specific decoration is an elegant method to label membrane components possessing certain chemical properties. Such investigations should improve our knowledge of the chemical composition of the intramembranous particles. When the first results on freeze-fracture electron microscopy of thylakoid membranes were reported (Mühlethaler *et al.*, 1965), we postulated that the membrane particles were of proteinaceous nature. This hypothesis was based on the fact that membrane particles are less evident in membranes with low protein concentrations, such as myelin. Additional evidence came from a study by Pinto da Silva *et al.* (1971), who showed that the binding sites of ferritin-labeled antibodies against erythrocyte ghost surface receptors are correlated with that of the intramembranous particles exposed by freeze-cleavage. It is now widely agreed that the particles seen on fracture faces are the structural representation of integral membrane proteins. The most direct evidence for the protein nature of intramembrane particles comes from reconstitution experiments with liposomes. The completely smooth surface faces of fractured phospholipid vesicles become covered with particles when amphipathic proteins are integrated into the liposome (Segrest *et al.*, 1974). The structural and functional identification of the intramembranous particles is not possible from freeze-fractured specimens, since the particles are not distinctive enough to enable a specific composition or physiological activity to be ascribed to them. According to their sizes, they may represent multimeric aggregates of a single type of protein or associations of various types of polypeptides. Whether all integral membrane particles are composed of proteins was recently questioned by Verkleij and Ververgaert (1978). They investigated the protein composition in the outer (exoplasmic) membrane of *E. coli* and tried to relate it to the particle distribution. They found proteins which they called b, c, and d, which were considered as possible candidates for the composition of the intramembranous particles. In order to test this hypothesis, mutants have been developed that lack these proteins. It was found that the absence of either b, c, d, or b + c did not change the particle density. In mutants lacking protein d, the particles occupy about 75% of the outer membrane face. If all three proteins are absent, the particulated surface is reduced to about 25%. These data are consistent with the view that the particles are of proteinaceous nature. Additional results obtained with lipopolysaccharide mutants of *E. coli* could, however, contradict this idea. Heptose-deficient lipopolysaccharide mutants also exhibit a drastic reduction in particle density (Verkleij *et al.*, 1976). Additional evidence that some of the particles seen after freeze-fracturing are composed of lipopolysaccharides came from an experiment in which the cellular lipopolysaccharides of wild-type *E. coli* were extracted with ethylenediamine tetraacetate (EDTA). After this treatment, which released half of the lipopolysaccharides but not the protein, the number of particles was reduced by about 50%. This led to the hypothesis that lipopolysaccharides are involved in the outer membrane patches. In this study, the complementarity between the two fracture faces was also investigated. It was found that some particles formed imprints or pits on the complementary membrane face and others did not. The complementary pits were assigned to the lipopolysaccharide complexes, which means that the fracture plane will run along the hydrophobic edge of the acyl chains. In conclusion, Verkleij and Ververgaert (1978) postulated from their study on *E. coli* membranes that particles which lack complementary pits are determined predominantly by proteins spanning the membrane and that particles which show complementary impressions are determined by lipid micelles.

In the case of the human erythrocyte membrane, a number of reports have suggested that the intramembranous particles contain two major penetrating proteins, band 3 and the major sialoglycoprotein (glycophorin A; Bretscher, 1973; Steck, 1974). As shown by Bächi *et al.* (1977), erythrocytes of a rare homozygous condition involving a complete lack of the major sialoglycoprotein still carry the same number of particles as normal erythrocytes. As shown by Steck (1974), an erythrocyte ghost contains 9.4×10^5 copies of band 3 molecules (mol. wt. approximately 100,000) which are probably in a dimeric form in the membrane, and 5×10^5 copies of the sialoglycoprotein (mol. wt. 31,000). As shown by Bächi *et al.* (1977), erythrocyte membranes contain 3000 particles per μm^2, or a total of 5×10^5 intramembranous particles, present on the plasma fracture face. This would be compatible with the assumption that each particle is composed of two band 3 molecules and one sialoglycoprotein (total mol. wt. 230,000). In this case, the sialoglycoprotein would contribute only 13% to the weight of the particle and its absence could not be detected in the electron microscope as a change in particle diameter. If the sialoglycoprotein forms its own population of intramembranous particles, distinct from those containing band 3 protein, the absence of the sialoglycoprotein would result in a 13% reduction in the particle density of the fracture face. The study by Bächi *et al.* (1977) indicates that membrane particles may be aggregates of several polypeptides and that the lack of one constituent must not, as a consequence, be detectable in a reduced number of particles. Therefore, the heterogeneity in size and diameter of intramembrane particles may reflect the association of a variable number of polypeptides. Identification of the nature of these particles is complicated because the freeze-fracture microscopy provides no information on the biochemical composition of the molecular organization of plasma membranes. A first attempt to overcome this obstacle is, as discussed above, the decoration technique as introduced by Gross *et al.* (1978b). Other routes leading to a better identification of morphological details are the "split membrane" technique and the antibody-labeling technique.

D. Split Membrane Technique

Based on the concept that frozen membranes are cleaved into two halves (Branton, 1966, 1971), new methods have been developed for a chemical analysis of each half (Fisher, 1975, 1976, 1978, 1982). In order to obtain large areas of fractured membranes, the biomembranes to be cleaved must be attached to a smooth, flat surface. One convenient method is the attachment by electrostatic interaction. Most biomembranes possess a net negative charge at physiological pH and will readily adsorb polycations. Glass slides are convenient as a smooth surface suitable for attachment. The surface, however, possesses a net negative charge. Therefore, the glass must be treated with polycations to positively charge the surface. Mazia *et al.* (1975) and Fisher (1975) used polylysine for the adhesion of cells. Care must be taken to clean the glass meticulously with detergents and acids in order to avoid a defective polylysine coat. Substitutes for polylysine are alcian blue, ruthenium red, and cetylpyridinium (Sommer, 1977). Alcian blue seems to be the cation of choice over all the others. The adsorbed cells even remain on the slide after washing with distilled water but, if required, can be removed from the glass surface with buffered saline. As shown by Büechi and Bächi (1979), a good attachment of cells to a substratum can be achieved with a 2% solution of 3-aminopropyltriethoxysilane in acetone followed by glutaraldehyde (1%) treatment to achieve covalent binding. With this method, Büechi and Bächi (1979) were able to study the plasmatic surface of human erythrocytes. The cells were attached to derivatized cover slip surfaces and then opened by squirting the monolayer with PBS, leaving behind fragments of membranes with their plasmatic surfaces exposed. Using a similar procedure, the cytoplasmic surface of the yeast plasma membrane was visualized. The membrane vesicles were adsorbed to cationized glass squirted with a hypotonic buffer stream, freeze-dried, and then platinum–carbon shadowed (Maurer and Mühlethaler, 1981). Subsequently, the adsorbed membranes were removed and solubilized in order to study the protein composition. For localization studies, the adsorbed membranes can also be labeled with antibodies. This enables the protein distribution at the cytoplasmic inner surface of the plasmalemma to be studied. The use of polylysine, alcian blue, or aminopropylsilane for membrane adsorption has allowed careful quantitative studies of split membranes (Nermut, 1982) to be carried out.

It has been shown that both lipid and polypeptide components of split membranes can be analyzed. The lipids can be extracted and the polypeptides characterized by sodium dodecyl sulfate (SDS)–polyacrylamide gel electrophoresis (SDS-PAGE). In addition, the transmembrane distribution of radioisotopic lipids and proteins can be analyzed (Fisher, 1982). In combination with the freeze-fracture technique, the "cell monolayer technique" allows us to study both the external and the internal surfaces of the plasma membrane, as well as the interior hydrophobic fracture faces. If the cells are adsorbed on carbon films, then lysed and squirted, they can be negatively stained, labeled with antibodies, or used for biochemical studies.

Chemical analysis of fractured membranes have been carried out on photosynthetic membranes (Park and Pfeifhofer, 1974), red blood cells (Fisher, 1976, 1982; Edwards *et al.*, 1979), and on yeast plasma membranes (Maurer and Mühlethaler, 1981). In these studies, the asymmetrical distribution of lipids and proteins across the plane of the membranes was established. In his analysis on the distribution of cholesterol in human erythrocyte membranes, Fisher (1976) found that cholesterol was more prevalent on the exterior side than on the interior portion of the membrane. A similar study of the transmembrane distribution of radioiodinated concanavalin A (Fisher, 1982) indicated that this protein is located in the exterior half of the same membrane. Although sialoglycoproteins are associated with the outer half of the membrane, the anion transport protein (band 3) remained predominantly with the inner half of the membrane (Edwards *et al.*, 1979). In this study, it was also noted that these transmembrane proteins were cleaved during fractionation. In the yeast plasma membranes, two glycoproteins of molecular weight 160,000 and 240,000 were found, both of which exhibited invertase activity (Maurer and Mühlethaler, 1981, 1982). The cytoplasmic surface showed paracrystalline arrays of particles which resembled in size and lattice constant the intramembranous particles generally seen after freeze-fracturing. SDS-PAGE of the adsorbed membranes showed that the glycoproteins and two hydrophobic polypeptides predominated in these arrays. In conclusion, it may be said that this so-called monolayer technique is a very useful procedure for membrane studies.

E. Immunoelectron Microscopy

From the very beginning of electron microscopy, efforts were made to relate the structural details seen in electron micrographs with their functional activities and biochemical composition. Various attempts to develop heavy-metal stains for specific reactions with biological molecules were of only limited success. The great selectivity of binding of antibody molecules and their high specificity for biological molecules offered a potential reagent for the identification and visualization of cellular components. The handicap which had to be overcome, before they could be used for electron microscopy, was their lack of contrast. In order to be detected against the background of the surrounding material, the antibody molecules had to be labeled with a marker of sufficient contrast. This was achieved by Singer (1959), who coupled ferritin as a marker to the antibody molecule. This protein has an iron-rich core which can be easily visualized in the electron microscope. Subsequently, other markers such as peroxidase, which is capable of giving an electron-dense reaction product (Nakane and Kawaoi, 1974), hemocyamin (Karnovsky *et al.*, 1972), or small viruses (Hämmerling *et al.*, 1969) were used. Experiments with colloidal iron–dextran complexes (Marshall and Rutherford, 1971) and colloidal gold (Faulk and Taylor, 1971) were also very successful. These immunolabeling techniques can be used with all the major electron microscopic techniques, such as transmission electron microscopy (TEM), replica and freeze-etching techniques, as well as in scanning electron microscopy (SEM). Thin-section labeling can be handicapped by the fact that due to fixation, dehydration, and embedding, the macromolecules become denatured. Since binding can only occur if the matching antigenic determinant is accessible to the antibody, changes in conformation or shrinking phenomena may give negative results.

Antiserum can be produced by injecting either an isolated component, or complex particles, or even entire membranes into animals. Even after injection of hydrophobic proteins, isolated with SDS, antibodies can be found which cross react with its native polypeptide in the membrane.

The general procedure for antibody labeling has been described in a number of papers (de Petris, 1978; Singer, 1978; Roth, 1982, 1983). For a long time the immunoferritin labeling technique was preferentially used. It can be used for the investigation of cell surface distribution of membrane polypeptides or transmembrane glycoproteins. As an example of membrane labeling, our procedure for the localization of *Rhodopseudomonas viridis* reaction center and light-harvesting proteins using ferritin-labeled antibodies will be briefly described (Jay *et al.*, 1983). In these photosynthetic membranes, four polypeptides with apparent molecular weights of 38,000, 33,000, 27,000, and 24,000 (reaction centre) and three low molecular weight polypeptides, 11,000, 8000, and 6000 (light-harvesting proteins), were identified. As a first step, antibodies were produced against these polypeptides which had been eluted from SDS-gels, and tested for specificity by an immunoblotting assay. With the available antibodies two labeling techniques are in use, the direct and the indirect method. The first is a one-step procedure in which the specimens are directly labeled with an antibody-marker conjugate. It is the simplest method as it involves only a single

Table I. Indirect Antibody Labeling of Membranes for Electron Microscopy[a]

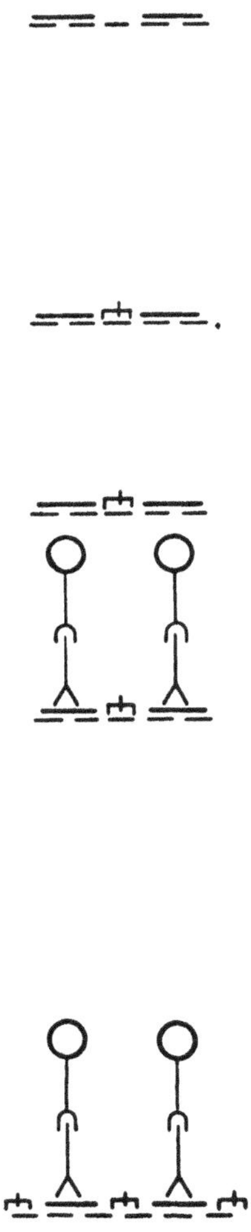

Copper grid coated with Scotch, carbon film and glow-discharged

↓

Grids inverted onto membrane suspension (0.5 mg/ml in PBS)
Incubated 15 min at 4°C

↓

Grids washed by inverting onto PBS (ca. 10 ml)

↓

Incubated 15 min on PBS containing 0.1 M glycine (BPS/glycine)

↓

Incubated 30 min, normal goat serum in PBS/glycine (2 ug/ml)

↓

Short wash in PBS/glycine

↓

Antiserum incubation using specific rabbit serum in PBS/glycine (0.005 mg/ml) — IgG

↓

Washed on PBS (ca. 1 ml each grid in an individual dish)

↓

Incubated in goat anti-rabbit-ferritin in PBS/glycine (60 min 1/100 diluted conjugate). Each grid in individual dishes. — Ferritin-αIgG

↓

Washed on PBS (ca. 1 ml)

↓

Glutaraldehyde-fixed (freshly diluted to 1% in PBS), 15 min

↓

Water-washed (left overnight at 4°C if necessary)

↓

Frozen in liquid nitrogen

↓

Freeze-dried and shadowed

[a] Procedure for indirect labeling of photosynthetic membrane proteins with goat anti-rabbit ferritin conjugate. Abbreviations: PBS, phosphate-buffered saline; αIgG, anti-immunoglobulin G; Scotch, extracted glue from Scotch Tape in chloroform.

incubation and washing. This reduces the often unwanted nonspecific staining and allows the use of whole immunoglobulin fractions instead of the purified antibodies which are required in most of the indirect techniques. Using ferritin as a marker, the one-step procedure also has the highest resolution for the localization of the membrane antigens. However, steric hindrance problems may arise due to the close proximity of the large electron-dense marker to the antigen binding site. The technique can also be used to label different antigens in the same membrane by using conjugates with different markers.

In the indirect technique, the specimens are first incubated with an unmodified antibody and subsequently exposed to an antibody–marker conjugate directed against the attached ligand. As a first label, often rabbit antibodies are used which are then visualized with, for example, goat anti-rabbit–marker conjugate. In this technique, the use of unmodified antibodies may avoid some of the problems encountered in the preparation and use of conjugates, such as the low yield of active conjugated molecules (important when, for example, precious monoclonal antisera are used) or the formation of aggregates. The indirect or two-step technique is not only convenient (one marker conjugate can be used for a multitude of different antisera) but is often more sensitive than direct labeling due to the introduction of an additional, magnifying step. In Table 1 our procedure for indirect labeling of photosynthetic membranes of *Rhodopseudomonas viridis* is shown (Jay *et al.*, 1983). The presence of glycine in the washing

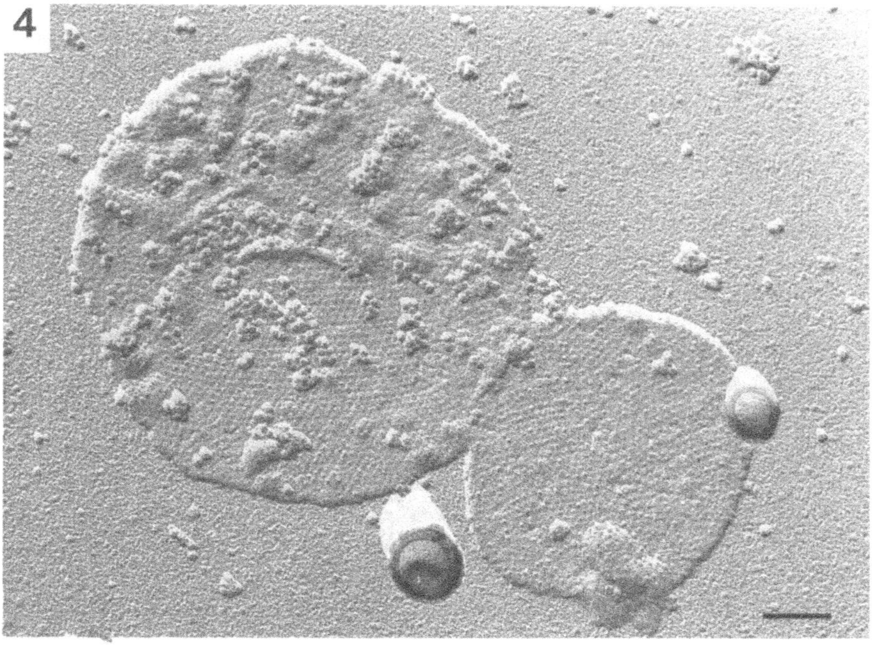

Figure 4. Photosynthetic membrane of *Rps. viridis* labeled with anti-reaction center serum and visualized with a ferritin conjugate. Scale bar: 0.1 μm. From Jay *et al.* (1983).

steps and incubation in normal goat serum was found to be important in order to avoid nonspecific adhesion of the ferritin conjugate. An example of ferritin-labeling is shown in Figure 4. In this sample, a protein (27,000) from the reaction center was labeled according to the procedure given in the scheme. The ferritin marker tends to form clusters and does not exhibit the uniform distribution one would expect on such a regular lattice structure. As the orientation of the proteins and not their lateral distribution is being studied, this does not interfere with the interpretation of the results. If the concentration of the label is raised, the underlying membrane is fully saturated with ferritin conjugate and the binding sites disguised. Since the lattice of the underlying hexagonal unit has a size of 13 nm, a precise localization using ferritin marker (12.5 nm diameter; Bielig *et al.*, 1966) and IgG (ca. 25 × 3.5 nm; Hall *et al.*, 1959) is not possible.

F. Colloidal Gold Marker

In 1971, Faulk and Taylor introduced colloidal gold as a high-contrast marker. In their experiments, they used a rabbit anti-*Salmonella* antiserum to coat colloidal gold particles. They succeeded in a specific labeling of the surface of bacteria by incubating them with the immunoglobulin–gold complex. Improvements in this technique were made by Horisberger *et al.* (1975, 1978), Horisberger and Rosset (1977), Geoghegan and Ackermann (1977), Wagner and Wagner (1976, 1977), Roth and Wagner (1977a,b), Roth *et al.* (1978) and Roth (1982, 1983). The colloidal gold markers have a wide application and can be used as markers for scanning electron microscopy (Horisberger, *et al.*, 1975), as well as for freeze-etch electron microscopy and light microscopy. At present, a variety of ligands such as immunoglobulins, toxins, protein A, avidin, and lectins are in use as labels together with colloidal gold as a marker. According to Roth (1983), several methods of preparing colloidal gold have been worked out enabling particle sizes between 2 and 20 nm. If different size particles are used, double-labeling studies can also be carried out. The protein adsorption to gold particles is due to electrostatic interactions between the negatively charged surface of gold and positively-charged groups of the polypeptides. In general, the coat macromolecules form a stable complex with gold particles if the pH value is close to the isoelectric point of a given protein. At these pH values, the zwitterion form of the polypeptides is dominant and the interfacial tension is maximal (Roth, 1983). Colloidal gold, like ferritin, can be applied for both direct and indirect labeling. Protein A from *Staphylococcus aureus* bound to colloidal gold has proved to be particularly useful (Roth *et al.*, 1978). With this complex, antigen-bound antibodies can be localized via their Fc portion in a manner similar to that described earlier with goat anti-rabbit-ferritin conjugates. As an example of its application, the same photosynthetic membrane of *Rhodopseudomonas viridis* thylakoid has been used. In Figure 5, the regular lattice of reaction center and light-harvesting complexes is visible, but in this preparation a small antibody–gold conjugate (≈ 50 Å) has been used as marker for one of the light-harvesting polypeptides (apparent mol. wt. of 11,000). Due to its high atomic number, the gold particle exhibits an extremely high contrast in the transmission microscope. Even very small particles of 2–3 nm in diameter can easily be seen on

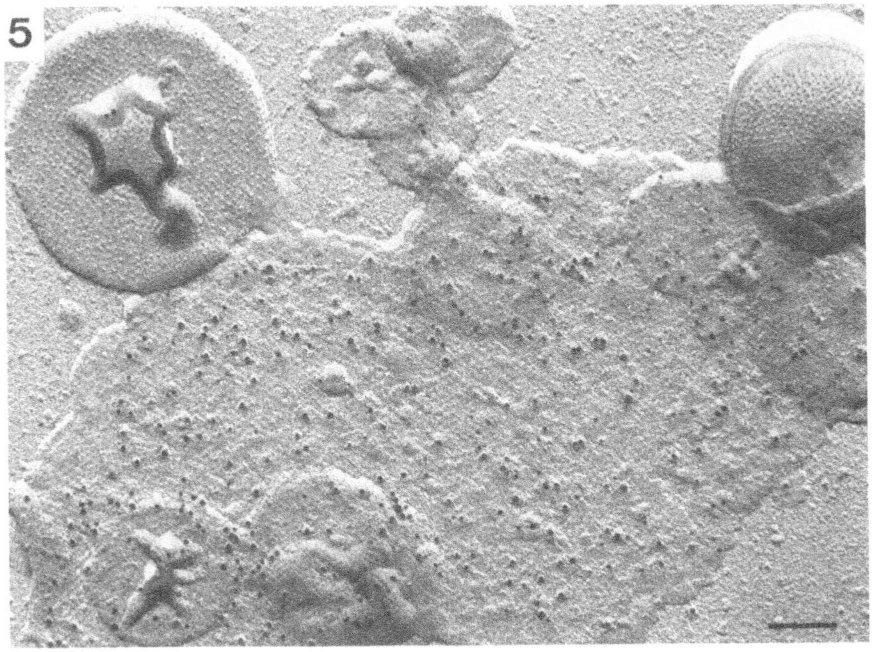

Figure 5. Photosynthetic membrane of *Rps. viridis* labeled with antibody–gold conjugate against light-harvesting polypeptides. Scale bar: 0.1 μm.

the structural background of the membrane. Compared with the normal antibody–ferritin conjugates, colloidal gold complexes with Fab fragments are considerably smaller, and therefore allow better resolution and reduced steric hindrance problems. Colloidal gold as a marker is now generally used because it is easy to make in a variety of sizes, coating with biological macromolecules is simple, and less technical problems arise with the conjugates. After staining, the specimen can be studied either in the light microscope, where the colloidal gold appears either purple or blue, or in the electron microscope, where the particles are visible with high contrast as shown in Figure 5. The application of colloidal gold conjugates has revolutionized labeling techniques for microscopy.

C. Cryoelectron Microscopy

Since the early days of electron microscopy, biologists were concerned with the problem of dehydration of living matter. If the aqueous environment is removed during specimen preparation, structural deformations must be expected. In order to prevent such artifacts, various attempts were made to preserve the natural environment even during beam exposure. All the efforts to construct an "environment cell" (Dupouy *et al.*, 1960) in a high-voltage microscope were not successful. High resolution is prevented by radiation damage, Brownian motion of the particles in the cells and the large cross section thus amplifying inelastic scattering. Recently, another combination of methods was attempted in order to reduce beam damage and dehydration artifacts. This was the application of the cryomicroscope with superconducting lenses as developed and applied by Dietrich *et al.* (1977, 1979) in combination with frozen hydrated specimens (Dubochet *et al.*, 1981, 1982). This cryomicroscope, operating at liquid helium temperature, offers some advantages over the more conventional types. The electronmagnetic lenses have an improved time stability and operate in a ultrahigh cryogenic vacuum at a temperature of 4.2°K. As a consequence, a low rate of etching and low specimen contamination result and in addition radiation damage seems to be reduced (Dietrich *et al.*, 1979). This reduction at low temperature is important because beam damage is the primary limiting factor for high-resolution studies. Radiation damage is mainly caused by the ionization, which results from inelastic scattering of electrons in the specimen. These scattered electrons have enough energy to split chemical bonds and these molecular fragments, in the form of free radicals, may diffuse away or form new complex arrangements with the macromolecules present in the specimen (Dubochet *et al.*, 1981). As shown by Knapek and Dubochet (1980) biological specimens may lose up to 20–80% of their mass with the electron dose required for high-resolution images. This mass loss does not take place at 4°K because the split fragments are not lost by sublimation. If frozen hydrated specimens are examined in the cryomicroscope, the object must always be kept below a temperature of − 135°C; otherwise the amorphous water may go through phase transition and form ice crystals. The state of the hydrated sample can be checked by electron diffraction which indicates unambiguously if the specimen is in the vitrified or crystalline state. Vitrified samples are also a prerequisite for sectioning in cryomicrotomes but this state can only be achieved to a limited extent. According to Talmon (1982) and Walter *et al.* (1982), vitrification of suspended biological objects in water can be achieved by spreading a thin film of a suspension onto a strongly hydrophilic grid and immediate quench-freezing by propulsion into liquid propane or ethane cooled by liquid nitrogen. Vitrification cannot be achieved if the water layer is thicker than 1 μm because crystallization in this case is inevitable. An additional difficulty lies in the transfer of the specimens from the quenching solution into the microscope. For this crucial step, a new cryotransfer system has been developed by Hax and Lichtenegger (1982) for the conventional microscope. Another complication with respect to standard microscopy is the very low contrast of frozen hydrated and unstained specimens. As shown by Lepault and Dubochet (1982), vitrified catalase crystals diffract with a resolution better than 0.1–0.2 nm.

As outlined above, this new technique is not yet developed for routine work, but already in this state of performance it offers some advantages over the standard techniques. Probably the crucial thing in the application of this technique is the low contrast. This is due to the small mass density difference between aqueous and organic domains of the specimen. Unstained material leads to very unsatisfactory and blurred images. Even with dark-field illumination it is difficult to see the details of the specimens. As shown by Carlemalm and Kellenberger (1982), the application of Z-contrast could be a method for better imaging. Unfortunately, this mode of imaging requires a STEM. Here, in contrary to the conventional transmission electron microscope, the elastically and inelastically scattered electrons can be analyzed independently. Imaging by the ratio of elastically to inelastically scattered electrons depends primarily on the atomic composition of the material. With this mode of imaging, Carlemalm and Kellenberger (1982) were able to observe integral membrane proteins in unstained thin sections of septate junctions in the testis of *Drosophila*. If good frozen hydrated cryosections become available, the combination of this method with Z-contrast imaging offers an alternative to conventional microscopy. For membrane research, frozen hydrated specimens studied in a cryomicroscope may become another interesting technique. For both methods, the contrast problem may be overcome by resins containing metallo-organic constituents (Armbruster *et al.*, 1982; Carlemalm *et al.*, 1982) or by adding heavy-atom solutes before freezing. For membrane studies, both techniques offer new insights into the structure of this complex system.

H. Image Processing and Three-Dimensional Structure Determination

The most rapidly growing field in electron microscopy within the past few years is concerned with image processing and three-dimensional structure determination. As the resolution obtained with the microscope improved, the effects of phase contrast, radiation damage, or preparation artifacts became pronounced. Many details present in structural arrangements are obscured by "noise." In order to visualize the native structure, methods have been developed to enhance this image relative to the noise. In a regular lattice, for instance, the structure of the subunits repeats at each lattice point, whereas the noise varies randomly. A simple procedure for enhancing the periodic image details has been introduced by Markham *et al.* (1963). In this technique, a photographic superposition is made in which the image is moved by an amount corresponding to some presumed symmetry between each exposure in the set. In order to determine the structural details, a number of different superposition images are compared. The construction showing the strongest enhancement of periodic detail is then selected. This technique is very time-consuming because each periodicity examined must be analyzed by this trial and error process. As shown by Klug and Berger (1964), optical diffraction is a better and more reliable technique for determining translational periodicities and symmetry in electron micrographs. The basic principle of this method is the selection of the periodic components from the aperiodic ones. The optical diffraction pattern displays at once all of the periodic fractures of the image, and the position of each diffraction spot gives an accurate measure of the periodicity. The noise, on the other hand, produces a background spread over the whole diffraction plane. The technique of optical filtering as developed by Klug and de Rosier (1966) is based on the principle of "filtering" the diffraction pattern of the image, which is produced by directing a laser beam through the photographic plate. The selected periodic components in the diffraction pattern can be filtered out with an opaque mask containing small holes in the positions of the diffraction spots. From these selected components, the image can be synthesized without the undesired random noise components. A detailed description of this optical filtering procedure has been given by Erickson *et al.* (1978). Since the preparation of masks for filtering is a very difficult and time-consuming process, methods have been developed to simulate optical filtering using computers (Klug, 1979). As an example, a micrograph of the yeast plasmalemma membrane freeze-fractured in UHV at liquid nitrogen temperatures is shown (Figure 3a and b). The processed images of the plasmic and exoplasmic fracture faces show the structure of the particles more pronounced than in the original micrograph (Kübler and Baumeister, 1978; Kübler *et al.*, 1978). The crater-like particles on the plasmatic fracture face have caused ring-like depressions on the complementary face. If the original micrograph is compared with the digitized and reconverted image, the improvement is obvious for the exoplasmic face. The depression of the rim, which is not visible in the original, is clearly revealed and, in addition, the hexagonal substructure seen on the plasmatic face is portrayed. As shown by Maurer and Mühlethaler (1981), these regular patches of membrane particles can be isolated and characterized. It was found that the particles contain glycoproteins which show invertase activity and two hydrophobic proteins. As shown in this example, noise-filtered images are very helpful for the determination of the substructure of membranes. With the improvement of image analysis, the optical diffraction method is generally used for a rapid assessment of the quality and resolution of an electron micrograph. Instead of the optical or computer filtering, a crystallographic analysis is now generally applied. As first shown by de Rosier and Klug (1968), the main advantage of the numerical transforms is the preservation of the phases. The procedure generally used is schematically shown in Figure 6. As in image filtering, the electron micrograph is checked in an optical diffractometer in order to find the best-ordered area for further processing. This area is then scanned on a microdensitometer which transforms the raster points into density values. The sampling frequency and the raster spot size depend on the expected resolution. The optical densities of the raster points are subsequently transformed into a set of Fourier coefficients. Amplitudes and phases of the Fourier components are printed out separately and their values at reciprocal lattice points are read off the numerical transforms. After lattice and phase refinement, a contour map of the structure can be obtained by inverse Fourier transformation. Based on this procedure, the three-dimensional structure of an object can also be reconstructed. The different two-dimensional projections are obtained by tilting the specimen relative to the electron beam. The Fourier transforms of these projections can then be combined to form a three-dimensional transform and an inverse Fourier transformation of these data gives the map of the structure in three dimensions. The first object investigated this way was the purple membrane in *Halobacterium halobium* by Unwin and Henderson (1975). Since then, a number of other objects such as microorganisms, cell constituents, and membrane systems of various plant and animal origin were screened for regular patterns. Regular arrays were found in the so-called HPI-layer, a wall component of *Micrococcus (Deinococcus) radiodurans* with a hexagonally ordered protein layer (Baumeister and Kübler, 1978), ribosomal crystals in the Italian lizard (*Lacerta sicula*) formed during hibernation (Kühlbrandt and Unwin, 1980), cytochrome *c*-reductase (Leonard *et al.*, 1980), cytochrome oxidase (Henderson *et al.*, 1977; Frey *et al.*, 1978) and photosynthetic membranes of *Rhodopseudomonas viridis* (Wehrli and Kübler, 1980; Miller, 1982; Welte and Kreutz, 1982), and of pea chloroplast light-harvesting complex (Kühlbrandt *et al.*, 1982, 1983). The thylakoids of *Rhodopseudomonas viridis* consist of flattened vesicles which are stacked like the grana regions in chloroplasts.

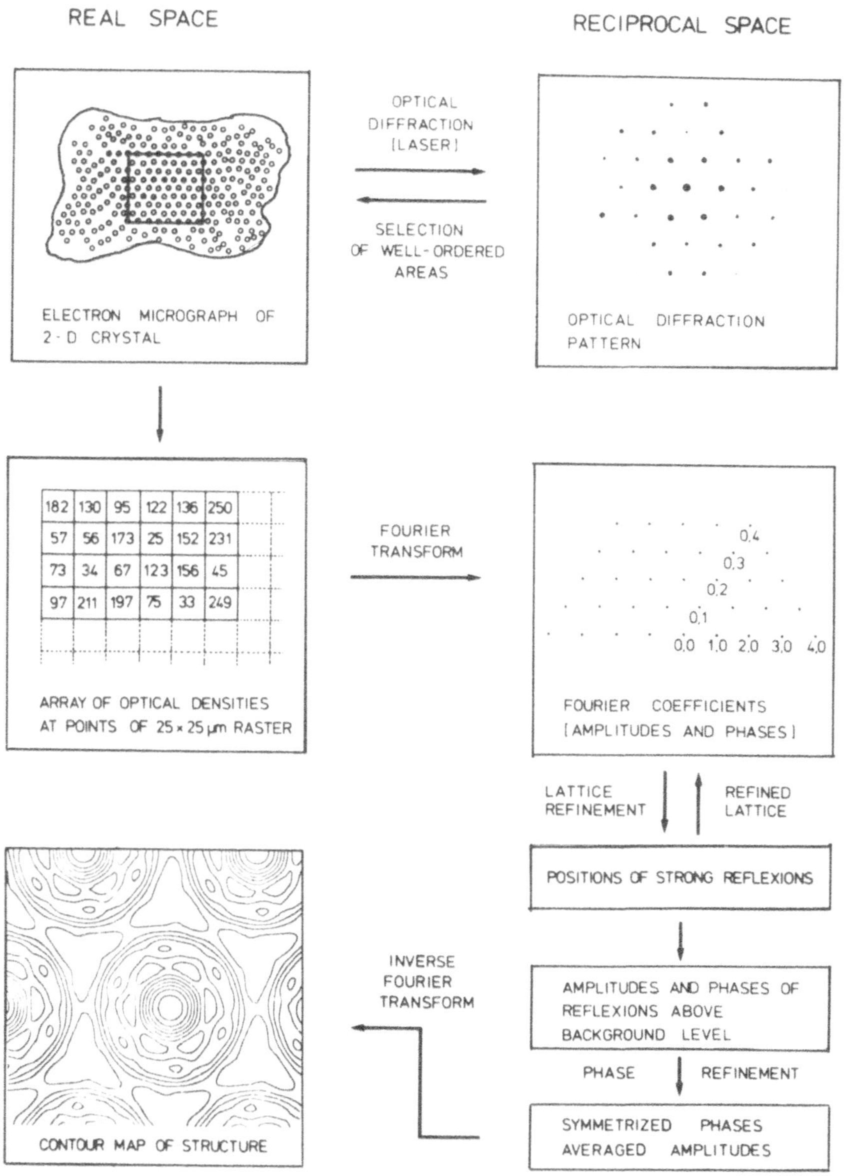

Figure 6. Scheme of successive steps in image analysis by crystallographic methods. By courtesy of W. Kühlbrandt and T. Thaler.

If the membrane polypeptides are characterized by SDS gels, four polypeptides with apparent molecular weights of 38,000, 33,000, 27,000, and 24,000, which according to Pucheu *et al.* (1976) are forming the reaction center complex and three low-molecular-weight proteins with molecular weights of 11,000, 8000, and 6000, are seen (Jay *et al.*, 1983). There are a number of indications that the latter correspond to the light-harvesting complex. The crystallinity of this membrane *in vivo* allows image processing techniques to be applied. Regular crystalline sheets can be formed *in vitro* when only the light-harvesting polypeptides (mol. wt. 27,000 and 25,000) of pea chloroplasts are crystallized (Kühlbrandt *et al.*, 1983; Figure 7). Upon precipitation with mono- and divalent cations in the presence of low concentrations of the detergent Triton X-100, the light-harvesting polypeptides form stacked arrays similar to the thylakoids in the grana region (Li and Hollingshead, 1982; Mullet and Arntzen, 1980; Kühlbrandt *et al.*, 1983). The antennae polypeptides appear to have a dual function, namely, the harvesting of light and to mediate thylakoid adhesion. The contour map of the crystalline LHC reveals that the structural units have roughly a triangular outline arranged in a hexagonal lattice (Figure 8). Kühlbrandt *et al.* (1983) calculated a unit mass of about 170,000. This would allow the packing of three polypeptides each of

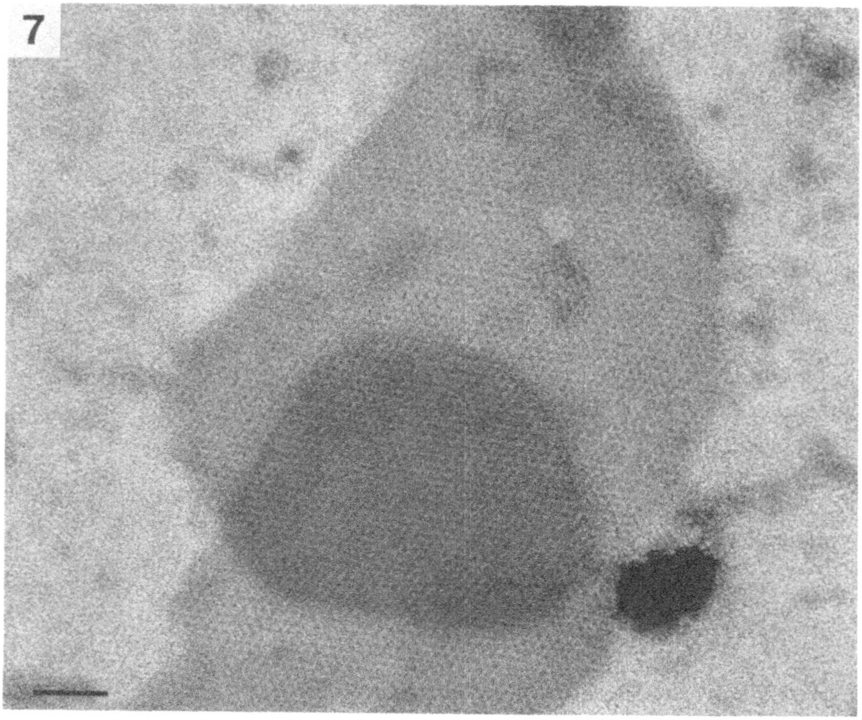

Figure 7. Membrane crystals of light-harvesting complex (LHC) polypeptides negatively stained with uranyl acetate. The stain-excluding regions form a hexagonal mesh. Scale bar: 0.1 μm. From Kühlbrandt *et al.* (1983).

8

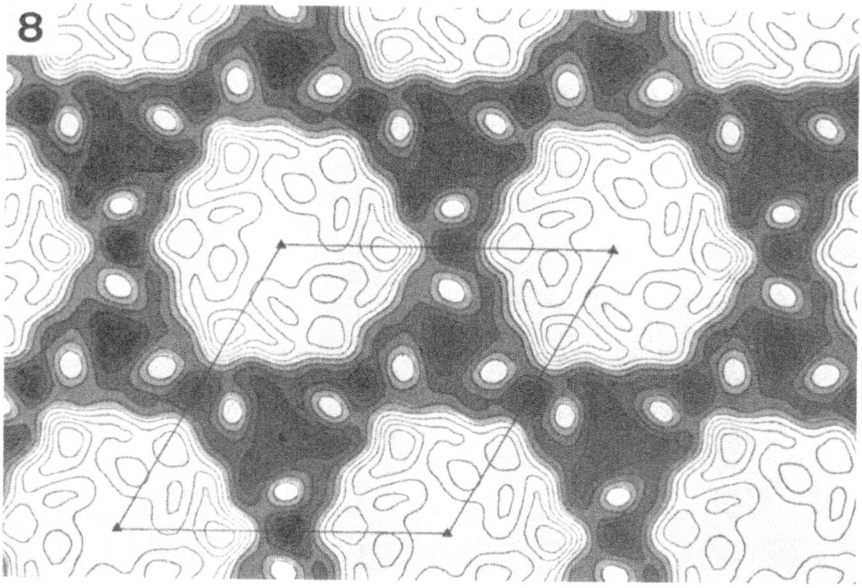

Figure 8. Projection map of negatively stained LHC membrane crystals. Stain-excluding regions are shown in shades of grey. The crystallographic unit cell (rhombus) contains two structural units of three-fold symmetry. Within each triangular unit, three elliptical staining regions are resolved. The stain distribution in the unit cell suggests that the two triangular units exhibit mirror symmetry in projection. From Kühlbrandt *et al.* (1983).

25,000 and 27,000 molecular weight in one triangular unit. The three-fold symmetry of the structural unit implies that it is composed of three identical subunits; the monomeric light-harvesting complex probably contains both polypeptides.

As demonstrated by these examples, the procedure of image reconstruction, in combination with biochemical analysis, is especially suited to advance our present knowledge on biological membranes.

ACKNOWLEDGMENTS

We express our thanks to Miss Erika Abächerli and Mrs. Margrit Leuzinger for the preparation of the manuscript.

REFERENCES

Armbruster, B. L., Carlemalm, E., Chiovetti, R., Garavito, R. M., Hobot, J. A., Kellenberger, E., and Villiger, W., 1982, Specimen preparation for electron microscopy using low temperature embedding resins, *J. Microscopy* **126**:77–85.

Bächi, Th., Whiting, K., Tanner, M. J. A., Metaxas, M. N., and Aristee, D. J., 1977, Freeze-fracture electron microscopy of human erythrocytes lacking the major membrane sialoglycoprotein, *Biochim. Biophys. Acta* **464**:635–639.

Bachmann, L., and Schmitt, W. W., 1971, Improved cryofixation applicable to freeze-etching, *Proc. Natl. Acad. Sci. USA* **68:**2149–2152.
Baumeister, W., and Kübler, O., 1978, Topographic study of the cell surface of Micrococcus radiodurans, *Proc. Natl. Acad. Sci. USA* **75:**5525–5528.
Berzborn, R., Kopp, F., and Mühlethaler, K., 1974, Mobility of chloroplast coupling factor 1 (CF_1) at the thylakoid surface as revealed by freeze-etching after antibody labelling, *Z. Naturforsch.* **29c:**694–699.
Bielig, H.-J., Kratky, O., Rohns, G., and Wawra, H., 1966, Small-angle scattering of apoferritin in solution, *Biochim. Biophys. Acta* **112:**110–118.
Branton, D., 1966, Fracture faces of frozen membranes, *Proc. Natl. Acad. Sci. USA* **55:**1048–1056.
Branton, D., 1971, Freeze-etching studies of membrane structure, *Philos. Trans. R. Soc. London Ser.* **261:**133–138.
Brenner, S., and Horne, R. W., 1959, A negative staining method for high resolution electron microscopy of viruses, *Biochim. Biophys. Acta* **34:**103–110.
Bretscher, M. S., 1973, Membrane structure: Some general principles, *Science* **181:**622–629.
Büechi, M., and Bächi, Th., 1979, Immunofluorescence and electron microscopy of the cytoplasmic surface of the human erythrocyte membrane and its interaction with Sendai virus, *J. Cell Biol.* **83:**338–347.
Carlemalm, E., and Kellenberger, E., 1982, The reproducible observation of unstained embedded cellular material in thin sections: Visualisation of an integral membrane protein by a new mode of imaging for STEM, *EMBO J.* **1:**63–68.
Carlemalm, E., Garavito, R. M., and Villiger, W., 1982, Resin development for electron microscopy and an analysis of embedding at low temperature, *J. Microsc.* **126:**123–143.
Cherry, R. J., 1979, Rotational and lateral diffusion of membrane proteins, *Biochim. Biophys. Acta* **559:**289–327.
Danielli, J. F., and Davson, H., 1935, A contribution to the theory of permeability of thin films, *J. Cell. Comp. Physiol.* **5:**495–508.
Deamer, D. W., and Branton, D., 1967, Fracture planes in an ice-bilayer model membrane system, *Science* **158:**655–657.
de Petris, S., 1978, Immunoelectron microscopy and immunofluorescence in membrane biology, in: *Methods in Membrane Biology,* Vol. 9 (E. D. Korn, ed.), Plenum Press, New York, pp. 1–201.
de Rosier, D. J., and Klug, A., 1968, Reconstruction of three dimensional structures from electron micrographs, *Nature* **217:**130–134.
Dietrich, I., Fox, F., Knapek, E., Lefranc, G., Nachtrieb, K., Weyl, R., and Zerbst, H., 1977, Improvements in the electron microscopy by application of superconductivity, *Ultramicroscopy* **2:**241–249.
Dietrich, I., Formanek, H., Fox, F., Knapek, E., and Weyl, R., 1979, Reduction of radiation damage in an electron microscope with a superconducting lens system, *Nature (Lond.)* **277:**380–381.
Dubochet, J., Knapek, E., and Dietrich, I., 1981, Reduction of beam damage by cryoprotection at 4°K, *Ultramicroscopy* **6:**77–80.
Dubochet, J., McDowall, A., Freeman, R., and Lepault, J., 1982, Cryoprotection on organic specimens, in: *Proc. 10th Int. Congr. on EM,* Vol. 1, Hamburg, pp. 19–23.
Dupouy, G., Perrier, F., and Durrieu, L., 1960, Microscopie electronique. L'observation de la matière vivante au moyen d'un microscope électronique fonctionnant sous très haute tension, *C.R. Acad. Sci.* **251:**2836–2841.
Edidin, M., 1982, Lateral diffusion of membrane proteins, in: *Membranes and Transport,* Vol. 1 (A. N. Martonosi, ed.), Plenum Press, New York, pp. 141–144.
Edwards, H. H., Mueller, Th. J., and Morrison, M., 1979, Distribution of transmembrane polypeptides in freeze fracture, *Science* **203:**1343–1345.
Erickson, H. P., Voter, W. A., and Leonard, K., 1978, Image reconstruction in electron microscopy: Enhancement of periodic structure by optical filtering, *Meth. Enzymol.* **49:**36–43.
Escaig, J., and Nicolas, G., 1976, Cryo-fractures de matériel biologique réalisées à très basses températures en ultra-vide, *C. R. Acad. Sci.* **283D:**1245–1248.
Faulk, W. P., and Taylor, G. M., 1971, An immunocolloid method for the electron microscope, *Immunochemistry* **8:**1081–1083.
Fernández-Morán, H., 1962, Cell-membrane ultrastructure, low-temperature electron microscopy and X-ray diffraction studies of lipoprotein components in lamellar systems, *Circulation* **26:**1039–1065.
Fisher, K. A., 1975, "Half" membrane enrichment: Verification by electron microscopy, *Science* **190:**983–985.

Fisher, K. A., 1976, Analysis of membrane halves: Cholesterol, *Proc. Natl. Acad. Sci. USA* **73:**173–177.

Fisher, K. A., 1978, Split membrane lipids and polypeptides, in: *Proc. 9th Int. Congr. on Electron Microscopy*, Vol. 3, Toronto, pp. 521–532.

Fisher, K. A., 1982, Monolayer freeze-fracture autoradiography: Quantitative analysis of the transmembrane distribution of radioiodinated Concanavalin A, *J. Cell Biol.* **93:**155–163.

Frey, T. G., Chan, S. H. P., and Schatz, G., 1978, Structure and orientation of cytochrome C oxidase in crystalline membranes. Studies by electron microscopy and by labelling with subunit-specific antibodies, *J. Biol. Chem.* **253:**4389–4395.

Frye, C. D., and Edidin, M., 1970, The rapid intermixing of cell surface antigens after formation of mouse-human heterokaryons, *J. Cell Sci.* **7:**319–335.

Garber, M., and Steponkus, P. L., 1974, Identification of chloroplast coupling factor by freeze-etching and negative staining techniques, *J. Cell Biol.* **63:**24–34.

Geoghegan, W. D., and Ackermann, G. A., 1977, Adsorption of horseradish peroxidase, ovomucoid and anti-immunoglobulin to colloidal gold for the indirect detection of Concanavalin A, wheat germ agglutinin and goat antihuman immunoglobulin G on cell surfaces at the electron microscopic level: A new method, theory and application, *J. Histochem. Cytochem.* **25:**1187–1200.

Gigg, R., and Payne, S., 1969, The reaction of glutaraldehyde with tissue lipids, *Chem. Phys. Lip.* **3:**292–295.

Gorter, E., and Grendel, F., 1925, On bimolecular layers of lipids on the chromocytes of the blood, *J. Exp. Med.* **41:**439–443.

Gross, H., Bas, E., and Moor, H., 1978a, Freeze-fracturing in ultrahigh vacuum at $-196°C$, *J. Cell Biol.* **76:**712–728.

Gross, H., Kübler, O., Bas, E., and Moor, H., 1978b, Decoration of specific sites on freeze-fractured membranes, *J. Cell Biol.* **79:**646–656.

Gulik-Krzywicki, T., and Costello, M. J., 1978, The use of low temperature X-ray diffraction to evaluate freezing methods used in freeze-fracture electron microscopy, *J. Microsc.* **112:**102–113.

Hake, T., 1965, Studies on the reactions of OsO_4 and $KMnO_4$ with amino acids, peptides and proteins, *Lab. Invest.* **14:**470–474.

Hall, C. E., Nisonoff, A., and Slyter, H. S., 1959, Electron microscopic observations of rabbit antibodies, *J. Biophys. Biochem. Cytol.* **6:**407–411.

Hämmerling, U., Aoki, T., Wood, H. A., Old, L. J., Boyse, E. A., and de Harven, E., 1969, New visual markers of antibody for electron microscopy, *Nature* **223:**1158–1159.

Hax, W. M. A., and Lichtenegger, S., 1982, A cryo-transfer system for EM 400 electron microscopes, *Electron Opt. Bull.* **117:**1–5.

Henderson, R., Capaldi, R. A., and Leigh, J. S., 1977, Arrangement of cytochrome oxidase molecules in two-dimensional vesicle crystals, *J. Mol. Biol.* **112:**631–648.

Höchli, M., 1974, Entwicklung einer neuen Druckgefrier-Technik und ihre Anwendung in der Elektronenmikroskopie, *Diss. ETH* No. 5304.

Höchli, M., and Hackenbrock, C. R., 1976, Fluidity in mitochondrial membranes: Thermotropic lateral translational motion of intramembrane particles, *Proc. Natl. Acad. Sci. USA* **73:**1636–1640.

Horisberger, M., and Rosset, J., 1977, Colloidal gold, a useful marker for transmission and scanning electron microscopy, *J. Histochem. Cytochem.* **25:**295–305.

Horisberger, M., Rosset, J., and Bauer, H., 1975, Colloidal gold granules as markers for cell surface receptors in the scanning electron microscope, *Experientia* **31:**1147–1148.

Horisberger, M., Rosset, J., and Vonlanthen, M., 1978, Location of lectin receptors on rat hepatocytes by transmission and scanning electron microscopy, *Experientia* **34:**274–276.

Jay, F., Lambillotte, M., and Mühlethaler, K., 1983, Localisation of Rhodopseudomonas viridis reaction centre and light harvesting proteins using ferritin-antibody labelling, *Eur. J. Cell Biol.*, **30:**1–8.

Karnovsky, M. J., Unanue, E. R., and Leventhal, M., 1972, Ligand-induced movement of lymphocyte membrane macromolecules. II. Mapping of surface moieties, *J. Exp. Med.* **136:**905–930.

Klug, A., 1979, Image analysis and reconstruction in the electron microscopy of biological macromolecules, *Chem. Scripta* **14:**245–256.

Klug, A., and Berger, J. E., 1964, An optical method for the analysis of periodicities in electron micrographs and some observations on the mechanics of negative staining, *J. Mol. Biol.* **10:**565–569.

Klug, A., and de Rosier, D. J., 1966, Optical filtering of electron micrographs, reconstruction of one sided images, *Nature* **212**:29–32.

Knapek, F., and Dubochet, J., 1980, Beam damage to organic material is considerably reduced in cryo-electron microscopy, *J. Mol. Biol.* **141**:147–161.

Kotowycz, G., and Suzuki, O., 1973, A carbon-13 nuclear magnetic resonance study of binding of manganese (II) to purine and pyrimidine nucleosides and nucleotides, *Biochemistry* **12**:3434–3439.

Kübler, O., and Baumeister, W., 1978, The structure of a periodic cell wall component (HPI-layer) of Micrococcus radiodurans, *Cytobiology* **17**:1–9.

Kübler, O., Gross, H., and Moor, H., 1978, Complementary structures of membrane fracture faces obtained by ultrahigh vacuum freeze-fracturing at −196°C and digital image processing, *Ultramicroscopy* **3**:161–168.

Kühlbrandt, W., and Unwin, P. N. T., 1980, Structural analysis of stained and unstained two-dimensional ribosomal crystals, in: *Electron Microscopy at Molecular Dimensions* (W. Baumeister and W. Vogell, eds.), Springer, Heidelberg, pp. 108–116.

Kühlbrandt, W., Wehrli, E., Thaler, Th., and Mühlethaler, K., 1982, Structures of two regular arrays of photosynthetic complexes, in: *Proc. 10th Int. Congr. on EM*, Vol. 3, Hamburg, pp. 45–46.

Kühlbrandt, W., Thaler, Th., and Wehrli, E., 1983, The structure of membrane crystals of the light-harvesting chlorophyll a/b-protein complex, *J. Cell Biol.*, **96**:1414–1424.

Latta, H., and Hartmann, J. F., 1950, Use of a glass edge in thin sectioning for electron microscopy, *Proc. Soc. Exp. Biol. Med.* **74**:436–439.

Lehninger, A. L., Wadkins, C. L., Cooper, C., Devlin, T. M., and Gamble, J. L., 1958, Oxidative phosphorylation, *Science* **128**:450–456.

Lenard, J., and Singer, S. J., 1968, Alteration of the conformation of proteins in red blood cell membranes and in solution by fixatives used in electron microscopy, *J. Cell Biol.* **37**:117–121.

Leonard, K., Wingfield, P., Arad, T., and Weiss, H., 1980, Membrane crystals of ubiquinone: Cytochrome C reductase from Neurospora mitochondria, in: *Electron Microscopy at Molecular Dimensions* (W. Baumeister and W. Vogell, eds.), Springer, Heidelberg, pp. 101–107.

Lepault, J., and Dubochet, J., 1982, Electron microscopy of frozen-hydrated specimens, in: *Proc. 10th Int. Congr. on EM*, Vol. 3, Hamburg, pp. 111–112.

Li, J., and Hollingshead, C., 1982, Formation of crystalline arrays of chlorophyll a/b-light harvesting protein by membrane reconstitution, *Biophys. J.* **37**:363–370.

Luft, J. H., 1956, Permanganate—a new fixative for electron microscopy, *J. Biophys. Biochem. Cytol.* **2**:799–801.

Luftig, R. B., Wehrli, E., and McMillan, P. N., 1977, The unit membrane image: A reevaluation, *Life Sci.* **21**:285–300.

Markham, R., Frey, S., and Hills, G. J., 1963, Methods for the enhancement of image detail and accentuation of structure in electron microscopy, *Virology* **20**:88–102.

Marshall, P. R., and Rutherford, D., 1971, Physical investigations on colloidal iron-dextran complexes, *J. Coll. Interface Sci.* **37**:390–402.

Maurer, A., and Mühlethaler, K., 1981, Isolation and characterization of paracrystalline arrays of the plasma membrane of baker's yeast Saccharomyces cerevisiae, *Eur. J. Cell Biol.* **24**:216–225.

Maurer, A., and Mühlethaler, K., 1982, Isolation and localization of plasma membrane-bound invertase in yeast (Saccharomyces cerevisiae), *Eur. J. Cell Biol.* **26**:219–227.

Mazia, D., Schatten, G., and Sale, W., 1975, Adhesion of cells to surfaces coated with polylysine, *J. Cell Biol.* **66**:198–200.

Miller, K. R., 1982, Three-dimensional structure of a photosynthetic membrane, *Nature* **300**:53–55.

Moor, H., and Mühlethaler, K., 1963, Fine structure in frozen-etched yeast cells, *J. Cell Biol.* **17**:609–628.

Moor, H., Mühlethaler, K., Waldner, H., and Frey-Wyssling, A., 1961, A new freezing-ultramicrotome, *J. Biophys. Biochem. Cytol.* **10**:1–13.

Mühlethaler, K., 1971a, Studies on freeze-etching of cell membranes, *Int. Rev. Cytol.* **31**:1–19.

Mühlethaler, K., 1971b, The ultrastructure of plastids, in: *Structure and Function of Chloroplasts* (M. Gibbs, ed.), Springer, Heidelberg, pp. 7–34.

Mühlethaler, K., Moor, H., and Szarkowski, J. M., 1965, The ultrastructure of the chloroplast lamellae, *Planta* **67**:305–323.

Müller, H. R., 1957, Die Gefriertrocknung als Fixierungsmethode für licht- und elekronenmikroskopische Untersuchungen an Pflanzenzellen, *J. Ultrastruct. Res.* **1**:109–137.

Müller, M., Meister, N., and Moor, H., 1980, Freezing in a propane jet and its application in freeze-fracturing, *Mikroskopie (Wien)* **36**:129–140.

Mullet, J. E., and Arntzen, Ch. J., 1980, Simulation of grana stacking in a model membrane system. Mediation by a purified light-harvesting pigment-protein complex from chlorplasts, *Biochim. Biophys. Acta* **589**:100–117.

Nägeli, C., and Cramer, C., 1855, *Pflanzenphysiologische Untersuchungen*, Heft 1, F. Schulthess, Zürich.

Nakane, P. K., and Kawaoi, A., 1974, Peroxidase-labelled antibody. A new method of conjugation, *J. Histochem. Cytochem.* **22**:1084–1091.

Nermut, M. V., 1982, The "cell monolayer technique" in membrane research, *Eur. J. Cell Biol.* **28**:160–172.

Newman, S. B., Borysko, E., and Swerdlow, M., 1949, Ultramicroscopy by a new method, *J. Res. Natl. Bur. Stand.* **43**:183–199.

Oleszko, S., and Moudrianakis, E. N., 1974, The visualization of photosynthetic coupling factor in embedded spinach chloroplasts, *J. Cell Biol.* **63**:936–948.

Overton, E., 1899, Ueber die allgemeinen osmotischen Eigenschaften der Zelle, ihre vermutlichen Ursachen und ihre Bedeutung für die Physiologie, *Viertelj. Schr. Naturf. Ges. Zürich* **44**:88–135.

Park, R. B., and Pfeifhofer, A., 1974, Chemical composition of fractured membrane halves, in: *Membrane Proteins in Transport and Phosphorylation* (G. F. Azzone, M. E. Klingenberg, E. Quagliariello, and N. Siliprandi, eds.), North-Holland, Amsterdam, pp. 97–102.

Pike, C. S., and Berry, J. A., 1980, Membrane phospholipid phase separations in plants adapted to or acclimated to different thermal regimes, *Plant Physiol.* **66**:238–241.

Pinto da Silva, P., Douglas, S. D., and Branton, D., 1971, Localization of A antigen sites on human erythrocyte ghosts, *Nature* **232**:194–195.

Porter, K., Claude, A., and Fullam E., 1945, A study of culture cells by electron microscope, *J. Exp. Med.* **81**:233–246.

Pucheu, N. L., Kerber, N. L., and Garcia, A. F., 1976, Isolation and purification of reaction centre from Rhodopseudomonos viridis NHTC 133 by means of LDAO, *Arch. Mikrobiol.* **109**:301–305.

Racker, E., Tyler, D. D., Estabrook, R. W., Conover, T. E., Parsons, D. F., and Chance, B., 1965, Correlations between electron-transport activity, ATP-ase and morphology of submitochondrial particles, in: *Oxidases and Related Systems*, Vol. 2 (T. E. King, H. S. Mason, and M. Morrison, eds.), John Wiley, New York, pp. 1077–1094.

Richards, F. M., and Knowles, J. R., 1968, Glutaraldehyde as a protein cross-linking reagent, *J. Mol. Biol.* **37**:231–233.

Riehle, M., 1968, Ueber die Vitrifizierung verdünnter wässériger Lösungen. *Thesis ETH*, No. 4271.

Robertson, J. D., 1958, A molecular theory of cell membrane structure, in: *Proc. 4th Int. Congr. EM*, Berlin, pp. 159–171.

Roth, J., 1982, The protein A-gold (pAg) technique—a qualitative and quantitative approach for antigen localisation on thin sections, in: *Techniques in Immunocytochemistry*, Vol. 1 (G. R. Bullock and P. Petrusz, eds.), Academic Press, London, pp. 107–134.

Roth, J., 1983, The colloidal gold marker system for light and electron microscopic cytochemistry (theory and application), in: *Techniques in Immunocytochemistry*, Vol. 2 (G. R. Bullock and P. Petrusz, eds.), Academic Press, London, pp. 217–284.

Roth, J., and Wagner, M., 1977a, Peroxidase and gold complexes of lectins for double labelling of surface-binding sites by electron microscopy, *J. Histochem. Cytochem.* **25**:1181–1184.

Roth, J., and Wagner, M., 1977b, Redistribution and internalisation of anti-A_{Hel} (anti-$_{HP}$)- and Concanavalin-A-binding sites, *Exp. Pathol.* **14**:311–320.

Roth, J., Bendayn, M., and Orci, L., 1978, Ultrastructural localization of intracellular antigens by the use of protein A-gold complex, *J. Histochem. Cytochem.* **26**:1074–1081.

Schneider, H., Lemasters, J. J., Höchli, M., and Hackenbrock, Chr. R., 1980, Liposome-mitochondrial inner membrane fusion (lateral diffusion of integral electron transfer components), *J. Biol. Chem.* **255**:3748–3756.

Segrest, J. P., Gulik-Krywicki, T., and Sardet, C. H., 1974, Association of the membrane-penetrating polypeptide segment of the human erythrocyte MN-glycoprotein with phospholipid bilayers, *Proc. Natl. Acad. Sci. USA* **71**:3294–3298.

Singer, S. J., 1959, Preparation of an electron-dense antibody conjugate, *Nature* **183:**1523–1524.

Singer, S. J., 1978, Immunoferritin electron microscopy and the structure of membranes, in: *Proc. 9th Int. Congr. EM*, Vol. 3, Toronto, pp. 533–539.

Sleytr, U. B., 1970, Die Gefrierätzung korrespondierender Bruchhälften: Ein neuer Weg zur Aufklärung von Membranstrukturen, *Protoplasma* **70:**101–117.

Sleytr, U. B., and Umrath, W., 1970, Freeze-etching: Technical developments and general interpretation problems, in: *Proc. 6th Eur. Congr. EM*, Vol. 2, Jerusalem, pp. 50–55.

Sommer, J. R., 1977, To cationize glass, *J. Cell Biol.* **75:**245a.

Sowers, A. E., and Hackenbrock, Ch. R., 1981, Rate of lateral diffusion of intramembrane particles: Measurement by electrophoretic displacement and rerandomization, *Proc. Natl. Acad. Sci. USA* **78:**6246–6250.

Staehelin, L. A., and Bertaud, W. S., 1971, Temperature and contamination dependent freeze-etch images of frozen water and glycerol solutions, *J. Ultrastruct. Res.* **37:**146–168.

Steck, T. L., 1974, The organization of proteins in the human red blood cell membrane, *J. Cell Biol.* **62:**1–19.

Steere, R. L., 1957, Electron microscopy of structural detail in frozen biological specimens, *J. Biophys. Biochem. Cytol.* **3:**45–60.

Steere, R. L., 1973, Preparation of high resolution freeze-etch, freeze-fracture, frozen surface and freeze-dried replicas in a single freeze-etch module, and the use of stereo electron microscopy to obtain maximum information from them, in: *Freeze-Etching Techniques and Application* (E. L. Benedetti and P. Favard, eds.), Soc. française de Micr. Electr., Paris, pp. 223–255.

Steere, R. L., and Moseley, M., 1970, Modified freeze-etch equipment permits simultaneous preparation of 2–10 double replicas, in: *Proc. 7th Int. Congr. EM*, Vol. 1, Grenoble, pp. 451–452.

Talmon, Y., 1982, Frozen hydrated specimens, in: *Proc. 10th Int. Congr. EM*, Vol. 1, Hamburg, pp. 25–32.

Unwin, P. N. T., and Henderson, R., 1975, Molecular structure determination by electron microscopy of unstained crystalline specimens, *J. Mol. Biol.* **94:**425–440.

Verkleij, A. J., and Ververgaert, P. H. J. Th., 1978, Freeze-fracture morphology of biological membranes, *Biochim. Biophys. Acta* **55:**303–327.

Verkleij, A. J., Lugtenberg, E. J. J., and Ververgaert, P. H. J. Th., 1976, Freeze-etch morphology of outer membrane mutants of Escherichia coli K12, *Biochim. Biophys. Acta* **426:**581–586.

Wagner, M., and Wagner, B., 1976, Electron microscopic detection of blood group antigen A on human erythrocytes by means of ferritin- and gold-labelled protectin of Helix pomatia, *Z. Immunol. Forsch.* **151:**117–125.

Wagner, M., and Wagner, B., 1977, Electron microscopic detection of the cryptantigen A_{HP}^{hum} (Friedenreich antigen) on human erythrocytes by means of gold-labelled agglutinin from Helix pomatia, *Z. Immunol. Forsch.* **153:**450–456.

Walter, C. A., Chang. J.-J., McDowall, A. W., and Dubochet, J., 1982, Freezing of biological material for electron microscopy, in: *Proceed. 10th Int. Congr. EM*, Vol. 3, Hamburg, pp. 115–116.

Walzthöny, D., Moor, H., and Gross, H., 1981, Ice crystals specifically decorate hydrophilic sites on freeze-fractured model membranes, *Ultramicroscopy* **6:**259–266.

Wehrli, E., and Kübler, O., 1980, The two-dimensional lattice of the photosynthetic membrane of Rhodopseudomonas viridis, in: *Electron Microscopy at Molecular Dimensions* (W. Baumeister and W. Vogell, eds.), Springer, Heidelberg, pp. 48–56.

Wehrli, E., Mühlethaler, K., and Moor, H., 1970, Membrane structure as seen with a double replica method for freeze-fracturing, *Exp. Cell Res.* **59:**336–339.

Welte, W., and Kreutz, W., 1982, Formation, structure and composition of a planar hexagonal lattice composed of specific protein-lipid complexes in the thylakoid membranes of Rhodopseudomonas viridis, *Biochim. Biophys. Acta* **692:**479–488.

Winkelmann, H., 1981, Problems of physical fixation by freezing, *Acta Histochem.* **23(Suppl.):**11–19.

2

Associations of Cytoskeletal Proteins with Plasma Membranes

Carl M. Cohen and Deborah K. Smith

I. INTRODUCTION

Eukaryotic cells contain an extensive network of self-associated, interconnected filaments and tubules referred to collectively as the cytoskeleton. The proteins of the cytoskeleton are responsible for the generation of nearly all types of cellular movement, from gross amoeboid migration to subtle intracellular motions of organelles or chromosomes. Cytoskeletons may be defined teleologically as the intracellular filaments or proteins which combine to generate cell movement and to control cell shape. The cytoskeleton can also be defined operationally as the protein residue which remains when cells are extracted with non-ionic detergents such as Triton X-100. Examination of these detergent-resistant cellular residues by electron microscopy shows them to consist of several types of filaments. Each filament type has a distinct diameter and each is distributed within the cell in a characteristic pattern.

Treatment of cells with non-ionic detergents solubilizes all membrane lipids and removes the cytoplasm, leaving behind the detergent-insoluble proteins and filaments of the cytoskeleton. However, detergent-extracted cells retain to a remarkable degree the shape of the original unextracted cell. This elementary observation shows that the cytoskeleton is responsible in large part for stabilizing cell shape. Further, under appropriate conditions, detergent-extracted, or lysed, cells can manifest certain cytoplasmic activities similar to those found in intact cells (Taylor and Condeelis, 1979).

While isolated cytoskeletons retain their integrity and perhaps certain of their functions, it is clear that some of the most important aspects of cytoskeletal organization involve contacts with the plasma membrane. Direct evidence for such contacts has

Carl M. Cohen • Department of Biomedical Research, St. Elizabeth's Hospital, Boston, Massachusetts 02135; and Departments of Medicine and of Molecular Biology and Microbiology, Tufts University School of Medicine, Boston, Massachusetts 02135. *Deborah K. Smith* • Department of Biomedical Research, St. Elizabeth's Hospital, Boston, Massachusetts 02135. *Present address:* Department of Physiology, Duke University Medical Center, Durham, North Carolina 27710.

been provided by electron microscopy, which shows that the various cytoskeletal filaments frequently terminate near or run parallel to the plasma membrane. In several instances the functional importance of these contacts is understood, while in many others it has only been hypothesized. This review will consider several relatively well-characterized examples of cytoskeletal–membrane contacts. However, the scope of cytoskeletal phenomena is broad and there are numerous examples of cytoskeletal activities which should logically involve the cell membrane but for which the evidence is still meager. With some exceptions we will restrict our detailed discussions to cytoskeletal–membrane associations which have been clearly documented both biochemically and morphologically.

II. THE COMPONENTS OF THE CYTOSKELETON

As the cytoskeleton has received more and more attention over the past decade, the list of cytoskeletal or cytoskeleton-associated proteins has grown. However, the bulk of the cytoskeleton is made up of three filament systems, microfilaments, microtubules, and intermediate filaments. The characteristics and membrane associations of each of these will be considered in the following sections. While the filamentous components of the cytoskeleton are by far its most visually prominent constituents, there are a large number of filament binding or connecting proteins, such as actin-binding proteins, which also contribute to cellular function. These will also be discussed in the following sections.

A. Actin

Actin is the single most prominent component of cytoskeletons and of whole cells, accounting for as much as 30% of total cellular protein in some cell types (Pollard and Weihing, 1974). The properties of this protein have been extensively reviewed (Oosawa and Kasai, 1971; Pollard and Weihing, 1974; Korn, 1978, 1982), and only some essential characteristics will be considered here. Monomeric or G-actin has a molecular weight of about 42,000 daltons and generally contains a tightly bound molecule of ATP. Upon the addition of salt, and warming to room temperature, actin monomers polymerize into double helical 5-nm-diameter microfilaments which can be many microns long. While the amino acid composition and isoelectric points of actins from different cell types vary somewhat, all actins seem to share similar biochemical features. The most notable of these, aside from forming filaments, is the ability of actin to interact with myosin filaments to generate motion or force by utilizing energy from the splitting of ATP. While myosin is present in nonmuscle cells, its role in the generation of movement is still, to a certain extent, conjectural.

The forms of actin in a typical eukaryotic cell can conveniently be divided into three categories: (1) filaments or filament bundles, (2) gels or networks, and (3) nonpolymerized.

1. Actin Filaments

While individual actin filaments can be seen only by electron microscopy, filament bundles or cables can be detected by phase contrast microscopy as well as by immunofluorescence microscopy using anti-actin antibodies or other labels. Actin filaments exist in every type of eukaryotic cell, but filament bundles or stress fibers are most characteristic of static, well-spread cells such as fibroblasts. Although certain cells have been shown to contain stress fibers *in vivo* (Byers and Fujiwara, 1982), stress fibers have been most extensively studied in cultured cells. For our purposes stress fibers are particularly interesting structures because, as will be discussed in Section V, they frequently terminate (or originate, it is unclear which) at the plasma membrane. The pattern of cellular stress fibers is a function of the cell type as well as the stage of the cell cycle in which the cell is observed. For example, cultured fibroblasts just after plating appear rounded and have polygonal networks of actin bundles just beneath their membrane. Well-spread fibroblasts, on the other hand, tend to have highly extended parallel arrays of stress fibers (Lazarides, 1976a,b). Thus, the shape of the cell correlates closely with the appearance of the cytoskeleton (see also Marek *et al.*, 1982). While some stress fibers contact the plasma membrane, the function of such contacts, particularly in cells which are relatively static, is not apparent. Indeed, stress fibers are most characteristic of cells which are relatively nonmotile, being sparse or absent altogether from dynamic cells (Herman *et al.*, 1981). Highly motile cells seem to be characterized by a subcortical meshwork of cytoskeletal proteins often characterized as being gel-like. This constitutes the second major type of actin structure within cells.

2. Actin Gels or Networks

The submembrane region of cultured cells has been visualized in striking detail by quick-freeze deep-etch electron microscopy (Heuser and Kirschner, 1980) as well as by other methods (Mesland *et al.*, 1981). This cortical region is rich in actin and contains a disorganized array of highly interconnected filaments. This region of the cytoplasm was also called the hyaline ectoplasm by early investigators and is distinguished by the fact that organelles are excluded from it. More recently, it has been suggested that the cell cortex is composed of a gel-like layer of cytoskeletal proteins which has been called the cortical plasma-gel (Stossel, 1978). Isolated cell membranes frequently retain some of this cortical material which can be eluted and can be shown by gel electrophoresis to be rich in actin. It is thought that this gel-like network is constructed by the joining or crosslinking of short actin filaments by actin-binding proteins.

Actin-binding proteins comprise an extensive class of macromolecules which share the common property of binding to F-actin (see Weeds, 1982 for review). Those proteins which have been examined to date have similar structures, being elongated (~2000 Å), thin, relatively flexible molecules having actin-binding sites at each of their ends. This bivalent property explains how actin-binding proteins can crosslink actin filaments in solution to form highly viscous gel-like networks similar to those found in the cell cortex. With the exception of the red cell actin-binding protein, spectrin, and its close relative fodrin, from brain, and vinculin (to be discussed below), no convincing association of an actin-binding protein with a cell membrane has been demonstrated. It has been suggested that actin gels may be precursors of stress fibers (Lazarides, 1976b; Gordon and Bushnell, 1979). When a gel is pulled or extended, its constituent filaments may align, giving a bundle-like appearance. Whether or how such a mechanism may operate *in vivo* is not known.

3. Nonpolymerized Actin

When a cell is homogenized or lysed and subjected to ultracentrifugation, a proportion of the cellular actin invariably remains in the supernatant. Some of this nonsedimentable actin may reflect a pool of nonpolymerized intracellular actin. It is unclear, however, whether the conditions of cell lysis can perturb the intracellular G–F-actin equilibrium and lead to the depolymerization of F-actin. It is known, however, that cells contain proteins which bind to actin monomers and prevent them from participating in filament formation. The most well-characterized of these proteins is profilin, which has been isolated from several cell types and forms a 1:1 complex with G-actin (see Korn, 1982; Weeds, 1982 for reviews). Profilin may help regulate actin filament formation or monomer exchange into filaments. However, except in special cases (Tilney, 1978; Tilney *et al.*, 1978), the significance of this type of regulation, or how it is controlled, is unknown.

4. Actin–Membrane Associations

Evidence for the association of actin with cell membranes includes electron microscopic observations of F-actin contacting the cytoplasmic surface of plasma membranes as well as biochemical identification of actin on plasma membranes. Several selected examples of specialized actin–membrane contacts will be discussed in the following sections (see also Weatherbee, 1981). Here we will review some general aspects of actin–membrane associations.

Isolated plasma membranes frequently contain a 43,000-dalton protein which is usually, and with good reason, assumed to be actin (Gruenstein *et al.*, 1975; Wickus *et al.*, 1975; Bray and Thomas, 1975; McNutt, 1978; Moore *et al.*, 1978; Phillips *et al.*, 1981; Luna *et al.*, 1981; Mescher *et al.*, 1981; Allore and Barber, 1982; Carraway *et al.*, 1982). Also, actin filaments have been clearly identified in association with plasma membranes by electron microscopy (Pollard and Korn, 1973; Clark *et al.*, 1975; McNutt, 1978; Moore *et al.*, 1978; Kelley *et al.*, 1980; Mesland *et al.*, 1981; Phillips *et al.*, 1981; Sugrue and Hay, 1981; Hirokawa and Tilney, 1982). Specific regions of the cytoplasm may become highly enriched in actin as a result of specialized cellular activities or structural requirements. During phagocytosis, for example, an amorphous gel of actin accumulates in pseudopodia which advance around the engulfed particle. Although electron microscopy shows actin to be close to pseudopodial membranes (Reaven and Axline, 1973; Hartwig *et al.*, 1977), little is known about the nature of its membrane association there. By comparison, the linear bundles of actin filaments in the intestinal brush border are more organized and form well-visualized connections with the membrane, both at their ends and along their length (see Section VI). Bundles of actin filaments are also closely aligned with the cell membrane along the cleavage furrow during cytokinesis, although whether or how the actin is associated with the membrane is unknown (Schroeder, 1975). Actin binds to plasma membranes from several cell types including Dictyostelium (Jacobson, 1980; Luna *et al.*, 1981) and red cells, but only in the latter case is the binding well understood (see Section IV).

In a few cases, actin is associated with specific membrane components. Koch and Smith (1978) have shown that in the murine mastocytoma line P815, membrane-associated actin can be co-isolated with an integral membrane component bearing the H-2 histocompatibility antigen. Similarly, a variety of integral membrane proteins, when clustered by multivalent ligands, appear to make direct transmembrane contact with submembranous actin filaments (Ash *et al.*, 1977, 1980; Singer *et al.*, 1978). It has been proposed (Ash *et al.*, 1980) that this interaction is mediated by a class of integral membrane proteins (as yet unidentified) which are normally connected to, or aligned over, stress fibers. Actin has also been reported to be exposed on the surface of lymphocytes (Owen *et al.*, 1978) and platelets (after activation) (George *et al.*, 1980). These findings are paradoxical since it is unclear how actin can gain access to the extracellular cell surface.

B. Microtubules

Microtubules are hollow, extended, tubular structures 25 nm in diameter and are composed of 110,000-dalton subunits. These subunits are themselves composed of two similar tubulin monomers, α- and β-tubulin, 55,000 daltons each. Microtubules can be induced to assemble from tubulin monomers in solution; the conditions which favor assembly and disassembly have been studied extensively (Scheele and Borisy, 1979). Just as the organization of actin within cells is controlled by a wide variety of actin-binding proteins, the state and associations of microtubules are also influenced by a group of proteins referred to as microtubule associated proteins (MAPs). These consist of high-molecular-weight proteins (>250,000 daltons) as well as lower-molecular-weight "tau" proteins. Evidence suggesting that microtubules can interact with cell membranes has been recently reviewed by Weatherbee (1981).

Perhaps one of the clearest examples of the association of microtubules with a cell membrane is evident in the marginal band of nucleated erythrocytes. Here, a tightly packed bundle of 10–15 microtubules circles the periphery of the cell making close contact with the plasma membrane (Wolfe, 1967; Behnke, 1970; Cohen, 1978). Removal of the microtubules by sonication or depolymerization reveals on the membrane a series of remnant "tracks" (Granger and Lazarides, 1982) which may be related to microtubule–membrane attachment sites. Other studies suggest an important relationship between cell surface components and microtubules (reviewed by Berlin *et al.*, 1979). However, it remains unknown whether a direct association of microtubules with the membrane or its constituent proteins explains these phenomena. For example, phagocytosis of zymosan by polymorphonuclear leukocyte cells is correlated with the appearance of microtubule growth from centrioles, but the microtubules are conspicuously absent from the immediate region of phagocytosis. Similarly, in other cell types, immunofluorescence using antitubulin generally fails to show any accumulation of tubulin near cell membranes, although microtubules sometimes run parallel to the cell periphery.

Efforts to document direct tubulin–membrane associations have also focused on model systems. Several groups have investigated the association of microtubule proteins with synthetic liposomes and found that tubulin binds avidly to lipid bilayers (Caron and Berlin, 1979; Kumar *et al.*, 1981; Klausner *et al.*, 1981). Whether a direct tubulin–bilayer association is related to the putative association of tubulin with cell membranes or whether proteins are involved is speculative. However, in the most well-documented cases of tubulin–membrane association (cilia and flagella) there seem to be specific entities mediating the lateral and end-on attachment to membranes. Filamentous bridges having a periodicity of about 220 nm connect microtubules to ciliary and flagellar membranes (Dentler, 1981). Other evidence also supports such lateral connections (Ringo, 1967; Allen, 1968; Dentler, 1981; Sattler and Staehelin, 1979). Structural evidence for the association of the distal tips of microtubules with ciliary or flagellar membranes has also been published (Dentler, 1980). The central pair of microtubules generally terminates in one or more plate-like structures which are apparently attached to the membrane. Other types of cilia appear to contain a membrane-associated cap consisting of a ball-like structure above two smaller plates (Dentler, 1980). In no case have the individual proteins connecting the microtubules to membranes been identified. Biochemical evidence for membrane-associated tubulin in cilia and flagella (Stephens, 1977) as well as other cells (Bhattacharyya and Wolff, 1975) has also been presented.

Numerous reports have documented membrane-associated tubulin in brain and synaptosomal membranes (Blitz and Fine, 1974; Kornguth and Sunderland, 1975; Bhattacharyya and Wolff, 1975, 1976; Feit *et al.*, 1977). However, the functional capacity of this membrane-associated tubulin is uncertain. Although ultrastructural demonstrations of close associations between microtubules and cell membranes are common in plants and protozoans, visualization of such associations are rare in mammalian cells with the exception of nervous tissue such as synaptic vesicles (Gray, 1975). Microtubules have, however, been observed to associate *in situ* with intracellular organelles, particularly mitochondria (Bradley and Satir, 1979). A specific association between purified tubulin and isolated rat mitochondria has also been shown by Bernier-Valentin and Rousset (1982).

C. Intermediate Filaments

Intermediate filaments are a heterogeneous group of intracellular filaments which share the limited feature of being about 10 nm in diameter, intermediate in size between microfilaments and microtubules. Five separate classes of intermediate filaments have been identified within eukaryotic cells (Lazarides, 1980). These are (1) keratin filaments which are 8 nm in diameter, are composed of 6–7 polypeptides of 47,000–58,000 daltons, and are found primarily in epithelial cells, (2) vimentin filaments, composed of the 56,000–58,000-dalton monomer, vimentin, found in a variety of cell types including muscle, nervous tissue, and epithelial cells, (3) desmin filaments, having a diameter of 10 nm, composed of the 50,000–55,000-dalton protein, desmin, and found in muscle cells, (4) neurofilaments which are about 10 nm in diameter, are composed of several different protein constituents, and are found exclusively in neurons, and (5) glial filaments, 8 nm in diameter, and composed primarily of a 51,000-dalton protein referred to as the glial fibrillary acidic protein. Glial filaments are found in the cytoplasm of astrocytes and other types of glial cells.

The function of intermediate filaments is still under active investigation, but it has been suggested that they play an important role in organizing the cytoplasm in as yet undefined ways (Lazarides, 1980). Desmin filaments in smooth muscle cells apparently contact cell membranes through membrane-bound dense plaques, which also serve as attachment sites for actin filaments (Uehara *et al.*, 1971; Cooke and Chase, 1971; Cooke, 1976). In this sense, dense plaques seem analogous to the Z-lines of striated muscle. In fact, desmin is a prominent component of Z-lines, being localized at the periphery of each Z-disk (Lazarides, 1980). Desmin has also been localized to regions of contact between Z-lines and plasma membranes, and it has been suggested that desmin acts to link Z-disks to the plasma membrane and to other organelles (Lazarides, 1980). Desmin may also play a role in mediating the attachment of actin to membrane sites such as dense bodies on intercalated disks of cardiac muscle (Lazarides, 1980). Vimentin filaments may also be associated with plasma and intracellular membranes, and have been observed to terminate at the nuclear membrane as well as at focal adhesions of cultured fibroblasts (Lehto *et al.*, 1978; Small and Celis, 1978; Virtanen *et al.*, 1979).

Intermediate filaments have also been shown to be associated with plasma membranes of nucleated avian erythrocytes, probably by some type of lateral association. These filaments traverse the cytoplasm and attach to the nucleus (Granger and Lazarides, 1982). While the marginal band of microtubules in these cells is confined to a narrow track around the periphery of the cell, the membrane-associated intermediate filaments seem to be confined to the faces of the cell, away from the microtubules. Finally, extensive three-dimensional associations between neurofilaments and membranous organelles have been clearly observed by the quick-freeze deep-etch method (Hirokawa, 1982) applied to axons. Others have also noted associations of intermediate filaments with organelles such as mitochondria (Bernstein and Wollman, 1975).

D. The Role of α-Actinin

α-Actinin is a 95,000-dalton rod-like molecule approximately 300 Å long. In striated muscle, this protein is confined exclusively to the Z-line where actin filaments terminate. In solution, purified α-actinin forms regular crosslinks between actin filaments, and induces the formation of filament bundles (Holmes *et al.*, 1971; Podlubnaya *et al.*, 1975). Because of its localization to sites of filament anchorage in muscle and its presence in nonmuscle cells, α-actinin has been suggested to mediate the association of actin filaments with cell membranes. In nonmuscle cells, α-actinin is distributed periodically along stress fibers as well as at the vertices of polygonal actin networks (Lazarides and Burridge, 1975; Feramisco, 1979; Geiger *et al.*, 1981). Distinct accumulations of α-actinin at the termini of stress fibers in fibroblasts or at actin–membrane junctions in other cells have also been clearly documented (Geiger, 1982; Geiger *et al.*, 1981).

Whether α-actinin mediates actin–membrane attachment is still subject to question. α-Actinin has been identified as a component of the membranes of a variety of cells (Burridge and McCullough, 1980; Hoessli *et al.*, 1980; Burridge *et al.*, 1980; Yeltman *et al.*, 1981), but Burridge and McCullough (1980) have shown that when membrane-associated α-actinin is eluted, nearly all of the membrane-associated actin remains behind. Geiger *et al.* (1981) have shown that α-actinin can be identified near the sites of filament–membrane contact in smooth muscle and intestinal brush border cells. However, the α-actinin molecules are rather dispersed relative to the actual filament terminus (see Section V). Similarly, a careful study of focal adhesion sites in fibroblasts has shown that α-actinin is not concentrated at the site of closest filament–membrane contact (Chen and Singer, 1982). On the other hand, α-actinin is enriched at regions of the fibroblast cell membrane which make so-called "close contacts" with the substratum (see Section V) and may participate in lateral actin filament–membrane association. α-Actinin could mediate such attachment via an unidentified integral membrane protein (Chen and Singer, 1982), although other alternatives are possible. Meyer *et al.* (1982) have recently shown that purified α-actinin can associate specifically with model membranes containing glycerides and fatty acids. Whether this type of association is relevant to the situation *in vivo* remains to be established.

The evidence for an α-actinin-mediated attachment of actin filaments to cell membranes remains equivocal. This protein may organize filaments into bundles which are then anchored to membranes by other proteins (see below), but it may also promote attachment of filaments to membranes in other ways. It seems unlikely, however, that α-actinin is the universal promoter of actin–membrane contacts that it was once suspected of being.

III. CYTOSKELETAL FUNCTIONS

There is a wide variety of evidence demonstrating the involvement of cytoskeletal proteins in cell movement (Taylor and Condeelis, 1979). The evidence includes correlations between the state of the cytoskeleton and the motile state of the cell as well as observation of the effects on cell movement of microfilament-disrupting agents like cytochalasin B. Further, the major contractile proteins, actin, myosin, tropomyosin, and α-actinin, have all been purified from a wide variety of motile cells ranging from amoebas to human leukocytes, and have been localized within these cells by immunofluorescence and electron microscopy. An important role for membrane–cytoskeletal associations has frequently been assumed to be a logical requirement for amoeboid movement. But there is a lack of experimental data to provide sound documentation for this assumption. It is possible that attachment of actin filaments to membranes, and thus to an underlying substrate, can serve as an anchor by which a cell can pull forward. Recent evidence suggests, however, that highly motile cells possess few or no membrane-attached filament bundles or stress fibers (Herman *et al.*, 1981). The amorphous actin gel in the cell cortex may be more important for this type of movement. While the identity of some of the proteins which comprise the gel is known, the factors, if any, responsible for their association with membranes are almost entirely unknown. Further, how the cortical actin gel generates pushing or ruffling movements at the periphery of cells is also not known. It has been emphasized by Izzard and Lochner (1980) that the classical role of actin filaments in striated muscle is to pull one part of a cell towards another via myosin ATPase. Such a mechanism cannot easily explain protrusive movement of cell processes into the surrounding medium.

Mitosis and cytokinesis also involve the intracellular contractile apparatus. While microtubules are the major structural elements involved in chromosome separation, actin filaments and myosin have also been localized in the mitotic spindle. Although the role of actin filaments in mitosis is still uncertain, actin–membrane associations may be important during cytokinesis (Schroeder, 1981). During cytokinesis, the development of the cleavage furrow is coincidental with the appearance of a circumferential ring of actin filaments. These filaments are located close to the plasma membrane and have been postulated to be involved in the contraction of the cleavage furrow, but whether they are membrane-associated is unknown. The occasional appearance in the cleavage furrow of the actin-binding protein filamin as well as α-actinin and myosin has also been noted (Fujiwara and Pollard, 1976; Mabuchi and Ikuno, 1977; Nunnally *et al.*, 1980).

Cytoskeletal proteins play a major role in cellular import and export functions, and the role of cytoskeletal–membrane contacts in phagocytosis and endocytosis is discussed in Section VII-B. There is also substantial evidence for the involvement of microfilaments, and possibly other contractile proteins, in secretion. Associations of actin filaments (Burridge and Phillips, 1975; Ostlund *et al.*, 1977; Howell and Tyhurst, 1979; Wilkins and Lin, 1981) or α-actinin (Jockusch *et al.*, 1977) with secretory vesicles have been described. Some evidence exists for a specific *in vitro* association between chromaffin granules and F-actin (Fowler and Pollard, 1982).

Nerve axons are known to have extensive, relatively ordered cytoskeletons containing actin, myosin, tropomyosin, actin-binding proteins, neurofilaments, and microtubules. The movement of vesicles or molecules along the length of the axon is known to involve some or all of these proteins. The slowest of these movements involves the gross advancement of the entire cytoskeleton from the cell body to the axon terminus (Lasek and Hoffman, 1976), while more rapid movements may involve translocation of particles along the cytoskeleton. Connections between the various filament systems and intracellular axonal membranes have recently been visualized by the rapid-freeze deep-etch method (Hirokawa, 1982).

While the above processes represent some of the most interesting examples of cytoskeletal involvement in cellular processes they are at the same time the most complex and least understood. In several of the examples discussed, a role for membrane-associated actin or other cytoskeletal elements has been suggested. In none of these cases, however, has unambiguous evidence proving the involvement of membrane–cytoskeletal associations been obtained. The reconstitution of actin onto isolated membranes provides some support for the notion that contacts observed in electron micrographs are not simply fortuitous. However the identity of the membrane components involved in such associations are still largely unknown.

By contrast with our understanding of these dynamic events, there is a considerable amount of data relating to cytoskeletal–membrane associations which are more static or structural in nature. In several instances, proteins which mediate the attachment of actin to membranes have been identified and purified. The next several sections will consider three examples of such cytoskeletal–membrane contacts.

IV. THE ERYTHROCYTE MEMBRANE SKELETON: A COMPLETELY MEMBRANE-ASSOCIATED CYTOSKELETON

The human erythrocyte contains the simplest and most well defined cytoskeleton of any mammalian cell. While the cytoskeletons of all other eukaryotic cells are composed of extensive transcellular arrays of filaments and tubules which make only limited contact with the plasma membrane, the red cell cytoskeleton is entirely membrane-associated. Electron microscopy has clearly shown that all of the major cytoskeletal proteins are located within about 10 nm of the lipid bilayer (Ziparo *et al.*, 1978; Tsukita *et al.*, 1980, 1981). In recognition of this unique property, the red cell cytoskeleton has also been referred to as a membrane skeleton.

A. Composition of the Erythrocyte Membrane Skeleton

As with other cells, the cytoskeleton of the mature erythrocyte is operationally defined as the residue of interconnected proteins which remains when cells are extracted with non-ionic detergents such as Triton X-100. In the case of the red cell, this residue, also called a Triton "shell," retains the shape of the original cell to a remarkable degree (Yu *et al.*, 1973). Electron microscopy shows that the shells consist of a highly disordered, interconnected network of filamentous proteins. Analysis of cytoskeletons on SDS polyacrylamide gels reveals four major components (Yu *et al.*, 1976). Spectrin, the major constituent, is, in its simplest form, a heterodimer composed of a single α chain (mol. wt. 240,000) and a single β chain (mol. wt. 220,000) and comprises about 75% of the total cytoskeleton mass. Actin comprises about 5% of the cytoskeleton and the proteins referred to as polypeptides 4.1 and 4.9 comprise about 5% each. Each of these proteins has been shown to be bound to the cytoplasmic membrane surface (Steck, 1974). These are peripheral proteins because they can be eluted from the membrane by manipulation of ionic strength or pH, or by addition of salts, without disrupting the membrane bilayer (Steck, 1974).

B. Spectrin

Spectrin is the best studied of the cytoskeletal proteins. In solution, spectrin dimers ($\alpha\beta$) can self-associate to form tetramers ($\alpha_2\beta_2$) which is apparently the form of the molecule on the membrane (Ralston, 1976; Ungewickell and Gratzer, 1978; Shotton *et al.*, 1979; Middaugh and Ji, 1980). Purified spectrin has been visualized by electron microscopy of shadowed specimens. In the $\alpha\beta$ dimeric form, spectrin appears as an elongated, highly flexible, double-stranded molecule with an average contour length of 97 nm. Spectrin tetramers are exactly twice as long and arise from head-to-head association of two dimers (Shotton *et al.*, 1979). It was thought that spectrin was unique to red blood cells, but it is now known that spectrin is an actin-binding protein (for review see Cohen, 1983) and is similar to other actin-binding proteins such as brain fodrin which may also be membrane-associated (Burridge *et al.*, 1982; Bennett *et al.*, 1982).

C. Actin

Actin in the red cell membrane skeleton is primarily of the β isotype (Pinder *et al.*, 1978). Because the functional form of actin in cytoskeletons is certainly the F or filamentous form, many attempts have been made to identify F-actin in the red cell membrane skeleton. While early attempts to identify actin filaments based solely on their morphology gave ambiguous results, there is now substantial biochemical evidence for the presence of F-actin in red cell membrane skeletons. Much of this evidence is based upon the fact that preformed actin filaments can nucleate the polymerization of G-actin under conditions in which spontaneous polymerization would be slow or nonexistent. Similarly, red cell membranes, and an oligomeric complex of spectrin, actin, and band 4.1 extracted from membrane skeletons, can nucleate actin polymerization in a manner identical to purified F-actin (Cohen and Branton, 1979; Lin and Lin, 1979; Pinder *et al.*, 1979; Brenner and Korn, 1980). Further, the plant alkaloid cytochalasin B blocks membrane (Cohen *et al.*, 1978) or spectrin-actin-4.1 complex-induced actin polymerization (Lin, 1981) in a manner identical to its action on F-actin.

The above observations provide indirect evidence for the presence of F-actin in the membrane skeleton. However, to date it has been impossible to visualize actin filaments in skeletons by electron microscopy. Some estimates of the number of actin filaments which may be present in skeletons have been based upon cytochalasin B binding. The data suggest that the filaments are short, perhaps containing fewer than 10–13 monomers each (about 25 nm long), and may be uniformly dispersed throughout the skeleton (Lin and Lin, 1979; Brenner and Korn, 1980). In contrast, other studies suggest that skeletal actin filaments may be longer, possibly containing 30–60 monomers, and being up to 100 nm in length. This length would make them similar in size to spectrin (Atkinson *et al.*, 1982).

D. Polypeptides 4.1 and 4.9

Band 4.1 is a globular protein of molecular weight between 78,000 and 82,000 daltons and comprises about 5% of the red cell cytoskeletal mass. When red cell ghosts or membrane skeletons are electrophoresed in continuous buffer gel systems, band 4.1 migrates as a single band, while in discontinuous buffer gel systems (Laemmli, 1970), it migrates as a closely spaced doublet designated 4.1a and b (Mueller and Morrison, 1977). Partial peptide mapping of 4.1a and b suggests that they have related peptide sequences (Goodman *et al.*, 1982).

Band 4.9, a 48,000-dalton phosphoprotein, is an often neglected constituent of the red cell membrane skeleton, yet is apparently present at a concentration equimolar with spectrin tetramers (Ralston and Crisp, 1981; Siegel and Branton, 1982). Little is known about the function of this protein, but recent work suggests that it may be involved in maintaining membrane skeletal structure (Ralston and Crisp, 1981). A recent report (Siegel and Branton, 1982) indicates that purified polypeptide 4.9 can interact with actin filaments to reduce the rate of actin polymerization, and possibly stabilize short actin oligomers. Such functions would obviously have important consequences for the regulation of actin filament length in the membrane skeleton.

E. Ankyrin

The attachment of the entire complex of cytoskeletal proteins to the plasma membrane is achieved largely by a single protein, ankyrin, a 200,000-dalton globular protein (Bennett, 1982a). This protein was originally designated band 2.1 (Steck, 1974), and it gives rise to a series of lower-molecular-weight membrane-associated proteolytic products which, along with band 2.1, have also been referred to collectively as syndeins (Yu and Goodman, 1979). Triton shells can be prepared which lack ankyrin but which still remain self-assembled with the general shape of the red cell (Sheetz, 1979). Nevertheless, ankyrin is central to the function of the membrane skeleton. The considerable work which went into its identification and characterization has recently been reviewed (Bennett, 1982a).

Ankyrin is located at the cytoplasmic membrane surface and is responsible for the high affinity, saturable binding of spectrin to red cell membranes (Bennett, 1982a). Purified ankyrin, or proteolytic products of ankyrin, effectively compete for and inhibit spectrin binding to membranes (Bennett, 1982a). The association of ankyrin with spectrin has been visualized by electron microscopy which shows that ankyrin binds to spectrin about 200 Å from the head of the dimer (that end which participates in dimer–dimer contacts to make tetramers; Tyler *et al.*, 1979, 1980). Other studies (Litman *et al.*, 1980; Morrow *et al.*, 1980) have shown that it is the spectrin β chain which binds to ankyrin and that, on the average, only one ankyrin binding site is occupied per spectrin tetramer *in vivo* (Bennett, 1979).

Although ankyrin mediates the attachment of spectrin to the membrane, it is not itself an integral membrane protein since it can be extracted from membranes by media containing high salt (Tyler *et al.*, 1979) without disrupting the lipid bilayer. However, ankyrin is tightly associated with the cytoplasmic portion of the major red cell transmembrane protein band 3 (Bennett, 1982a,b). In solution, ankyrin binds to band 3 with a 1:1 stoichiometry (Bennett, 1982a), but the molecular details of the ankyrin–band 3 complex are still uncertain. It is unclear for example whether all band 3 molecules in the membrane can bind ankyrin. While there are 10^6 copies of band 3 per cell, there are only about 110,000–125,000 binding sites for ankyrin per cell (Bennett, 1982a). Yet, studies in which purified band 3 molecules were incorporated into liposomes showed that the bulk of the purified band 3 molecules could bind ankyrin (Hargreaves *et al.*, 1980). Even if band 3 exists as a tetramer in the membrane as has been suggested by Nigg and Cherry (1979) and Weinstein *et al.* (1980), there would still be only one ankyrin binding site per every 2 or 2.5 tetramers. Recent work (Bennett, 1982b) has shown that band 3 molecules isolated in a 1:1 complex with ankyrin have peptide maps and other characteristics which are virtually identical to non-ankyrin-associated band 3. This further limits the possible distinctions between ankyrin and non-ankyrin-associated band 3 molecules.

F. Associations between Spectrin, Actin, and Band 4.1

The studies outlined above show how the major cytoskeletal protein spectrin is attached to the lipid bilayer by its association with band 3 via ankyrin. However, the cytoskeletal proteins, spectrin, actin, and bands 4.1 and 4.9, remain together in a self-associated shell even upon extraction of all band 3 and ankyrin. This shows that contacts other than those between spectrin and the membrane are important to cytoskeletal integrity. Many of these intermolecular associations have been defined and have recently been reviewed elsewhere (Branton *et al.*, 1981; Gratzer, 1981; Cohen, 1983).

One of the key interconnections in the membrane skeleton is that between spectrin and actin, a connection which is mediated by band 4.1. Measurement of the association of spectrin and F-actin has been made by various techniques including viscometry, binding studies, sedimentation analysis, and measurements of spectrin–actin complex formation on membranes (reviewed in Branton *et al.*, 1981; Cohen, 1983). The general conclusion from these studies is that without band 4.1, spectrin can associate only weakly with F-actin while the presence of band 4.1 promotes, strengthens, or stabilizes this association considerably. Visualization of these molecules by electron microscopy shows that band 4.1 binds at or very close to the tail of spectrin dimers or tetramers (Branton *et al.*, 1981). Further, electron microscopy of spectrin–F-actin complexes (with or without band 4.1) shows that spectrin's F-actin binding site is also at its tail (Branton *et al.*, 1981). It is unclear whether band 4.1 forms a link between spectrin and actin or promotes their association in some other, possibly allosteric, way. Studies of abnormal red cells deficient in or lacking band 4.1 show that band 4.1, and presumably the spectrin–actin complex it stabilizes, is an important determinant of red cell shape and cytoskeletal stability (reviewed in Cohen and Branton, 1981).

Although band 4.1 binds to spectrin, we know that spectrin is not required for band 4.1 to bind to red cell membranes because spectrin and actin can be entirely eluted from membranes without removing the bulk of band 4.1 (Steck and Yu, 1973). Moreover, purified band 4.1 will bind selectively to red cell inside-out vesicles, which lack spectrin and actin, in an apparently nonsaturable manner (Cohen and Foley, 1982). It has also been suggested that the association between band 4.1 and the membrane may involve the integral membrane protein PAS 2, which has been called glycoconnectin (Owens *et al.*, 1980; Mueller and Morrison, 1981). These and other observations (Cohen and Foley, 1982) show that band 4.1 may be capable of binding to both spectrin and membranes at the same time. Thus, band 4.1 could serve as a secondary membrane attachment site for the lengthy spectrin tetramer, securing spectrin's tail to the membrane in a manner similar to the way in which ankyrin secures the head. However, membranes depleted of ankyrin but containing band 4.1 have only a limited capacity

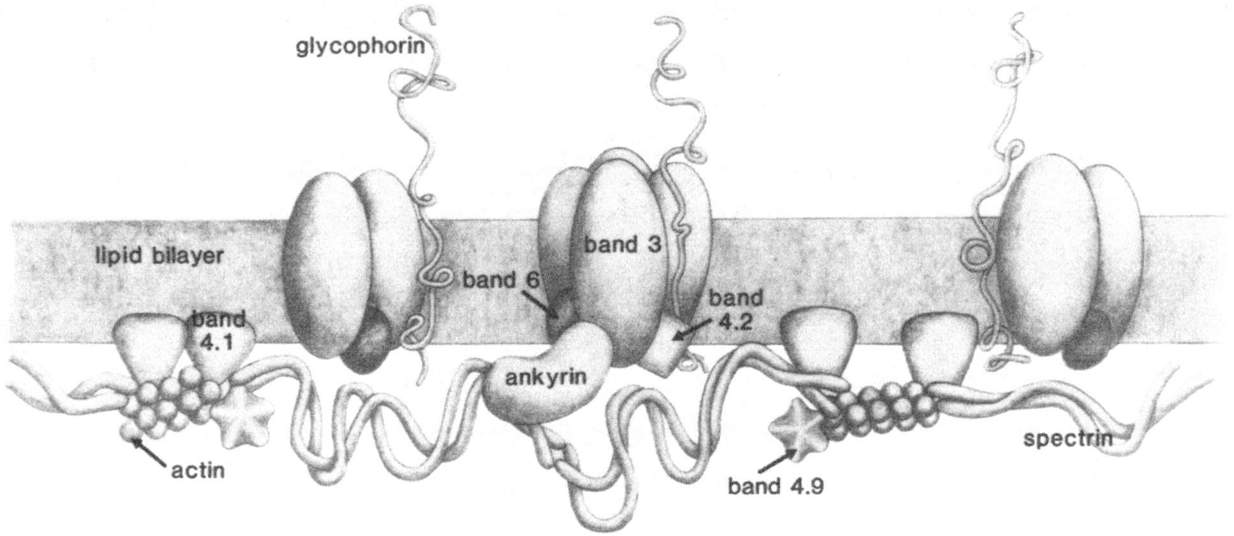

Figure 1. Model of the red cell membrane and its associated membrane skeleton illustrating some of the known protein–protein associations.

to bind spectrin (Bennett, 1982a). Thus, even though the spectrin–4.1 complex has a high K_d, in the range of 10^{-7} M, this association is not sufficient by itself to bind spectrin to membranes. In addition to the protein–protein associations discussed above, it is likely that the cytoskeletal proteins have important associations with membrane lipids. It is possible, for example, that some of the binding of band 4.1 to red cell vesicles may be accounted for by association with lipids.

C. Ultrastructure of the Membrane Skeleton

Recent work examining thin sections of red cell membranes fixed with tannic acid-glutaraldehyde has provided detailed views of the red cell membrane skeleton *in situ* (Tsukita *et al.*, 1980, 1981). Obliquely cut sections of membrane contain a homogeneously distributed but disorganized meshwork consisting of spectrin molecules connected to 25-nm-diameter nodes. The 25-nm-diameter nodes were similar in size and general appearance to structures seen by low-angle shadowing of mixtures of spectrin, actin and band 4.1 *in vitro* (Cohen *et al.*, 1980). Spectrin molecules radiated from these nodes which were proposed to consist of short aggregates or protofilaments of actin, possibly in association with band 4.1 (Cohen *et al.*, 1980). Micrographs of thin sections have also shown that the cytoskeletal proteins are confined to within about 10 nm of the bilayer (Ziparo *et al.*, 1978) even on inside-out vesicles with extremely small radii of curvature (Tsukita *et al.*, 1981). This may show, as suggested above, that the spectrin molecule is multiply tethered to the membrane via ankyrin (at the middle of tetramers) and band 4.1–actin complexes (at the ends of tetramers).

In summary, the biochemical and ultrastructural observations outlined above suggest a relatively simple picture of the membrane skeleton (see Figure 1). It seems likely that the unit cell of the membrane skeleton contains a core of actin molecules, forming a protofilament containing 5–30 monomers. Attached to this core are 2–12 spectrin dimers with associated molecules of band 4.1 (one per dimer) and band 4.9 (one per 2 dimers). This unit cell can be used to build a two-dimensional cytoskeleton by allowing all dimers to self-associate with adjacent unit cell dimers to form tetramers. Attachment of the complex to the membrane occurs through ankyrin and, to a lesser extent, band 4.1 (see Figure 1).

V. CYTOSKELETAL INVOLVEMENT IN CELL–SUBSTRATUM ASSOCIATIONS

A. Focal Adhesions

Focal adhesions are one of the most visually striking examples of the association of actin filaments with cell membranes. Focal adhesions, or adhesion plaques, are discrete regions found on the ventral surface of fibroblasts and other cells where the cell membrane makes its closest approach to the substratum (Curtis, 1964; Abercrombie and Dunn, 1975; Izzard and Lochner, 1976, 1980). Both electron microscopy and interference reflection microscopy suggest that the cell membrane approaches within

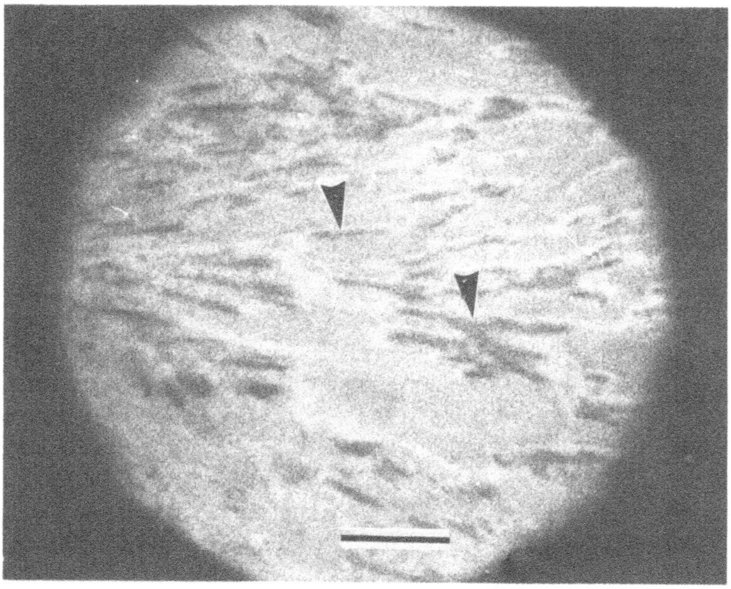

Figure 2. Interference reflection micrograph illustrating focal contacts (arrowheads) in Nil 8 hamster fibroblasts. Bar = 10 μm. (Photo courtesy of I. I. Singer).

10–15 nm of the substratum in these regions (Abercrombie *et al.*, 1971; Izzard and Lochner, 1980). Interference reflection microscopy is a simple method for the visualization of adhesion plaques *in situ*. The technique in its simplest form involves illuminating substrate-associated cells with light of a particular wavelength (Curtis, 1964). Using an appropriate optical detection system, focal adhesions appear as discrete black patches due to reflection interference of the illuminating light between the cell membrane and the substratum (Figure 2). Close associations appear as extended gray areas. Ultrastructural studies have shown that the intracellular sides of focal adhesions contain bundles of actin filaments which apparently terminate end-on at the membrane.

Efforts to understand the relationship between actin filaments, cell membranes, and the extracellular matrix in the region of focal adhesions can be divided into three categories: (1) efforts to define the role of vinculin, α-actinin, and other proteins thought to be involved in the attachment of actin filaments to the cytoplasmic membrane surface, (2) efforts to identify integral or transmembrane proteins involved in the attachment, and (3) studies designed to explore the role of extracellular fibronectin or other extracellular matrix proteins such as collagen in the attachment of actin to focal adhesions (see also Hynes, 1981).

B. The Role of α-Actinin and Vinculin in Focal Adhesions

As indicated in Section II-D, α-actinin has been a favorite candidate for mediating the attachment of actin filaments to cell membranes because of its role in striated muscle. Thus, the discovery of α-actinin in nonmuscle cells led to the suggestion that it may be involved in the association of actin filaments with cell membranes (Lazarides and Burridge, 1975; Mooseker and Tilney, 1975). In support of this idea, it was found that the polarity of actin filaments emanating from focal adhesions is identical to the polarity of actin filaments which terminate at the Z-line. (The polarity of an actin filament is defined by the direction of "arrowheads" formed on the filament when molecules of myosin fragment S-1 bind to it. In muscle, the barbed ends of the arrowheads are towards the Z-line, and in cultured cells the barbed ends are towards the focal adhesion.) However, increasing evidence has suggested that α-actinin is unlikely to be directly involved in mediating the end-on attachment of actin filaments to cell membranes (see Section II-D).

Despite the uncertain role of α-actinin in nonmuscle cells there is now strong evidence that the protein vinculin is involved with securing actin filaments to focal adhesions and other intracellular sites. Vinculin is a 130,000-dalton protein first isolated from chicken gizzard smooth muscle (Geiger, 1979; Geiger *et al.*, 1980). Geiger (1979) observed that while vinculin was present throughout the cytoplasm of fibroblasts, it appeared to be concentrated in patches which were coincident with focal adhesions. Subsequently, Geiger *et al.* (1980) showed by immunoelectron microscopy of frozen sections that vinculin was concentrated at several intracellular sites of actin–membrane contact. Vinculin was localized to the zonula adherens at the junctional complex of intestinal epithelial cells, to the membrane-associated dense plaques of smooth muscle cells, and to the fascia adherens of the intercalated disk membranes of cardiac muscle cells. Using a different approach, Burridge and Feramisco (1980) microinjected FITC-labeled vinculin into living fibroblasts and observed that the fluorescent protein had two patterns of intracellular distribution. One pattern was bright focal patches at the ends of actin filaments (likely to be focal adhesions) and the other was a fibrillar or streaky pattern most commonly found on the dorsal cell surface. Burridge and Feramisco (1980) also found that α-actinin appeared to be codistributed with vinculin.

Vinculin has been identified as a component of substrate-attached membranes of cultured cells whose upper surface has been removed by shearing (Avnur and Geiger, 1981). Vinculin was seen only in regions of the substrate-attached membrane identified by reflection interference microscopy as focal contacts. α-Actinin was also detected on these membranes in a punctate distribution along stress fibers and at focal adhesions. α-Actinin as well as filamin were also found on the exposed membrane surface of sheared cells at sites corresponding to the vertices of polygonal actin filament networks. These sites may be another type of actin–membrane contact distinct from focal contacts. Geiger (1982) also showed that fluorescent α-actinin added to demembranated cells was incorporated into actin bundles along their length and particularly at their termini, near sites of vinculin incorporation.

The studies cited above used light microscopy and showed that vinculin was highly enriched at focal adhesions. Electron microscopy of frozen sections also showed that vinculin is concentrated within 50 nm of filament–membrane junctions (Geiger *et al.*, 1980, 1981). It is not known, however, whether vinculin is itself attached to the cell membrane (via an integral membrane protein for example), or whether it accumulates in regions of actin–membrane contact for other reasons. Vinculin found in focal adhesions is apparently resistant to extraction by non-ionic detergents but can be removed by media of slightly alkaline pH or by exposure to media containing 1% deoxycholate or 1 M KI (Avnur and Geiger, 1981). It may be that other proteins which are tightly associated with the cell membrane at sites of membrane–substrate contact mediate the binding of vinculin to focal adhesions (Lehto *et al.*, 1980; Chen and Singer, 1982; Bayley and Rees, 1982).

C. Association of Vinculin with F-Actin

While the association of vinculin with the cell membrane is largely uncharacterized, there is evidence that vinculin binds directly to actin filaments. Wilkins and Lin (1982) and Burridge and Feramisco (1982) demonstrated that purified vinculin binds to actin filaments in solution with a stoichiometry of one vinculin per 1500–2000 actin monomers, and a dissociation constant of 20–45 nM. The stoichiometry suggests that vinculin may be interacting only with filament ends, a suggestion which would be consistent with the localization of vinculin to the ends of actin filament bundles in focal adhesions. However, vinculin has a number of effects upon both preformed and growing actin filaments which suggest that perhaps more than one type of actin–vinculin interaction is possible. Thus, vinculin increases the critical monomer concentration necessary for actin polymerization (Burridge and Feramisco, 1982; Wilkins and Lin, 1982). Further, vinculin decreases the viscosity of F-actin and delays the onset of salt- or nuclei-induced polymerization (Jockusch and Isenberg, 1981; Burridge and Feramisco, 1982; Wilkins and Lin, 1982). In addition, Jockusch and Isenberg (1981, 1982) and Wilkins and Lin (1982) have observed that vinculin induces actin filaments to form bundles in solution. Finally, *in vitro* studies leave open the question of whether vinculin actually does bind to the membrane-attached end of actin filaments (compare Isenberg *et al.*, 1982; Wilkins and Lin, 1982). At the time of this writing, vinculin has been visualized *in vitro* by electron microscopy (using ferritin-conjugated anti-vinculin antibodies) only along the length of actin filaments (Isenberg *et al.*, 1982).

The *in vitro* studies cited above suggest that vinculin may interact with actin filaments in two separate ways. First, vinculin may bind to the membrane-associated ends of filaments and actually link these ends to the membrane. Second, vinculin may bring actin filaments into a configuration favorable for membrane attachment by causing filaments to form bundles.

D. Effects of Cell Transformation on Focal Adhesions

The study of transformed cells has also shed some light on the role of vinculin. Transformed cells frequently have morphologies strikingly different from those of their normal counterparts. Further, the cytoskeletons of transformed cells exhibit fewer stress fibers, and in general appear less ordered than normal cytoskeletons (Pollack *et al.*, 1975; Wang and Goldberg, 1976; Pollack and Rifkin, 1976; Verderame *et al.*, 1980; Carley *et al.*, 1981). Whether these alterations are related to tumorigenicity *per se* or simply to cell shape is still, however, open to question (Der *et al.*, 1981).

Transformed cells are frequently rounded, and have fewer contacts with the substratum and with adjacent cells (for review see Vasiliev and Gelfand, 1977). Because cell–substratum contacts are known to be mediated by focal adhesions, these structures have been of interest to investigators seeking to understand cell transformation. It now appears that the actin–membrane contact at focal adhesions may be one of the prime loci of transformation-induced changes.

In cells transformed by Rous sarcoma virus (RSV), transformation has been attributed to the effects of a single gene product designated pp 60^{src}, a 60,000-dalton phosphoprotein possessing a unique capability to phosphorylate proteins at tyrosine residues (Collett and Erikson, 1978; Hunter and Sefton, 1980). The organization of actin, α-actinin, and vinculin is dramatically altered in cells transformed by RSV (David-Pfeuty and Singer, 1980; Carley *et al.*, 1981; Shriver and Rohrschneider, 1981). In RSV-transformed cells, focal adhesions are not oval shaped and dispersed under the ventral cell surface as in normal cells. Instead, they are clustered close together, leading to loss of the typical spread-out cell morphology. The discovery that vinculin was one of the few proteins which was phosphorylated by pp 60^{src} (Sefton *et al.*, 1981; Shriver and Rohrschneider, 1981; Hynes, 1982) lent support to the idea that vinculin was involved in focal adhesion function. Sefton *et al.* (1981) showed that vinculin contained 8-fold more phosphotyrosine in cells transformed by RSV than in uninfected cells. Other cytoskeletal proteins such as filamin, myosin, α-actinin, and vimentin appeared to be phosphorylated on tyrosine very little or not at all.

Further evidence that vinculin is one of the few targets of pp 60^{src} was provided by Shriver and Rohrschneider (1981). They showed that pp 60^{src} was concentrated within adhesion plaques of infected cells in a distribution coincident with that of vinculin. Vinculin is not, however, entirely localized to adhesion plaques, and it is not known whether different classes of vinculin are phosphorylated to different degrees. Moreover, Hynes (1982) has noted that even at the elevated phosphotyrosine levels found in transformed cells, there is only about 0.01 phosphotyrosine per vinculin. Nevertheless, it is possible either that there is a rapid turnover of phosphate and a large fraction of vinculin is phosphorylated briefly or that phosphorylation of only a small fraction of the vinculin is sufficient to exert a large effect (Hynes, 1982). The distribution of α-actinin, another protein which is located at or near adhesion plaques (David-Pfeuty and Singer, 1980), is also altered in RSV-infected cells but this protein seems not to be affected by pp 60^{src}. While the organization of α-actinin or even actin may be affected by changes in the degree of vinculin phosphorylation, it is also likely that other intracellular changes contribute to the cytoskeletal alterations associated with RSV.

E. The Relationship of Focal Adhesions with the Extracellular Matrix

Actin filament bundles at focal adhesions may have important transmembrane contacts with proteins of the extracellular matrix. Thus, several lines of evidence suggest that there is continuity between cortical actin filaments and extracellular fibronectin. For example, disorganization of the cytoskeleton by cytochalasin B induces the release or loss of cell surface fibronectin (Mautner and Hynes, 1977; Ali and

Hynes, 1977; Kurkinen *et al.*, 1978). Further, addition of fibronectin to fibronectin-poor cells results in the organization of cytoplasmic actin into bundles (Ali *et al.*, 1977; Willingham *et al.*, 1977). Comparison of the distribution of actin and fibronectin by immunofluorescence during cell spreading reveals that the appearance of stress fibers is followed closely by the appearance of fibronectin strands exactly underneath them (Hynes, and Destree, 1978; Heggeness *et al.*, 1978; Der *et al.*, 1981). These observations have led to the suggestion that intracellular actin filament bundles make some kind of direct transmembrane connection with extracellular fibronectin.

While observations using immunofluorescence have established the close proximity of intracellular actin fibers and extracellular fibronectin, the limited resolution of this technique (0.2 μm) prevents a clear demonstration of coincidence. Electron microscopic observations have, however, shown that some actin bundles which contact the cell membrane appear to make a transmembrane connection with fibronectin filaments at the corresponding extracellular side (Singer, 1979; Singer and Paradiso, 1981). The junction of these filamentous systems has been termed the fibronexus (Singer, 1979; Figure 3). Such transmembrane connections are seen in a variety of cell types (Trotter, 1981). The fibronexus may be identical to focal adhesions even though the fibronexus has been most frequently seen on the dorsal cell surface which lacks or is poor in vinculin (Singer, 1979, 1982). In addition to the morphological evidence for actin–fibronectin connections, other studies have shown that fibronectin will bind actin *in vitro* (Keski-Oja *et al.*, 1980), although with low affinity (Koteliansky *et al.*, 1981). However, the relevance of these observations is unclear since there is no evidence that either actin or fibronectin can tranverse the plasma membrane.

Whether fibronectin is present at all focal adhesions on the ventral cell surface has been the subject of controversy. In contrast to observations by I. Singer and co-workers of close transmembrane relationships between actin bundles and fibronectin filaments, several groups have reported that fibronectin is absent from some or all focal adhesions (Chen and Singer, 1980; Birchmeier *et al.*, 1980; Fox *et al.*, 1980; Badley *et al.*, 1980; Avnur and Geiger, 1981). Chen and Singer (1982) concluded that focal adhesions had no underlying fibronectin, while close contacts (regions of 30–50 nm spacing between the membrane and substrate) were rich in fibronectin at the extracellular side. The appearance of fibronectin in close, but not focal, contacts has also been noted by others (Norton and Izzard, 1982). Further, Virtanen *et al.* (1982) reported that cells which are blocked from secreting fibronectin fail to form focal adhesions altogether.

Singer (1982) has shown that the presence of fibronectin in focal adhesions is determined, at least in part, by the concentration of serum in which cells are cultured. Singer postulates that at high (5% v/v) levels of serum fibroblasts exhibit heightened surface activity and that this results in the failure of cells to form stable, highly adhesive, fibronectin-positive attachments. Conversely, cells which have been cultured at low serum concentrations tend to be well spread and exhibit less activity. These cells are characterized by a high frequency of fibronectin- and vinculin-positive focal adhesions (Hynes, 1981; Singer, 1982). These conclusions are complicated by the fact that fibroblasts can actually remove extracellular fibronectin from the substratum (Avnur and Geiger, 1981), and that antibodies to substrate-associated molecules may have

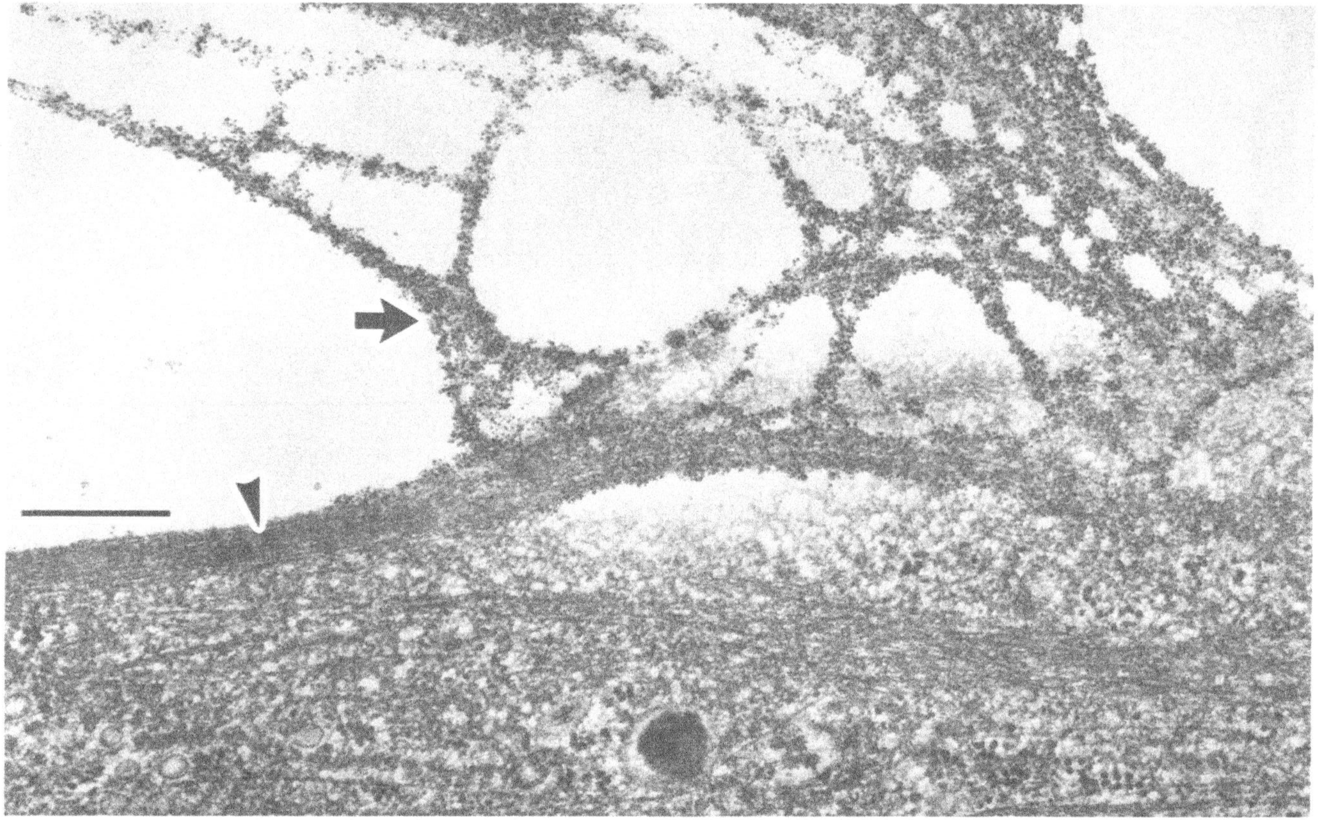

Figure 3. Electron micrograph of Nil 8 hamster fibroblasts stained for fibronectin by an indirect immunoferritin method. Note the extensive ferritin label in the fibronectin-containing extracellular matrix (arrow) and the apparent continuity between fibronectin and the intracellular actin bundle (arrowhead). Bar = 0.5 μm. (Photo courtesy of I. I. Singer).

difficulty penetrating focal adhesions (Grinell, 1980). Nevertheless, it seems likely that some focal contacts are fibronectin positive and others are fibronectin negative and that the relative proportion of each may be a function of growth conditions.

F. The Formation of Focal Adhesions

The question of how focal contacts form and how actin filaments and vinculin become associated with them remains unanswered. Geiger (1982) proposed that focal adhesion formation may be initiated by local patching of membrane receptors in areas of transient cell–substratum contact. At these regions, vinculin may become attached to the cytoplasmic membrane surface (possibly to a specific membrane receptor) and initiate the attachment of actin bundles. α-Actinin may also be involved in the organization of the associated filaments somewhat more distant from the actual membrane contact (Chen and Singer, 1982). Figure 4 summarizes some of the possible associations discussed above.

The origin of actin filaments which terminate at focal adhesions is also not known. Do these filaments result from rearrangement of the relatively disordered network of actin filaments located in the cell cortex or do they arise *de novo* in response to some initiating stimulus at the focal adhesion? It is particularly difficult to envision a mechanism whereby actin filaments actually grow from the membrane into the cytoplasm. Such a mechanism would require that actin monomers be added to the end of the filament anchored to the membrane at vinculin-containing sites (see for example Tilney

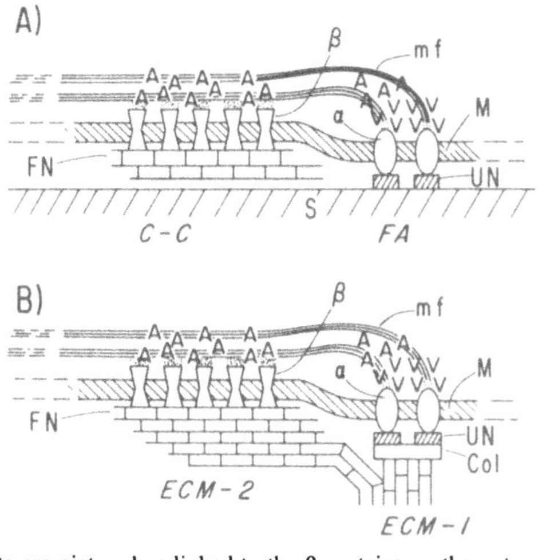

Figure 4. (A) Highly schematic representation of the molecular ultrastructure and transmembrane interactions involved at fibroblast–substratum focal adhesion (FA) and close contact (C-C) sites. M designates the ventral surface membrane of the cell; S, the substratum; mf, microfilaments; A, α-actinin; V, vinculin; FN, extracellular fibronectin; UN, unknown factors thought to play a role in cell–substratum adhesion; α and β are two distinct but unidentified transmembrane proteins. At FA sites, microfilament bundles are known to terminate at the membrane, and vinculin may mediate the attachment by way of the α proteins. Although fibronectin is shown absent from the FA sites here, this depends upon growth conditions (see text). At C-C sites, it is proposed that microfilaments are laterally associated with the cytoplasmic face of the membrane, perhaps linked to the β integral membrane proteins by α-actinin; fibronectin filaments are pictured as linked to the β proteins on the outer membrane surface. (B) Transmembrane associations with the extracellular matrix (ECM). Here two types of association are shown; ECM-1 type association is analogous to a focal adhesion and would be identified as a fibronexus (see text). These may be mediated by collagen (Col; Chen and Singer, 1982) and other unknown factors (UN). ECM-2 type associations are analogous to close contacts. Reprinted by permission from Chen and Singer (1982).

et al., 1981). Arguing against this possibility is evidence suggesting that vinculin in solution may inhibit actin monomer addition to the membrane-attached ends (barbed ends) of filaments (Wilkins and Lin, 1982), although this is still controversal (Isenberg *et al.*, 1982). Alternatively, actin filaments could grow by addition of monomers to the "pointed" end of the membrane-attached filament. Although the pointed end has a lower affinity for actin monomers than the barbed end *in vitro*, capping of the barbed end by the membrane could make it the preferred end *in vivo*.

C. The Function of Focal Adhesions

Our understanding of membrane-associated actin is further hampered by our ignorance of the function of filament bundles. It is tempting to speculate that filament bundles which are anchored to the membrane exert a force between one part of the cell and another, as has been proposed for filaments of the contractile ring (Schroeder, 1975). A force could, for example, be exerted between opposing sets of filament bundles via myosin interconnections in a fashion similar to striated muscle. Ultrastructural evidence for such a phenomenon is, however, lacking. Myosin is known to be a component of stress fibers (Fujiwara and Pollard, 1976) but how it functions in that configuration is unknown.

It may be that membrane-associated actin filament bundles serve a structural rather than dynamic role. As discussed in Section II-A.2, the most dynamic regions of cells contain few or no easily recognizable filament bundles (Herman *et al.*, 1981) but seem to be rich in a gel-like disorganized array of actin filaments and associated proteins. Stress fibers have been detected *in vivo* (Byers and Fujiwara, 1982), but there also they may promote cellular adhesion rather than produce movement. It may be that stress fibers and membrane-associated filament bundles are a way for the cell to anchor itself to the extracellular matrix via specialized membrane sites.

VI. CYTOSKELETAL–MEMBRANE INTERACTIONS IN MICROVILLI OF THE INTESTINAL BRUSH BORDER

The lining of the intestine consists of highly polarized epithelial cells whose apical (luminal) surface is covered with numerous uniform microvilli 0.1–2 mm in length and approximately 0.1 mm in diameter. This highly organized structure has been termed a brush border and mediates absorption of nutrients and water from luminal contents. Electron microscopic studies have revealed that these microvilli contain microfilament bundles of approximately 20–30 actin filaments (see Figure 5) which extend lengthwise through the microvilli to form a core (Ishikawa *et al.*, 1969; Brunser and Luft, 1970; Mukherjee and Staehelin, 1971; Mooseker and Tilney, 1975). These bundles extend down past the microvillar base and splay out into a region of more sparsely arranged microfilaments and intermediate filaments called the terminal web, which is arranged roughly perpendicular to the microvilli (Mukherjee and Staehelin, 1971; Mooseker and Tilney, 1975; Hirokawa and Heuser, 1981). Myosin filaments have also been seen in the terminal web by electron microscopy of thin sections

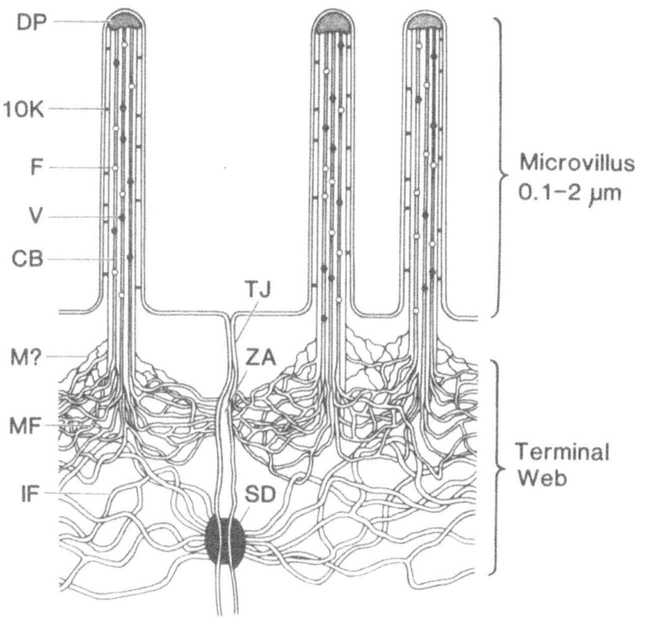

Figure 5. Schematic illustration of the arrangement of cytoskeletal components in the intestinal brush border based on biochemical and morphologic data detailed in the text. Structures are not necessarily to scale. Terminal web structure is largely derived from morphology of quick-freeze deep-etch samples as described by Hirokawa and Heuser (1981) and Hirokawa *et al.* (1982). DP, electron-dense patch at the microvillus tip where core bundle microfilaments may terminate (or originate); 110K, 110,000-dalton protein which crossbridges the core bundle and plasma membrane; F, fimbrin, an actin-bundling protein; V, villin, an actin-bundling protein; CB, microfilament core bundle; M?, crossfilaments which may contain myosin, suggested from studies of Hirokawa *et al.* (1982); MF, microfilament; IF, intermediate filament; TJ, tight junction between adjacent epithelial cells; ZA, zonula adherens, junctional region with which terminal web microfilaments are associated; SD, spot desmosome complex, site of membrane interaction with terminal web intermediate filaments.

(Mooseker and Tilney, 1975). The presence of myosin exclusively in the terminal web has also been confirmed by immunofluorescence microscopy (Mooseker *et al.*, 1978; Bretscher and Weber, 1978b) and electron microscopy of quick-frozen deep-etched samples (Hirokawa and Heuser, 1981; Hirokawa *et al.*, 1982).

The terminal web microfilaments appear to insert into the membrane at the zonula adherens, an intermediate junction that rings the microvillar-covered apical surface of the cell at the point of cell–cell contact (Rodewald *et al.*, 1976; Hull and Staehelin, 1979). The protein vinculin, which may be involved in microfilament bundle–membrane interactions at focal adhesions (see Section V), is selectively distributed in the cell near the zonula adherens, and may play a role in terminal web microfilament association with this junctional complex (Geiger *et al.*, 1981). The intermediate filaments interact with spot desmosomes at the junctional complex (Franke *et al.*, 1978,

1979; Hull and Staehelin, 1979) and may be connected to terminal web actin filaments by fine fibrils which can be displaced by high concentrations of myosin S-1 fragment (Hirokawa *et al.*, 1982). Whether these fibrils are myosin-related remains in question.

The microvillus core bundle consists of actin filaments of uniform polarity as illustrated by myosin S-1 decoration, which reveals arrowhead complexes pointing away from the membrane (Mooseker and Tilney, 1975; Begg *et al.*, 1978). At the tip of each core bundle is an electron-dense patch that appears to be membrane-associated and into which the core microfilament bundles insert (Mooseker and Tilney, 1975; Mukherjee and Staehelin, 1971; Brunser and Luft, 1970). This dense patch may be involved in microfilament polymerization, which occurs at this end of the core bundle (Tilney and Cardell, 1970; Mooseker *et al.*, 1982). Along its length, the core bundle is attached to the plasma membrane by fine crossbridges which extend laterally from the core in a helical pattern (Mukherjee and Staehelin, 1971; Mooseker and Tilney, 1975; McNutt, 1978; Matsudaira and Burgess, 1979, 1982b; Hirokawa and Heuser, 1981). These lateral arms exhibit a 33-nm periodicity and point away from the microvillus tip in a manner similar to the direction of S-1 decoration. The bridges may interact with particular sites on the actin filaments (suggested by the periodicity) and appear to interact with dense patches on the microvillar membrane (Matsudaira and Burgess, 1982b). Freeze-fracture studies of microvillar membrane particle distribution, ostensibly in search of correlated helical particle patterns, have generally revealed random distributions, although orderly arrangements have been reported (Mukherjee and Staehelin, 1971; Tilney and Mooseker, 1976; McNutt, 1978; Neutra, 1979).

A series of elegant experiments from a number of laboratories have recently led to the identification of the major protein components involved in microvillar filament organization. Bretscher and Weber (1978a) developed a method to isolate pure microvilli and cores from the brush border, which enabled detailed biochemical studies of the microvillar constituents. In addition to actin filaments, microvilli cores contain three major proteins with molecular weights of 110,000, 95,000, and 68,000 daltons (Bretscher and Weber, 1979; Matsudaira and Burgess, 1979). Matsudaira and Burgess (1979) studied the arrangement of these proteins in the core by correlating selective extraction of individual proteins with structural changes in the microfilament core. Addition of Mg^{2+}-ATP to isolated cores at 4°C resulted in the loss of the lateral arms connecting the core bundle to the membrane, but had no effect upon the integrity of the core bundle itself. That this treatment also led to a great reduction in the amount of microvillus-associated 110,000-dalton protein suggested that this polypeptide was part of the arm. Conversely, when isolated cores were treated with the detergent deoxycholate, the core filaments became unbundled, yet retained their lateral arms. Concomitantly, the 95,000- and 68,000-dalton proteins were released from the core, suggesting that these proteins may be involved in core filament bundling. These suggestions have been confirmed by subsequent isolation and characterization of the 95,000-dalton protein which has been named villin (Bretscher and Weber, 1979) and the 68,000-dalton protein now named fimbrin (Bretscher and Weber, 1980a). Villin and fimbrin have both been shown to bind F-actin and calcium and to induce filament bundling (Bretscher and Weber, 1979; Bretscher and Weber, 1980a; Bretscher and

Weber, 1980b; Mooseker *et al.*, 1980; Craig and Powell, 1980; Glenney *et al.*, 1981; Matsudaira and Burgess, 1982a). The bundles formed do not have lateral arms, but otherwise resemble the microfilament core bundle of intestinal microvilli.

The 110,000-dalton (110K) protein therefore appears to be the component which mediates microfilament–membrane attachment in the microvilli. In calmodulin gel-overlay experiments, this protein was shown to bind calmodulin independently of calcium (Glenney and Weber, 1980) as do intact cores (Howe *et al.*, 1980). The 110K protein was originally believed to be α-actinin, but further immunologic and peptide mapping studies have shown that the two proteins are distinct (Schollmeyer *et al.*, 1974; Mooseker and Tilney, 1975; Bretscher and Weber, 1978b; Geiger *et al.*, 1979; Mooseker and Stephens, 1980). Although preparative gel electrophoresis has been used to isolate this protein for antibody production, it has not yet been isolated in a native form suitable for biochemical studies. The protein as released by Mg^{2+}-ATP treatment of isolated cores (Matsudaira and Burgess, 1979) was reported to be unstable and subject to rapid degradation (Coudrier *et al.*, 1981). Likewise, the protein was sensitive to proteolysis in isolated cores (Glenney and Weber, 1980; Mooseker and Stephens, 1980; Glenney *et al.*, 1982c). Although immunofluorescence studies showed the protein to be located primarily in the microvilli portion of the brush border, staining of some terminal web filaments was also reported (Coudrier *et al.*, 1981; Glenney *et al.*, 1982c).

Recently, Coudrier *et al.* (1981, 1982) have described another major protein which was found to be associated with microvillus cores prepared from Triton X-100 extracted microvilli. This protein of 140,000 daltons (140K) behaves as an integral membrane protein and appears to be a glycoprotein, as shown by its binding to concanavalin A and inhibition of this binding by α-methyl-D-glucopyranoside. By immunofluorescence and immunoferritin labeling of thin sections, this protein appears to be localized only in the microvilli of the brush border and not in the terminal web region. Furthermore, the ferritin–antibody complexes appear to be localized on the outside of the microvilli. These data suggest that the 140K protein may be a membrane glycoprotein which spans the membrane to interact with the microfilament core. Whether the 140K protein is similar to red cell band 3 protein in this regard is an important question for future research in this area.

In summary, the microvillus is a fascinating biological structure exhibiting several types of associations between cytoskeletal proteins and the plasma membrane. While the lateral arms (110K protein) connecting the filament core bundle to the microvillus membrane have been partially defined, much remains to be learned about this connection. It will be particularly important to determine how the lateral arms attach to the membrane, and whether this attachment is mediated by other, possibly integral, membrane proteins. In addition, it will be important to identify the dense material at microvilli tips and to characterize any actin–membrane contacts which occur there. The firm identification of the cytoskeletal and membrane proteins involved in the associations discussed above will be an important contribution to our understanding of cytoskeletal–membrane contacts.

VII. MEMBRANE-ASSOCIATED CYTOSKELETAL ELEMENTS AND THE CONTROL OF CELL SURFACE RECEPTOR DYNAMICS

The cytoskeletal–membrane associations of erythrocytes, adhesion plaques, and brush border microvilli described above are relatively static and structural in nature. However, there is mounting evidence that the cytoskeleton may also function in dynamic membrane processes involving cell surface components. Such processes include ligand-induced capping or clustering of surface receptors, pinocytosis and phagocytosis, and modulation of lateral diffusion rates of membrane components. Specific associations of cytoskeletal proteins and membrane components involved in these processes have not been demonstrated, but indirect evidence suggests that the state of the cytoskeleton is important. Such evidence includes studies with agents which perturb microfilaments of microtubules, morphologic observations of cytoskeletal structures with both electron and light microscopy, and isolation of cytoskeletal–membrane complexes. Possible associations of cytoskeletal and membrane components in three dynamic processes are detailed below.

A. Capping

Since the phenomenon of ligand-induced capping was first described in lymphocytes, similar phenomena have been observed in a number of other systems (for reviews see dePetris, 1977; or Weatherbee, 1981). In the classic definition of capping, binding of multivalent ligands (such as antibodies to surface determinants) to the cell surface is followed by temperature-dependent clustering and energy-dependent movement of the ligand–receptor complexes to one pole of the cell to form a cap. However, further studies have revealed the capping phenomenon to be very heterogeneous, depending on the cell or the ligand involved. For example, ligand-independent capping has been described (Yahara and Kamimoto-Sameshima, 1977, 1979; Bourguignon *et al.*, 1981; Yahara *et al.*, 1979), as well as capping of receptors induced by ligands which are not multivalent (Oliver and Berlin, 1982). Drastic cell shape changes may occur during cap formation and such shape changes may be a prerequisite for capping (see Oliver and Berlin, 1982). Cell locomotion may be stimulated during capping or the capping process may occur on locomoting cells (Braun *et al.*, 1978; Schreiner *et al.*, 1976; Schreiner and Unanue, 1975a,b). Some ligands induce cap formation rapidly while others proceed slowly or require secondary ligands to cap, possibly reflecting separate mechanisms (Unanue, 1976; Braun *et al.*, 1978). However, the involvement of the cytoskeleton in capping has been well established, although the specific mechanisms remain undetermined.

The energy dependence and the rapidity of the capping process suggested that the lateral movement of ligand–receptor complexes in the membrane was not simple diffusion, an idea which has been supported by studies on rates of ligand–receptor movement (Unanue and Schreiner, 1977; Koppel *et al.*, 1982). The involvement of cytoskeletal elements was first suggested by studies showing that cytoskeletal-perturbing drugs could alter the capping process (Taylor *et al.*, 1971; Yahara and Edelman, 1972, 1973; Edelman *et al.*, 1973; dePetris, 1975). To date, a wealth of reports have been published which provide some evidence for the involvement of cytoskeletal structures in capping and suggest that redistribution of surface ligands is accompanied by redistribution of cytoskeletal proteins.

For example, binding of antibodies to surface immunoglobulins, T-25 antigens, or H-2 antigens of lymphocytes induces rapid capping of these molecules, which is accompanied by concentration of myosin in the cytoplasm just under the cap, as detected by immunofluorescence microscopy (Bourguignon and Singer, 1977; Schreiner *et al.*, 1977; Braun *et al.*, 1978). Similarly, high concentrations of actin and dense microfilament networks have been detected in the cell cortex immediately under caps induced by a variety of ligands in a number of cell types (Albertini and Clark, 1975; Sundqvist and Ehrnst, 1976; Bourguignon and Singer, 1977; Albertini *et al.*, 1977; Oliver *et al.*, 1977; Toh and Hard, 1977; Gabbiani *et al.*, 1977; Albertini and Anderson, 1977; Mousa *et al.*, 1978; Sundqvist *et al.*, 1978; Bourguignon *et al.*, 1978; Bourguignon and Rozek, 1980; Butman *et al.*, 1980). Such concentrations of actin and myosin are not present in the resting cell and seem to correlate with receptor capping. Accumulation of the actin-binding protein α-actinin has also been observed associated with surface immunoglobulin, concanavalin A, or H-2 caps on lymphocytes (Geiger and Singer, 1979).

Recently, Dellagi and Brouet (1982) have described redistribution of vimentin intermediate filaments accompanying capping of β_2-microglobulin or surface immunoglobulin on lymphocytes. When cells were treated with colcemid, which results in microtubule depolymerization and coiling of intermediate filaments (see Lazarides, 1980), caps were localized on the membrane in the region of intermediate filament coils. Herman and Albertini (personal communication) have recently observed similar phenomena in concanavalin A capping of colchicine-treated ovarian granulosa cells. Aggregation of intermediate filaments has also been observed between the cap and nucleus of concanavalin A-capped lymphocytes (Bourguignon and Bourguignon, 1981b).

While cocapping of tubulin and surface receptors has been reported (Albertini and Clark, 1975; Gabbiani *et al.*, 1977), other evidence has suggested that microtubule *absence* may actually stimulate capping. Yahara and Edelman (1972, 1973) reported that excessive concentrations of concanavalin A bound to lymphocytes inhibited capping of surface immunoglobulin or antigens. Furthermore, this restriction could be released by the microtubule-disrupting drug colchicine, suggesting that microtubules anchored the ligand–receptor complexes in some manner and that release from this restraint allowed lateral redistribution of complexes (Edelman *et al.*, 1973). Capping of concanavalin A receptors on leukocytes or macrophages is also stimulated by microtubule-disrupting agents such as nocodozole, colcemid, colchicine, or diamide (Oliver *et al.*, 1976, 1980; Albertini *et al.*, 1977; Pick and Wilner, 1979). Evidence suggests that such microtubule disassembly allows reorganization of other cytoskeletal components, and results in asymmetry of cell structure and membrane functions (see Oliver and Berlin, 1982).

The cocapping of cytoskeletal proteins or structures and ligand–receptor complexes points to functional associations of surface and cytoplasmic components, but leaves open the question of whether these associations are direct or indirect. While clear-cut data on this point are scarce, several conceptual schemes have been proposed to account for these phenomena. Some of these models have been reviewed elsewhere (Weatherbee, 1981; Oliver and Berlin, 1982), and their merits will not be discussed here. For our purposes, such models fall into two categories. The first type postulates some manner of direct interaction of cytoskeletal proteins with membrane receptors. For example, Edelman and co-workers proposed a model in which microtubules served to anchor receptors, while microfilaments could attach to and move receptors laterally in the membrane (Edelman *et al.*, 1973; Edelman, 1976). Bourguignon and Singer (1977) suggested a similar model in which microfilaments were attached to ligand-bound receptors by some unidentified protein.

The second type of model postulates a more indirect influence of cytoskeletal components on receptor topography. Bretscher (1976) and Harris (1976) suggested that oriented receptor movement was a result of membrane flow from one pole of a cell to the other, e.g., from the leading edge to the uropod or cap of a motile cell. Recently, Oliver and Berlin (1982) have described a more complex model based in part on a concept described by Hewitt (1979). They suggest that reorganization of microfilaments generates asymmetric tension in the nearby membrane, resulting in formation and translation of directional waves in the membrane, which can carry certain receptors to a common destination.

There is some data to support the notion of direct connections between cytoskeletal proteins and membrane receptors. For example, Flanagan and Koch (1978) described a complex of surface immunoglobulin and actin, which could be isolated from NP-40 detergent lysates of lymphocytes whose surface immunoglobulin had been cross-linked by anti-immunoglobulin antibodies. Isolation of the complex was dependent on such crosslinking, suggesting that the actin–membrane association was induced by ligand–receptor crosslinking.

A slightly different approach was used by Condeelis (1979), who isolated concanavalin A-induced caps from *Dictyostelium discoideum* amoebae by making use of the resistance of such caps to disruption by the detergent Triton X-100. He found that when concanavalin A bound to its receptors on the cell, significant amounts of actin and myosin were found in a detergent-resistant plasma membrane fraction. These amounts remained the same during the time course of cap formation, until the cap was complete. After that, the amounts of actin and myosin associated with the isolated caps decreased. These data suggested that ligand binding to the cell surface triggered associations of these cytoskeletal proteins with the ligand–receptor complexes, whose movement into a cap was accomplished by an actomyosin sliding filament mechanism. Although no specific actin-binding membrane proteins were described, the myosin was thought to be associated with the membranes indirectly through its binding to actin, since most myosin could be removed by addition of Mg^{2+}-ATP. This was similar to actin and myosin interactions on Dictyostelium membranes described by Spudich (1974). Condeelis (1981) also found that no gelation factors were associated with isolated membranes and that actin and myosin remained associated with membranes even under conditions which solate actin gels. Therefore, actin gelation was not required for actin–myosin–membrane interactions.

The possibility of direct associations between actin and Dictyostelium membranes was further supported by Luna *et al.* (1981), who showed that isolated Dictyostelium membrane vesicles (presumably with inside-out orientation) could be crosslinked by F-actin. This actin–membrane association seemed to be mediated by membrane proteins, since protease-treated membranes no longer crosslinked F-actin. These investigators also extracted isolated amoebae membranes with Triton X-100 and found that actin and myosin were associated with membrane proteins in a Triton-insoluble residue.

Myosin has also been found in purified plasma membranes of murine T-lymphoma cells, and myosin 20K light chains were shown to be preferentially associated with the membranes of cells capped by colchicine treatment (Bourguignon and Bourguignon, 1981a; Bourguignon *et al.*, 1981). These myosin light chains were phosphorylated to a greater degree in membranes from capped vs. uncapped cells. Phosphorylation of myosin light chains was proposed as a regulatory mechanism for myosin attachment to the membrane. Such attachment might, in turn, activate an actomyosin system which could collect membrane receptors into a cap. In addition, a phosphorylated 28K membrane protein was identified which coprecipitated with myosin in antimyosin immunoprecipitation. Whether this may represent a "linkage" protein between the membrane and cytoskeleton remains to be seen.

Evidence has also accumulated suggesting that intracellular Ca^{2+} and calmodulin may have regulatory roles in capping through influences on cytoskeletal organization. Altering intracellular Ca^{2+} levels by use of the Ca^{2+} ionophore A23187 or by local anesthetics inhibits capping and results in disruption of preformed caps (Braun *et al.*, 1978; Poste *et al.*, 1975; Ryan *et al.*, 1974; Schreiner and Unanue, 1976a,b). Disruption of caps is concomitant with dispersal of myosin concentrated near the cap (Braun *et al.*, 1978). It was suggested that calcium might cause such effects by inducing a disorganized contraction of cortical cytoskeletal structures over the entire cell, thus hindering cap-associated filaments, or by disrupting filament–membrane interactions. It is noteworthy that Ca^{2+} (via calmodulin) has been shown to regulate myosin light chain kinase and consequently phosphorylation of myosin, which was increased in T-lymphoma cell caps (Bourguignon *et al.*, 1981; see above). Further evidence for the role of calmodulin in capping was described by Salisbury *et al.*, (1980, 1981), who showed that calmodulin becomes concentrated in the cell cortex under the cap. However, their studies suggest that this calmodulin is necessary for the endocytosis of ligand–receptor complexes that occurs subsequent to cap formation rather than for cap formation *per se*.

In summary, cytoskeletal reorganizations occurring during the capping process may provide driving or directional forces for ligand–receptor redistribution. Associations of contractile and cytoskeletal proteins with the plasma membrane have been suggested by experiments such as those described above, although specific membrane components which may interact with cytoskeletal elements have not been identified. Identification of such components will lead to a better understanding of the role of the cytoskeleton in the dynamic process of receptor capping.

B. Endocytosis

Binding of particles or ligands to their respective receptors on the cell surface triggers invagination of the membrane and internalization of the attached particles or ligands in plasma-membrane-derived endocytic vesicles. Small ligands such as hormones, plasma transport proteins, asialoglycoproteins, and lysosomal enzymes bind to individual receptors on the cell surface which become clustered into specialized areas of the plasma membrane known as coated pits (Goldstein *et al.*, 1979; Kaplan, 1981). The ligand–receptor complexes in coated pits are rapidly internalized in a process termed receptor-mediated endocytosis (see Pastan and Willingham, 1981). Larger particles can also be internalized by the process of phagocytosis, in which the cell extends pseudopods around the particle, forming a large membrane-bound vesicle (Stossel and Hartwig, 1976; Silverstein *et al.*, 1977; Silverstein and Loike, 1980). Evidence for the involvement of cytoskeletal proteins in both receptor-mediated endocytosis and phagocytosis is largely indirect, being suggested by the inhibitory effects of cytoskeletal-perturbing drugs on these processes (Allison and Davies, 1974; Axline and Reaven, 1974; Hartwig *et al.*, 1980; Salisbury *et al.*, 1980). In some cases however, colocalization of cytoskeletal proteins with areas of endocytic activity has suggested a functional relationship between the two.

The involvement of cytoskeletal proteins with membrane components related to endocytosis has been suggested by several studies. Bloom *et al.* (1980) showed that coated pits were aligned on the membrane over actin stress fibers, although a direct connection between the two was not demonstrated. There is, however, some structural evidence for direct association between coated vesicles and cytoskeletal filaments (Salisbury *et al.*, 1980; Heuser, 1980; see Figure 6). In addition, *in vivo* studies have shown that G- and F-actin as well as α-actinin and tropomyosin can all bind to the protein clathrin (Schook *et al.*, 1979; Puszkin *et al.*, 1979), which is the major constituent of the coat of coated vesicles and pits (Pearse, 1975, 1976). Clathrin can itself bind to plasma membranes. While the specific mechanism of association is not yet known, Unanue *et al.*, (1981) have described *in vitro* reassociation of clathrin with endocytic vesicles stripped of endogenous clathrin. Clathrin binding was diminished in vesicles in which a 110,000-dalton polypeptide had been selectively degraded by the protease elastase, suggesting a role for this polypeptide in clathrin association with the membrane. A similar 110,000-dalton protein which was extractable from vesicles in high-salt media was previously shown to promote clathrin assembly into "baskets" (Keen *et al.*, 1979).

The ingestion of larger particles via phagocytosis also involves the cytoskeletal apparatus. Immunofluorescence and electron microscopic studies of phagocytosing cells have shown actin to be concentrated in the cell cortex near the site of particle ingestion. Dense arrays of microfilaments are also present in the pseudopods surrounding the particle (Reaven and Axline, 1973; Allison and Davies, 1974; Korn *et al.*, 1974; Stossel and Hartwig, 1976; Hartwig *et al.*, 1977; Berlin and Oliver, 1978; Painter and McIntosh, 1979; Painter *et al.*, 1981). Myosin and actin-binding protein have also been detected near the membrane at sites of phagocytosis in macrophages

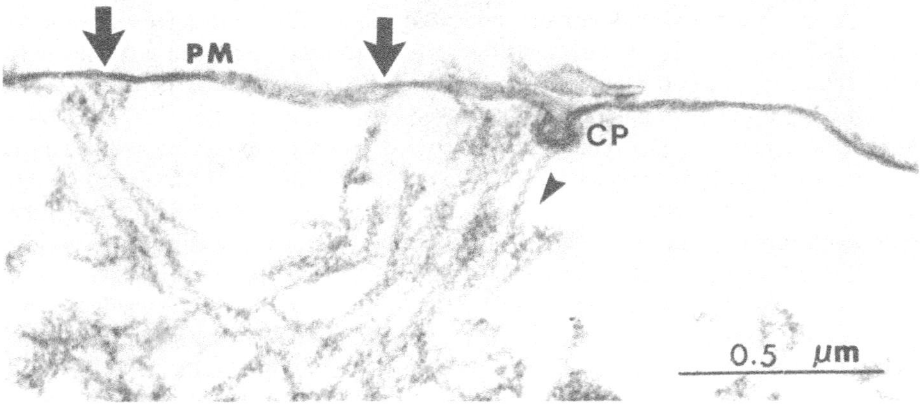

Figure 6. Microfilaments associated with a coated pit (CP) in WiL 2 lymphoblastoid cells. In this cell, the actin filaments have been decorated by heavy meromyosin (HMM); the polarity of the arrowheads on the filament (small arrowhead) point away from the coated pit. Also shown are several instances of actin filaments attached to the plasma membrane (PM; large arrows). (Photo courtesy of J. L. Salisbury, J. S. Condeelis, and P. Satir).

(Hartwig *et al.*, 1977; Painter and McIntosh, 1979; Stendahl *et al.*, 1980; Painter *et al.*, 1981). In an attempt to define the three-dimensional structure of actin–membrane contacts during phagocytosis, Boyles and Bainton (1981) examined cell interiors by scanning electron microscopy. These investigators observed a network of branched filaments radiating from globular centers on the cytoplasmic surface of the plasma membrane near the site of particle attachment. Similar structures were also previously seen associated with the plasma membranes of adhering leukocytes (Boyles and Bainton, 1979). As the membrane invaginated during particle ingestion, these filaments seemed to disappear, leaving a bare area of membrane with which lysosomal membranes could fuse. Recently, Aggeler and Werb (1982), using the rapid-freeze deep-etch technique, reported similar denuded regions on invaginated membranes, but they also noted large patches of clathrin-coated membrane at sites of particle attachment, similar to large clathrin sheets found in substrate-associated HeLa cells (Maupin and Pollard, 1981). Thus, the involvement of cytoskeletal and membrane components in phagocytosis may be analogous to their roles in substrate attachment and spreading.

It has been noted that capping of ligand–receptor complexes (described in Section VII-A) is followed by rapid internalization of the complexes (see dePetris, 1977). Furthermore, Walter *et al.* (1980) noted that this adsorptive endocytosis, as well as phagocytosis of opsonized erythrocytes and other particles, was restricted to the cap region of the cell. In a comparison of capping and phagocytosis in leukocytes, Berlin and Oliver (1978) noted that both processes involve reorganizations of microfilaments and disappearance of microtubules, although such skeletal changes in phagocytosis are limited to areas of particle attachment. The interactions of skeletal components with the membrane at phagocytic sites could therefore be similar to skeletal involvement in capping.

C. *Cell Surface Receptor Topography and Mobility*

As described previously, the lateral movement of ligand–receptor complexes during clustering, capping, or endocytosis appears to involve major cytoskeletal rearrangements. However, cytoskeletal elements may have other, more subtle, roles in regulating cell surface receptor mobility and topography. The development of the technique of fluorescence photobleaching recovery (FPR; see Cherry, 1979; Webb *et al.*, 1981; Peters, 1981) has allowed direct study of the mobility of surface-exposed proteins of plasma membranes in a variety of cells.

Fluorescence photobleaching studies have shown that many membrane proteins are restricted in their rates of movement, and that within the limits of the technique some proteins appear immobile. For example, the lateral diffusion constants for proteins on a variety of cell surfaces have been found to be $\leq 10^{-10}$ cm^2/sec, which is significantly less than the diffusion rates of lipids in the same membranes ($\sim 10^{-8}$ cm^2/sec; see Cherry, 1979 or Peters, 1981 for thorough compilations of such data). In addition, only a fraction of these membrane proteins was mobile ($\leq$80%; see Webb *et al.*, 1981). On the other hand, lipid diffusion rates were found to approach theoretical values ($\sim 10^{-8}$ cm^2/sec), being limited only by membrane viscosity (Saffman and Delbrück, 1975; Saffman, 1976; Peters, 1981; Webb *et al.*, 1981). However, when proteins such as M-13 phage coat protein, erythrocyte glycophorin, or band 3 were reconstituted into bare lipid vesicles, they were found to have unrestricted mobility, similar to that of vesicular lipid (Wu *et al.*, 1978; Smith *et al.*, 1979; Vaz *et al.*, 1981; Chang *et al.*, 1981). This finding implied that diffusion of proteins in a biological membrane is influenced by factors other than the intrinsic properties of the lipid bilayer, and suggested that other membrane-associated elements may be involved. While involvement of cytoskeletal elements in modulation of integral protein mobility has been suggested in a number of cells, a clear example of this type of interaction is the role of the red cell cytoskeleton in restricting the mobility of transmembrane proteins.

1. *Mobility of Red Cell Membrane Proteins*

The restriction of protein lateral mobility in red cell membranes became evident when it was found that concanavalin A (con A) receptors [now known to be primarily band 3 protein (Tanner and Anstee, 1976; West *et al.*, 1978)] could not be capped by the ligand (Loor *et al.*, 1972). In early experiments, Peters *et al.* (1974), using the technique of fluorescence microphotolysis (a synonym of fluorescence photobleaching recovery), found that the lateral diffusion constant (D) of fluorescein isothiocyanate (FITC)-labeled red cell membrane proteins was $\leq 3 \times 10^{-12}$ cm^2/sec in ghosts at 20–23°C. Fowler and Branton (1977) found a similar value of 6×10^{-12} cm^2/sec at 23°C for FITC-labeled proteins, by measuring the time for equilibration of label after fusion of labeled and unlabeled cells. Such slow rates for protein lateral diffusion have since been confirmed by a number of laboratories (Schindler *et al.*, 1980; Golan and Veatch, 1980; Sheetz *et al.*, 1980; Chang *et al.*, 1981; Sheetz *et al.*, 1982; Smith and Palek, 1982). Lipid molecules, however, appear to diffuse much faster ($D \cong 10^{-9}$ cm^2/sec at 23°C) in the erythrocyte membrane (Kapitza and Sackmann, 1980; Thompson and Axelrod, 1980), although these diffusion rates are 10-fold slower than in model membranes (Peters, 1981). Involvement of cytoskeletal elements in restriction of integral protein mobility was suggested by freeze-fracture experiments which revealed (1) that intramembrane particles could only be aggregated in vesicles with reduced amounts of spectrin, and (2) that particle aggregation was related to aggregation of residual spectrin (Elgsaeter and Branton, 1974; Elgsaeter *et al.*, 1976; Yu and Branton, 1976; Shotton *et al.*, 1978). In addition, Nicolson and Painter (1973) found that antibodies to spectrin, when sealed into ghosts, caused clustering of membrane glycoproteins. In related experiments, multivalent ligands which crosslinked external saccharide moieties caused rearrangement of cytoskeletal proteins on the internal face of the red cell membrane (Ji and Nicolson, 1974).

Removal of spectrin from the membrane by low-ionic-strength extraction (Golan and Veatch, 1980) or displacement of spectrin from the membrane by a proteolytic fragment of ankyrin containing its spectrin binding site (Fowler and Bennett, 1978) was shown to result in increased rates of lateral diffusion of integral membrane proteins. In addition, lateral diffusion of integral proteins was approximately 50 times faster ($D = 2.5 \times 10^{-9}$ cm^2/sec) in spherocytic mouse erythrocytes (Sheetz *et al.*, 1980), which lack major cytoskeletal proteins including spectrin (Greenquist *et al.*, 1978). Therefore, the interaction of spectrin with the membrane clearly restricts lateral motion of integral proteins.

Evidence also suggests that the state of the cytoskeletal structure can influence the lateral diffusion of transmembrane proteins. Schindler *et al.* (1980) and Sheetz *et al.* (1982) showed that lateral mobility of integral membrane proteins was increased by addition of polyphosphates (polyanions) such as ATP, 2,3-diphosphoglycerate (2,3-DPG), or triphosphoinositide to erythrocyte membranes. ATP and 2,3-DPG were shown to cause disruption of isolated cytoskeletons but these compounds did not affect cytoskeletal attachment to the membrane (Sheetz and Casaly, 1980), suggesting that polyphosphates increase membrane protein lateral mobility by perturbing cytoskeletal associations other than spectrin–membrane interaction. Conversely, diffusion rates were shown to be decreased by polyamines (polycations) such as neomycin or spermine (Schindler *et al.*, 1980), or by mild oxidative intermolecular crosslinking of spectrin (Smith and Palek, 1982), suggesting that stabilization of the cytoskeletal structure restricts membrane protein lateral movement.

Golan and Veatch (1980), using FPR on ghosts, demonstrated the existence of both mobile and immobile fractions of band 3 protein, which varied in quantity with temperature and ionic strength. However, the percentages of band 3 which were either mobile or immobile did not show the same dependence on temperature or ionic strength as diffusion rates. This suggested that band 3 lateral movement was controlled by at least two separate mechanisms affecting either size or rate of movement of the mobile fraction of band 3. Such mechanisms were discussed by Koppel *et al.* (1981), who pointed out that direct linkage of band 3 to the cytoskeleton would result in immobilization of the protein, while steric hindrance of band 3 by the cytoskeleton would reduce the diffusion rates of unbound (and therefore mobile) proteins. Such steric hindrance was originally described by Cherry *et al.* (1976) as trapping of cytoplasmic portions of integral proteins which protrude from the bilayer into a cytoskeletal "net." Through this indirect mechanism, cytoskeletal elements could presumably modulate mobility (and function?) of membrane proteins without requiring specific direct associations between skeletal and integral proteins.

In addition to diffusing laterally in the membrane, integral proteins may also rotate in the membrane. Using flash photolysis of triplet probes, Cherry and co-workers have described rotational mobility of band 3 about an axis perpendicular to the plane of the membrane. Initial observations with ghosts and spectrin-depleted vesicles suggested that spectrin does not restrict rotation of band 3 (Cherry *et al.*, 1976), since extraction of spectrin did not change rotational diffusion coefficients for eosin-labeled band 3. Also, addition of antispectrin antibodies to leaky ghosts had no effect on rotation of band 3 (Nigg *et al.*, 1980). However, more recent studies have suggested that high-ionic-strength extraction of ankyrin and/or polypeptide 4.1 from spectrin-depleted vesicles or tryptic cleavage of the cytoplasmic portion of band 3 results in enhancement of band 3 rotational mobility (Nigg and Cherry, 1980; Cherry and Nigg, 1981). This suggests that band 3 does have restricted rotation due to interaction with the cytoskeleton. Recently, Sakaki *et al.* (1982) have shown that rotational diffusion of band 3, which was reconstituted in phospholipid vesicles, was reduced by binding of cytoskeletal proteins (spectrin, actin, polypeptide 4.1) and ankyrin. The specific nature of such restriction, however, remains unclear.

2. *Membrane Protein Mobility in Other Cell Types*

The role of the cytoskeleton in restricting membrane protein mobility in a variety of cells was suggested by alterations in mobility induced by cytoskeletal-disrupting agents. Schlessinger *et al.* (1976a,b, 1977a), using FPR, examined lateral diffusion of cell surface proteins in myoblasts and mast cells and found that a fraction of these proteins was immobile. Diffusion rates of mobile proteins were $\leq 10^{-10}$ cm^2/sec, which is at least two orders of magnitude slower than rates for lipid probes. Furthermore, cytochalasin B addition reduced the rate of diffusion and, in one case, colchicine treatment resulted in immobilization of a greater fraction of membrane proteins. Thus, a relationship between the cytoskeleton and membrane protein diffusion was suggested.

In several studies, binding of concanavalin A (con A) to fibroblasts or lymphocytes was correlated with reduced lateral mobility of membrane proteins (Schlessinger *et al.*, 1977b; Henis and Elson, 1981; Leuther *et al.*, 1981). It had been shown previously that the clustering of receptors which precedes capping could be inhibited by binding large amounts of con A to cells, an effect which appeared to involve microtubules and microfilaments (see Edelman, 1976). In order to test whether this restriction of lateral mobility was due to indirect effects of con A binding on the cytoskeleton, Schlessinger *et al.* (1977b) bound an aggregate of con A-clumped platelets to one region of a fibroblast and measured lateral mobility in another area of the cell by FPR.

They showed that platelets bound to one region of the membrane (covering 4–30% of the cell surface) decreased lateral diffusion rates of surface proteins in other areas of the membrane by a factor of six. This reduction in mobility was partially reversed by addition of microtubule-disrupting agents. In a later study using the same technique, Henis and Elson (1981) showed that con A-platelet binding to lymphocytes inhibited mobility of surface immunoglobulins. This inhibition was partially reversed by addition of cytochalasin B or colchicine, and was completely reversed by addition of both drugs. However, reversal of inhibition only resulted in diffusion rates equal to the original value, which was still much less than the rate predicted for free diffusion in the bilayer. Therefore, this ligand-induced cytoskeletal involvement occurred independently of restriction of membrane proteins in the unperturbed cell.

Immobilization of cell surface receptors during differentiation has been reported and cytoskeletal-perturbing agents were shown to interfere with these processes (Fishman *et al.*, 1981; Herman and Fernandez, 1982). Differences in membrane protein mobilities in normal and SV40-transformed fibroblasts have also been suggested to be related to differences in cytoskeletal organization resulting from transformation (Eldridge *et al.*, 1980). Immobilization of surface proteins also occurs in specialized cell structures such as coated pits, cell junctions, or synaptic junctions (see for example Geiger *et al.*, 1982); however, these are not discussed in this review.

Most studies implicating the cytoskeleton in modulation of membrane protein mobility have relied on effects produced by cytoskeletal-perturbing drugs. However, an alternate approach has been taken recently by Webb and co-workers, who have measured diffusion rates in areas of plasma membrane which have pulled away from and have apparently separated from the cytoskeleton (Webb *et al.*, 1981; Tank *et al.*, 1982; Wu *et al.*, 1982). Diffusion of acetylcholine receptors and con A receptors on such blebs of muscle cells and myoblasts was shown to be 10- to 100-fold faster than diffusion of these same proteins on intact cell membranes (Tank *et al.*, 1982). Similarly, mobility of con A receptors on lymphocytes and lymphoma cells was also enhanced on areas of the membrane detached from the cortical cytoplasm by swelling (Wu *et al.*, 1982). Also low-density-lipoprotein receptors appeared to diffuse faster in such blebs on fibroblasts (Webb *et al.*, 1981). These faster rates approach values for diffusion limited only by hydrodynamic drag in the membrane, suggesting that constraint of receptor mobility is due to associations between the membrane and cortical cytoplasm.

Whether such associations involve cytoskeletal–membrane interactions analogous to those in the erythrocyte is presently unknown, although membrane-associated cytoskeletal structures have been suggested from studies of detergent extraction of cells (Ben-Ze'ev *et al.*, 1979; Mescher *et al.*, 1981). Mounting evidence for the existence of erythrocyte-like skeletal proteins in other cell types may suggest similar cytoskeletal influences on transmembrane protein mobilities.

VIII. SUMMARY

Conceptual schemes explaining the various motile, protrusive or contractile events which nearly all cells undergo invariably postulate one or several types of cytoskeletal–membrane contact. However, in only a few cases such as the intestinal microvillus, fibroblast adhesion plaques, and the red cell membrane skeleton have cytoskeletal–membrane contacts been clearly demonstrated. Moreover, in fewer cases still (probably only the red cell) have such contacts been defined at the molecular level.

An important goal of studies of cytoskeletal–membrane contacts is to define the protein molecules responsible for them. Studies of the red cell have led to the suggestion that a special class of connecting molecules may be responsible for mediating certain types of actin–membrane contacts. Indeed, spectrin-like proteins have now been found in a wide variety of cells (Goodman *et al.*, 1981; Davis and Bennett, 1982; Repasky *et al.*, 1982; Glenney *et al.*, 1982a,b), and in some cases appear to be on or close to plasma membranes (Burridge *et al.*, 1982; Bennett *et al.*, 1982). Ankyrin-like proteins have also been found to be widely distributed in nonerythroid cells (Bennett, 1979; Bennett and Davis, 1981) although it is not known whether these proteins have the same function as red cell ankyrin. Analogues to red cell band 4.1 have been found in nonerythroid cells (Cohen *et al.*, 1982) but, as is the case for ankyrin, it is not known what role, if any, these proteins play in nonerythroid cytoskeletal–membrane attachments. It is our expectation that studies of these red cell protein analogues may lead to an understanding of cytoskeletal contacts in nonerythroid cells which equals that of the red cell.

Another area in which rapid progress can be expected on a molecular level is the study of actin–membrane contacts in focal adhesions. Here, two proteins, α-actinin and vinculin, have been identified as being involved with filament–membrane attachment. Recently, a variant of vinculin, referred to as metavinculin, has been identified (Siliciano and Craig, 1982). Metavinculin behaves in some respects like an integral membrane protein and may be directly involved in actin–membrane binding. Further work on the effects of phosphorylation of vinculin by the pp 60^{src} kinase may also yield important information about how actin–membrane contacts are regulated.

Finally, the intestinal brush border microvillus has recently become an accessible and well-defined system for the study of cytoskeleton–membrane contacts. Here too, particular proteins which may mediate the attachment of actin filaments to the microvillus membrane have been proposed, and it is only a matter of time before they are isolated and characterized in detail.

It is logical to expect that there are multiple types of contacts between cytoskeletal proteins and membranes, each suited to the cell in which they are found. End-on attachment of actin filaments to membranes may be characteristic of static structures such as adhesion plaques or microvilli. Lateral association, or multipoint attachments through actin gels, may be characteristic of more dynamic states. It is entirely possible as well that the most dynamic cellular regions such as ruffles may be poor or lacking in cytoskeletal membrane contacts, enabling the greatest degree of structural remodeling to occur rapidly. Similarly, each type of contact may be reflected in distinct, transmembrane effects on cell surface receptors and proteins, some being more restrictive than others. The merging of our developing knowledge of cytoskeletal and membrane structure with that of cell surface topography and dynamics is one of the challenges for future research.

The literature search for this review was completed in the fall of 1982. Since then considerable progress has been made, especially in the characterization of nonerythroid spectrin-like proteins.

ACKNOWLEDGMENTS

This work was supported in part by NIH grants HL 24382, HL 15157, and HL 27215. We wish to thank Drs. David Albertini, John Condeelis, Brian Herman, Jeffrey Salisbury, Peter Satir, Irwin Singer, and S. J. Singer for providing micrographs or access to unpublished data. We also wish to express our sincere thanks to Lucille Paul for manuscript preparation and to Cela Libeskind for her artwork.

REFERENCES

Abercrombie, M., and Dunn, G. A., 1975, Adhesions of fibroblasts to substratum during contact inhibition observed by interference reflection microscopy, *Exp. Cell Res.* **92**:57–62.

Abercrombie, M., Heaysman, J. E. M., and Pegrum, S. M., 1971, The locomotion of fibroblasts in culture: IV. Electron microscopy of the leading lamella, *Exp. Cell Res.* **67**:359–369.

Aggeler, J., and Werb, Z., 1982, Initial events during phagocytosis by macrophages viewed from outside and inside the cell: Membrane-particle interactions and clathrin, *J. Cell Biol.* **94**:613–623.

Albertini, D. F., and Anderson, E., 1977, Microtubule and microfilament rearrangments during capping of concanavalin A receptors on cultured ovarian granulosa cells, *J. Cell Biol.* **73**:111–127.

Albertini, D. F., and Clark, J. I., 1975, Membrane-microtubule interactions, *Proc. Natl. Acad. Sci. USA* **72**:4976–4980.

Albertini, D. F., Berlin, R. D., and Oliver, J. M., 1977, The mechanism of concanavalin A cap formation in leukocytes, *J. Cell Sci.* **26**:57–75.

Ali, I. U., and Hynes, R. O., 1977, Effects of cytochalasin B and colchicine on attachment of a major surface protein of fibroblasts, *Biochim. Biophys. Acta* **471**:16–24.

Ali, I. U., Mautner, V. M., Lanta, R. P., and Hynes, R. O., 1977, Restoration of normal morphology, adhesion and cytoskeleton in transformed cells by addition of a transformation-sensitive surface protein, *Cell* **11**:115–126.

Allen, R. D., 1968, A reinvestigation of cross sections of cilia, *J. Cell Biol.* **37**:825–831.

Allison, A. C., and Davies, P., 1974, Mechanisms of endocytosis and exocytosis, *Symp. Soc. Exp. Biol.* **28**:419–446.

Allore, R. J., and Barber, B. H., 1982, Structural studies of purified pig lymph node plasma membrane: Association of cytoskeletal components with the plasma membrane and the effect of detergent solubilization, *Can. J. Biochem.* **60**:57–70.

Ash, J. F., Louvard, D., and Singer, S. J., 1977, Antibody-induced linkages of plasma membrane proteins to intracellular actomyosin-containing filaments in cultured fibroblasts, *Proc. Natl. Acad. Sci. USA* **74**:5584–5588.

Ash, J. F., Louvard, D., and Singer, S. J., 1980, Interactions between the plasma membrane and cytoskeleton of culture fibroblasts, *Prog. Clin. Biol. Res.* **41**:925–930.

Atkinson, M., Morrow, J., and Marchesi, V., 1982, The polymeric state of actin in the human erythrocyte cytoskeleton, *J. Cellular Biochem.* **18**:493–505.

Avnur, Z., and Geiger, B., 1981, The removal of extracellular fibronectin from areas of cell-substrate contact, *Cell* **25**:121–132.

Axline, S. G., and Reaven, E. P., 1974, Inhibition of phagocytosis and plasma membrane mobility of the cultured macrophage by cytochalasin B. Role of subplasmalemmal microfilaments, *J. Cell Biol.* **62**:647–659.

Badley, R. A., Woods, A., Smith, C. G., and Rees, D. A., 1980, Actomyosin relationships with surface features in fibroblast adhesions, *Exp. Cell Res.* **126**:263–272.

Bayley, S. A., and Rees, D. A., 1982, Analysis of the proteins, glycoproteins and glycosaminoglycans of fibroblast adhesions to substratum, *Biochim. Biophys. Acta* **689**:351–362.

Begg, D. A., Rodewald, R., and Rebhun, L. I., 1978, The visualization of actin filament polarity in thin sections. Evidence for the uniform polarity of membrane associated filaments, *J. Cell Biol.* **79**:846–852.

Behnke, O., 1970, Microtubules in disk-shaped blood cells, *Int. Rev. Exp. Pathol.* **9**:1–92.

Bennett, V., 1979, Immunoreactive forms of human erythrocyte ankyrin are present in diverse cells and tissues, *Nature* **281**:597–599.

Bennett, V., 1982a, The molecular basis for membrane-cytoskeleton association in human erythrocytes, *J. Cellular Biochem.* **18**:49–65.

Bennett, V., 1982b, Isolation of an ankyrin-band 3 oligomer from human erythrocyte membranes, *Biochim. Biophys. Acta* **689**:475–484.

Bennett, V., and Davis, J., 1981, Erythrocyte ankyrin: Immunoreactive analogues and associated with mitotic structures in cultured cells and with microtubules in brain, *Proc. Natl. Acad. Sci. USA* **78**:7550–7554.

Bennett, V., Davis, J., and Fowler, W. E., 1982, Brain spectrin. A membrane-associated protein related in structure and function to erythrocyte spectrin, *Nature* **299**:126–131.

Ben-Ze'ev, A., Duerr, A., Solomon, F., and Penman, S., 1979, The outer boundary of the cytoskeleton: A lamina derived from plasma membrane proteins, *Cell* **17**:859–865.

Berlin, R. D., and Oliver, J. M., 1978, Analogous ultrastructure and surface properties during capping and phagocytosis in leukocytes, *J. Cell Biol.* **77**:789–804.

Berlin, R. D., Caron, J. M., and Oliver, J. M., 1979, Microtubules and the structure and function of cell surfaces, in: *Microtubules* (K. Roberts and J. S. Hyans, eds.), Academic Press, New York, pp. 443–486.

Bernier-Valentin, F., and Rousset, B., 1982, Interaction of tubulin with rat liver mitochondria, *J. Biol. Chem.* **257**:7092–7099.

Bernstein, L. H., and Wollman, S. H., 1975, Association of mitochondria with desmosomes in the rat thyroid gland, *J. Ultrastruct. Res.* **53**:87–92.

Bhattacharyya, B., and Wolff, J., 1975, Membrane-bound tubulin in brain and thyroid tissue, *J. Biol. Chem.* **250**:7639–7646.

Bhattacharyya, B., and Wolff, J., 1976, Polymerization of membrane tubulin, *Nature* **264**:576–577.

Birchmeier, C., Kreis, T. E., Eppenberger, H. M., Winterhalter, K. H., and Birchmeier, W., 1980, Corrugated attachment membrane in WI-38 fibroblasts: Alternating fibronectin fibers and actin-containing focal contacts, *Proc. Natl. Acad. Sci. USA* **77**:4108–4112.

Blitz, A. L., and Fine, R. E., 1974, Muscle-like contractile proteins and tubulin in synaptosomes, *Proc. Natl. Acad. Sci USA* **74**:4472–4476.

Bloom, W. S., Fields, K. L., Yen, S. H., Haver, K., Schook, W., and Puszkin, S., 1980, Brain clathrin: Immunofluorescent patterns in cultured cells and tissues, *Proc. Natl. Acad. Sci. USA* **77**:5520–5524.

Bourguignon, G. J., and Bourguignon, L. Y. W., 1981a, Isolation and initial characterization of a lymphocyte cap structure, *Biochim. Biophys. Acta* **646**:109–118.

Bourguignon, L. Y. W., and Bourguignon, G. J., 1981b, Immunocytochemical localization of intermediate filament proteins during lymphocyte capping, *Cell Biol. Int. Rep.* **5**:783–789.

Bourguignon, L. Y. W., and Rozek, R. J., 1980, Capping of concanavalin A receptors and their association with microfilaments in monolayer grown human fibroblastoid cells, *Cell Tissue Res.* **205**:77–84.

Bourguignon, L. Y. W., and Singer, S. J., 1977, Transmembrane interactions and the mechanism of capping of surface receptors by their specific ligands, *Proc. Natl. Acad. Sci USA* **74**:5031–5035.

Bourguignon, L. Y. W., Tokoyasu, K. T., and Singer, S. J., 1978, The capping of lymphocytes and other cells, studied by an improved method for immunofluorescence staining of frozen sections, *J. Cell Physiol.* **95**:239–258.

Bourguignon, L. Y. W., Nagpal, M. L., and Hsing, Y.-C., 1981, Phosphorylation of myosin light chain during capping of mouse T-lymphoma cells, *J. Cell Biol.* **91**:889–894.

Boyles, J., and Bainton, D. F., 1979, Changing patterns of plasma membrane-associated filaments during the initial phases of polymorphonuclear leukocyte adherence, *J. Cell Biol.* **82**:347–368.

Boyles, T., and Bainton, D. F., 1981, Changes in plasma membrane associated filaments during endocytosis and exocytosis in polymorphonulcear leukocytes, *Cell* **24**:905–914.

Bradley, T. J., and Satir, P., 1979, Evidence of microfilament-associated mitochondrial movement, *J. Supramol. Struct.* **12**:165–175.

Branton, D., Cohen, C., and Tyler, J., 1981, Interaction of cytoskeletal proteins on the human erythrocyte membrane, *Cell* **24**:24–32.

Braun, J., Fujiwara, K., Pollard, T. D., and Unanue, E., 1978, Two distinct mechanisms for redistribution of lymphocyte surface macromolecules. I. Relationship to cytoplasmic myosin, *J. Cell Biol.* **79:**409–418.

Bray, D., and Thomas, C., 1975, The actin content of fibroblasts, *J. Biochem.* **147:**221–228.

Brenner, S., and Korn, E., 1980, Spectrin/actin complex isolated from sheep erythrocytes accelerates actin polymerization by simple nucleation, *J. Biol. Chem.* **255:**1670–1676.

Bretscher, M. S., 1976, Directed lipid flow in cell membranes, *Nature* **260:**21–23.

Bretscher, A., and Weber, K., 1978a, Purification of microvilli and an analysis of the protein components of the microfilament core bundle, *Exp. Cell Res.* **116:**397–407.

Bretscher, A., and Weber, K., 1978b, Localization of actin and microfilament-associated proteins in the microvilli and terminal web of the intestinal brush border by immunofluorescence microscopy, *J. Cell Biol.* **79:**839–845.

Bretscher, A., and Weber, K., 1979, Villin: The major microfilament-associated protein of the intestinal microvillus, *Proc. Natl. Acad. Sci. USA* **76:**2321–2325.

Bretscher, A., and Weber, K., 1980a, Fimbrin, a new microfilament-associated protein present in microvilli and other cell surface structures, *J. Cell Biol.* **86:**335–340.

Bretscher, A., and Weber, K., 1980b, Villin is a major protein of the microvillus cytoskeleton which binds both G and F actin in a calcium-dependent manner, *Cell* **20:**839–847.

Brunser, O., and Luft, J. H., 1970, Fine structure of the apex of absorptive cells from rat small intestine, *J. Ultrastruct. Res.* **31:**291–311.

Burridge, K., and Feramisco, J. R., 1980, Microinjection and localization of a 130K protein in living fibroblasts: A relationship to actin and fibronectin, *Cell* **19:**587–595.

Burridge, K., and Feramisco, J. R., 1982, α-Actinin and vinculin from nonmuscle cells: Calcium-sensitive interactions with actin, in: *Cold Spring Harbor Symposia on Quantitative Biology,* Volume 46, Cold Spring Harbor Laboratory, Cold Spring Harbor, New York, pp. 587–598.

Burridge, K., and McCullough, L., 1980, The association of α-actinin with the plasma membrane, *J. Supramol. Struct.* **13:**53–65.

Burridge, K., and Phillips, J. H., 1975, Association of actin and myosin with secretory granule membranes, *Nature* **254:**526–529.

Burridge, K., Feramisco, J., and Blose, S., 1980, The association of α-actinin and clathrin with the plasma membrane, *Prog. Clin. Biol. Res.* **41:**907–924.

Burridge, K., Kelly, T., and Mangeat, P., 1982, Nonerythrocyte spectrins: Actin-membrane attachment proteins occurring in many cell types, *J. Cell Biol.* **95:**478–486.

Butman, B., Bourguignon, G. J., and Bourguignon, L. Y. W., 1980, Lymphocyte capping induced by polycationized ferritin, *J. Cell Physiol.* **105:**7–15.

Byers, H. R., and Fujiwara, K., 1982, Stress fibers in cells *in situ:* Immunofluorescence visualization with antiactin, antimyosin and anti-alpha-actin, *J. Cell Biol.* **93:**804–811.

Carley, W. W., Barak, L. S., and Webb, W. W., 1981, F-actin aggregates in transformed cells, *J. Cell Biol.* **90:**797–802.

Caron, J. M., and Berlin, R. D., 1979, Interaction of microtubule proteins with phospholipid vesicles, *J. Cell Biol.* **81:**665–671.

Carraway, K. L., Cerra, R. F., Jung, G., and Carothers Carraway, C. A., 1982, Membrane-associated actin from the microvillar membranes of ascites tumor cells, *J. Cell Biol.* **94:**624–630.

Chang, C., Takeuchi, H., Ito, T., Machida, K., and Ohnishi, S., 1981, Lateral mobility of erythrocyte membrane proteins studied by the fluorescence photobleaching recovery technique, *J. Biochem.* **90:**997–1004.

Chen, W.-T., and Singer, S. J., 1980, Fibronectin is not present in the focal adhesions formed between normal cultured fibroblasts and their substrata, *Proc. Natl. Acad. Sci. USA* **77:**7318–7322.

Chen, W.-T., and Singer, S. J., 1982, Immunoelectron microscopic studies of the sites of cell–substratum and cell–cell contacts in cultured fibroblasts, *J. Cell Biol.* **95:**205–222.

Cherry, R. J., 1979, Rotational and lateral diffusion of membrane proteins, *Biochim. Biophys. Acta* **559:** 289–327.

Cherry, R. J., and Nigg, E. A., 1981, Rotational diffusion of erythrocyte membrane proteins, in: *The Function of Red Blood Cells: Erythrocyte Pathobiology* (D. F. H. Wallach, ed.), Alan R. Liss, New York, pp. 59–77.

Cherry, R. J., Bürkli, A., Busslinger, M., Schneider, G., and Parish, G. R., 1976, Rotational diffusion of band 3 proteins in the human erythrocyte membrane, *Nature* **263:**389–393.

Clarke, M., Schatten, G., Mazia, D., and Spudich, J. A., 1975, Visualization of actin fibers associated with the cell membrane in amoebae of Dictyostelium discoideum, *Proc. Natl. Acad. Sci. USA* **72:**1758–1762.

Cohen, C. M., 1983, The molecular organization of the red cell membrane skeleton, *Semin. Haematol.* **20:**141–158.

Cohen, C. M., and Branton, D., 1979, The role of spectrin in erythrocyte membrane-stimulated actin polymerisation, *Nature* **279:**163–165.

Cohen, C. M., and Branton, D., 1981, The normal and abnormal red cell cytoskeleton: A renewed search for molecular defects, *Trends Biochem. Sci.* **6:**266–268.

Cohen, C. M., and Foley, S., 1982, The role of band 4.1 in the association of actin with erythrocyte membranes, *Biochim. Biophys. Acta* **688:**691–701.

Cohen, C. M., Jackson, P., and Branton, D., 1978, Actin-membrane interactions: Association of G-actin with the red cell membrane, *J. Supramol. Struct.* **9:**113–124.

Cohen, C. M., Tyler, J., and Branton, D., 1980, Spectrin-actin associations studied by electron microscopy of shadowed preparations, *Cell* **21:**875–883.

Cohen, C. M., Foley, S. F., and Korsgren, C., 1982, A protein immunologically related to erythrocyte band 4.1 is found on stress fibers of non-erythroid cells, *Nature* **299:**648–650.

Cohen, W. D., 1978, Observations on the marginal band system of nucleated erythrocytes, *J. Cell Biol.* **78:**260–273.

Collett, M. S., and Erikson, R. L., 1978, Protein kinase activity associated with ovarian sarcoma virus src gene product, *Proc. Natl. Acad. Sci. USA* **75:**2021–2024.

Condeelis, J., 1979, Isolation of concanavalin A caps during various stages of formation and their association with actin and myosin, *J. Cell Biol.* **80:**751–758.

Condeelis, J. S., 1981, Reciprocal interactions between the actin lattice and cell membrane, *Neurosci. Res. Prog. Bull.* **19:**83–99.

Cooke, P. H., 1976, A filamentous cytoskeleton in vertebrate smooth muscle fibers, *J. Cell Biol.* **68:**539–556.

Cooke, P. H., and Chase, R. H., 1971, Potassium chloride-insoluble myofilaments in vertebrate smooth muscle cells, *Exp. Cell Res.* **66:**417–425.

Coudrier, E., Reggio, H., and Louvard, D., 1981, Immunolocalization of the 110,000 molecular weight cytoskeletal protein of intestinal microvilli, *J. Mol. Biol.* **152:**49–66.

Coudrier, E., Reggio, H., and Louvard, D., 1982, The cytoskeleton of intestinal microvilli contains two polypeptides immunologically related to proteins of striated muscle, in: *Cold Spring Harbor Symposia on Quantitative Biology,* Volume XLVI, Part 2, Cold Spring Harbor Laboratory, Cold Spring Harbor, New York, pp. 881–892.

Craig, S., and Powell, L., 1980, Regulation of actin polymerization by villin, a 95,000 dalton cytoskeletal component of intestinal brush borders, *Cell* **22:**739–746.

Curtis, A. S. G., 1964, The mechanism of adhesion of cells to glass. A study by interference reflection microscopy, *J. Cell Sci.* **20:**199–218.

David-Pfeuty, T., and Singer, S. J., 1980, Altered distributions of the cytoskeletal proteins vinculin and α-actinin in cultured fibroblasts transformed by Rous sarcoma virus, *Proc. Natl. Acad. Sci. USA* **77:**6687–6691.

Davis, J., and Bennett, V., 1982, Microtubule-associated protein 2, a microtubule-associated protein from brain, is immunologically related to the α subunit of erythrocyte spectrin, *J. Biol. Chem.* **257:**5816–5820.

Dellagi, K., and Brouet, J. C., 1982, Redistribution of intermediate filaments during capping of lymphocyte surface molecules, *Nature* **298:**284–286.

Dentler, W. L., 1980, Structures linking the tips of ciliary and flagellar microtubules to the membrane, *J. Cell Sci.* **42:**207–220.

Dentler, W. L., 1981, Microtubule-membrane interactions in cilia and flagella, *Int. Rev. Cytol.* **72:**1–47.

dePetris, S., 1975, Concanavalin A receptors, immunoglobulins and O antigen of the lymphocyte surface. Interactions with concanavalin A and with cytoplasmic structures, J. Cell Biol. **65:**123–146.

dePetris, S., 1977, Distribution and mobility of plasma membrane components on lymphocytes, in: *Dynamic Aspects of Cell Surface Organization* (G. Poste and G. L. Nicholson, eds.), North Holland Publishing Co., New York, pp. 643–728.

Der, C. J., Ash, J. F., and Stanbridge, E. J., 1981, Cytoskeletal and transmembrane interactions in the expression of tumorigenicity in human cell hybrids, *J. Cell Sci.*, **52:**151–166.

Edelman, G. M., 1976, Surface modulation in cell recognition and cell growth, *Science* **192:**218–226.

Edelman, G. M., Yahara, I., and Wang, J. L., 1973, Receptor mobility and receptor–cytoplasmic interactions in lymphocytes, *Proc. Natl. Acad. Sci. USA* **70:**1442–1446.

Eldridge, C. A., Elson, E. L., and Webb, W. W., 1980, Fluorescence photobleaching recovery measurements of surface lateral mobilities on normal and SV40-transformed mouse fibroblasts, *Biochemistry* **19:**2075–2079.

Elgsaeter, A., and Branton, D., 1974, Intramembrane particle aggregation in erythrocyte ghosts. I. The effects of protein removal, *J. Cell Biol.* **63:**1018–1030.

Elgsaeter, A., Shotton, D. M., and Branton, D., 1976, Intramembrane particle aggregation in erythrocyte ghosts. II. The influence of spectrin aggregation, *Biochim. Biophys. Acta* **426:**101–122.

Feit, H., Kelly, P., and Cotman, C. W., 1977, Identification of a protein related to tubulin in the post synaptic density, *Proc. Natl. Acad. Sci. USA* **74:**1047–1051.

Feramisco, J. R., 1979, Microinjection of fluorescently labeled α-actinin into living fibroblasts, *Proc. Natl. Acad. Sci. USA* **76:**3967–3971.

Fishman, M. C., Dragsten, P. R., and Spector, I., 1981, Immobilization of concanavalin A receptors during differentiation of neuroblastoma cells, *Nature* **290:**781–783.

Flanagan, J., and Koch, G. L. E., 1978, Cross-linked surface Ig attaches to actin, *Nature* **273:**278–281.

Fowler, V., and Bennett, V., 1978, Association of spectrin with its membrane attachment site restricts lateral mobility of human erythrocyte integral membrane proteins, *J. Supramol. Struct.* **8:**215–221.

Fowler, V., and Branton, D., 1977, Lateral mobility of human erythrocyte integral membrane proteins, *Nature* **268:**23–26.

Fowler, V. M., and Pollard, H. B., 1982, In vitro reconstitution of chromaffin granule-cytoskeleton interactions: Ionic factors influencing the association of F-actin with purified chromaffin granule membranes, *J. Cell. Biochem.* **18:**295–311.

Fox, C. H., Cottler-Fox, M. H., and Yamada, K. M., 1980, The distribution of fibronectin in attachment sites of chick fibroblasts, *Exp. Cell Res.* **130:**477–481.

Franke, W. W., Weber, K., Osborn, M., Schmid, E., and Freudenstein, C., 1978, Antibody to prekeratin. Decoration of tonofilament-like arrays in various cells of epithelial character, *Exp. Cell Res.* **116:**429–445.

Franke, W. W., Appelhans, B., Schmid, E., Freudenstein, C., Obborn, M., and Weber, K., 1979, The organization of cytokeratin filaments in the intestinal epithelium, *Eur. J. Cell Biol.* **19:**255–268.

Fujiwara, K., and Pollard, T. D., 1976, Fluorescent antibody localization of myosin in the cytoplasm, cleavage furrow, and mitotic spindle of human cells, *J. Cell Biol.* **71:**848–875.

Gabbiani, G., Chaponnier, C., Zumbe, A., and Vassalli, P., 1977, Actin and tubulin co-cap with surface immunoglobulins in mouse B lymphocytes, *Nature* **269:**697–698.

Geiger, B., 1979, A 130K protein from chicken gizzard: Its localization at the termini of microfilament bundles in cultured chicken cells, *Cell* **18:**193–205.

Geiger, B., 1982, Involvement of vinculin in contact-induced cytoskeletal interactions, in: *Cold Spring Harbor Symposia on Quantitative Biology,* Volume 46, Cold Spring Harbor Laboratory, Cold Spring Harbor, New York, pp. 671–682.

Geiger, B., and Singer, S. J., 1979, The participation of α-actinin in the capping of cell membrane components, *Cell* **16:**213–222.

Geiger, B., Tokuyasu, K. T., and Singer, S. J., 1979, Immunocytochemical localization of α-actinin in intestinal epithelial cells, *Proc. Natl. Acad. Sci. USA* **76:**2833–2837.

Geiger, B., Tokuyasu, K. T., Dutton, A. H., and Singer, S. J., 1980, Vinculin, an intracellular protein localized at specialized sites where microfilament bundles terminate at cell membranes, *Proc. Natl. Acad. Sci. USA* **77:**4127–4131.

Geiger, B., Dutton, A. H., Tokuyasu, K. T., and Singer, S. J., 1981, Immunoelectron microscopy studies of membrane-microfilament interactions: Distributions of α-actinin, tropomyosin, and vinculin in intestinal epithelial brush border and chicken gizzard smooth cells, *J. Cell Biol.* **91:**614–628.

Geiger, B., Avnur, Z., and Schlessinger, J., 1982, Restricted mobility of membrane constituents in cell-substrate focal contacts of chicken fibroblasts, *J. Cell Biol.* **93:**495–500.

George, J. N., Lyons, R. M., and Morgan, R. K., 1980, Membrane changes associated with platelet activation, *J. Clin. Invest.* **66**:1–9.

Glenney, J. R., and Weber, K., 1980, Calmodulin-binding proteins of the microfilaments present in isolated brush borders and microvilli of intestinal epithelial cells, *J. Biol. Chem.* **255**:10551–10554.

Glenney, J. R., Jr., Glenney, P., Osborn, M., and Weber, K., 1982a, An F-actin- and calmodulin-binding protein from isolated intestinal brush borders has a morphology related to spectrin, *Cell* **28**:843–854.

Glenney, J. R., Jr., Glenney, P., and Weber, K., 1982b, Erythroid spectrin, brain fodrin, and intestinal brush border proteins (TW-260/240) are related molecules containing a common calmodulin-binding subunit bound to a variant cell type-specific subunit, *Proc. Natl. Acad. Sci. USA* **79**:4002–4005.

Glenney, J. R., Jr., Osborn, M., and Weber, K., 1982c, The intracellular localization of the microvillus 110K protein, a component considered to be involved in side-on membrane attachment of F-actin, *Exp. Cell Res.* **138**:199–205.

Glenney, V. R., Kaulfus, P., Matsudaira, P., and Weber, K., 1981, F-actin binding and bundling properties of fimbrin, a major cytoskeletal protein of microvillus core filaments, *J. Biol. Chem.* **256**:9283–9288.

Golan, D. E., and Veatch, W., 1980, Lateral mobility of band 3 in the human erythrocyte membrane studied by fluorescence photobleaching recovery: Evidence for control by cytoskeletal interactions, *Proc. Natl. Acad. Sci. USA* **77**:2537–2541.

Goldstein, J. L., Anderson, R. G. W., and Brown, M. S., 1979, Coated pits, coated vesicles, and receptor-mediated endocytosis, *Nature* **279**:679–685.

Goodman, S. R., Zagon, I. S., and Kulikowski, R. R., 1981, Identification of a spectrin-like protein in nonerythroid cells, *Proc. Natl. Acad. Sci. USA* **78**:7570–7574.

Goodman, S. R., Yu, J., Whitfield, C., Culp, E., and Posnak, E., Erythrocyte membrane skeletal protein bands 4.1a and b are sequence-related phosphoproteins, *J. Biol. Chem.* **257**:4564–4569.

Gordon, W. E., III, and Bushnell, A., 1979, Immunofluroescent and ultrastructural studies of polygonal microfilament networks in respreading non-muscle cells, *Exp. Cell Res.* **120**:335–348.

Granger, B. L., and Lazarides, E., 1982, Structural associations of synemin and vimentin filaments in avian erythrocytes revealed by immunoelectron microscopy, *Cell* **30**:263–275.

Gratzer, W., 1981, The red cell membrane and its cytoskeleton, *J. Biochem.* **198**:1–8.

Gray, E. G., 1975, Presynaptic microtubules and their association with synaptic vesicles, *Proc. R. Soc. Lond. (Biol.)* **190**:367–372.

Greenquist, A. C., Shohet, S. G., and Bernstein, S. E., 1978, Marked reduction of spectrin in hereditary spherocytosis in the common house mouse, *Blood* **51**:1149–1155.

Grinnell, F., 1980, Visualization of cell substratum adhesion plaques by antibody exclusion, *Cell Biol. Int. Rep.* **4**:1031–1036.

Gruenstein, E., Rich, A., and Weihing, R. R., 1975, Actin associated with membranes from 3T3 mouse fibroblast and HeLa cells, *J. Cell Biol.* **64**:223–234.

Hargreaves, W., Giedd, K., Verkleij, A., and Branton, D., 1980, Reassociation of ankyrin with band 3 in erythrocyte membranes and in lipid vesicles, *J. Biol. Chem.* **255**:11965–11972.

Harris, A., 1976, Recycling of dissolved plasma membrane components as an explanation of the capping phenomenon, *Nature* **263**:781–783.

Hartwig, J. H., Davies, W. A., and Stossel, T. P., 1977, Evidence for contractile protein translocation in macrophage spreading, phagocytosis and phagolysosome formation, *J. Cell Biol.* **75**:956–967.

Hartwig, J. H., Yin, H. L., and Stossel, T. D., 1980, Contractile proteins and the mechanism of phagocytosis in macrophages, in: *Mononuclear Phagocytes,* Vol. II (R. van Furth, ed.), Martinus Nijhoff, The Hague, pp. 971–996.

Heggeness, M. H., Ash, J. F., and Singer, S. J., 1978, Transmembrane linkage of fibronectin to intracellular actin-containing filaments in cultured human fibroblasts, *N.Y. Acad. Sci.* 414–417.

Henis, Y. I., and Elson, E. L., 1981, Inhibition of the mobility of mouse lymphocyte surface immunoglobulins by locally bound concanavalin A, *Proc. Natl. Acad. Sci. USA* **78**:1072–1076.

Herman, B. A., and Fernandez, S. M., 1982, Dynamics and topographical distribution of surface glycoproteins during myoblast fusion: A resonance energy transfer study, *Biochemistry* **21**:3275–3283.

Herman, I. M., Crisona, N. J., and Pollard, T. D., 1981, Relation between cell activity and the distribution of cytoplasmic actin and myosin, *J. Cell Biol.* **90**:84–91.

Heuser, J. E., 1980, Three-dimensional visualization of coated vesicle formation in fibroblasts, *J. Cell Biol.* **84**:560–583.

Heuser, J. E., and Kirschner, M. W., 1980, Filament organization revealed in platinum replicas of freeze-dried cytoskeletons, *J. Cell Biol.* **86**:212–234.

Hewitt, J. A., 1979, Surf-riding model for cell capping, *J. Theoret. Biol.* **80**:115–127.

Hirokawa, N., 1982, Cross-linker system between neurofilaments, microtubules, and membranous organelles in frog axons revealed by the quick-freeze, deep-etching method, *J. Cell Biol.* **94**:129–142.

Hirokawa, N., and Heuser, J. E., 1981, Quick-freeze, deep-etch visualization of the cytoskeleton beneath surface differentiations of intestinal epithelial cells, *J. Cell Biol.* **91**:399–409.

Hirokawa, N., and Tilney, L. G., 1982, Interactions between actin filaments and between actin filaments and membranes in quick-frozen and deeply etched hair cells of the chick ear, *J. Cell Biol.* **95**:249–261.

Hirokawa, N., Tilney, L. G., Fujiwara, K., and Heuser, J. E., 1982, Organization of actin, myosin, and intermediate filaments in the brush border of intestinal epithelial cells, *J. Cell Biol.* **94**:425–443.

Hoessli, D., Rungger-Brandle, E., Jockusch, B. M., and Gabbiani, G., 1980, Lymphocyte α-actinin: Relationship to cell membrane and co-capping with surface receptors, *J. Cell Biol.* **84**:305–314.

Holmes, G. R., Goll, E. E., and Suzuki, A., 1971, Effect of α-actinin on actin viscosity, *Biochim. Biophys. Acta* **253**:240–253.

Howe, C. L., Mooseker, M. S., and Graves, T. A., 1980, Brush-border calmodulin: A major component of the isolated microvillus core, *J. Cell Biol.* **85**:916–923.

Howell, S. L., and Tyhurst, M., 1979, Interaction between insulin-storage granules and F-actin in vitro, *J. Biochem.* **178**:367–371.

Hull, B. E., and Staehelin, L. A., 1979, The terminal web. A re-evaluation of its structure and function, *J. Cell Biol.* **81**:67–82.

Hunter, T., and Sefton, B. M., 1980, The transforming gene product of Rous sarcoma virus phosphorylates tyrosine, *Proc. Natl. Acad. Sci. USA* **77**:1311–1315.

Hynes, R., 1982, Phosphorylation of vinculin by $pp60^{src}$: What might it mean? *Cell* **28**:437–438.

Hynes, R. O., 1981, Relationships between fibronectin and the cytoskeleton; in: *Cytoskeletal Elements and Plasma Membrane Organization* (G. Poste and G. L. Nicolson, eds.) North Holland Pub. Co., New York, pp. 100–137.

Hynes, R. O., and Destree, A. T., 1978, Relationships between fibronectin (LETS protein) and actin, *Cell* **15**:875–886.

Isenberg, G., Leonard, K., and Jockusch, B. M., 1982, Structural aspects of vinculin-actin interactions, *J. Mol. Biol.* **158**:231–249.

Ishikawa, H., Bischoff, R., and Holtzer, H., 1969, The formation of arrowhead complexes with heavy meromyosin in a variety of cell types, *J. Cell Biol.* **43**:312–328.

Izzard, C. S., and Lochner, L. R., 1976, Cell-to-substrate contacts in living fibroblasts: An interference reflexion study with an evaluation of the technique, *J. Cell Sci.* **21**:129–159.

Izzard, C. S., and Lochner, L. R., 1980, Formation of cell-to-substrate contacts during fibroblast motility: An interference-reflexion study, *J. Cell Sci.* **42**:81–116.

Jacobson, B. S., 1980, Actin binding to the cytoplasmic surface of the plasma membrane isolated from *Dictyostelium discoideum*, *Biochem. Biophys. Res. Commun.* **97**:1493–1498.

Ji, T. H., and Nicolson, G. L., 1974, Lectin binding and perturbation of the outer surface of the cell membrane induces a transmembrane organizational alteration at the inner surface, *Proc. Natl. Acad. Sci. USA* **71**:2212–2216.

Jockusch, B. M., and Isenberg, G., 1981, Interaction of α-actinin and vinculin with actin: Opposite effects on filament network formation, *Proc. Natl. Acad. Sci. USA* **78**:3005–3009.

Jockusch, B. M., and Isenberg, G., 1982, Vinculin and α-actinin: Interaction with actin and effect on microfilament network formation, in: *Cold Spring Harbor Symposia on Quantitative Biology*, Vol. 46, Cold Spring Harbor, New York, pp. 613–623.

Jockusch, B. M., Burger, M. M., DaPrada, M., Richards, J. G., Chaponnier, C., and Gabbiani, G., 1977, α-Actinin attached to membranes of secretory vesicles, *Nature* **270**:628–629.

Kapitza, H.-G., and Sackmann, E., 1980, Local measurement of lateral motion in erythrocyte membranes by photobleaching techniques, *Biochim. Biophys. Acta* **595**:56–64.

Kaplan, J., 1981, Polypeptide-binding membrane receptors: Analysis and classification, *Science* **212**:14–20.

Keen, J. H., Willingham, M. C., and Pastan, I. H., 1979, Clathrin-coated vesicles: Isolation, dissociation, and factor-dependent reassociation of clathrin baskets, *Cell* **16:**303–312.

Kelley, R. O., Trotter, J. A., Marek, L. F., Perdue, B. D., and Taylor, C. B., 1980, Variation in cytoskeletal assembly during spreading of progressively subcultivated human embryo fibroblasts (IMR-90), *Mech. Aging Dev.* **13:**127–141.

Keski-Oja, J., Sen, A., and Todaro, G. J., 1980, Direct association of fibronectin and actin molecules in vitro, *J. Cell Biol.* **85:**527–533.

Klausner, R. D., Kumar, N., Weinstein, J. N., Blumenthal, R., and Flavin, M., 1981, Interaction of tubulin with phospholipid vesicles, *J. Biol. Chem.* **256:**5879–5885.

Koch, G. L. E., and Smith, M. J., 1978, An association between actin and the major histocompatibility antigen H-2, *Nature* **273:**274–278.

Koppel, D. E., Sheetz, M. P., and Schindler, M., 1981, Matrix control of protein diffusion in biological membranes, *Proc. Natl. Acad. Sci. USA* **78:**3576–3580.

Koppel, D. E., Oliver, J. M., and Berlin, R. D., 1982, Surface functions during mitosis III. Quantitative analysis of ligand-receptor movement into the cleavage furrow: Diffusion vs. flow, *J. Cell Biol.* **93:**950–960.

Korn, E. D., 1978, Biochemistry of actomyosin-dependent cell motility (a review), *Proc. Natl. Acad. Sci. USA* **75:**588–599.

Korn, E. D., 1982, Actin polymerization and its regulation by proteins from nonmuscle cells, *Phys. Rev.* **62:**672–737.

Korn, E. D., Bowers, B., Batzri, S., Simons, S. R., and Victoria, E. J., 1974, Endocytosis and exocytosis: Role of microfilaments and involvement of phospholipids in membrane fusion, *J. Supramol. Struct.* **2:**517–528.

Kornguth, S. E., and Sunderland, E., 1975, Isolation and partial characterization of a tubulin-like protein from human and swine synaptosomal membrane, *Biochim. Biophys. Acta* **393:**100–114.

Koteliansky, V. E., Glukhova, M. A., Morozkin, A. D., Musatov, A. P., Shirinsky, V. P., Tskhovrebova, L. A., and Smirnov, V. N., 1981, A study of actin-fibronectin interaction, *FEBS Lett.* **133:**31–35.

Kumar, N., Klausner, R. D., Weinstein, J. N., Blumenthal, R., and Flavin, M., 1981, Interaction of tubulin with phospholipid vesicles. II. Physical changes of the protein, *J. Cell Biol.* **256:**5886–5889.

Kurkinen, M., Wartiovaaara, J., and Vaheri, A., 1978, Cytochalasin B releases a major surface associated glycoprotein, fibronectin, from cultured fibroblasts, *Exp. Cell Res.* **111:**127–137.

Laemmli, U. K., 1970, Cleavage of structural proteins during the assembly of the head of bacteriophage T4, *Nature* **227:**680–685.

Lasek, R. J., and Hoffman, P. N., 1976, The neuronal cytoskeleton, axonal transport and axonal growth, in: *Cell Motility, Book A* (R. Goldman, T. Pollard, and J. Rosenbaum, eds.), Cold Spring Harbor Laboratory, Cold Spring Harbor, New York, pp. 1021–1050.

Lazarides, E., 1976a, Actin, alpha-actinin and tropomyosin interaction in the structural organization of actin filaments in nonmuscle cells, *J. Cell Biol.* **68:**202–219.

Lazarides, E., 1976b, Aspects of the structural organization of actin filaments in tissue culture cells, in: *Cell Motility, Book A* (R. Goldman, T. Pollard, and J. Rosenbaum, eds.), Cold Spring Harbor Laboratory, Cold Spring Harbor, New York, pp. 347–360.

Lazarides, E., 1980, Intermediate filaments as mechanical integrators of cellular space, *Nature* **283:**249–256.

Lazarides, E., and Burridge, K., 1975, α-Actinin: Immunofluorescent localization of a muscle structural protein in nonmuscle cells, *Cell* **6:**289–298.

Lehto, V. P., Virtanen, I., and Kurki, P., 1978, Intermediate filaments anchor the nuclei in nuclear monolayers of cultured human fibroblasts, *Nature* **272:**175–177.

Lehto, V. P., Vartio, T., and Virtanen, I., 1980, Enrichment of a 140 Kd surface glycoprotein in adherent, detergent-resistant cytoskeletons of cultured human fibroblasts, *Biochem. Biophys. Res. Commun.* **95:**909–916.

Leuther, M. D., Peacock, J. S., Krakaver, H., and Barisas, B. G., 1981, Changes in lectin receptor lateral mobilities accompany lymphocyte stimulation, *J. Immunol.* **127:**893–899.

Lin, D., 1981, Spectrin-4.1-actin complex of the human erythrocyte: Molecular basis of its ability to bind cytochalasins with high-affinity and to accelerate actin polymerization in vitro, *J. Supramol. Struct. Cellular Biochem.* **15:**129–138.

Lin, D., and Lin, S., 1979, Actin polymerization induced by a motility-related high-affinity cytochalasin binding complex from human erythrocyte membrane, *Proc. Natl. Acad. Sci. USA* **76:**2345–2349.

Litman, D., Hsu, C., and Marchesi, V., 1980, Evidence that spectrin binds to macromolecular complexes on the inner surface of the red cell membrane, *J. Cell Biol.* **42:**1–22.

Loor, F., Forni, F., and Pernis, B., 1972, The dynamic state of the lymphocyte membrane. Factors affecting the distribution and turnover of surface immunoglobulin, *Eur. J. Immunol.* **2:**203–212.

Luna, E. J., Fowler, V. M., Swanson, J., Branton, D., and Taylor, D. L., 1981, A membrane cytoskeleton from *Dictyostelium discoideum*. I. Identification and partial characterization of an actin-binding activity, *J. Cell Biol.* **88:**396–409.

Mabuchi, I., and Ikuno, M., 1977, The effect of myosin antibody on the division of starfish blastomeres, *J. Cell Biol.* **74:**251–263.

Marek, L. F., Kelley, R. O., and Perdue, B. D., 1982, Organization of the cytoskeleton in square fibroblasts, *Cell Motil.* **2:**115–130.

Matsudaira, P. T., and Burgess, D. R., 1979, Identification and organization of the components in the isolated microvillus cytoskeleton, *J. Cell Biol.* **83:**667–673.

Matsudaira, P. T., and Burgess, D. R., 1982a, Partial reconstruction of the microvillus core bundle: Characterization of villin as a Ca^{+}-dependent, actin-bundling/depolymerizing protein, *J. Cell Biol.* **92:**648–656.

Matsudaira, P. T., and Burgess, D. R., 1982b, Organization of the cross-filaments in intestinal microvilli, *J. Cell Biol.* **92:**657–664.

Maupin, P., and Pollard, T. D., 1981, Visualization of coated membrane regions, membrane specializations, and cytoplasmic filaments of cultured cells, *J. Cell Biol.* **91:**300a.

Mautner, V., and Hynes, R. O., 1977, Surface distribution of LETS protein in relation to the cytoskeleton of normal and transformed cells, *J. Cell Biol.* **75:**743–768.

McNutt, S. N., 1978, A thin-section and freeze-fracture study of microfilament-membrane attachments in choroid plexus and intestinal microvilli, *J. Cell Biol.* **79:**774–787.

Mescher, M. F., Jose, M. J. L., and Balk, S. P., 1981, Actin-containing matrix associated with the plasma membrane of murine tumor and lymphoid cells, *Nature* **289:**139–144.

Mesland, D. A. M., Spiele, H., and Roos, E., 1981, Membrane-associated cytoskeleton and coated vesicles in cultured hepatocytes visualized by dry-cleaving, *Exp. Cell Res.* **132:**169–184.

Meyer, R. K., Schindler, H., and Burger, M. M., 1982, α-Actinin interacts specifically with model membranes containing glycerides and fatty acids, *Proc. Natl. Acad. Sci. USA* **79:**4280–4284.

Middaugh, C. R., and Ji, T. H., 1980, A photochemical crosslinking study of the subunit structure of membrane-associated spectrin, *Eur. J. Biochem.* **110:**567–592.

Moore, P. B., Ownby, C. L., and Carraway, K. L., 1978, Interactions of cytoskeletal elements with the plasma membrane of sarcoma 180 ascites tumor cells, *Exp. Cell Res.* **115:**331–342.

Mooseker, M. S., and Stephens, R. E., 1980, Brush-border alpha actinin? Comparison of two proteins of the microvillus core with alpha-actinin by two-dimensional peptide mapping, *J. Cell Biol.* **86:**466–474.

Mooseker, M. S., and Tilney, L. G., 1975, The organization of an actin filament-membrane complex: Filament polarity and membrane attachment in the microvilli of intestinal epithelial cells, *J. Cell Biol.* **67:**725–743.

Mooseker, M. S., Pollard, T. D., and Fujiwara, K., 1978, Characterization and localization of myosin in the brush border of intestinal epithelial cells, *J. Cell Biol.* **79:**444–453.

Mooseker, M. S., Graves, T. A., Wharton, K. A., Falco, N., and Howe, C. L., 1980, Regulation of microvillus structure: Calcium-dependent solation and cross-linking of actin filaments in the microvilli of intestinal epithelial cells, *J. Cell Biol.* **87:**809–822.

Mooseker, M. S., Pollard, T. D., and Wharton, K. A., 1982, Nucleated polymerization of actin from the membrane-associated ends of microvillar filaments in the intestinal brush border, *J. Cell Biol.* **95:**223–233.

Morrow, J., Speicher, D., and Knowles, W., 1980, Identification of functional domains of human erythrocyte spectrin, *Proc. Natl. Acad. Sci. USA* **77:**6592–6596.

Mousa, G. Y., Trevithick, J. R., Bechberger, J., and Blain, D. G., 1978, Cytohalasin D induces the capping of both leukaemia viral proteins and actin in infected cells, *Nature* **274:**808–809.

Mueller, T., and Morrison, M., 1977, Detection of a variant of protein 3, the major transmembrane protein of the human erythrocyte, *J. Biol. Chem.* **252:**6573–6576.

Mueller, T., and Morrison, M., 1981, Glycoconnectin (PAS 2), A membrane attachment site for the human erythrocyte cytoskeleton, in: *Erythrocyte Membranes 2: Recent Clinical and Experimental Advances* (W. Kruckeberg, J. Eaton, and G. Brewer, eds.), Alan R. Liss, New York, pp. 95–112.

Mukherjee, T. M., and Staehelin, L. A., 1971, The fine structural organization of the brush border of intestinal epithelial cells, *J. Cell Sci.* **8**:573–599.

Neutra, M. R., 1979, Linear arrays of intramembrane particles on microvilli in primate large intestine, *Anat. Rec.* **193**:367–381.

Nicolson, G. L., and Painter, R. G., 1973, Anionic sites of human erythrocyte membranes. II. Transmembrane effects of anti-spectrin on the topography of bound positively charged colloidal particles, *J. Cell Biol.* **59**:395–406.

Nigg, E. A., and Cherry, R. J., 1979, Influence of temperature and cholesterol on the rotational diffusion of band 3 in the human erythrocyte membrane, *Biochemistry* **18**:3457–3465.

Nigg, E. A., and Cherry, R. J., 1980, Anchorage of a band 3 population at the erythrocyte cytoplasmic membrane surface: Protein rotational diffusion measurements, *Proc. Natl. Acad. Sci. USA* **77**:4702–4706.

Nigg, E. A., Bron, C., Girardet, M., and Cherry, R. J., 1980, Band 3-glycophorin A association in erythrocyte membranes demonstrated by combining protein diffusion measurements with antibody-induced cross-linking, *Biochemistry* **19**:1887–1893.

Norton, E. K., and Izzard, C. S., 1982, Fibronectin promotes formation of the close cell-to-substrate contact in cultured cells, *Exp. Cell Res.* **139**:463–467.

Nunnally, M. H., D'Angelo, J. M., and Craig, S. W., 1980, Filamin concentration in cleavage furrow and midbody region: Frequency of occurrence compared with that of alpha-actinin and myosin, *J. Cell Biol.* **87**:219–226.

Oliver, J. M., and Berlin, R. D., 1982, Mechanisms that regulate the structural and functional architecture of cell surfaces, *Int. Rev. Cytol.* **74**:55–94.

Oliver, J. M. Albertini, D. F., and Berlin, R. D., 1976, Effects of glutathione-oxidizing agents on microtubule assembly and microtubule-dependent surface properties of human neutrophils, *J. Cell Biol.* **71**:921–932.

Oliver, J. M., Lalchandani, R., and Becker, E. L., 1977, Actin redistribution during concanavalin A cap formation in rabbit neutrophils, *J. Reticuloendothel. Soc.* **21**:359–364.

Oliver, J. M., Gelfand, E. W., Pearson, C. B., Pfeiffer, J. R., and Dosch, H-M., 1980, Microtubule assembly and concanavalin A capping in lymphocytes: Reappraisal using normal and abnormal human peripheral blood cells, *Proc. Natl. Acad. Sci. USA* **77**:3499–3503.

Oosawa, F., and Kasai, M., 1971, Actin, in: *Subunits in Biological Systems*, Part A (S. N. Timasheff and G. D. Fasman, eds.), Marcel Dekker, New York, pp. 261–321.

Ostlund, R., Leung, J., and Kipnis, D., 1977, Muscle actin filaments bind pituitary secretory granules in vitro, *J. Cell Biol.* **73**:78–87.

Owen, M. J., Auger, J., Barber, B. H., Edwards, A. J., Walsh, F. S., and Crumpton, M. J., 1978, Actin may be present on the lymphocyte surface, *Proc. Natl. Acad. Sci. USA* **75**:4484–4488.

Owens, J., Mueller, T., and Morrison, M., 1980, A minor sialoglycoprotein of the human erythrocyte membrane, *Arch. Biochem. Biophys.* **204**:247–254.

Painter, R. G., and McIntosh, A. T., 1979, The regional association of actin and myosin with sites of particle phagocytosis, *J. Supramol. Struct.* **12**:369–384.

Painter, R. G., Whisenand, J., and McIntosh, A. T., 1981, Effects of cytochalasin B on actin and myosin association with particle binding sites in mouse macrophages: Implications with regard to the mechanism of action of the cytochalasins, *J. Cell Biol.* **91**:373–384.

Pastan, I. H., and Willingham, M. C., 1981, Receptor-mediated endocytosis of hormones in cultured cells, *Annu. Rev. Physiol.* **43**:239–250.

Pearse, B. M. F., 1975, Coated vesicles from pig brain: Purification and biochemical characterization, *J. Mol. Biol.* **97**:93–98.

Pearse, B. M. F., 1976, Clathrin: A unique protein associated with intracellular transfer of membranes by coated vesicles, *Proc. Natl. Acad. Sci. USA* **73**:1255–1259.

Peters, R., 1981, Translational diffusion in the plasma membrane of single cells as studied by fluorescence microphotolysis, *Cell Biol. Int. Rep.* **5**:733–760.

Peters, R., Peters, J., Tews, K. H., and Bähr, W., 1974, A microfluorimetric study of translational diffusion in erythrocyte membranes, *Biochim. Biophys. Acta* **367**:282–294.

Phillips, M. J., Oda, M., Yousef, I. M., and Funatsu, K., 1981, Effects of cytochalasin B on membrane-associated microfilaments in a cell-free system, *J. Cell Biol.* **91**:524–527.

Pick, E., and Wilner, I., 1979, Cytoskeletal control of concanavalin A receptor mobility in peritoneal macrophages, *Exp. Cell Res.* **118**:151–158.

Pinder, J., Ungewickell, E., and Bray, D., 1978, The spectrin–actin complex and erythrocyte shape, *J. Supramol. Struct.* **8**:439–445.

Pinder, J., Ungewickell, E., and Calvert, R., 1979, Polymerization of G-actin by spectrin preparations: Identification of the active constituent, *FEBS Lett.* **104**:396–400.

Podlubnaya, Z. A., Tskhovrebova, L. A., Zualishtsbvili, M. M., and Stefanenko, G. A., 1975, Electron microscopic study of alpha actinin, *J. Mol. Biol.* **92**:357–359.

Pollack, R., and Rifkin, D. B., 1976, Modification of mammalian cell shape: Redistribution of intracellular actin by SV40 virus, proteases, cytochalasin B and dimethyl sulfoxide, in: *Cell Motility* (R. Goldman, T. Pollard, and J. Rosenbaum, eds.), Cold Spring Harbor Laboratory, Cold Spring Harbor, New York, pp. 389–401.

Pollack, R., Osborn, M., and Weber, K., 1975, Patterns of organization of actin and myosin in normal and transformed cultured cells, *Proc. Natl. Acad. Sci. USA* **72**:994–998.

Pollard, T. D., and Korn, E. D., 1973, Electron microscopic identification of actin associated with isolated amoeba plasma membranes, *J. Cell Chem.* **248**:448–450.

Pollard, T. D., and Weihing, R. R., 1974, Actin and myosin and cell movement, *CRC Crit. Rev. Biochem.* **2**:1–65.

Poste, G., Papahadjopoulos, D., Jacobson, K., and Vail, W. J., 1975, Local anesthetics increase susceptibility of untransformed cells to agglutination by con A, *Nature* **253**:552–554.

Puszkin, S., Maimon, J., and Schook, W. J., 1979, Clathrin association with low-molecular weight proteins acting as cofactors for the assembly/disassembly of baskets, *J. Cell Biol.* **83**:293a.

Ralston, G., 1976, The influence of salt on the aggregation state of spectrin from bovine erythrocyte membranes, *Biochim. Biophys. Acta* **443**:387–393.

Ralston, G., and Crisp, E., 1981, The action of organic mercurials on the erythrocyte membrane, *Biochim. Biophys. Acta* **649**:98–104.

Reaven, E. P., and Axline, S. G., 1973, Subplasmalemmal microfilaments and microtubules in resting and phagocytizing cultivated macrophages, *J. Cell Biol.* **59**:12–27.

Repasky, E. A., Granger, B. L., and Lazarides, E., 1982, Wide-spread occurrence of avian spectrin in nonerythroid cells, *Cell* **29**:821–833.

Ringo, D. L., 1967, Flagellar motion and fine structure of the flagellar apparatus in Chlamydomonas, *J. Cell Biol.* **33**:543–571.

Rodewald, R., Newman, S. B., and Karnovsky, M. J., 1976, Contraction of isolated brush borders from the intestinal epithelium, *J. Cell Biol.* **70**:541–554.

Ryan, G. B., Unanue, E. R., and Karnovsky, M. J., 1974, Inhibition of surface capping of macromolecules by local anesthetics and tranquilisers, *Nature* **250**:56–57.

Saffman, P. G., 1976, Brownian motion in thin sheets of viscous fluid, *J. Fluid Mech.* **73**:593–602.

Saffman, P. G., and Delbrück, M., 1975, Brownian motion in biological membranes, *Proc. Natl. Acad. Sci. USA* **72**:3111–3113.

Sakaki, T., Tsuji, A., Chang, C.-H., and Ohnishi, S. I., 1982, Rotational mobility of an erythrocyte membrane integral protein band 3 in dimyristoylphosphatidylcholine reconstituted vesicles and effect of binding of cytoskeletal proteins, *Biochemistry* **21**:2366–2372.

Salisbury, J. L., Condeelis, J. S., and Satir, P., 1980, Role of coated vesicles, microfilaments, and calmodulin in receptor-mediated endocytosis by cultured B lymphoblastoid cells, *J. Cell Biol.* **87**:132–141.

Salisbury, J. L., Condeelis, J. S., Maihle, N. J., and Satir, P., 1981, Calmodulin localization during capping and receptor-mediated endocytosis, *Nature* **294**:163–164.

Sattler, C. A., and Staehelin, L. A., 1974, Ciliary membrane differentiations in *Tetrahymena pyriformis*. *Tetrahymena* has four types of cilia, *J. Cell Biol.*, **62**:473–490.

Scheele, R. B., and Borisy, G. C., 1979, *In vitro* assembly of microtubules, in: *Microtubules* (J. S. Hyams and K. Roberts, eds.), Academic Press, New York, pp. 175–254.

Schindler, M., Koppel, D. E., and Sheetz, M. P., 1980, Modulation of membrane protein lateral mobility by polyphosphates and polyamines, *Proc. Natl. Acad. Sci. USA* **77**:1457–1461.

Schlessinger, J., Koppel, D. E., Axelrod, D., Jacobson, K., Webb, W. W., and Elson, E. L., 1976a, Lateral transport on cell membranes: Mobility of concanavalin A receptors on myoblasts, *Proc. Natl. Acad. Sci. USA* **73**:2409–2413.

Schlessinger, J., Webb, W. W., Elson, E. L., and Metzger, H., 1976b, Lateral motion and valence of Fc receptors on rat peritoneal mast cells, *Nature* **264**:550–552.

Schlessinger, J., Axelrod, D., Koppel, D. E., Webb, W. W., and Elson, E. L., 1977a, Lateral transport of a lipid probe and labeled proteins on a cell membrane, *Science* **195**:305–309.

Schlessinger, J., Elson, E. L., Webb, W. W., Yahara, I., Rutishauser, U., and Edelman, G. M., 1977b, Receptor diffusion on cell surfaces modulated by locally bound concanavalin A, *Proc. Natl. Acad. Sci. USA* **74**:1110–1114.

Schollmeyer, J. V., Goll, D. E., Tilney, L. G., Mooseker, M. S., Robson, R. M., and Stromer, M. H., 1974, Localization of α-actinin in non-muscle material, *J. Cell Biol.* **63**:304a.

Schook, W. J., Puszkin, S., Bloom, W. S., Ores, C., and Kochwa, S., 1979, Mechanochemical properties of brain clathrin: Interactions with actin and α-actinin and polymerization into basketlike structures or filaments, *Proc. Natl. Acad. Sci. USA* **76**:116–120.

Schreiner, G. F., and Unanue, E. R., 1975a, The modulation of spontaneous and anti-Ig-stimulated motility of lymphocytes by cyclic nucleotides and adrenergic and cholinergic agents, *J. Immunol.* **114**:802–808.

Schreiner, G. F., and Unanue, E. R., 1975b, Anti-Ig-triggered movement of lymphocytes. Specificity and lack of evidence for directional migration, *J. Immunol.* **114**:809–814.

Schreiner, G. F., and Unanue, E. R., 1976a, Calcium-sensitive modulation of Ig capping: Evidence supporting a cytoplasmic control of ligand-receptor complexes, *J. Exp. Med.* **143**:15–31.

Schreiner, G. F., and Unanue, E. R., 1976b, The disruption of immunoglobulin caps by local anesthetics, *Clin. Immunol. Immunopathol.* **6**:264–269.

Schreiner, G. F., Braun, J., and Unanue, E. R., 1976, Spontaneous redistribution of surface immunoglobulin in the motile B lymphocyte, *J. Exp. Med.* **144**:1683–1688.

Schreiner, G. F., Fujiwara, K., Pollard, T., and Unanue, E. R., 1977, Redistribution of myosin accompanying capping of surface Ig, *J. Exp. Med.* **145**:1393–1398.

Schroeder, T. E., 1975, Dynamics of the contractile ring, in: *Molecules and Cell Movement* (S. Inoue and R. E. Stephens, eds.), Raven Press, New York, pp. 305–334.

Schroeder, T. E., 1981, Interrelations between the cell surface and the cytoskeleton in cleaving sea urchin eggs, in: *Cytoskeleton Elements and Plasma Membrane Organization* (G. Poste and G. L. Nicolson, eds.), North-Holland, New York, pp. 170–209.

Sefton, B. M., Hunter, T., Ball, E. H., and Singer, S. J., 1981, Vinculin: A cytoskeletal target of the transforming protein of rous sarcoma virus, *Cell* **24**:165–174.

Sheetz, M. P., 1979, Integral membrane protein interaction with Triton cytoskeletons of erythrocytes, *Biochim. Biophys. Acta* **557**:122–134.

Sheetz, M. P., and Casaly, J., 1980, 2,3-Diphosphoglycerate and ATP dissociate erythrocyte membrane skeletons, *J. Biol. Chem.* **255**:9955–9960.

Sheetz, M. P., Schindler, M., and Koppel, D. E., 1980, Lateral mobility of integral membrane proteins is increased in spherocytic erythrocytes, *Nature* **285**:510–512.

Sheetz, M. P., Febbroriello, P., and Koppel, D. E., 1982, Triphosphoinositide increases glycoprotein lateral mobility in erythrocyte membranes, *Nature* **296**:91–93.

Shotton, D. M., Thompson, K., Wofsy, L., and Branton, D., 1978, Appearance and distribution of surface proteins of the human erythrocyte membrane: An electron microscope and immunochemical labeling study, *J. Cell Biol.* **76**:512–531.

Shotton, D., Burke, B., and Branton, D., 1979, The molecular structure of human erythrocyte spectrin, *J. Mol. Biol.* **131**:303–329.

Shriver, K., and Rohrschneider, L., 1981, Organization of $pp60^{src}$ and selected cytoskeletal proteins within adhesion plaques and junctions of rous sarcoma virus-transformed rat cells, *J. Cell Biol.* **89**:525–533.

Siegel, D., and Branton, D., 1982, Human erythrocyte band 4.9, *J. Cell Biol.* **95**:265a.

Siliciano, J. D., and Craig, S. W., 1982, Meta-vinculin—a vinculin-related protein with solubility properties of a membrane protein, *Nature* **300**:533–535.

Silverstein, S. C., and Loike, J. D., 1980, Phagocytosis, in: *Mononuclear Phagocytes*, Vol. II (R. van Furth, ed.), Martinus Nijhoff, The Hague, pp. 895–917.

Silverstein, S. C., Steinman, R. M., and Cohn, Z. A., 1977, Endocytosis, *Annu. Rev. Biochem.* **46**:669–722.

Singer, I. I., 1979, The fibronexus: a transmembrane association of fibronectin-containing fibers and bundles of 5 nm microfilaments in hamster and human fibroblasts, *Cell* **16**:675–685.

Singer, I. I., 1982, Association of fibronectin and vinculin with focal contacts and stress fibers in stationary hamster fibroblasts, *J. Cell Biol.* **92**:398–408.

Singer, I. I., and Paradiso, P. R., 1981, A transmembrane relationship between fibronectin and vinculin (130 kd protein): Serum modulation in normal and transformed hamster fibroblasts, *Cell* **24**:481–492.

Singer, S. J., Ash, J. F., Bourguignon, L. Y. W., Heggeness, M. H., and Louvard, D., 1978, Transmembrane interactions and the mechanisms of transport of proteins across membranes, *J. Supramol. Struct.* **9**:373–389.

Small, J. V., and Celis, J. E., 1978, Direct visualization of the 10 nm (100 Å) filament network in whole and enucleated cultured cells, *J. Cell Sci.* **31**:393–409.

Smith, D. K., and Palek, J., 1982, Modulation of lateral mobility of band 3 in the erythrocyte membrane by oxidative cross-linking of spectrin, *Nature* **297**:424–425.

Smith, L. M., Smith, B. A., and McConnell, H. M., 1979, Lateral diffusion of M-13 coat protein in model membranes, *Biochemistry* **18**:2256–2259.

Spudich, J. A., 1974, Biochemical and structural studies of actomyosin-like proteins from non-muscle cells. II. Purification, properties, and membrane association of actin from amoebae of *Dictyostelium discoideum*, *J. Biol. Chem.* **249**:6013–6020.

Steck, T., 1974, The organization of proteins in the human red cell membrane, *J. Cell Biol.* **62**:1–19.

Steck, T., and Yu, J., 1973, Selective solubilization of proteins from red blood cell membranes by protein perturbants, *J. Supramol. Struct.* **1**:220–232.

Stendahl, O. I., Hartwig, J. H., Brotschi, E. A., and Stossel, T. P., 1980, Distribution of actin-binding protein and myosin in macrophages during spreading and phagocytosis, *J. Cell Biol.* **84**:215–224.

Stephens, R. E., 1977, Major membrane protein differences in cilia and flagella: Evidence for a membrane-associated tubulin, *Biochemistry* **16**:2047–2048.

Stossel, T. P., 1978, Contractile proteins in cell structure and function, *Annu. Rev. Med.* **29**:427–457.

Stossel, T. P., and Hartwig, J. H., 1976, Phagocytosis and the contractile proteins of pulmonary macrophages, in: *Cell Motility* (R. Goldman, T. Pollard, and J. Rosenbaum, eds.), Cold Spring Harbor Laboratory, New York, **vol. B.**, pp. 529–544.

Sugrue, S. P., and Hay, E. D., 1981, Response of basal epithelial cell surface and cytoskeleton to solubilized extracellular matrix molecules, *J. Cell Biol.* **91**:45–54.

Sundqvist, K.-G., and Ehrnst, A., 1976, Cytoskeletal control of surface membrane mobility, *Nature* **264**:226–231.

Sundqvist, K.-G., Oheskog, P., and Ege, T., 1978, Cytochalasin B induces polarisation of plasma membrane components and actin in transformed cells, *Nature* **274**:915–917.

Tank, D. W., Wu, E.-S., and Webb, W. W., 1982, Enhanced molecular diffusibility in muscle membrane blebs: Release of lateral constraints, *J. Cell Biol.* **92**:207–212.

Tanner, M. J. A., and Anstee, D. J., 1976, A method for the direct demonstration of the lectin-binding components of the human erythrocyte membrane, *Biochem. J.* **153**:265–270.

Taylor, D. L., and Condeelis, J. S., 1979, Cytoplasmic structure and contractility in amoeboid cells, *Int. Rev. Cytol.* **56**:57–143.

Taylor, R. B., Duffus, P. H., Raff, M. C., and dePetris, S., 1971, Redistribution and pinocytosis of lymphocyte surface immunoglobulin molecules induced by anti-immunoglobulin antibody, *Nature New Biol.* **233**:225–229.

Thompson, N. L., and Axelrod, D., 1980, Reduced lateral mobility of a fluorescent lipid probe in cholesterol-depleted erythrocyte membrane, *Biochim. Biophys. Acta* **597**:155–165.

Tilney, L. G., 1978, Polymerization of actin. V. A new organelle, the actomere, that initiates the assembly of actin filaments in thyone sperm, *J. Cell Biol.* **47**:408–422.

Tilney, L. G., and Cardell, R. R., Jr., 1970, Factors controlling the reassembly of the microvillus border of the small intestine of the salamander, *J. Cell Biol.* **47**:408–422.

Tilney, L. G., and Mooseker, M. S., 1976, Actin filament-membrane attachment: Are membrane particles involved? *J. Cell Biol.* **71**:402–416.

Tilney, L. G., Kiehart, D. P., Sardet, C., and Tilney, M., 1978, Polymerization of actin. IV. Role of Ca^{++} and H^+ in the assembly of actin and in membrane fusion in the acrosomal reaction of echinoderm sperm, *J. Cell Biol.* **77**:536–550.

Tilney, L. G., Bonder, E. M., and DeRosier, D. J., 1981, Actin filaments elongate from their membrane-associated ends, *J. Cell Biol.* **90**:485–494.

Toh, B. H., and Hard, G. C., 1977, Actin co-caps with concanavalin A receptors, *Nature* **269**:696–697.

Trotter, J. A., 1981, The organization of actin in spreading macrophages, *Exp. Cell Res.* **132**:235–248.

Tsukita, S., Tsukita, S., and Ishikawa, H., 1980, Cytoskeletal network underlying the human erythrocyte membrane, *J. Cell Biol.* **85**:567–576.

Tsukita, S., Tsukita, S., and Ishikawa, H., 1981, Electron microscopic study of reassociation of spectrin and actin with the human erythrocyte membrane, *J. Cell Biol.* **90**:70–77.

Tyler, J., Hargreaves, W., and Branton, D., 1979, Purification of two spectrin-binding proteins: Biochemical and electron microscopic evidence for site-specific reassociation between spectrin and bands 2.1 and 4.1, *Proc. Natl. Acad. Sci. USA* **76**:5192–5196.

Tyler, J., Reinhardt, B., and Branton, D., 1980, Associations of erythrocyte membrane proteins. Binding of purified bands 2.1 and 4.1 to spectrin, *J. Biol. Chem.* **255**:7034–7039.

Uehara, Y., Campbell, G. R., and Burnstock, G., 1971, Cytoplasmic filaments in developing and adult vertebrate smooth muscle, *J. Cell Biol.* **50**:484–497.

Unanue, E. R., 1976, Cytological analysis of histocompatibility molecules, in: *The Role of the Products of the Histocompatibility Gene Complex in Immune Response* (D. H. Katz and B. Benacerraf, eds.), Academic Press, New York, pp. 603–642.

Unanue, E. R., and Schreiner, G. F., 1977, Structure and function of surface immunoglobulin of lymphocytes, in: *Dynamic Aspects of Cell Surface Organization* (G. Poste and G. L. Nicolson, eds.), North-Holland Publishing Co., New York, pp. 619–641.

Unanue, E. R., Ungewickell, E., and Branton, D., 1981, The binding of clathrin triskelions to membranes from coated vesicles, *Cell* **26**:439–446.

Ungewickell, E., and Gratzer, W., 1978, Self-association of human spectrin, *Eur. J. Biochem.* **88**:379–385.

Vasiliev, J. M., and Gelfand, I. M., 1977, Mechanisms of morphogenesis in cell cultures, *Int. Rev. Cytol.* **50**:159–274.

Vaz, W. L. C., Kapitza, H. G., Stümpel, J., Sackmann, E., and Jovin, T. M., 1981, Translational mobility of glycophorin in bilayer membranes of dimyristoylphosphatidyl choline, *Biochemistry* **20**:1392–1396.

Verderame, M., Alcorta, D., Egnor, M., Smith, K., and Pollack, R., 1980, Cytoskeletal F-actin patterns quantitated with fluorescein isothiocyanate-phalloidin in normal and transformed cells, *Proc. Natl. Acad. Sci. USA* **77**:6624–6628.

Virtanen, I., Kurkinen, M., and Lehto, V.-P., 1979, Nucleus anchoring cytoskeleton in chicken red blood cells, *Cell Biol. Int. Rep.* **3**:157–162.

Virtanen, I., Vartio, T., Badley, R. A., and Lehto, V.-P., 1982, Fibronectin in adhesion, spreading and cytoskeletal organization of cultured fibroblasts, *Nature* **298**:660–663.

Walter, R. J., Berlin, R. D., Pfeiffer, J. R., and Oliver, J. M., 1980, Polarization of endocytosis and receptor topography on cultured macrophages, *J. Cell Biol.* **86**:199–211.

Wang, E., and Goldberg, A. R., 1976, Changes in microfilament organization and surface topography upon transformation of chick embryo fibroblasts with Rous sarcoma virus, *Proc. Natl. Acad. Sci. USA* **73**:4065–4069.

Weatherbee, J. A., 1981, Membranes and cell movements: Interactions of membranes with the proteins of the cytoskeleton, in: *International Review of Cytology*, Supplement 12, Academic Press, New York, pp. 113–176.

Webb, W. W., Barak, L. S., Tank, D. W., and Wu, E.-S., 1981, Molecular mobility on the cell surface, *Biochem. Soc. Symp.* **46**:191–205.

Weeds, A., 1982, Actin-binding proteins—regulators of cell architecture and motility, *Nature* **296**:811–816.

Weinstein, R., Khodadad, J., and Steck, T., 1980, The band 3 protein intramembrane particle of the human red blood cell, in: *Membrane Transport in Erythrocytes* (U. Lassen, H. Ussing; and J. Wieth, eds.), Munksgaard, Copenhagen, pp. 35–50.

West, C. M., McMahon, D., and Molday, R. S., 1978, Identification of glycoproteins, using lectins as probes, in plasma membranes from *Dictyostelium discoideum* and human erythrocytes, *J. Biol. Chem.* **253:**1716–1724.

Wickus, G., Gruenstein, E., Robbins, P. W., and Rich, A., 1975, Decrease in membrane-associated actin of fibroblasts after transformation by Rous sarcoma virus, *Proc. Natl. Acad. Sci. USA* **72:**746–749.

Wilkins, J. A., and Lin, S., 1981, Association of actin with chromaffin granule membranes and the effect of cytochalasin B on the polarity of actin filament elongation, *Biohim. Biophys. Acta* **642:**55–66.

Wilkins, J. A., and Lin, S., 1982, High-affinity interaction of vinculin with actin filaments in vitro, *Cell* **28:**83–90.

Willingham, M. C., Yamada, K. M., Yamada, S. S., Pouyssegur, J., and Pastan, I., 1977, Microfilament bundles and cell shape are related to adhesiveness to substratum and are dissociable from growth control in cultured fibroblasts, *Cell* **10:**375–380.

Wolfe, S. L., 1967, Microtubular elements and cell membranes isolated from nucleated erythrocytes, *J. Ultrastruct. Res.* **17:**588–597.

Wu, E.-S., Jacobson, K., Szoka, F., and Portis, A., Jr., 1978, Lateral diffusion of a hydrophobic peptide, N-4-nitrobenz-2-oxa-1,3-diazole gramicidin S, in phospholipid multibilayers, *Biochemistry* **17:**5543–5550.

Wu, E.-S., Tank, D. W., and Webb, W. W., 1982, Unconstrained lateral diffusion of concanavalin A receptors on bulbous lymphocytes, *Proc. Natl. Acad. Sci. USA* **79:**4962–4966.

Yahara, I., and Edelman, G. M., 1972, Restrictions of the mobility of lymphocyte immunoglobulin receptors by concanavalin A, *Proc. Natl. Acad. Sci. USA* **69:**608–612.

Yahara, I., and Edelman, G. M., 1973, The effects of concanavalin A on the mobility of lymphocyte surface receptors, *Exp. Cell Res.* **81:**143–155.

Yahara, I., and Kamimoto-Sameshima, F., 1977, Ligand-independent cap formation: Redistribution of surface receptors on mouse lymphocytes and thymocytes in hypertonic medium, *Proc. Natl. Acad. Sci. USA* **74:**4511–4515.

Yahara, I., and Kamimoto-Sameshima, F., 1979, Analysis of ligand-independent cap formation induced in hypertonic medium, *Exp. Cell Res.* **119:**237–252.

Yahara, I., Iwashita, S., Ebina, T., Satake, M., and Ishida, N., 1979, Inhibition of ligand-independent cap formation of mouse lymphocytes and Raji cells by neocarzinostatin, *Cancer Res.* **39:**4687–4793.

Yeltman, D. R., Jung, G., and Carraway, K. L., 1981, Isolation of α-actinin from sarcoma 180 ascites cell plasma membranes and comparison with smooth muscle α-actinin, *Biochim. Biophys. Acta* **668:**201–208.

Yu, J., and Branton, D., 1976, Reconstitution of intramembrane particles in recombinants of erythrocyte protein band 3 and lipid: Effects of spectrin-actin association, *Proc. Natl. Acad. Sci. USA* **73:**3891–3895.

Yu, J., and Goodman, S., 1979, Syndeins: The spectrin-binding protein(s) of the human erythrocyte membrane, *Proc. Natl. Acad. Sci. USA* **76:**2340–2344.

Yu, J., Fischman, D., and Steck, T. L., 1973, Selective solubilization of proteins and phospholipids from red blood cell membranes by non-ionic detergents, *J. Supramol. Struct.* **1:**233–248.

Ziparo, E., Lemay, A., and Marchesi, V., 1978, The distribution of spectrin along the membranes of normal and echinocytic human erythrocytes, *J. Cell Sci.* **34:**91–101.

3

Cell Coupling

Camillo Peracchia

I. INTRODUCTION

A fascinating hypothesis on the origin of Metazoa envisions some flagellated Protozoa of the Cambrian period aggregating into small communities where food procurement and mutual protection against the environment could be more efficiently provided (Hyman, 1940). The simplest such colonies may have been the ancestors of modern Porifera, Ctenophora, and Coelenterata.

Porifera are simple, stationary organisms in which flagellated cells, the choanocytes, form the walls of complex canals which empty into a common chamber open to the outside via the osculum. Water flows through the canals and out the osculum at a slow rate by the beating of the flagella. Ctenophora and Coelenterata, on the contrary, have been more complex organisms from the beginning, their cells having acquired early the ability to differentiate for various functions.

In these early cell communities, several functions fundamental for survival could not be efficiently performed without synchronous or sequential activation of groups of cells. An example is locomotion in "comb jellies" (phylum Ctenophora). These transparent walnut-shaped creatures with a sense organ (the statocyst) at the aboral end and the mouth at the other end are equipped with eight rows of ciliated bodies, the comb plates, which extend along the length of the animal from the statocyst to the mouth. The comb plates are swimming organs which by the beating strokes of their cilia, directed toward the aboral end, move the animal mouth-foremost. Effective swimming could be achieved only if the ciliated cells were activated sequentially from the aboral end of the comb row (Tamm, 1973). To coordinate their beating function, the cells had to be provided with a system for fast signal transmission. A system of channels for direct cell-to-cell communication became available. The cells became electrically coupled; the structural correlate of such a system is a junction generally known as a gap junction. Indeed, the comb plates of these early Ctenophora provide

Camillo Peracchia • Department of Physiology, School of Medicine, University of Rochester, Rochester, New York 14642.

one of the earliest, if not the earliest, examples of gap junctions (Satterlie and Case, 1978).

In contrast with the successful synchronization of cell activity in Ctenophora is the largely inefficient water movement in Porifera. Here the flagella do not beat in coordination, and how any water current at all can be produced is still unclear. It is not surprising to learn that Porifera do not have gap junctions (Gilula, 1978) and, although the choanocytes may have some form of electrical interaction (Loewenstein, 1967), this is obviously insufficient to coordinate their function. It could very well be that the lack of gap junctions was the very reason why these organisms did not evolve, remaining at a level of structure which has been termed "cellular," i.e., loose aggregates of cells hardly formed into tissue (Hyman, 1940). In any event, there is little doubt that the capacity of these early cells to establish direct lines of communication has represented one of the most important steps in the development of complex multicellular organisms.

Direct cell-to-cell communication (cell coupling) is a feature of virtually all multicellular organisms from the lower invertebrates to mammals. The most obvious role of cell coupling is to coordinate the electrical activity of groups of cells, for example, in heart contraction, uterine contracture at labor, and simultaneous firing of groups of neurons. However, many other nonelectrical cooperative phenomena are also known to depend on direct cell-to-cell communication, as not only small ions are exchanged but also larger molecules including amino acids, oligosaccharides, nucleotides, and second messengers (Hooper and Subak-Sharpe, 1981). Cell coupling is believed to allow cells to synchronize hormonal effects, equilibrate metabolic and ionic pools, and exchange regulatory molecules for cell growth and differentiation.

II. WHICH MOLECULES DIFFUSE FROM CELL TO CELL?

A. Electrical Coupling

A variety of protoplasmic molecules are presently known to be exchanged between coupled cells, but for several years after cell-to-cell coupling was discovered, neighboring cells were believed to communicate only electrically with each other. Evidence for electrical coupling between cells was first obtained in the Purkinje fibers of sheep heart (Weidmann, 1952) and soon confirmed in a variety of excitable cells (reviewed in Peracchia, 1980). Electrical coupling is understood as the capacity of cells to transmit passive (electrotonic) potentials without significantly reducing their amplitude. There are several ways in which electrical coupling can be detected and measured. Most often, information on electrical coupling is needed in complex multicellular systems. In these cases, the coupling ratio or coupling coefficient (V_2/V_1) is usually measured.

A standard set up for measuring the coupling ratio employs three microelectrodes (Figure 1). Two of them, one for injecting current (I) and the other for recording the change in voltage (V_1) caused by current injection, are inserted in a cell (C_1). Another microelectrode is inserted in a neighboring cell (C_2) for recording the simultaneous change in voltage (V_2) in this cell. From the current and voltage records, the input

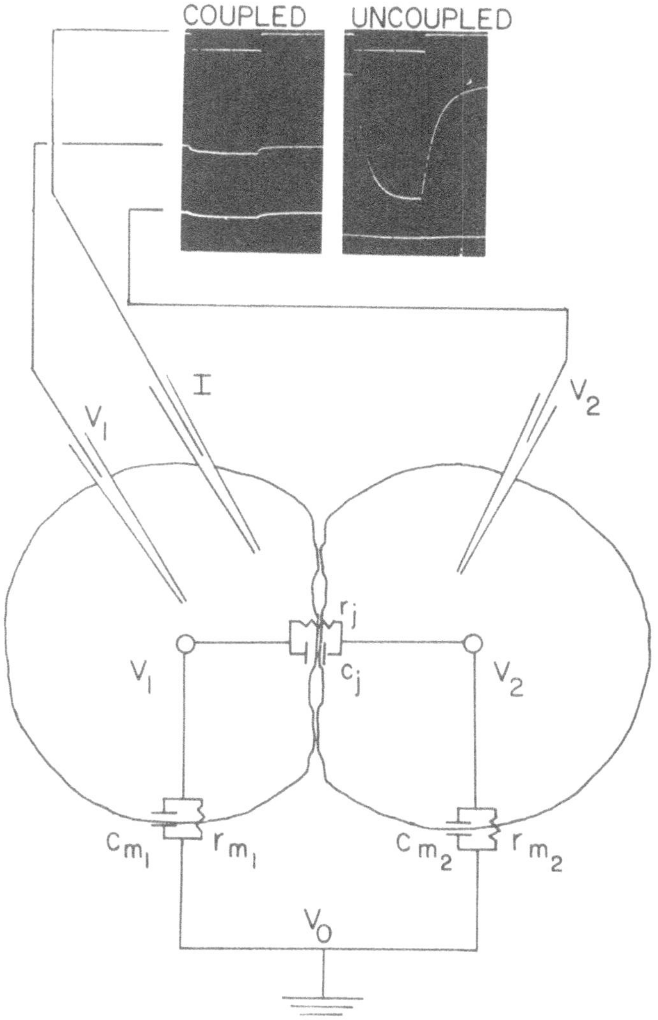

Figure 1. Measurement of electrical coupling. A standard set up involves a three-microelectrode system. Two microelectrodes, one for injecting current (I) and one for recording voltage (V_1), are inserted in cell 1 (left). Another microelectrode is inserted in a neighboring cell (right) to record the simultaneous change in voltage (V_2) in this cell. In well-coupled cells, V_1 and V_2 have similar amplitudes because the junctional resistance (r_j) is small; thus, the coupling ratio (V_2/V_1) is close to one. Uncoupling occurs when r_j increases significantly. This is reflected by a large increase in V_1 as the current injected is restricted to cell 1, and a decrease in V_2, as less current spreads to cell 2. Thus, V_2/V_1 in uncoupled cells is close to zero. The upper panel shows current and voltage records as seen on the oscilloscope screen in coupled and uncoupled conditions. r_m and c_m are the nonjunctional membrane resistivity and capacitancé respectively. c_j is the junctional membrane capacitance. V_0 is the voltage of the bath (ground).

resistance (r_i), the transfer resistance (r_t), and the coupling ratio (α) can be calculated (Watanabe and Grundfest, 1961; Bennett, 1966):

$$r_i = \frac{V_1}{I} \tag{1}$$

$$r_t = \frac{V_2}{I} \tag{2}$$

$$\alpha = \frac{V_2}{V_1} \tag{3}$$

In well-coupled cells, the coupling ratio is close to one, indicating that the electrotonic potential (V_1) evoked in C_1 spreads to C_2 with minimal decrement, while in uncoupled cells, it is close to zero. Although the coupling ratio does not allow one to calculate absolute values of junctional membrane resistance (r_j), it provides a sensitive test for electrical coupling and its relative changes, as a decrease in coupling is manifested by a decrease in V_2, as the amount of current passing from C_1 to C_2 diminishes, and an increase in V_1 as r_j increases (Figs. 1,18, and 19). In simple systems such as cell pairs (daughter cells; Figure 1), the nonjunctional membrane resistances (r_m) of the two cells are assumed to be equal, and thus the coupling ratio assumes its simplest form (Bennett, 1966):

$$\frac{V_2}{V_1} = \frac{r_m}{r_j + r_m} \tag{4}$$

In this case, the junctional resistance (r_j) can easily be calculated as

$$V_2 - V_1 = r_j \cdot i_j \tag{5}$$

$$V_0 - V_1 = r_m \cdot i_m \tag{6}$$

$$V_0 - V_2 = r_m \cdot i_j \tag{7}$$

$$I = i_j + i_m \tag{8}$$

where i_j and i_m are the junctional and membrane currents respectively, and V_0 is the potential of the bath (ground). Solving Eqs. (5–8) for r_j gives

$$r_j = \frac{V_2^2 - V_1^2}{I \cdot V_2} \tag{9}$$

The junctional resistance can also be calculated in multiple cell aggregates such as linear arrays or thin sheets of cells, but unless the multicellular systems have simple geometry, this task becomes quite complex (Socolar and Loewenstein, 1979).

While the three-microelectrode arrangement for measuring electrical coupling can easily be applied to tissues with large cells such as insect salivary glands, amphibian embryos, crayfish nervous system, etc., this method can hardly be used with cells a few micrometers in diameter such as those of most vertebrate organs. In these cases, a procedure which employs only two microelectrodes can be used. This method allows one to inject current and record electrotonic potentials through the same microelectrode (reviewed by Eisenberg and Johnson, 1970). The microelectrode is connected to a current generator and a voltmeter, and the potential produced by the electrode, which artifactually adds up to the "true" membrane electrotonic potential, can be eliminated from the final recording by a balanced bridge circuit.

In cells in which the junctional resistance (r_j) can be determined, efforts have been made to calculate the specific electrical resistance of the junctional membrane. To do so, one needs to estimate the junctional area of the coupled cells. Early studies used estimates of the entire area of cell-to-cell apposition, rather than just the surface area occupied by gap junctions; thus, the values obtained, ranging from 1 to 10 $\Omega \cdot cm^2$ (Woodbury and Crill, 1961; Wiener *et al.*, 1964; Weidmann, 1966; Borek *et al.*, 1969; Siegenbeck Van Henkelom *et al.*, 1972), were probably too high, as they represented hypothetical cases in which the whole region of cell-to-cell apposition was a low-resistance junction. On the contrary, the area of low resistance represents only a small percentage of the total area of membrane apposition, and is probably restricted to the gap junction area.

Studies in which the intercellular resistance has been related exclusively to the gap junction area reported values of specific junctional resistance ranging from $3 \cdot 10^{-3}$ to 1.4 $\Omega \cdot cm^2$ (Kriebel, 1968; Spira, 1971; Jongsma and Van Rijn, 1972; Ito *et al.*, 1974; Brink and Barr, 1977; Brink and Dewey, 1978; Sheridan *et al.*, 1978). Resistance values of the order of 10^{-2} $\Omega \cdot cm^2$ (Ito *et al.*, 1974; Sheridan *et al.*, 1978) are probably the most reliable because the cell systems in which they were obtained are architecturally most suitable for this type of study.

Using a value of 10^{-2} $\Omega \cdot cm^2$ for the specific junctional resistance and assuming that the intercellular channels are hexagonally aggregated at a center-to-center distance of 10.0 nm, the resistance of each channel has been estimated to be 10^{10} Ω (Loewenstein, 1975) and thus its conductance 100 ps. A similar value has been reported for the conductance of the postsynaptic channels in neuromuscular junctions (Katz and Miledi, 1972), while the conductance of the channels of excitability is approximately one order of magnitude lower (Begenisich and Stevens, 1975; Conti *et al.*, 1975, 1976). Since the length of the intercellular channels is approximately twice that of the postsynaptic channels, their diameter must be 50–75% larger (Sheridan *et al.*, 1978), as channel conductance is inversely proportional to channel length. The channel conductance (g_c) is given (Loewenstein, 1975) by:

$$g_c = \frac{\pi a^2}{\tau l} + \frac{l}{\tau d f(2a/d)} \qquad (10)$$

where a is the channel radius; l, the channel length; τ, the resistivity of the channel core; d, the channel spacing; and $f(2a/d)$, the ratio of the resistance of the channel system to that of an isolated channel. In Eq. (10), the first term represents the conductance of an isolated channel and the second term the additional conductance resulting from the electrostatic interaction between channels. In special cases, like that of junction formation, the conductance of individual channels can be calculated more directly by measuring the quantum displacement of V_2 taking place when a new channel forms (Loewenstein *et al.*, 1978). However, in this case, the influence of neighboring channels on individual channel conductance is difficult to evaluate, as the newly formed channels may not be aggregated into typical gap junction particle arrays.

The junctional contact sufficient to synchronize the beating of chick ventricular myocytes could be as small as an area 50 nm in diameter (De Haan, 1982). Such a small area, even if it were entirely junctional, could not contain more than 40 channels. How could so few channels pass enough current to activate a neighboring myocyte? This important question was addressed by D. Noble (De Haan, 1982) in an attempt to estimate the minimal junctional conductance necessary to synchronize two myocytes. By using the following equation:

$$i_c = -c\frac{dV}{dt} \tag{11}$$

which relates the rate of voltage change on the capacitance (c) of the cell to the net ionic currents (i_c) necessary to produce that rate of voltage change, and by assuming 0.1 *V*/sec as the rate of pacemaker depolarization during diastole and $5 \cdot 10^{-6}$ μF as the capacitance of a cell 10 μm in diameter, Noble estimated the ionic current flowing during diastole in a single cell to be of the order of 0.5 pA. He thought that if two cells with intrinsic rates different by 100% come in contact, one cell may generate 0.5 pA and the other 0.25 pA. Thus, to make the currents equal, at least 0.1 pA must flow from one cell to the other. Assuming from noise records (De Haan, 1982) that the difference in voltage between the cells is not greater than 1 mV, Noble concluded that the junctional conductance should be of the order of 100 ps (10^{-13} A/10^{-3} V). Because this is believed to be the conductance of just one channel (see above), it is a startling result! Thus, theoretically, two cells could be electrically synchronized by just one or, in any event, a very small number of channels.

This being the case, one may ask why most cells are coupled by thousands of channels, judging from the number of gap junction particles. A large number of channels provides the cells with a comfortable safety margin, especially in view of the fact that not all the channels may be in an open state at any given time (see Section IV-C). However, the principal reason for this apparent "waste" is probably metabolic rather than electrical. In fact, electrical coupling is only one of the functions of direct cell-to-cell communication, typical of excitable cells only. The principal function of cell coupling is probably that of enabling cells to exchange metabolites and regulatory molecules. This wider meaning of cell coupling became apparent in the early 1960's when electrical coupling (Loewenstein and Kanno, 1964; Kuffler and Potter, 1964) and cell-to-cell diffusion of dyes (Loewenstein and Kanno, 1964) were demonstrated in nonexcitable cells. Indeed, the channel permeance for various metabolic intermediates known to diffuse from cell to cell is considerably smaller than that for small ions. Thus, a few channels would only allow minimal exchange of molecules. A method for direct measurement of channel permeance is not available (Socolar and Loewenstein, 1979). Thus, channel permeability can only be studied by testing the cell-to-cell diffusion of probes of known molecular weight and shape.

B. Cell-to-Cell Diffusion of Ions

While the observation of a low coupling resistance between cells suggested from the beginning the cell-to-cell diffusion of charged molecules, unequivocal evidence for this came from data on the intercellular diffusion of $^{42}K^+$ (Weidmann, 1966; Bennett *et al.*, 1967), $^{22}Na^+$, $^{36}Cl^-$, $^{125}I^-$, $^{35}SO_4^{2-}$ (Bennett *et al.*, 1967), Co^{2+} (Politoff *et al.*, 1974), [1 − ^{14}C]TEA (Weingart, 1974), $AuCl_3OH^-$, and lead-EDTA^{2-}(Turin, 1977). An elegant, yet indirect, demonstration of the free exchange of Na^+ and K^+ came from a study on cells with a Na^+,K^+-ATPase resistant to ouabain (Corsaro and Migeon, 1977). When ouabain-resistant cells were grown in mixed cultures with cells of normal ouabain sensitivity, sensitive cells in contact with resistant cells were not affected by ouabain. This suggested that Na^+ passed from sensitive to resistant cells and K^+ the other way around, thus, the Na^+,K^+-ATPase of resistant cells was able to maintain a fairly normal $[Na^+]_i$ and $[K^+]_i$ in their ouabain sensitive neighbors. Na^+ and K^+ diffused via gap junctions because the sensitive cells did not become ouabain resistant when cocultured with resistant cells of the L line, a line usually incapable of forming gap junctions (Goshima, 1970; Gilula *et al.*, 1972) and of establishing electrical (Goshima, 1969; Gilula *et al.*, 1972) and metabolic (Pitts, 1971) coupling. With a similar approach, Ledbetter and Lubin (1979) demonstrated the diffusion of $^{86}Rb^+$, a tracer for K^+.

C. Molecular Probes of Cell-to-Cell Coupling

The cell-to-cell exchange of molecules larger than small ions has been investigated by testing the diffusion of dyes and tracer molecules of known molecular weight and size. In most cases, dyes such as fluorescein (mol. wt. 334), lucifer yellow CH (mol. wt. 476), carboxy-fluorescein, etc. (reviewed in Peracchia, 1980), have been employed primarily to test the capacity of certain cells to communicate with each other. Recently, however, dyes have been used in conjunction with short polymers specifically to determine the size limit of permeant molecules and the permeability characteristics of the intercellular channels.

A number of natural and synthetic peptides, with their primary amino groups coupled to fluorescent dyes such as fluorescein isothiocyanate (FITC), dansyl chloride (DANS), or lissamine rhodamine B (LRB), ranging in molecular weight from 250 to 4158, have been tested (Simpson *et al.*, 1977). In epithelial cells of Chironomus salivary gland, tracers of molecular weight 1664 or lower diffused from the injected cell into its neighbors, while those of molecular weight 1926 or greater did not (Loewenstein, 1979). Interestingly, when the same tracers were iontophoresed in various mammalian cells (primary cultures and lines) of epithelial or fibroblastic origin, compounds heavier than molecular weight 900 were not transferred (Flagg-Newton *et al.*, 1979). These channels also showed more charge discrimination than those in insect cells, as tracers with two negative charges were transferred more easily than similar ones with three charges. The conclusion was that the mammalian channels are smaller and/or more polar than those of insects, and thus more discriminating (Flagg-Newton *et al.*, 1979). Since the permeant molecules tested were all 1.4–1.6 nm in width, varying only in length, the lower limit estimate of channel bore was 1.4–1.6 nm (Flagg-Newton *et al.*, 1979). An upper limit could not be set, as insufficient data on the channel fixed charges and their effects on channel permeability were available.

Evidence for the presence of negative charges in the channels (Flagg-Newton *et al.*, 1979; Brink and Dewey, 1980) suggested that by using neutral rather than charged probes, a more reliable estimate of channel diameter and a distinction between polar and steric restraints could be made. Recently, a series of neutral fluorescent probes, linear oligosaccharides, and branched glycopeptides, have been tested in insect and mammalian cells (Schwarzmann *et al.*, 1981). With these probes the channel diameter has been estimated to be 1.6–2.0 nm in mammalian cells and 2.0–3.0 nm in insect cells. Branched glycopeptides with four terminal galactoses (3.0 nm in width) did not permeate the insect channels, while the same molecules with the galactoses clipped off (2.05 nm in width) did. Neither molecule permeated the mammalian channels.

D. Metabolic Coupling

Channels 2.0–3.0 nm in size are obviously expected to be permeable to a large variety of metabolites. Direct evidence for intercellular exchange of metabolites is not available. However, elegant studies on metabolic cooperation between cultured cells have strongly suggested that nucleotides, amino acids, oligosaccharides, and second messengers are directly transferred from cell to cell.

Early studies utilized mutant cells unable to incorporate certain nucleotides because they lacked the necessary enzymes. A classic example is that of hamster fibroblasts (BHK cells). These cells are unable to incorporate hypoxanthine into their nucleic acids because they lack hypoxanthine–guanine phosphoribosyl transferase (HGPRT; EC 2.4.2.8; Subak-Sharpe *et al.*, 1966, 1969). The capacity of these cells to incorporate [^{3}H]hypoxanthine (demonstrated by autoradiography), when allowed to come in contact with a wild-type cell, clearly indicated that labeled nucleotides, synthesized by the wild cell from [^{3}H]hypoxanthine, passed into the mutant cell and were consequently incorporated into the nucleic acids (Subak-Sharpe *et al.*, 1966, 1969). Similar results were obtained with a BHK mutant lacking thymidine kinase (TK; EC 2.7.1.21; Cox *et al.*, 1970), which confirmed the generality of the phenomenon. The use of mutant cells, however, is somewhat unsatisfactory because the limited number of mutant lines available allows one to test only a few metabolites.

A more powerful procedure for studying metabolic coupling consists in prelabeling cells with a tritiated precursor and coculturing them with unlabeled recipient cells after the precursor is converted into an intermediate metabolite not diffusible through the plasma membrane (Simms, 1973; Pitts, 1976). By this procedure, the intercellular diffusion of uridine, glucose, fucose, glucosamine, 2-deoxyglucose, choline, a number of amino acids, proline, and tetrahydrofolate has been demonstrated (Pitts and Simms, 1977; Pitts and Finbow, 1977; Pitts, 1977).

Recently, a procedure combining freeze-substitution and autoradiography has been employed to demonstrate the diffusion of lysine, glucose, uridine, and deoxyglucose between mouse lens fibers (Goodenough *et al.*, 1980). This is noteworthy because the lens fiber junctions are structurally and functionally an unusual type of gap junction (Goodenough, 1979; Peracchia and Peracchia, 1980b; Nicholson *et al.*, 1980; Hertzberg *et al.*, 1982) and, although they had been shown to pass current and dyes (Rae, 1974, 1979), they were only assumed to pass metabolites. Indeed, the eye lens, being an avascular organ, must rely on metabolic cooperation to transfer nutrients and waste products to and from deep fiber cells.

Diffusion of cAMP has been demonstrated by means of two ingenious techniques, one applied to heart muscle fibers and the other to myocardial cells coupled to ovarian granulosa cells. The former (Tsien and Weingart, 1974) consisted of pulling ventricular fibers through a hole in a rubber partition separating two perfusion chambers. [^{3}H]-cAMP was added to chamber one where the fibers had been cut in a Ca^{2+}-free solution to prevent cell uncoupling. After allowing sufficient time for [^{3}H]-cAMP uptake by the cut fibers, the solution of chamber one was replaced with a Ca^{2+}-containing solution to "heal over" the loaded fibers (see Section IV-A). The fiber-to-fiber diffusion of [^{3}H]-cAMP was determined by measuring the radioactivity of the fibers in chamber two, frozen and cut into segments at different distances from the rubber partition. The other technique (Lawrence *et al.*, 1978) capitalized on two cAMP-dependent phenomena. One is norepinephrine (NE)-induced increase in heart frequency and action potential amplitude, and decrease in action potential duration in myocardial cells. The other is the follicle-stimulating hormone (FSH)-induced synthesis of plasminogen activator (PA) in ovarian granulosa cells. Cocultures of myocardial and granulosa cells showed that NE induces synthesis of PA only in granulosa cells coupled to myocardial cells, and that FSH increases myocardial beat frequency and action potential amplitude and decreases action potential duration only in myocardial cells coupled to granulosa cells. This clearly indicated that cAMP, synthesized in one cell type, had diffused to the other cell type. With a similar approach, Blalock and Stanton (1980) have shown that the interferon-mediated protection against viruses can be transferred from cell to cell. However, it is not yet clear whether such transfer is mediated by intercellular channels or by the extracellular medium, as a similar phenomenon also takes place between L cells (Blalock and Baron, 1977).

E. Variability in Channel Permselectivity

The permeability of gap junctions may vary among tissues and even among cells of the same tissue. We have already reported (Section II-C) the differences in channel permeability between insect and mammalian junctions (Flagg-Newton *et al.*, 1979). There are also examples of vertebrate cells with channels capable of transferring small ions but not dyes and larger molecules. In early studies, lack of fluorescein exchange between electrically well-coupled cells was described in embryos of Xenopus (Slack and Palmer, 1969), Asterias (Tupper and Sanders, 1972), and Fundulus (Bennett *et al.*, 1972) at the blastula stage, but these data were later reevaluated because lucifer yellow CH readily diffused between Fundulus blastomeres (Bennett *et al.*, 1977). The lack of fluorescein transfer was then attributed to the rapid loss of this dye through nonjunctional membranes and to the small junctional area available between embryonic cells. Recently, however, solid evidence for electrical coupling in the absence of dye transfer has been obtained in mouse blastocysts (Lo and Gilula, 1979), in certain neuroblasts (Goodman and Spitzer, 1979), and in hybrids between human fibroblasts and L cells (Azarnia and Loewenstein, 1977), suggesting that some cells may in fact have a partial cell-to-cell coupling. Indeed, a number of studies indicate that the intercellular channels may be capable of undergoing gradual changes in permeability, rather than assuming either a fully open or a fully closed state (Délèze and Loewenstein, 1976; Rose *et al.*, 1977; Azarnia and Loewenstein, 1977; Flagg-Newton and Loewenstein, 1979). Thus, cells may be capable of modulating the degree of intercommunication by limiting the size of exchangeable molecules.

An example of differences in permeability among cells of the same tissue is provided by the insect epidermis (Warner and Lawrence, 1982). In this tissue, cells of different segments are electrically well coupled and exchange lead-EDTA (mol. wt. 374) but not lucifer yellow CH (mol. wt. 450), while cells of the same segment are both electrically and dye coupled. Since differences in gap junction surface area between different epidermal regions were not observed (Lawrence and Green, 1975), these data suggest intrinsic differences in channel permeability.

F. Asymmetry of Channel Permeability

Asymmetrical permeability has first been described in crayfish giant-to-motor low-resistance junctions (Furshpan and Potter, 1959). In these junctions, depolarizing current pulses spread from giant to motor but not from motor to giant fiber and the inverse is true for hyperpolarizing pulses, which indicates that these junctions function as one-way rectifiers (Furshpan and Potter, 1959). These junctions may also be impermeable to fluorescein (Keeters *et al.*, 1974), although this might be the case in certain functional states only (Bennett, 1978). Aside from the crayfish, only a few other rectifying junctions have been described in other invertebrates (Smith and Bauman, 1969; Nicholls and Purves, 1970) and in vertebrates (Auerbach and Bennett, 1969; Slack and Warner, 1975).

Asymmetrical permeability is also seen in junctions between pairs of cultured mammalian cells of different types (Flagg-Newton and Loewenstein, 1980). In pairs made of B-cells coupled to Balb/c cells, glutamic acid (mol. wt. 700) diffuses more readily from B to Balb/c cells. Since B cells are known to make more permeable channels than Balb/c cells in homocellular junctions (Flagg-Newton *et al.*, 1979), this directional permselectivity indicates that molecules have a better chance to permeate heterocellular junctions by entering the channels from their larger or less polar halves. This may have a physiological meaning as heterocellular gap junctions have indeed been described in a variety of organs (reviewed in Peracchia, 1980).

III. HOW MOLECULES DIFFUSE FROM CELL TO CELL

The capacity of neighboring cells to exchange with each other charged and neutral molecules as large as 2–3 nm clearly suggests the existence of channels that provide the cells with cytoplasmic bridges well insulated from the extracellular medium. In the last two decades, a variety of techniques, including electron microscopy, X-ray diffraction, low-dose electron microscopy, and membrane biochemistry, have been extensively applied to the channel problem in order to define the channel location, molecular architecture, and chemical composition.

Studies showing that electrical coupling between cells is lost following gap junction membrane separation (Barr *et al.*, 1965; Dreifuss *et al.*, 1966; Kawamura and Konishi, 1967; Barr *et al.*, 1968; Asada and Bennett, 1971; Pappas *et al.*, 1971) or structural change (Peracchia and Dulhunty, 1974, 1976; Dahl and Isenberg, 1980; Délèze and Hervé, 1983), and that gap junctions are necessary for synchronous contraction in heart (Goshima, 1969, 1970; Hyde *et al.*, 1969; De Haan and Hirakow, 1972; Lawrence *et al.*, 1978) and smooth muscle (Garfield *et al.*, 1977), as well as for electrical and metabolic coupling between cultured cells (Gilula *et al.*, 1972; Azarnia *et al.*, 1974), have strongly suggested that the gap junction is the channel-bearing structure (although not necessarily the only one).

A. Gap Junction Architecture

Gap junctions are regions of apposition between neighboring cells where the plasma membranes appear separated by a 2- to 3-nm-thick (3–5 nm in certain invertebrates) extracellular space (Figure 2). The extracellular space or gap, from which the term "gap junction" originates (Revel and Karnovsky, 1967; Revel, 1968), is, however, only an apparent feature of these junctions. Indeed, the membranes are in contact with each other at discrete spots where intramembrane particles, polygonally

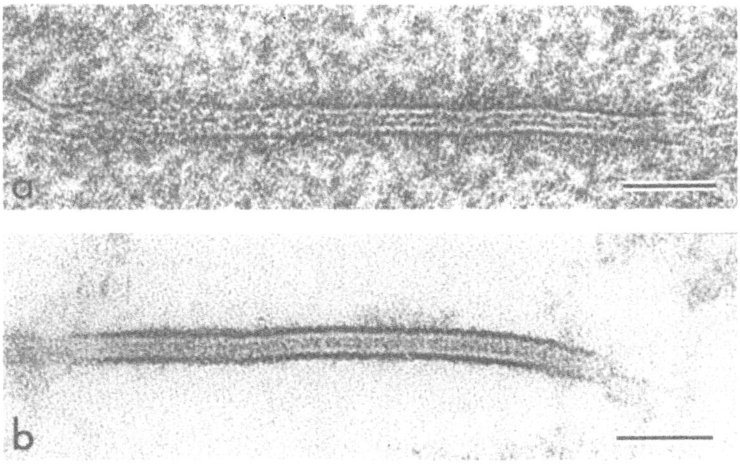

Figure 2. Vertebrate gap junctions. Thin sections. The cross-sectional profile of gap junctions displays two plasma membranes in close proximity, separated by a narrow (2–3 nm) extracellular space (gap). (a) Gap junction of rat stomach. (b) Gap junction isolated from rat liver. Bars = 50 nm. (a) From Peracchia (1980).

arranged in the plane of the membranes (Figures 3 and 4), protrude extracellularly, and bridge the gap by binding end to end to analogous particles of the adjoining membrane.

Gap junctions are usually pictured as round patches of closely apposed plasma membranes in which intramembrane particles, homogeneous in size and shape, are aggregated in disordered arrays at a center-to-center distance of approximately 10 nm (larger spacings are seen in most invertebrates). However, a great variability in gap junction architecture has been described. Gap junctions of various shapes, ranging in size from a few nanometers to several micrometers (Figures 3 and 4), and junctions

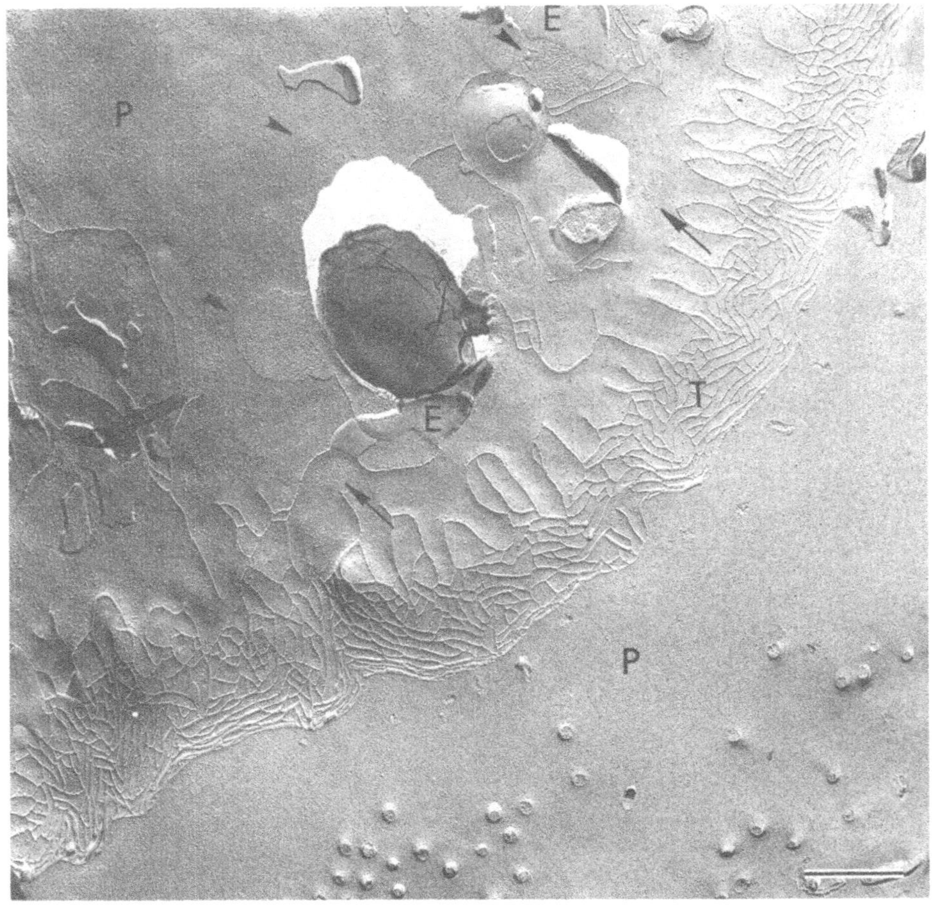

Figure 3. Junctional complexes in epithelial cells of rat stomach. Freeze-fracture replica. An extensive tight junction (zonula occludens; T) limits the lateral cell surface (upper left) from the luminal surface (lower right). The lateral surface displays several gap junctions variable in size and shape (arrowheads), a number of desmosomes (arrows) and scattered strands of tight-junctional elements. The circular profiles of cross-fractured microvilli are visible on the luminal surface. P = face P; E = face E; bar = 0.2 μM. From Peracchia (1977).

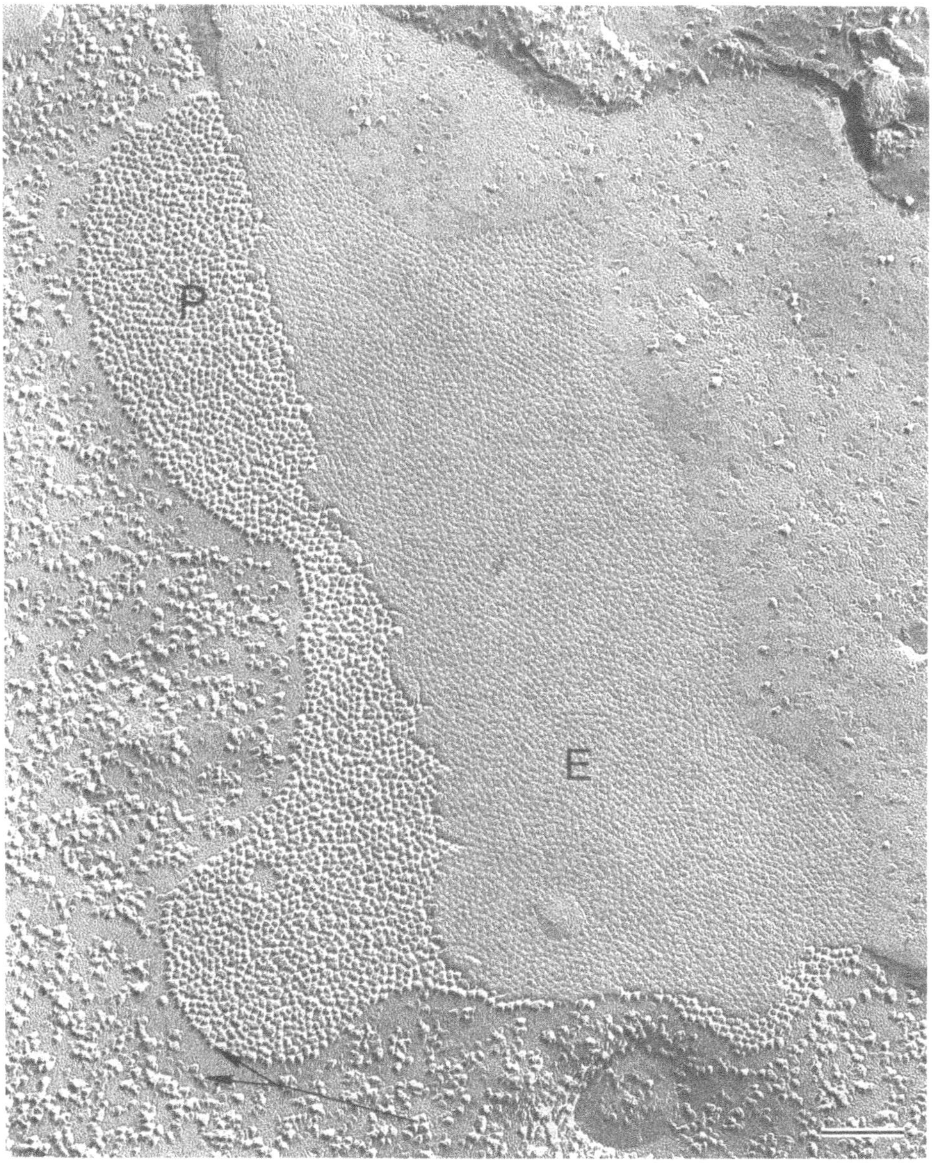

Figure 4. Rat liver gap junction. Freeze-fracture replica. Both the exoplasmic leaflet (E) of one membrane and the protoplasmic leaflet (P) of the adjacent membrane are exposed by the fracture. Face P displays 8.5-nm particles closely aggregated at a center-to-center distance of ~9.5 nm. Face E shows a similar array of pits, the complementary images of gap junction particles which have been fractured away. Notice that the junction is surrounded by a particle-free halo 25–30 nm thick (arrows). An interpretation of the nature of the halo is shown in Figure 12. Bar = 0.1 μm.

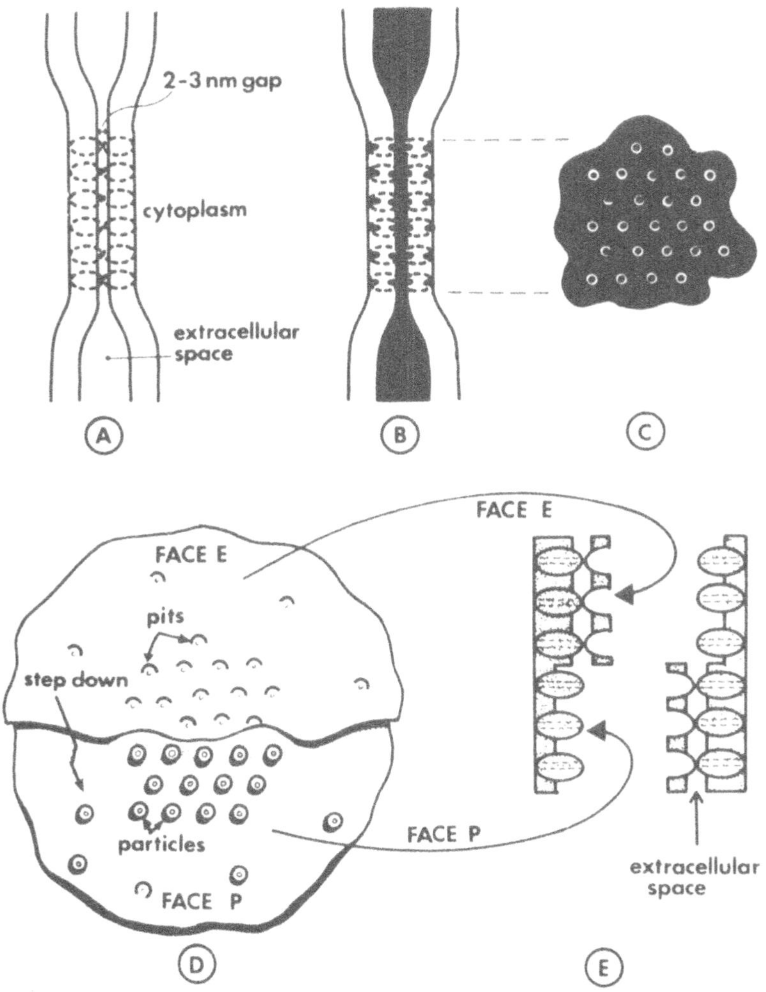

Figure 5. Schematic diagram of the ultrastructure of vertebrate gap junctions. (A) Cross-sectional profiles. (B) Profile after negative staining. (C) Face view after negative staining. (D) Freeze-fracture. The junction is fractured in a step-like fashion such that both the fractured surface of the exoplasmic leaflet (face E) and that of the protoplasmic leaflet (face P) are seen (as in Figure 4). (E) Profile of a junction fractured as in (D).

with different degrees of particle order, from total randomness (Figure 14a) to almost perfect crystallinity (Figure 14b), have been described. Crystalline, most often hexagonal, arrays are now believed to be typical of nonpermeable (uncoupled) junctions (Section IV-D).

The gap junction particles, usually about 8.5 nm in diameter in vertebrates (Figures 4 and 15a), are 10 nm or larger in most invertebrates (Figures 6 and 7). In vertebrates and some invertebrates, the particles usually fracture with the protoplasmic leaflet, appearing as bumps on P face (Figures 4 and 5D and E), but in most arthropods and some coelenterates, platyhelminthes, and annelids, they fracture mostly with the exoplasmic leaflet (Figures 6, 7, and 9).

Variable associations between gap junctions and cytoplasmic organelles such as mitochondria, endoplasmic reticulum, ribosomes, and vesicles have been reported, as well as gap junctions between different cell types called heterocellular gap junctions, to distinguish them from the more common homocellular junctions. Gap junctions

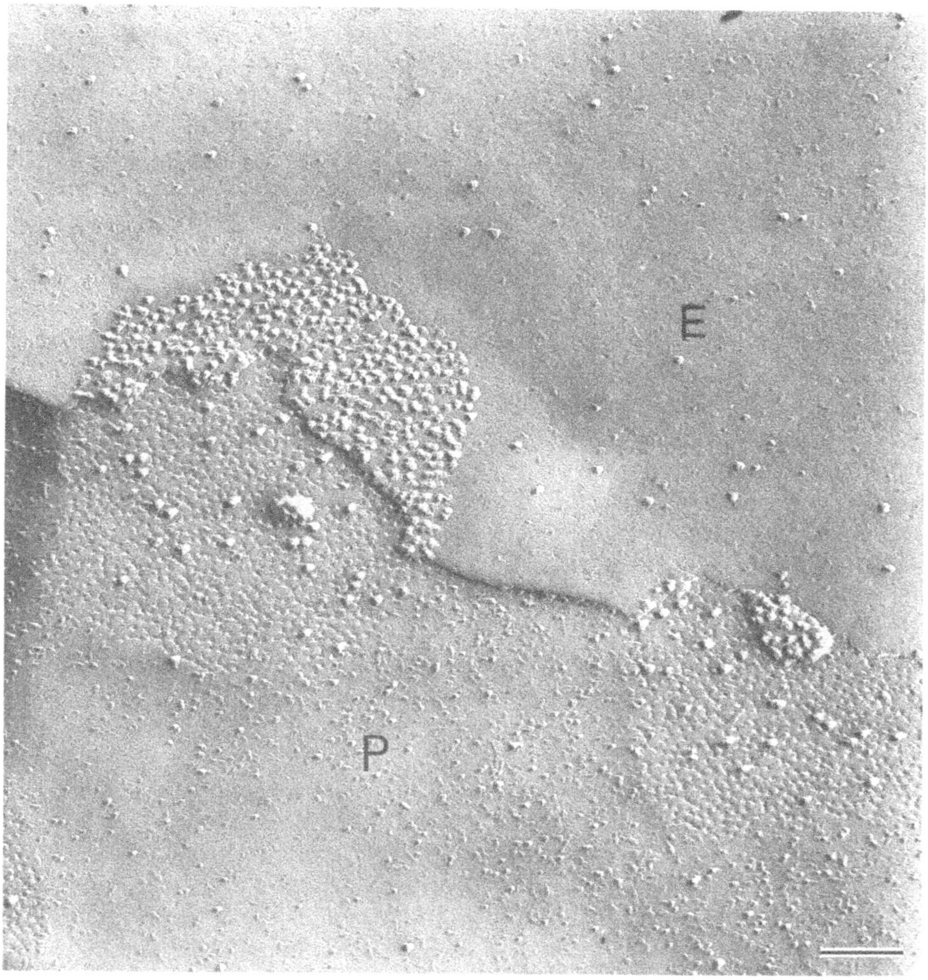

Figure 6. Arthropod gap junctions (crayfish glia). Freeze-fracture replica. In arthropods, as in some coelenterates, platyhelminthes, and annelids, the junctional particles fracture mostly with the exoplasmic leaflet (E). The junctions shown here are the most common gap junctions of arthropods, but they differ from crayfish neuronal junctions (Figure 7), as their particles are smaller and aggregate in a more disordered fashion (Peracchia and Dulhunty, 1976). P = face P; bar = 0.1 μm.

between different regions of the same cell, reflexive gap junctions, and intracellular gap junction structures, annular gap junctions, have also been described. The latter are believed to represent products of gap junction removal and degradation. Other generalities of gap junction architecture and data on the variability of gap junction ultrastructure have been extensively reviewed elsewhere (Staehelin, 1974; Larsen, 1977; Lane and Skaer, 1980; Peracchia, 1980).

B. Structure of Cell-to-Cell Channels

It is now generally accepted that the intramembrane particles of gap junctions constitute the molecular framework of the intercellular channels. This channel location, first suggested by the observation of electron-opaque dots at the center of particles in face-view images of lanthanum-stained (Payton *et al.*, 1969b; McNutt and Weinstein, 1970) or freeze-fractured (McNutt and Weinstein, 1970) gap junctions, was further supported by images obtained in crayfish low-resistance junctions studied by lanthanum or sodium phosphotungstate (NaPT) negative staining or by freeze-fracture techniques (Peracchia, 1973a,b; Peracchia, 1980). In freeze-fracture replicas, the electron-opaque dots were indeed shown to be small indentations (Figure 7). Since both intra- and extracellular ends of each particle displayed such indentations, they were interpreted as the internal and external mouth of a hemichannel, respectively (an intercellular channel resulting from the extracellular end-to-end fusion of two hemichannels; Figures 9 and 11a; Peracchia, 1973b, 1980).

A channel location at the center of the particles seems obvious as the particles are the only structures which bridge the extracellular space at gap junctions; however, unequivocal evidence for it is not yet available. A serious obstacle in the channel hunt is the channel-closing mechanism. This mechanism is activated by most preparative procedures; thus, most junctions employed for structural and chemical studies are believed to be in an uncoupled state, namely, with closed channels. In addition, the channel size is at the limit of or just below the resolving power of most of the available techniques, and electron-opaque tracers able to infiltrate and stain the channels from the cytoplasm, while in an open state, are not available. Indeed, an electron-density minimum, 2.0 nm in radius, was shown at the outer end of the particles in electron density maps derived from equatorial diffraction patterns of liver gap junctions (Makowski *et al.*, 1977). However, the interpretation of this low-density core as an aqueous channel was based on several assumptions and there was no evidence that indeed it spanned the whole membrane thickness to reach the cytoplasmic surface.

Studies with low-dose electron microscopy have confirmed the presence of a 2.0-nm-wide opening at the external end of the particles, but have not succeeded in defining the presumed channel within the membrane thickness in either of the two junctional configurations reported (Zampighi and Unwin, 1979; Unwin and Zampighi, 1980). This was attributed to the insufficient resolution of the technique, but whether or not the channels are in an open state in either of the two junctional configurations is debatable.

Data from three-dimensional electron density maps of gap junction profiles, derived from X-ray diffraction patterns, show the presumed channel extending through

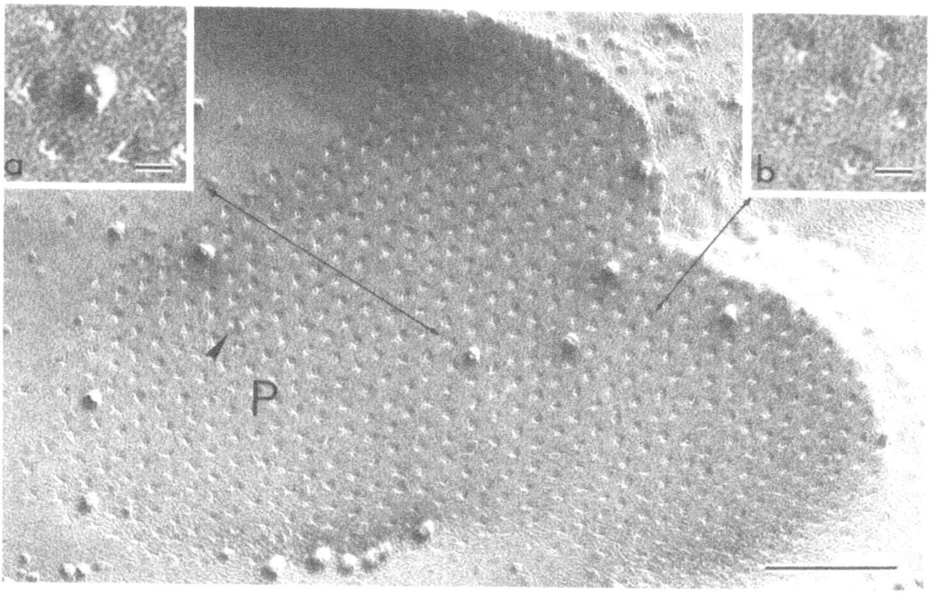

Figure 7. Neuronal low-resistance junction (crayfish). Freeze-fracture replica. Most of the particles fracture away with the exoplasmic leaflet leaving pits on face P (P). Some of the remaining particles display a central depression (inset a), believed to represent the extracellular mouth of a hemichannel (two hemichannels joined across the extracellular gap form a cell-to-cell channel). Many of the pits show central bumps (inset b). These are the complementary images of small depressions seen at the center of the particles in face E (not shown) and are believed to represent the cytoplasmic ends of the channels. The arrowhead points to a pit partially occupied by a small particle. The interpretation of the nature of this image is given in Figure 9c. Bar = 0.1 μm; bars in insets = 10 nm. From Peracchia (1973b).

the center of the particles (connexons) most of the way from the extracellular gap to the cytoplasmic surface (Makowski *et al.*, 1982). However, here also the channel patency could not be demonstrated as diffraction patterns from gap junctions exposed to 50% sucrose, a channel permeant (Bennett and Dunham, 1970), showed that sucrose fills the gap but does not enter the channel. Indeed, this confirms the hypothesis that the channel is closed at its cytoplasmic ends (Figure 11b; Peracchia, 1980). Makowski *et al.* (1982) also showed that the electron density near the cytoplasmic surface of the intramembrane particles decreases if the junctions are isolated in the presence of EGTA. This is an interesting result which suggests that EGTA may remove a component linked to the cytoplasmic surface of the junctions. In view of recent evidence for the binding of calmodulin to lens and liver junction proteins (Welsh *et al.*, 1981, 1982; Hertzberg and Gilula, 1981) and for a calmodulin participation in the uncoupling mechanism (Peracchia *et al.*, 1981, 1983; Peracchia, 1982b), it is possible that this lost component is calmodulin (see Section IV-F); indeed, calmodulin can be removed from membranes with EGTA (Kakiuchi *et al.*, 1978).

Evidence for channel occlusion at or near the cytoplasmic end of the particles has also been provided by ultrastructure studies. Edge views of negatively stained gap junctions frequently show electron-opaque lines extending from the gap into the membrane thickness at the intramembrane particles (Goodenough, 1976; Herzberg and Gilula, 1979; Peracchia, 1973a, 1980; Zampighi *et al.*, 1980). However, these lines do not span the membrane thickness, but stop abruptly at 4–6 nm from the center of the gap (Peracchia, 1980; Zampighi *et al.*, 1980), suggesting that the negative stain, having gained access to the channels probably from the gap, was stopped by the closed gates of the channels near their cytoplasmic ends (Peracchia, 1980).

While information on channel structure is still limited, some interesting data have been obtained on the architecture of gap junction particles. In crayfish junctions, estimates of particle height (Peracchia, 1973b; Hanna *et al.*, 1978), data on the particles' alternate fracture property (Figure 7; Peracchia, 1973b), and side view images of particles crossing the membrane core in thin sections (Figure 8; Peracchia, 1980), have suggested that the particles span the whole membrane thickness and protrude from both surfaces (Figures 9 and 11a; Peracchia, 1973b). This is well supported by X-ray diffraction studies on gap junctions isolated from mouse liver (Makowski *et al.*,

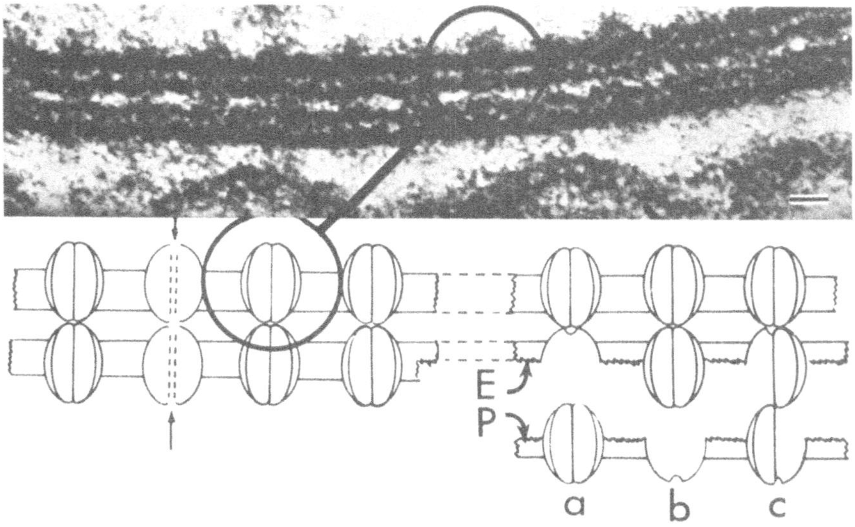

Figure 8. Neuronal gap junction (crayfish). Thin section. Transverse section through crayfish junctions often display images of intramembrane particles spanning the trilaminar profile of the membrane (encircled region). Bar = 10 nm. From Peracchia and Dulhunty (1976).

Figure 9. Schematic diagram of the structure and freeze-fracture behavior of crayfish neuronal junctions. Intramembrane particles composed of six subunits span the membrane thickness and bind to similar particles of the adjoining membrane. The cell-to-cell channels are pictured spanning the junctional thickness at the center of the particles (arrows). The right side of the diagram shows the fracture behavior of these junctions. Most particles fracture with the exoplasmic leaflet (b), some remain with the protoplasmic leaflet (a) and a few are fragmented (c) in such a way that some subunits remain on face P (P) and the other on face E (E). This feature is suggested by the observation of pits partially occupied by small particle fragments (Figure 7, arrowhead; Peracchia, 1974, 1980).

1977; Caspar *et al.*, 1977). The X-ray diffraction evidence is based on the observation that the electron-density minimum in the central stratum of the membrane is much higher than that expected from hydrocarbon chains alone, but is consistent with the presence of a mixture of hydrocarbons and proteins (Makowski *et al.*, 1977). The same study also showed that the electron density in the center of the gap is considerably greater than that of water which is consistent with ultrastructural evidence for an extracellular protrusion of the particle proteins (Peracchia, 1973a). Recently, high-resolution X-ray diffraction patterns showing strong near-meridional spots at about 0.47-nm spacing, and equatorial spots at about 1.0 nm, have suggested that a significant portion of the junctional protein is likely to be in a β-sheet conformation, with the polypeptide chains parallel to the junctional membrane surface (Makowski *et al.*, 1982).

Evidence for a subunit composition of the junctional particles has also been obtained. Particles displaying a six-pointed face-view image, seen both in invertebrate (crayfish; Peracchia, 1973a,b; Peracchia, 1980) and vertebrate (Hama and Saito, 1977; Henderson *et al.*, 1979; Sikerwar *et al.*, 1981) junctions, have suggested the presence of six subunits per particle. In crayfish junctions, particles with six subunits were seen in lanthanum (Peracchia, 1973a) or sodium phosphotungstate stained specimens (Figure 10; Peracchia, 1973b), as well as in freeze-fracture replicas (Peracchia, 1973b). In the same junctions, evidence for subunits was also provided by the ability of the particles to fracture along a plane perpendicular to the membrane surface (Figures 7 and 9c) (Peracchia, 1974, 1980). In vertebrates, particles displaying five or six subunits

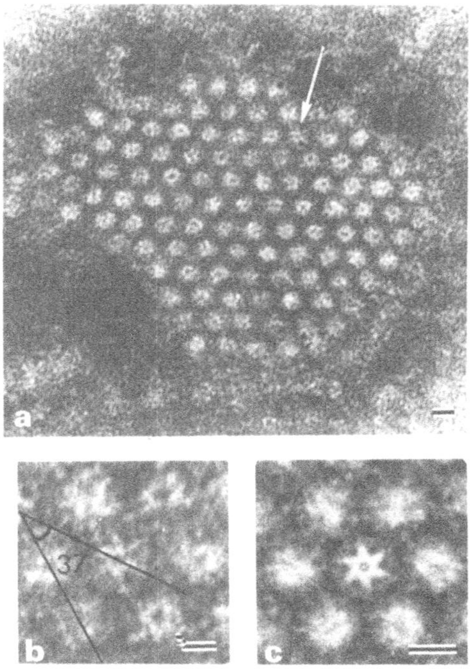

Figure 10. Isolated fragment of a crayfish neuronal junction negatively stained with sodium phosphotungstate (NaPT) (a). The diffusion of the electron opaque stain into the gap results in a negative opacization of the extracellular ends of the particles as shown in Figure 5B and C. The particle indicated by the arrow in (a) shows a six-subunit composition. The subunits are more clearly visible after a six-step photographic rotation (c) of this particle shown at higher magnification in (b). The hexameric particles are skewed in the plane of the membrane by an angle of approximately 37° (23° respectively) with respect to the axis passing through the particle centers (b). Recently, evidence for a skewed orientation of the particles has also been obtained in vertebrate junctions (Henderson *et al.*, 1979; Baker *et al.*, 1983). Bar = 10 nm. From Peracchia (1973b).

were seen in high resolution freeze-fractures of goldfish and guinea pig gap junctions (Hama and Saito, 1977) and in quick-frozen, fractured, and rotary shadowed mouse liver gap junctions (Sikerwar *et al.*, 1981). More convincing images were produced in isolated, negatively stained liver junctions studied by optical diffraction, filtering, and image reconstruction (Henderson *et al.*, 1979) and, more recently, in the same junctions studied with low-dose electron microscopy (Zampighi and Unwin, 1979; Unwin and Zampighi, 1980; Baker *et al.*, 1983). While in most gap junctions the particles seem to be composed of six subunits, the particles of lens fiber junctions may be composed of four subunits. This is suggested by the orthogonal or rhombic arrangement of these junctional particles after treatment with uncoupling agents (see Section IV-D; Peracchia and Peracchia, 1978, 1980a,b; Zampighi *et al.*, 1982) and by preliminary optical diffraction data (Zampighi *et al.*, 1982).

Three-dimensional image reconstructions of gap junction density distribution have been obtained from low-dose electron micrographs of negatively stained gap junctions (Unwin and Zampighi, 1980) and from X-ray diffraction (Makowski *et al.*, 1982). The former depicted the features of the junctional membrane surfaces and of the hydrophilic surfaces of the particle subunits, as these are the only regions accessible to the negative stain. The subunits were shown as structures 7.5 nm long protruding 0.5–1.0 nm from the cytoplasmic surface and 1.5–2.0 nm from the extracellular surface. They surrounded a narrow opening, the presumed channel, which widened to a diameter of 2.0 nm in the extracellular region. A similar map was obtained from X-ray diffraction (Makowski *et al.*, 1982).

The study with low-dose electron microscopy (Zampighi and Unwin, 1979; Unwin and Zampighi, 1980) also reported the apparent ability of the junctions to assume different configurations depending on the preparative procedure. In junctions negatively stained after a brief washing period (enough to remove only some of the detergent used for the isolation procedure), called A junctions, the particles had a well-defined annular appearance, while in junctions stained after extensive washing, B junctions, the particles were less distinct. Projection maps calculated from computer transforms of gap junction micrographs showed the particles of both junction configurations as hexagonally shaped rings arranged in a P6 lattice in the plane of the membrane. The rings appeared strongly contrasted and packed edge to edge in A junctions, but were weakly contrasted and packed vertex to vertex in B junctions, suggesting a rotation of the particles within the lattice plane in the conversion from A to B (Zampighi and Unwin, 1979). Three-dimensional maps of A and B configurations, calculated from micrographs of junctions tilted at angles from 0 to 79° (Unwin and Zampighi, 1980), indicated that the particle subunits were inclined with respect to an axis perpendicular to the membrane surface by angles of 14 and 9° in A and B junctions respectively, and that the particles may have rotated by approximately 23° within the lattice crystal during the transition from A to B. In addition, the apparent opening seen at the cytoplasmic end of the particles (Unwin and Zampighi, 1980; Makowski *et al.*, 1982) seemed to disappear in the transition from A to B, as if matter had moved toward the central axis of the junctional particle.

These findings prompted a hypothesis which envisioned a channel opening and closing mechanism based on subunit rotation and reduction of subunit inclination (Unwin and Zampighi, 1980). Recently, however, the basis of this hypothesis has been questioned by a study with low-dose microscopy testing the effects of different electron doses (Baker *et al.*, 1983). This study has indicated that the two junction configurations, A and B, may not represent structural changes in the junctional particles resulting in channel opening and closing, but rather an artifactual stain redistribution induced by electron irradiation. In fact, while low-irradiation micrographs showed hexagonally shaped particles with little stain at their centers (similar to B junctions) exposure to a higher electron dose, yet less than that used in conventional electron

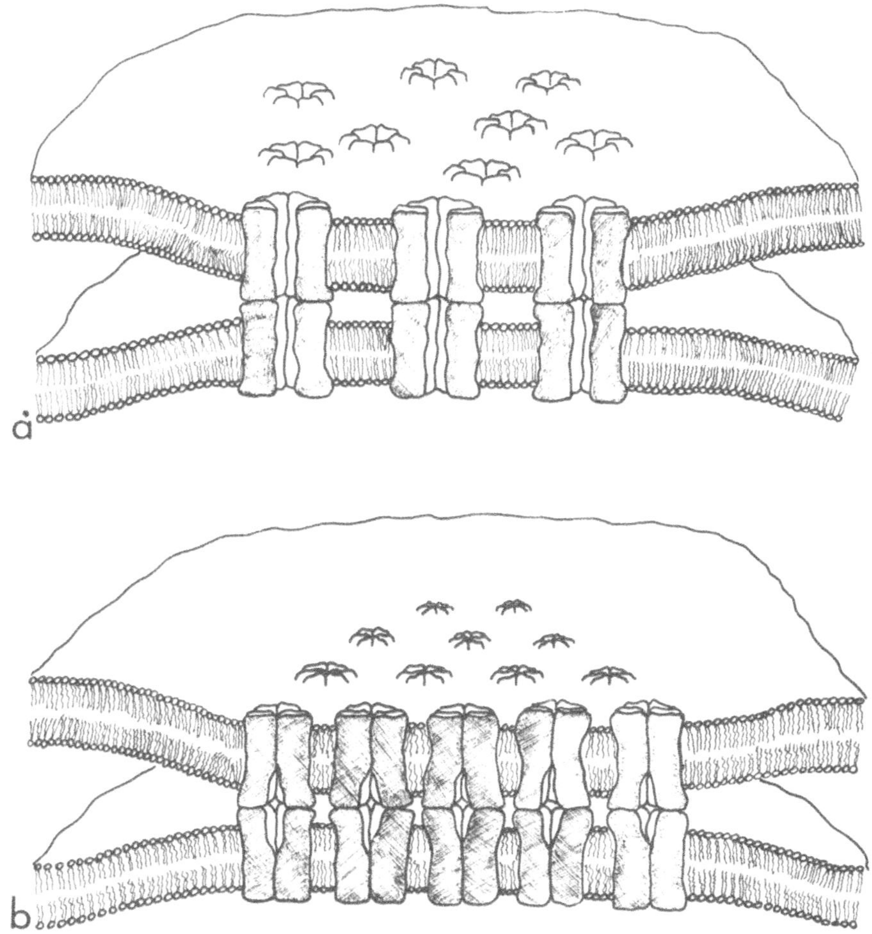

Figure 11. Tentative models of vertebrate gap junction architecture in coupled (a) and uncoupled (b) conditions. Intramembrane particles composed of six subunits span the membrane thickness and bind to similar particles of the adjoining membrane. Channels (1.5–2 nm wide), located at the center of the particles, provide cytoplasmic bridges for intercellular exchange of small molecules. The intramembrane particles form loose and disordered arrays in coupled junctions (a) and tighter and more ordered packings in uncoupled junctions (b). Cell-to-cell uncoupling is believed to result from the occlusion of the cytoplasmic ends of the channels (b). From Peracchia (1980).

microscopy, converted the particle images to those of hexagonally shaped rings with darkly stained centers (similar to A junctions).

In contrast to previous models which picture the particles as hexamers composed of identical subunits symmetrically paired with identical hexamers of the adjoining membrane (Makowski *et al.*, 1977; Zampighi and Unwin, 1979; Unwin and Zampighi, 1980), Baker *et al.* (1983) found the hexameric particles to be skewed in the plane of the membrane by an angle of approximately 8° with respect to the axis passing through the particle centers. Some evidence for a skewed orientation of the particles was previously obtained in the same junctions (Henderson *et al.*, 1979) and in crayfish junctions (Figure 10b; Peracchia, 1973b). A tentative model of gap junctions in coupled and uncoupled conditions is shown in Figure 11.

C. What Keeps Gap Junction Particles Aggregated?

In spite of the rapidly increasing knowledge on gap junction structure, there are general aspects of gap junction architecture still poorly understood. Indeed, little is known on the mechanisms by which the junctional particles are kept aggregated and about the nature of the interactions among junctional components. As a hypothesis it could be proposed that gap junction particle aggregation reflects an equilibrium among three forces (Figure 12): (1) a binding force (A), linking together the particles across the gap and thus keeping the membranes attached to each other at a 2–3-nm distance (vertebrates), (2) a repulsive force (B) between neighboring particles of the same membrane (possibly due to surface charges) which keeps the particles of coupled junctions in disordered arrays at a center-to-center distance of approximately 10 nm (vertebrates), and (3) a repulsive force (C) between the perijunctional plasma membranes, due to the extracellular glycoprotein chains, which maintains an extracellular space of approximately 15 nm. Since there is no evidence for interaction between junctional particles and cytoskeleton (Peracchia and Peracchia, 1980a; Hirokawa and Heuser, 1982), aside from the case of interiorizing gap junctions (Larsen *et al.*, 1979), it is reasonable to believe that the particles may be kept aggregated by force C, as close as force B permits. If this is true, the particles could migrate into the perijunctional regions only by eliminating force A and/or force C.

Consistent with this hypothesis may be the observation, in certain junctions, of smooth disks, approximately 65 nm in diameter, containing one or two particles at their center (Figure 13a). These may be regions from which the junctional particles were excluded by force C, imposed upon the membranes by the central particles (probably glycoproteins). These particles would tend to keep the membranes at a distance greater than 2–3 nm, which is supported by the fact that the disk regions are concave on the P face (Figure 13a) and convex on the E face (Figure 13b). This hypothesis may also explain the particle-free halo, often seen separating gap junctions from perijunctional regions (Figure 4). The halo is probably a region from which both glycoproteins and gap junction particles are excluded, the former because of the progressively decreasing width of the extracellular space (from 15 nm to 2–3 nm), and the latter because of force A.

For gap junction formation, the plasma membranes need to be brought close

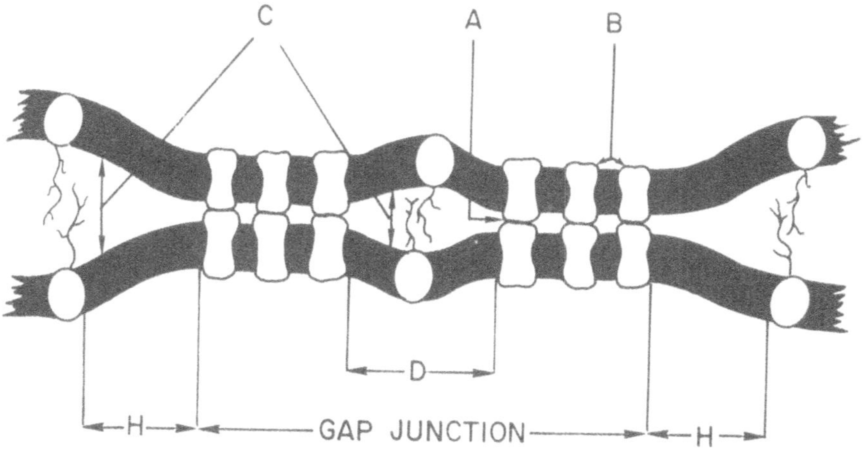

Figure 12. Hypothetical model of the potential energies determining particle aggregation in gap junctions. The junctional architecture is seen as the result of an equilibrium among three forces; a binding force (A), which links together opposite particles across the gap and keeps the membrane bilayers at a 2–3-nm distance; a repulsive force (B) between neighboring particles, probably due to protein surface charges, which keeps the particles of coupled junctions in disordered arrays at a center-to-center distance of ~10 nm (vertebrates); a repulsive force (C), between perijunctional membrane surfaces, by which an extracellular space of about 15 nm is maintained in nonjunctional regions. This force is probably due to hydration and charges of glycoprotein and glycolipid extracellular branches. In this model, the particles of coupled gap junctions are believed to be pushed against each other by force C as close as force B permits. The stronger the force (C), the tighter the particle packing. The very tight, pseudocrystalline, particle arrays of uncoupled junctions are believed to result from a reduction of force B. This may follow protein conformational changes and/or a change in protein surface charges. The crystalline arrays are likely to reflect the configuration of individual particles, hexagonal arrays from hexamers (Figures 10, 14b, and 15b), and orthogonal or rhombic arrays from tetramers (Figure 17). H = profile of the particle-free membrane region (halo) as seen in Figure 4. D = membrane profiles at the smooth disk-like regions seen in Figure 13.

enough to each other (2–3 nm) to allow the junctional proteins of neighboring cells to interact across the extracellular space. This may follow glycoprotein aggregation and formation of glycoprotein-free areas with a narrow extracellular space, in agreement with observations of particle-free membrane patches in developing gap junctions (Johnson *et al.*, 1974; Decker and Friend, 1974).

D. Gap Junction Composition

The gap junction is indeed a remarkable piece of membrane in many respects, but a most unusual characteristic of this membrane is its accessibility to research tools. We have already seen the wide variety of electrophysiological, ultrastructural, and crystallographic techniques that have been successfully applied to gap junctions. It will be shown next that gap junctions are also uniquely suitable for chemical studies.

The first and most important step for studying the composition of a membrane is our ability to isolate it in its purest form. Since the late 1960s, gap junctions have

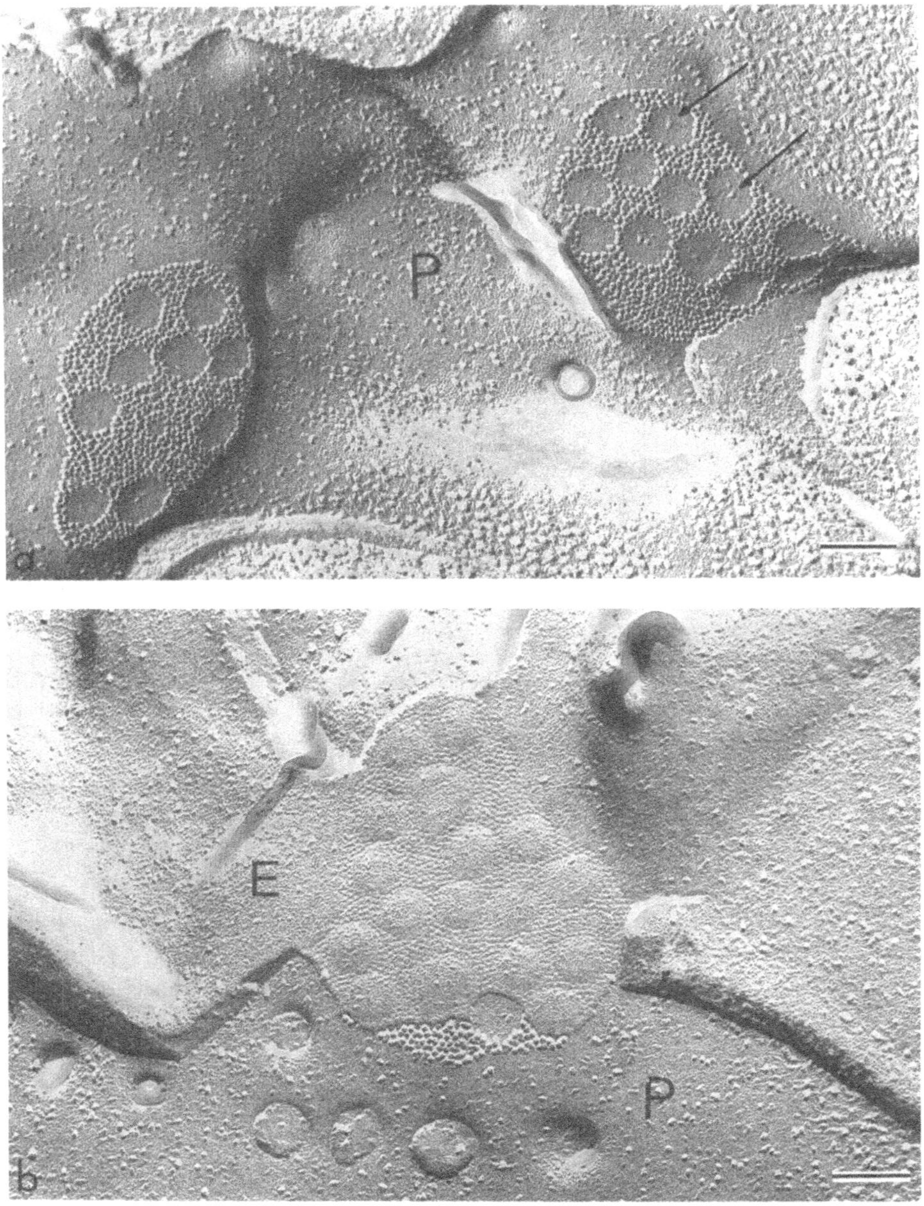

Figure 13. Vertebrate gap junctions (calf heart). Freeze-fracture replica. Smooth disks, 65–70 nm in diameter are frequently seen in calf heart gap junctions. The disks' surface is slightly concave on face P (P) and contains one or two particles near the center (a, arrows), likely to be glycoproteins trapped in the junction during junctional assembly. Complementary, convex, disk surfaces are seen in (b), on the exoplasmic fracture face (E). An interpretation of the nature of the disks is given in Figure 12. Bars = 0.1 μm. From Peracchia (1980).

been successfully isolated by capitalizing on their unusual resistance to the solubilizing action of certain detergents (Benedetti and Emmelot, 1968; Evans and Gurd, 1972; Goodenough and Stoeckenius, 1972). Enriched gap junction fractions were first produced by Benedetti and Emmelot (1968) with sodium deoxycholate, but later a variety of other detergents including *n*-lauryl sarcosine (Evans and Gurd, 1972; Goodenough and Stoeckenius, 1972), Lubrol WX (Zampighi and Robertson, 1977) and Brij 58 (Goodenough *et al.*, 1978) have been employed.

Gap junctions have been isolated from three vertebrate organs: liver, lens, and heart. The usual isolation procedure includes tissue homogenization, repeated wash in buffered solutions of low osmolarity, sucrose gradient centrifugation for isolation of plasma membrane-junction fractions, detergent treatment, and subsequently a second sucrose gradient centrifugation for isolation of gap junction fractions. This type of isolation procedure results in enriched gap junction fractions which are still partially contaminated with detergent-insoluble membrane vesicles, collagen, desmosomes (heart), and uricase crystals (rat liver). To isolate the gap junctions in highly purified forms, various enzymes such as collagenase, hyaluronidase, and trypsin have been used. With enzymatic treatment, the purest gap junction fractions have been isolated from the liver (Goodenough, 1974, 1976); however, a drawback of this method is the protein damage caused by the proteolytic enzymes (Duguid and Revel, 1975).

To study gap junction protein composition, enriched or purified junction fractions are dissolved in sodium dodecyl sulfate, and the protein components are separated by polyacrylamide gel electrophoresis (SDS-PAGE). During the past ten years, a variety of protein components ranging in molecular weight from 10,000 to 47,000 have been attributed to gap junctions (reviewed in Peracchia, 1980); however, now the general consensus is that there is only a main component of molecular weight approximately 28,000 in liver junctions (Duguid and Revel, 1975; Henderson *et al.*, 1979; Hertzberg and Gilula, 1979; Zampighi and Unwin, 1979; Finbow *et al.*, 1980; Nicholson *et al.*, 1981) and one of molecular weight approximately 26,000 (MIP 26) in lens junctions (Alcala *et al.*, 1975; Maisel *et al.*, 1976; Broekhuyse *et al.*, 1976; Bloemendal *et al.*, 1977; Hertzberg *et al.*, 1982; Bok *et al.*, 1982). In the heart, a variety of components of mol. wt. 31,000, 33,500, 38,000, and 47,000 (Kensler and Goodenough, 1980) and of molecular weight 19,000, 26,000, and 37,000 (Colaco and Evans, 1981) have been reported, but in view of the difficulty in completely eliminating desmosomes from these junctions, it is not yet clear which of these components belongs to gap junctions. Indeed, also in liver, direct evidence that the component of molecular weight 28,000 belongs to gap junctions is still lacking, but the purity of these gap junction fractions is such that one can feel quite confident about the relationship between this protein and gap junctions. Direct evidence, however, could only be provided by the binding of specific antibodies to gap junction membranes.

In spite of efforts from several laboratories, a specific antiserum to the liver component of molecular weight 28,000 has not yet been produced. In the lens, on the other hand, a number of laboratories have succeeded in preparing specific antisera against the main protein component (MIP 26; Acala and Maisel, 1978; Waggoner and Maisel, 1978; Broekhuyse *et al.*, 1979; Friedlander, 1980; Zigler and Horwitz, 1981), and recently, an elegant demonstration of the specific binding of rabbit anti-MIP 26 antibodies to lens junction membranes has been produced by indirect immunocytochemistry with ferritin-conjugated goat antirabbit IgGs (Bok *et al.*, 1982). Interestingly, the antibodies to the lens junction protein do not cross-react with the liver protein, indicating that they are immunologically different (Hertzberg, 1980; Zigler and Horwitz, 1981; Hertzberg *et al.*, 1982); indeed, these proteins have different peptide maps (Hertzberg, 1980; Nicholson *et al.*, 1980).

Recently, a remarkable study has produced the first partial sequencing of the liver component of molecular weight 28,000 (Nicholson *et al.*, 1981). Thus far, the sequence of the NH_2-terminal 52 amino acids has been determined. After 14 uncharged amino acids at the NH_2-terminus, Nicholson *et al.* (1981) found a strongly hydrophobic region of 18 residues, probably the intramembranous portion, interrupted by a single charged amino acid (Arg) and flanked at the NH_2-terminus by three basic residues and at the COOH-terminus by three acidic residues. These and other data suggested that the protein crosses the membrane more than once, is made of two major portions joined by a region accessible to proteases, and has its COOH-terminal at the cytoplasmic side of the junction.

Different from other membrane proteins, whose half-life is of the order of 30–60 hr, the liver gap junction protein has a half-life of 5–19 hr, suggesting that gap junctions may be capable of rapidly responding to the changes in physiological demands for cell-to-cell coupling (Yancey *et al.*, 1981; Fallon and Goodenough, 1981). An example of rapid gap junction formation is provided by the myometrium at the onset of parturition (Garfield *et al.*, 1978; Dahl and Berger, 1978). In this case, the rapid increase in gap junction number and size has been considered responsible for the development of well-synchronized activity in uterus during parturition.

By capitalizing on the rapid gap junction synthesis in uterine muscle, Dahl *et al.* (1981) have made the first successful attempt to isolate the mRNA of the gap junction protein. The isolated mRNA, incorporated via liposomes into communication-defective cells, induced electrical coupling in these cells. Although still preliminary, this study opens an exciting new chapter in gap junction research, that of gap junction biogenesis.

In spite of the numerous studies on gap junction chemistry, only few data are available on gap junction lipid composition. Junctions isolated with sarkosyl contain 36% (wt./wt.) lipids (Henderson *et al.*, 1979). The phospholipid (phosphatidylethanolamine, phosphatidylcholine, and phosphatidylserine) content is similar to that of plasma membranes, but gap junctions have a higher cholesterol to phospholipid ratio (C/PL; Evans and Gurd, 1972; Caspar *et al.*, 1977; Hertzberg and Gilula, 1979; Henderson *et al.*, 1979). The C/PL ratio is 1.5 after sarkosyl and 5.6 after Triton X-100 and urea treatment, suggesting a tight binding between cholesterol and the junction protein (Henderson *et al.*, 1979). An interesting finding is the apparent lack of sphingomyelin (Hertzberg and Gilula, 1979).

IV. HOW CELL-TO-CELL DIFFUSION OF MOLECULES IS REGULATED

The degree of communication between coupled cells can be decreased down to a virtually complete cell-to-cell uncoupling by a variety of treatments. Our knowledge of uncoupling finds its origin in some interesting observations of a nineteenth century physiologist. Engelmann (1877), while working on the effects of injury in cardiac muscle, noticed that upon mechanical injury, intact cells neighboring damaged cells suddenly depolarized (injury potential) but, within a few minutes, recovered their original membrane potential. Engelmann's conclusion was that "heart cells live together but die individually." We now know that the capacity of coupled cells to isolate themselves from damaged neighbors, healing-over, is a property of virtually all tissues where cell-to-cell coupling is present and is due to the occlusion of cell-to-cell channels.

A. Uncoupling Agents

In 1964, Délèze noticed that healing-over is a Ca^{2+}-dependent phenomenon. This important information came at the time when evidence for the involvement of Ca^{2+} in cell-to-cell uncoupling was accumulating (Loewenstein, 1966) and thus suggested that the healing-over process could very well be the result of uncoupling between injured and uninjured cells. Evidence for a close relationship between an increase in $[Ca^{2+}]_i$ and uncoupling was first provided by the observation that electrical uncoupling occurs in perforated cells (Loewenstein *et al.*, 1967) or cells made permeable to ions (Nakas *et al.*, 1966) only if the bathing medium contains Ca^{2+}. Indeed, most of the uncoupling treatments are known to increase the $[Ca^{2+}]_i$. Aside from Ca^{2+} injection, cell damage, or exposure to Ca^{2+} ionophores (Loewenstein and Rose, 1978), $[Ca^{2+}]_i$ increases when uncoupling is induced by metabolic inhibition, Ca^{2+} chelation, prolonged cooling, increase in $[Na^+]_i$, stimulation of secretion (reviewed in Peracchia, 1980), and probably hypoxia (Wojtczak, 1979). Rose and Loewenstein (1975, 1976) have elegantly demonstrated the simultaneous occurrence of an increase in $[Ca^{2+}]_i$ and uncoupling by monitoring the $[Ca^{2+}]_i$ with aequorin, a sensitive Ca^{2+} indicator, in insect salivary gland cells exposed to various uncoupling treatments. However, in some cases, uncoupling occurs without an aequorin-detectable increase in $[Ca^{2+}]_i$ (Rose and Loewenstein, 1976). Furthermore, in some cells, an increase in $[Ca^{2+}]_i$ causes a decrease in pH_i (Meech and Thomas, 1977). Evidence that uncoupling rapidly develops when pH_i decreases has suggested that H^+ may induce uncoupling independently from Ca^{2+} (Turin and Warner, 1977, 1980; Iwatsuki and Petersen, 1979; Weingart and Reber, 1979; Giaume *et al.*, 1980; Rink *et al.*, 1980; De Mello, 1980a; Spray *et al.*, 1981a, 1982; Reber and Weingart, 1982). However, in some systems, a decrease in pH_i has been shown to cause an increase in $[Ca^{2+}]_i$ (Lea and Ashley, 1978; Rose and Rick, 1978; Rink *et al.*, 1980), indicating that the uncoupling by low pH_i could also be Ca^{2+}-mediated. Rose and Rick (1978) took this stand in a study on insect cells in which pH_i, $[Ca^{2+}]_i$, and electrical uncoupling were monitored simultaneously. However, in Xenopus embryonic cells, only a small increase in $[Ca^{2+}]_i$ (from 0.2 to 0.35 μM) was measured during ucoupling by 100% CO_2 (Rink *et al.*, 1980). In this study, Ca^{2+} was not believed to be involved because a $[Ca^{2+}]_i$ of 40–70 μM had been reported to be necessary to electrically uncouple cells (Oliveira-Castro and Loewenstein, 1971).

Spray *et al.* (1981a) have supported the independent uncoupling action of H^+ by showing that the relationship between pH_i and junctional conductance is well fitted by a Hill curve with a $K = 5 \times 10^{-8}$ M (pK = 7.3) in Ambystoma and Fundulus embryonic cells. More recently, Spray *et al.* (1982) have shown that both H^+ and Ca^{2+} are independent uncoupling agents in blastomere pairs studied with an ingenious technique by which one of the two cells, drawn into a pipette, is cut open and internally perfused with EGTA-Ca^{2+} solutions buffered at different pH values. In this study, however, Ca^{2+} caused uncoupling only at concentrations greater than 200 μM.

The independent effect of H^+ has also been well supported in mammalian heart cells (Reber and Weingart, 1982). In these cells, in contrast to other systems (see above), intracellular acidosis sufficient to lower the coupling coefficient decreased, and alkalosis increased, the $[Ca^{2+}]_i$, when measured with a Ca^{2+}-sensitive microelectrode (Hess and Weingart, 1980). Thus, assuming that protonation of the channel-gating proteins does not significantly increase the Ca^{2+} affinity constant of their Ca^{2+} sites, one may conclude that both Ca^{2+} and H^+ uncouple cells independently. Consistent with this view is the finding of the independent effect of Ca^{2+} and H^+ on the structure of lens junctions (Peracchia and Peracchia, 1980b).

If Ca^{2+} and H^+ were acting on the same site, they could compete with each other. This would explain why Turin and Warner's (1977, 1980) uncoupling pH values vary from experiment to experiment (pH 6.4–6.85) and are consistently lower than those (pH 7.2–7.3) of Spray *et al.* (1981a). Different values have also been reported for the $[Ca^{2+}]_i$ capable of triggering uncoupling. Oliveira-Castro and Loewenstein (1971), and Spray *et al.* (1982) gave $[Ca^{2+}]_i$ ranging from 40 to 400 μM, although lower concentrations have been considered capable of causing partial uncoupling (Délèze and Loewenstein, 1976). These rather high values of $[Ca^{2+}]_i$, however, have been obtained only in ruptured cells or cells internally perfused, while estimates obtained in intact cells were considerably lower (Weingart, 1977; Dahl and Isenberg, 1980). Lower $[Ca^{2+}]_i$ was also found to affect junction structure (Peracchia and Peracchia, 1980a). It could be that a higher $[Ca^{2+}]_i$ is necessary to uncouple ruptured or internally perfused cells because they have partially lost a soluble intermediate. Indeed, internally perfused crayfish axons lose their capacity to uncouple (Johnston and Ramón, 1981). In view of recent data, the soluble intermediate factor, if present, could be calmodulin (Peracchia *et al.*, 1981, 1983; Peracchia, 1982b).

Aside from Ca^{2+} and H^+, the junctional conductance is decreased by glutaraldehyde (Bennett *et al.*, 1972), certain alcohols such as hexanol, heptanol, octanol, and nonanol (Johnston *et al.*, 1980; Bernardini *et al.*, 1982), 1–2% halothane (Hauswirth, 1968), large transjunctional potential differences (Spray *et al.*, 1981b; Harris *et al.*, 1981) and prolonged cooling (Payton *et al.*, 1969a). Interestingly, cytoplasmic acidification and glutaraldehyde do not seem to alter the voltage dependence of uncoupling, while glutaraldehyde appears to abolish the pH dependence of uncoupling (Spray *et al.*, 1981c), indicating that pH and voltage sensitivity of channel conductance may belong to different regulatory sites. Uncoupling is also believed to be induced by Mg^{2+}, Co^{2+}, Mn^{2+}, Sr^{2+}, Ba^{2+}, and La^{3+} (Oliveira-Castro and Loewenstein, 1971; Politoff *et al.*, 1974; De Mello, 1979); however, whether the effect of these ions is direct or mediated by Ca^{2+} or H^+ has not been established. Also unclear is the mechanism of action of glutaraldehyde, local anesthetics, and transjunctional potentials.

B. Is Uncoupling a Graded Phenomenon?

Délèze and Loewenstein (1976) have proposed that uncoupling is not all-or-nothing but rather a graded phenomenon in a study which has tested the changes in electrical coupling and fluorescein transfer in insect cells injected with various quantities of Ca^{2+}. Small Ca^{2+} injections, insufficient to depolarize the cell, prevented fluorescein transfer but did not electrically uncouple the cells. These data were later confirmed using other tracer molecules in insect (Rose *et al.*, 1977) and mammalian (Flagg-Newton and Loewenstein, 1979) cells. However, whether this phenomenon is due to a change in the channel size or to a change in the ratio between open and closed channels has not been established.

C. How to Enhance Coupling or Inhibit Uncoupling

In the past, most of the work on the regulation of cell coupling has been concerned with treatments that decrease coupling. Recently, some interest has also been shown in methods for improving cell coupling and for protecting cells from uncoupling. Weingart (1977) noticed an improvement in cell coupling during the early phase of the ouabain effect on heart sodium pump. The same effect was obtained with angiotensin II (Hermsmeyer, 1980) by increasing pH_i (Reber and Weingart, 1982) and, in insect cells, by treatment with β-ecdysone (Caveney and Blennerhassett, 1980) or 20-hydroxyecdysone (Caveney *et al.*, 1980). Preliminary experiments (Peracchia, 1982b) also show an increase in coupling ratio in Xenopus embryonic cells treated with calmidazolium (R 24571), a powerful calmodulin inhibitor (Van Belle, 1981). Uncoupling is inhibited by glucocorticoids in insect cells (Suzuki and Higashino, 1977; Higashino *et al.*, 1980), by internal perfusion in crayfish axons (Johnston and Ramón, 1981), and by calmodulin inhibitors in frog and Xenopus embryonic cells (Peracchia *et al.*, 1981, 1983; Peracchia, 1982b). Coupling improves in certain cultured cells and develops in other cells usually unable to communicate as a result of an increase in intracellular cAMP (Flagg-Newton *et al.*, 1981; Flagg-Newton and Loewenstein, 1981; Azarnia *et al.*, 1981).

D. Change in Junction Structure with Uncoupling

Since the mid-1960s, electron microscopists looked with interest at the possibility of detecting changes in junction structure with functional uncoupling. In early studies, uncoupling was always believed to result from separation of the junctional membranes (Barr *et al.*, 1965, 1968; Asada and Bennett, 1971; Pappas *et al.*, 1971). Although it is true that in cells subjected to hypertonic media the gap junction membranes separate (Goodenough and Gilula, 1974; Peracchia, 1977, 1980; Hirokawa and Heuser, 1982), in most cases uncoupling occurs without gross disruption of the junctional membranes.

A correlation between fine changes in junction structure and electrical uncoupling was first reported in crayfish neurons (Peracchia and Dulhunty, 1974, 1976). In this study, the junctional particles, loosely and irregularly packed in controls, became more tightly and regularly packed, occasionally forming crystalline arrays, in cells uncoupled

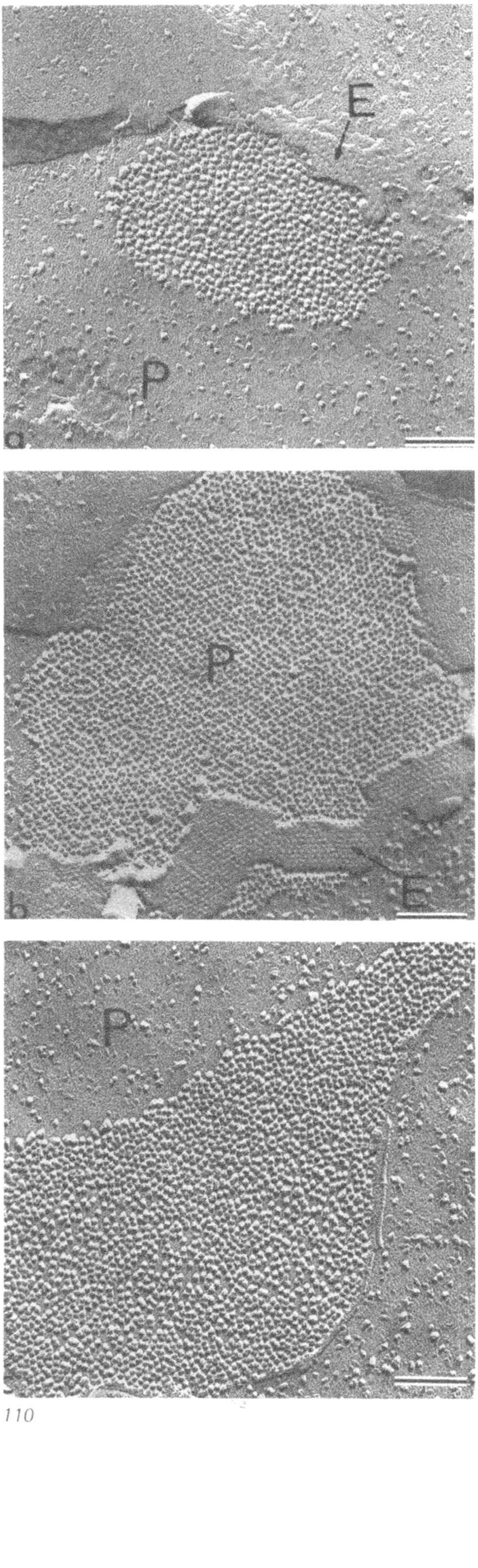

Figure 14. Vertebrate gap junctions (rat stomach). Freeze-fracture replicas. (a) A control junction. (b and c) Junctions from samples exposed to an uncoupling treatment (incubation in a Tyrode's solution containing 2,4-dinitrophenol). The intramembrane particles of control junctions are aggregated in a disordered array at a center-to-center distance of about 10 nm (a). With the uncoupling treatment the particles become more tightly and regularly aggregated (b and c). The change in particle packing is a gradual phenomenon which eventually results in gap junction crystallization (b). This is suggested by the observation of junctions with particle arrays of variable tightness and order (c). P = face P; E = face E; bars = 0.1 μm. From Peracchia (1977).

by Ca^{2+}-free solutions or metabolic inhibition. In addition, the particle diameter decreased by approximately 3 nm and the gap by approximately 2 nm. Similar changes in particle packing and particle diameter (approximately 1-nm decrease) were seen in gap junctions of rat stomach (Figure 14) and liver (Figure 15) epithelia exposed to uncoupling treatments such as inhibition of metabolism or hypoxia (Peracchia, 1977).

During the past five years, the increase in tightness and/or order of gap junction particles has been reported in a variety of vertebrate and invertebrate tissues after uncoupling treatments (Peracchia, 1978; Raviola *et al.*, 1978, 1980; Baldwin, 1979; Alcala *et al.*, 1979; Peracchia and Peracchia, 1980a,b; Dahl and Isenberg, 1980; Meda *et al.*, 1980; Kistler and Bullivant, 1980b; Bernardini and Peracchia, 1981; Shibata and Page, 1981; Berdan and Caveney, 1981; Hirokawa and Heuser, 1982; Benedetti *et al.*, 1981; Bernardini *et al.*, 1982; Délèze and Hervé, 1983). In just a few studies, the effect of the uncoupling treatments was monitored electrophysiologically (Peracchia and Dulhunty, 1974, 1976; Dahl and Isenberg, 1980; Délèze and Hervé, 1983), and only in the crayfish were the junctional changes seen in the cells in which the electrical uncoupling was detected electrophysiologically (Peracchia and Dulhunty, 1974, 1976).

In mammalian heart, Dahl and Isenberg (1980) reported an increase rather than a decrease in particle diameter in crystalline junctions of myocytes fixed soon after uncoupling, while smaller particles were seen only at later stages. Recently, however, a decrease in particle diameter similar to that reported in mammalian liver, stomach, and lens (Peracchia, 1977, 1978) has been seen in gap junctions of sheep heart fixed

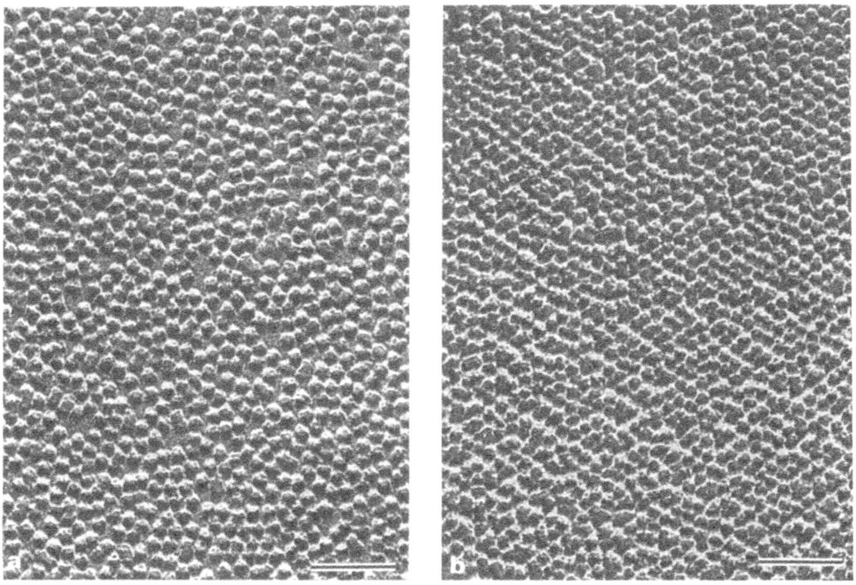

Figure 15. Vertebrate gap junctions (rat liver). Freeze-fracture replicas. The particles of control junctions (a) form a disordered array and are at a center-to-center distance of 9.5–10 nm. Those from samples subjected to an uncoupling treatment (hypoxia) (b) are hexagonally arranged at 8.5-nm distance. The average particle diameter is 8.5 ± 0.09 nm in (a) and 7.66 ± 0.08 nm (mean ± S.E.M.) in (b). Bars = 50 nm. From Peracchia (1977).

immediately after uncoupling (Délèze and Hervé, 1983). In any event, data on particle diameter should be taken with caution because small changes in particle size are difficult to measure in freeze-fracture replicas and could result from plastic deformation, surface contamination, and fluctuations in platinum deposition and platinum/carbon ratio.

Questions have been raised about the changes in gap junctions with functional uncoupling in view of the fact that glutaraldehyde, usually employed as a fixative, is itself an uncoupling agent (Politoff and Pappas, 1972; Bennett, 1973). Recently, however, similar changes in particle packing with uncoupling have been obtained in samples prepared for freeze-fracture without glutaraldehyde fixation (Raviola *et al.*, 1980; Baldwin, 1979; Peracchia and Peracchia, 1980b; Dahl and Isenberg, 1980), thus, glutaraldehyde may uncouple cells with a different mechanism not involving particle change. Raviola *et al.* (1978) have suggested that glutaraldehyde fixation indeed causes gap junction crystallization. In ciliary epithelium, the gap junctions appeared consistently crystalline unless they were rapidly frozen without glutaraldehyde. However, this is not the case in most junctions, as demonstrated quantitatively by several studies (Peracchia and Dulhunty, 1974, 1976; Hama and Saito, 1977; Peracchia, 1977, 1978; Baldwin, 1979; Dahl and Isenberg, 1980; Délèze and Hervé, 1983). In a recent study on liver gap junctions (Sikerwar and Malhotra, 1981), glutaraldehyde fixation has been claimed to induce a decrease in particle spacing (but not to change particle order), as the particles of junctions rapidly frozen in liquid nitrogen without fixation were more dispersed than those of junctions similarly frozen after glutaraldehyde. This particle clumping was believed to represent the structural correlate of uncoupling by glutaraldehyde. However, this study did not rule out the possibility that glutaraldehyde fixation, rather than clumping the particles, may have prevented particle dispersion during freezing in the unfixed junctions. Indeed, the freezing method (Van Harreveld *et al.*, 1965) used in this study has several limitations (Heuser *et al.*, 1975).

Raviola *et al.* (1980) have reported that tightening of particle packings up to the formation of crystalline arrays indeed occurs in samples rapidly frozen after exposure to various uncoupling treatments. But, in view of the slow time course and the apparent irreversibility of the morphological changes induced by prolonged anoxia or low pH, they felt that particle crystallization may not be the morphological correlate of electrical uncoupling. However, in this study, the morphological data were not quantitated and no attempt was made to electrophysiologically determine the effects of the prolonged uncoupling treatments. Prolonged anoxia or extended low-pH treatment may indeed induce irreversible electrical uncoupling.

Preliminary data of Hanna *et al.* (1981) indicate that crystallization does occur in heart gap junctions of a tunicate rapidly frozen in liquid helium (Heuser *et al.*, 1975) after a prolonged exposure to 100% CO_2, but not after a brief CO_2 treatment. This observation was taken to indicate that crystallization is not concomitant with uncoupling but lags behind. However, the data were not substantiated with quantitation and electrical recordings. One may object that, following brief exposures to CO_2, the cells may have been able to recouple before being frozen. It takes at least 1 min to mount tissues on the rapid freezing apparatus and while the tissue is exposed to room air, some CO_2 will leave the cells and pH_i will increase. Indeed, in samples uncoupled by 100% CO_2, recoupling occurs within 2–3 min of washing (Turin and Warner, 1977, 1980; Iwatsuki and Petersen, 1979; Spray *et al.*, 1981a).

In favor of the uncoupled state of crystalline gap junctions is a recent X-ray diffraction study (Makowski *et al.*, 1982) showing that sucrose does not gain access to the channels of crystalline gap junctions isolated from mouse liver, and also a study on insect embryos (Baldwin and Hakim, 1981) in which the gap junctions between cells of the midgut, which are normally electrically uncoupled, were always found with crystalline particle packings, while those between epidermal cells, which are usually coupled, had disordered particle arrays.

From these numerous and somewhat controversial data one may reach the conclusion that although there is little doubt that crystalline gap junctions are in a nonpermeable (uncoupled) state, junctions with disordered arrays may or may not be in an uncoupled state. Early stages of uncoupling are probably paralleled by a tightening of the particle arrays, with a possible decrease in particle diameter, while gap junction crystallization may be the latest stage of the particle-clumping process. Indeed, this was the message of the earliest studies (Peracchia and Dulhunty, 1976; Peracchia, 1977) where the structural changes were not described as an all-or-none phenomenon, but rather as gradual changes from loose and disordered particle packings to tighter and more regular arrays (Figure 14).

One may ask what causes the particles of gap junctions to increase in mutual affinity during uncoupling and whether the changes in particle array have anything to do with channel occlusion. An approach to this problem has been to test the direct effect on the junctions of uncoupling agents such as divalent cations and H^+. Indeed, most gap junctions were found to crystallize in epithelial cells of rat stomach treated with a calcium ionophore (Peracchia and Peracchia, 1980a), but this study did not prove that the structural changes are a direct effect of Ca^{2+} on junctional molecules. More direct answers were obtained from junctions isolated from eye lens fibers.

In the lens, while the surface epithelial cells have conventional gap junctions (Peracchia, 1978), the fiber cells are extensively joined (approximately 60% of their surface) by junctions which differ from typical gap junctions in a number of features; e.g., they are narrower (the gap is hardly detectable), crystallize in pseudohexagonal (Figure 16b), rhombic, and orthogonal (Figure 17) arrays with a 6.5-nm particle spacing (Peracchia and Peracchia, 1980a,b; Zampighi *et al.*, 1982), and their main protein (MIP 26) is unique in amino acid composition (Nicholson *et al.*, 1983) and immunological properties (Hertzberg *et al.*, 1982). Nevertheless, they are the only junctions between lens fibers, and since these cells are electrically well coupled (see Rae, 1979 for review) and pass dyes (Rae, 1974) and metabolites (Goodenough *et al.*, 1980), they are likely to be communicating junctions.

Unlike liver junctions, calf lens junctions can be isolated with the appearance of control (coupled) junctions, i.e., loose and disordered particle packings, if the lenses are homogenized and washed in EDTA solutions buffered to pH 7.4 (Figure 16a; Peracchia, 1978). If the junctions are subsequently incubated in Ca^{2+}-EDTA buffers (pH 7) with gradually increasing $[Ca^{2+}]_i$, the junctional particles assume crystalline arrangements at 0.5 μM or greater $[Ca^{2+}]$ (Figure 16b; Peracchia, 1978; Peracchia and Peracchia, 1980a). Similar results are obtained with a $[Mg^{2+}]$ greater than 1 mM

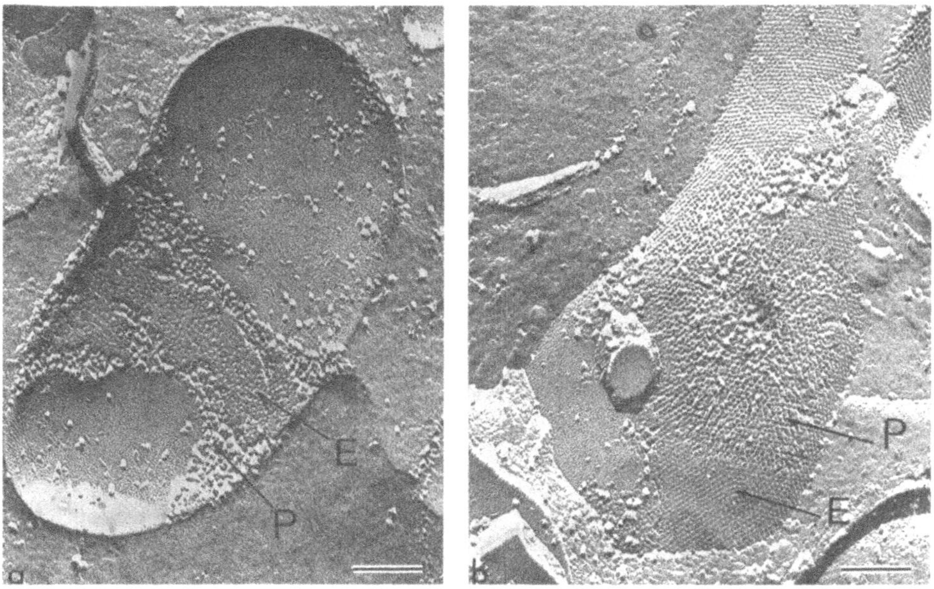

Figure 16. Isolated lens fiber junctions (calf). Freeze-fracture replicas. (a) A junction incubated in a Ca^{2+}-, Mg^{2+}-free solution. A similar junction incubated in 10 μM Ca^{2+}, 1 mM Mg^{2+} is shown (b). Note the change in particle packing, from disordered (a) to crystalline (pseudo-hexagonal; b), induced by divalent cations. Both junctions were previously treated with hexokinase–glucose solutions to hydrolyze endogenous ATP. P = face P; E = face E; bars = 0.1 μm. From Peracchia and Peracchia (1980a).

(Peracchia and Peracchia, 1980a) and with Ca^{2+}-, Mg^{2+}-free solutions buffered at pH 6.5 or lower (Figure 17; Peracchia and Peracchia, 1980b), suggesting that uncoupling agents affect directly the particle arrangement of lens junctions. These changes were reversible, did not require ATP (Figure 16), and also occurred in samples frozen without glutaraldehyde. Recently, a Ca^{2+}-induced particle crystallization has also been obtained in liver gap junctions (Hirokawa and Heuser, 1982). The ability of lens junctions to assume crystalline arrays has been confirmed in several studies (Alcala *et al.*, 1979; Kistler and Bullivant, 1980b; Benedetti *et al.*, 1981; Zampighi *et al.*, 1982). Kistler and Bullivant (1980a) have reported that in isolated lens junction fractions, only membranes with hexagonal particle arrays are junctions, while membranes with orthogonal arrays are individual membranes of a nonjunctional nature whose particles form crystalline arrays following trypsin digestion. However, the junctional nature of the orthogonal arrays has been clearly demonstrated in other studies (Peracchia and Peracchia, 1980b; Zampighi *et al.*, 1982). Thus, it is possible that the trypsin digestion and/or the different isolation procedure used by Kistler and Bullivant (1980a) have induced the crystalline junction membranes to separate.

Doubts have been raised on the validity of the studies on lens junction crystallization, in terms of uncoupling, on the basis of the apparent inability of lens junctions to crystallize in intact cells and in isolated fractions under certain experimental conditions (Goodenough, 1979; Kuszak *et al.*, 1981), and as a hypothesis, it has been proposed that lens fiber cells may indeed be unable to uncouple (Goodenough, 1979). However, recently, widespread crystallization has also been obtained in intact lens following an increase in $[Ca^{2+}]_i$ (Bernardini and Peracchia, 1981), and lens fibers have been shown to heal over, following a mechanical injury, only if Ca^{2+} is present (Bernardini *et al.*, 1981) and to uncouple after treatment with 2,4-dinitrophenol (Rae *et al.*, 1982). Further confirming the capacity of these cells to uncouple is the obser-

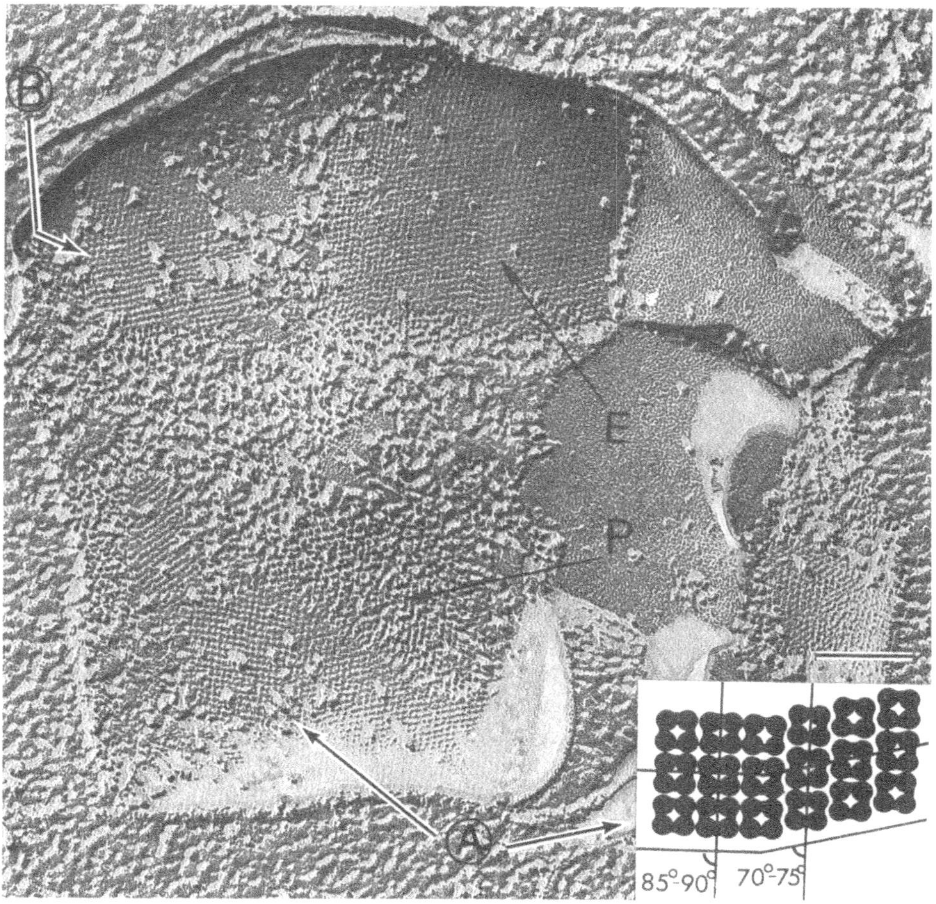

Figure 17. Isolated lens fiber junction (calf). Freeze-fracture replica. The junction was incubated at pH 6 in a Ca^{2+}-, Mg^{2+}-free solution. Particles and pits form orthogonal and rhombic arrays. The two types of arrays are often seen continuous with each other (areas A and B). Note that the change from one to the other array follows a shift in the orientation of only one of the two axes on which the particle rows are aligned. The inset shows an interpretation of the architecture of lens junction particles and their aggregation after crystallization. The particles are believed to be tetramers. P = face P; E = face E. Bar = 0.1 μm. From Peracchia and Peracchia (1980b).

vation that procion yellow, a junction-permeant molecule (Rae, 1974), does not diffuse from damaged to undamaged lens fiber cells (Bernardini *et al.*, 1981).

E. Hypotheses on Channel-Closing Mechanism

The mechanism by which uncoupling agents change the gap junction structure and close the channels is still poorly understood. As a hypothesis, both phenomena have been suggested to result from an interaction between the uncoupling agents and specific membrane sites, followed by conformational changes in the channel protein (Peracchia, 1980; De Mello, 1980b). In junctions with disordered arrays, the particles may be kept separated from each other by electrostatic repulsion (Figure 12), while a reduction in surface charges caused by the conformational changes could result in particle clumping and eventually in junction crystallization. However, recent evidence (Peracchia and Peracchia, 1983) for the existence of extracellular filamentous bridges (1 nm wide and 1.5 nm long) interlinking neighboring gap junction particles suggests that the mechanism driving the changes in particle distribution is probably more complex. The particle clumping must require lipid diffusion to perijunctional regions, which is likely to be a relatively slow process. This could explain why, in some studies, particle crystallization has been seen to lag behind functional uncoupling (see above). That particle clumping is caused by interaction with cytoskeletal filaments is unlikely, because deep-etching experiments have shown that the cytoplasmic surface of lens, stomach, and liver junctions is smooth (Peracchia and Peracchia, 1980a; Hirokawa and Heuser, 1982). Indeed, actin filaments have been seen associated with gap junctions, but only during the interiorization process (Larsen *et al.*, 1979).

In the past, the channels were believed to close at their extracellular end (Makowski *et al.*, 1977), but recent data indicate that the channel gating mechanism is probably located on the intracellular side (Figure 11b; Peracchia, 1980; Unwin and Zampighi, 1980; Makowski *et al.*, 1982). Unwin and Zampighi (1980) have suggested that the channels of liver junctions close at their cytoplasmic end as a result of a rotation and tilting of the six subunits composing the channel framework. This model was based on the observation of two conformational states in isolated junctions negatively stained with uranyl acetate and studied by low-dose electron microscopy (see Section III-B). Indeed, this would be a very exciting finding were it not for the rather unphysiological way in which the two conformational states were obtained, namely short vs. extensive junction wash following detergent treatment. Unfortunately, these authors did not attempt to test the effects on the junctions of known uncoupling agents such as Ca^{2+} or H^+; thus, until further evidence is provided, the physiological meaning of these data remains dubious. Moreover, questions have been raised on the very existence of the two conformational states, as preliminary data suggest that the two states may reflect an artifactual stain redistribution induced by electron irradiation rather than a change in junction structure (Baker *et al.*, 1983).

F. Is Calmodulin Involved in the Regulation of Cell-to-Cell Coupling?

The occlusion of the channels could either follow a direct interaction between the uncoupling ions and the channel proteins or result from the activation of inter-

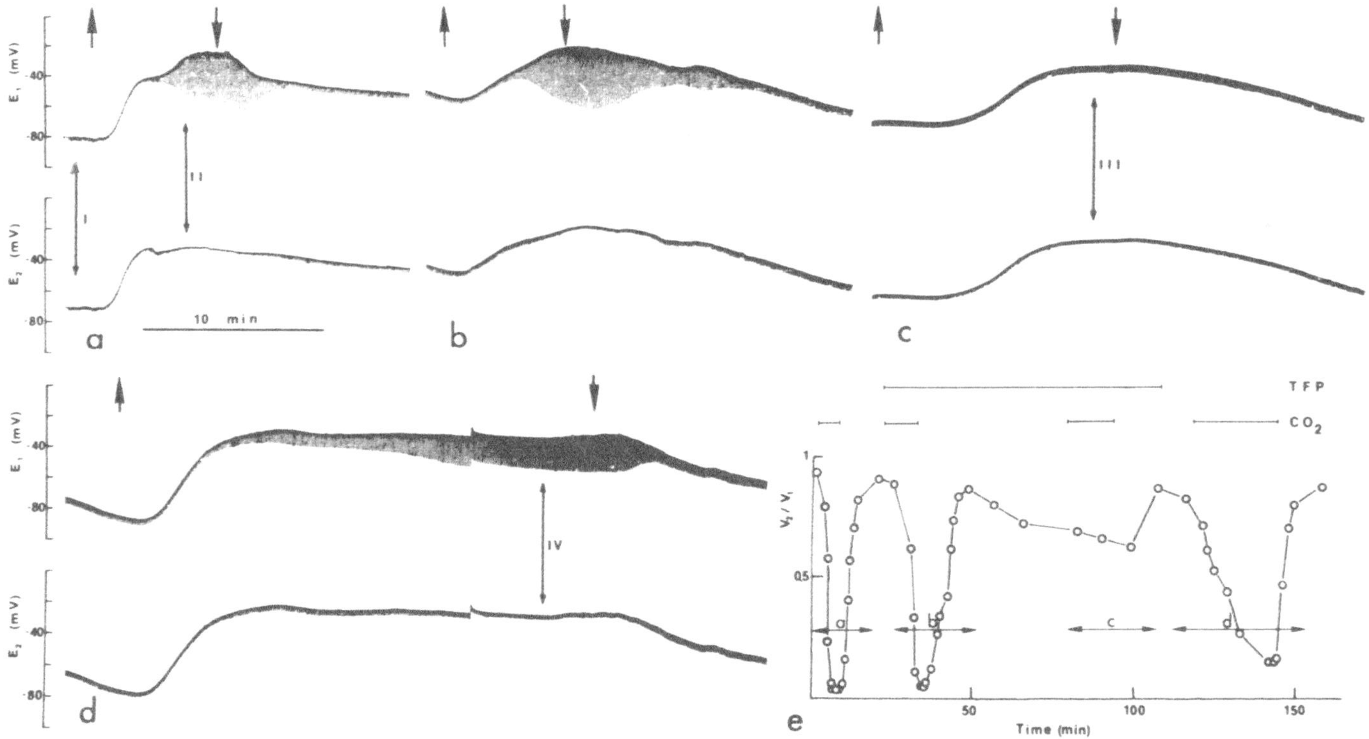

Figure 18. Changes of electrical coupling in neighboring cells of a 32-cell frog embryo exposed to 100% CO_2 in the presence and absence of trifluoperazine (TFP), a calmodulin inhibitor. Hyperpolarizing, rectangular, current pulses (5 nA, 300-msec duration) are injected in a cell (C_1) at 5-sec intervals. Membrane (E_1 and E_2) and electrotonic (V_1 and V_2) potentials are recorded in C_1 and in one of its neighboring cells (C_2). A low-speed chart recording of E and V in C_1 and C_2, upper and lower tracing respectively, is shown in (a–d). The large arrows mark beginning (↑) and end (↓) of exposure to 100% CO_2; the double-headed arrows labeled with roman numerals (a, c, and d) refer to the corresponding oscilloscope recordings shown in Figure 19. (e) Coupling ratio (V_2/V_1) sampled from the continuous chart recording seen in part in (a–d). The bars represent the periods of exposure to CO_2 and TFP (50 μM); the double-headed arrows labeled (a–d) indicate the periods of chart recording shown in the illustrations (a–d) respectively. Exposure to CO_2 in absence of TFP causes reversible depolarization of both cells and uncoupling, indicated by an increase of V_1 and a decrease of V_2 (a, e). Exposure to CO_2 6 min after superfusion with Ringer-TFP still causes uncoupling, although with a slower onset (b, e), but a subsequent exposure to CO_2 does not cause uncoupling (c, e). After a 10-min washing with Ringer's solution, the cells recover their ability to uncouple reversibly (d, e). From Peracchia *et al.* (1983).

mediary molecules. Recent data may support the latter hypothesis. The inability of internally perfused Ca^{2+} solutions to uncouple crayfish axons previously perfused with Ca^{2+}-free solutions induced Johnston and Ramón (1981) to suggest the existence of a soluble uncoupling intermediate. There is evidence that the intermediate may be calmodulin or a calmodulin-like protein (Peracchia *et al.*, 1981, 1983; Welsh *et al.*, 1981, 1982; Peracchia, 1982b).

The hypothesis of a calmodulin-mediated coupling regulation is based on evidence for the capacity of trifluoperazine (TFP), a calmodulin inhibitor, to reversibly inhibit lens junction crystallization by Ca^{2+} or H^+ (Peracchia *et al.*, 1981) and for the ability of both TFP (Figures 18 and 19) and calmidazolium (CDZ), a more specific calmodulin inhibitor (Van Belle, 1981), to reversibly inhibit the CO_2-induced electrical uncoupling of amphibian embryonic cells (Peracchia *et al.*, 1981, 1983; Peracchia, 1982b). With TFP (50 μM), uncoupling attempts were always unsuccessful after 45–60 min of incubation (Figures 18c and 19III); a lower TFP concentration (20 μM) also caused significant and partially reversible inhibition. With CDZ, the uncoupling inhibition developed after 30, 50, and 60 min of treatment with 100 nM, 70 nM, and 50 nM CDZ, respectively (Peracchia, 1982b). In the absence of CO_2, CDZ also caused an improvement in the electrical coupling as shown by an increase in V_2/V_1. Both TFP and CDZ did not significantly affect membrane potential and input resistance, indicating that the uncoupling inhibition is not the result of changes in nonjunctional membrane resistance. An involvement of calmodulin in the function of communicating junctions has also been suggested by Welsh *et al.* (1981, 1982) and by Hertzberg and Gilula (1981) on the basis of evidence for a specific binding of calmodulin to junction proteins. Interestingly, calmodulin interacts in a Ca^{2+}-independent manner with the lens junction protein (MIP 26; Welsh *et al.*, 1981, 1982; Hertzberg and Gilula, 1981), but requires Ca^{2+} to bind to the liver junction protein of molecular weight 28,000 (Hertzberg and Gilula, 1981). These findings are very interesting because they indicate a direct in-

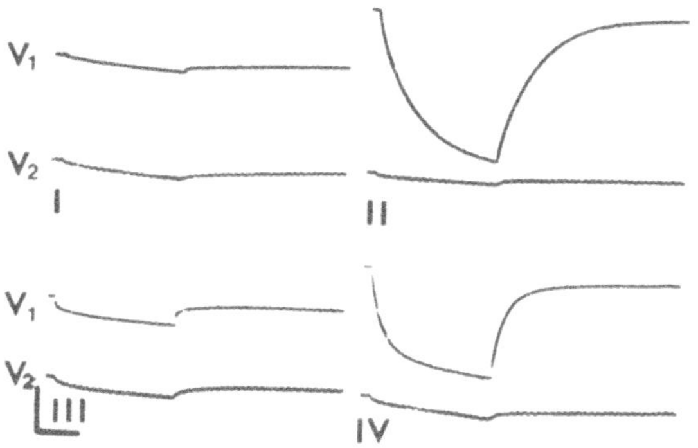

Figure 19. Oscilloscope recordings of the time course of electrotonic potentials in cell 1 (V_1) and cell 2 (V_2), sampled from the experiment shown in Figure 18. Well-coupled cells (I) are uncoupled by exposure to CO_2 (II). TFP prevents CO_2-uncoupling (III) but, after washing, the cells recover their ability to uncouple (IV). These recordings were made at the times indicated by the double arrows in Figures 18a, c, and d. Vertical bar = 10 mV. Horizontal bar = 150 msec. From Peracchia et al., 1983.

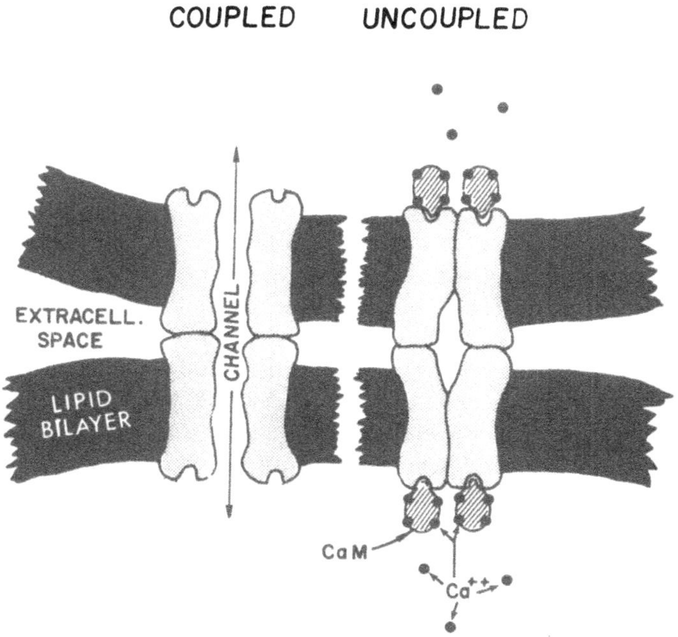

Figure 20. Hypothetical model of channel permeability regulation via calmodulin activation. The channel proteins are believed to have calmodulin sites. Calmodulin (CaM) would bind to the sites and induce protein conformational changes resulting in channel occlusion (uncoupled). In the lens, calmodulin may be attached to the junctions at all times, as it binds to the protein (MIP26) in a Ca^{2+}-independent manner, but it would affect the channels only when activated. In other junctions, calmodulin may bind and affect channel protein conformation only after being activated. This is suggested by evidence for a Ca^{2+}-dependent calmodulin binding to the liver junction protein.

teraction between calmodulin and channel proteins. In the lens, calmodulin may be bound to the junctions in its inactive form and affect the channel permeability when activated. In liver, on the contrary, calmodulin may interact with the junctions and affect the channels only after activation (Figure 20).

One is tempted to suggest that the activated calmodulin may directly induce a conformational change in the channel protein resulting in channel occlusion (Figure 20). Recently, this hypothesis has been supported by a preliminary study in which the intrinsic fluorescence of calmodulin, the lens junction protein (MIP26), and the calmodulin-MIP26 complex has been followed as a function of [Ca^{2+}] (Girsch and Peracchia, 1982). This method proved feasible because the two proteins have markedly different fluorescence emission spectra. Calmodulin contains no tryptophan and has two tyrosines, while MIP26 contains at least four tryptophans and little tyrosine. MIP26 exhibited no changes in fluorescence until above 100 μM Ca^{2+}, at which point scattering and precipitation occurred. In contrast, stepwise addition of Ca^{2+} to the calmodulin-MIP26 complex caused changes in fluorescence starting at 1 μM Ca^{2+}. The changes, which reached a maximum at 5 μM Ca^{2+}, were characterized by a 20% decrease in tryptophan emission at 342 nm and the appearance of a new peak at 319 nm (tyrosine). The quenching of tryptophan fluorescence indicates that Ca^{2+} induces a conformational change in the lens junction protein by means of calmodulin activation. Similar results have been obtained with circular dichroism experiments (Girsch and Peracchia, 1983a). Although still preliminary, these results are quite exciting, not only because they may shed some light on the mechanism of channel regulation, but also because they show that a great deal of information can be obtained by studying isolated junctional components. Ideally, one would like to study junction channels in reconstituted systems. A successful reconstitution of the lens junction protein (MIP26) in phospholipid vesicles has recently been obtained (Figure 21; Peracchia and Girsch, 1983). MIP26 self-assembles in the vesicle membrane, forming large channels, the permeability of which is regulated by Ca-calmodulin (Girsch and Peracchia, 1983b).

The hypothesis that envisions uncoupling as a calmodulin-mediated conformational change in channel protein does not exclude the possibility that cells can be uncoupled in a calmodulin-independent way. Indeed, the channel protein could be directly modified by certain uncoupling treatments. This may be the case in uncoupling induced by glutaraldehyde or transjunctional potential difference.

The chemical and structural differences among junctional proteins of different animals or different tissues of the same animal are not inconsistent with the hypothesis that all communicating junctions have a similar, calmodulin-mediated, regulatory

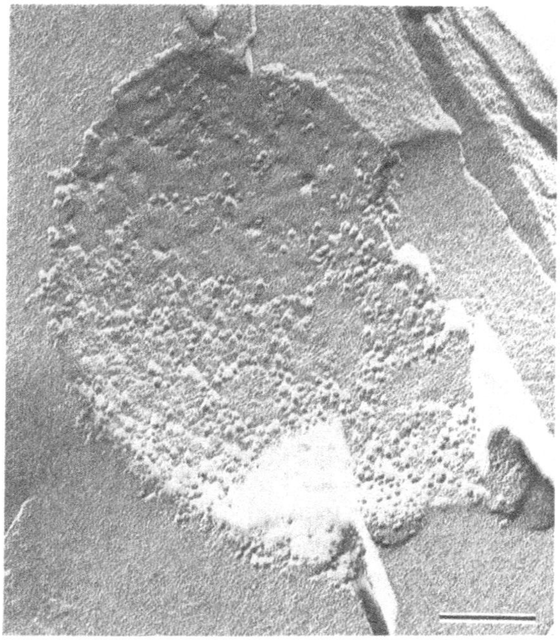

Figure 21. Reincorporation of isolated lens junction protein (MIP26) into phospholipid vesicles. MIP26 reassembles in the lipid bilayer forming intramembrane particles indistinguishable from lens junction particles (Figure 16a). Bar = 0.1 μm.

mechanism. Different types of cell-to-cell channel proteins may have developed to serve different cell-to-cell transport requirements, but the regulatory mechanism may have remained the same, as the physiological uncoupling conditions (cell death, hypoxia, hypercapnia, etc.) have similar effects on all cells.

ACKNOWLEDGMENTS

The author's work was supported by NIH Grant GM 20113. The author wishes to acknowledge Mrs. Lillian Peracchia for her technical assistance and her help in the preparation of this manuscript.

REFERENCES

Alcala, J., and Maisel, H., 1978, Specific antiserum to the main intrinsic polypeptide of chick lens fiber cell plasma membranes, *Exp. Eye Res.* **26**:219–221.

Alcala, J., Lieska, N., and Maisel, H., 1975, Protein composition of bovine lens cortical fiber cell membranes, *Exp. Eye Res.* **21**:581–595.

Alcala, J., Kuszak, J., Katar, M., Bradley, R. H., and Maisel, H., 1979, Relationship of intrinsic and peripheral proteins to chicken lens gap junction morphology, *J. Cell Biol.* **83**:269a.

Asada, Y., and Bennett, M. V. L., 1971, Experimental alteration of coupling resistance at an electrotonic synapse,, *J. Cell. Biol.* **49**:159–172.

Auerbach, A., and Bennett, M. V. L., 1969, A rectifying electrotonic synapse in the central nervous system of vertebrate, *J. Gen. Physiol.* **53**:211–237.

Azarnia, R., and Loewenstein, W. R., 1977, Intercellular communication and tissue growth. VIII. A genetic analysis of junctional communication and cancerous growth, *J. Membr. Biol.* **34**:1–28.

Azarnia, R., Larsen, W., and Loewenstein, W. R., 1974, The membrane junctions in communicating and non-communicating cells, their hybrids and segregants, *Proc. Natl. Acad. Sci. USA* **71**:880–884.

Azarnia, R., Dahl, G., and Loewenstein, W. R., 1981, Cell junction and cyclic-AMP. 3. Promotion of junctional membrane permeability and junctional membrane particles in a junction-deficient cell type, *J. Membr. Biol.* **63**:133–146.

Baker, T. S., Hollingshead, C., Caspar, D. L. D., and Goodenough, D. A., 1982, New gap junction structures revealed by low-irradiation electron microscopy, *Biophys. J.* **37**:285a.

Baker, T. S., Caspar, D. L. D., Hollingshead, C. J., and Goodenough, D. A., 1983, Gap junction structures. IV. Asymmetric features revealed by low-irradiation microscopy, *J. Cell Biol.* **96**:204–216.

Baldwin, K. M., 1979, Cardiac gap junction configuration after an uncoupling treatment as a function of time, *J. Cell Biol.* **82**:66–75.

Baldwin, K. M., and Hakim, R. S., 1981, Freeze-fracture analysis of gap and septate junctions in embryos of a moth, *Tissue Cell* **13**:691–699.

Barr, L., Dewey, M. M., and Berger, W., 1965, Propagation of action potentials and the structure of the nexus in cardiac muscle, *J. Gen. Physiol.* **48**:797–823.

Barr, L., Berger, W., and Dewey, M. M., 1968, Electrical transmission at the nexus between smooth muscle cells, *J. Gen. Physiol.* **51**:347–368.

Begenisich, T., and Stevens, C. F., 1975, How many conductance states do potassium channels have? *Biophys. J.* **15**:843–846.

Benedetti, E. L., and Emmelot, P., 1968, Hexagonal array of subunits in tight junctions separated from isolated rat liver plasma membranes, *J. Cell Biol.* **38**:15–24.

Benedetti, E. L., Dunia, I., Ramaekers, F. C. S., and Kibbelaar, M. A., 1981, Lenticular plasma membranes and cytoskeleton, in: *Molecular and Cellular Biology of the Eye Lens* (H. Bloemendal, ed.), John Wiley and Sons, New York, pp. 137–188.

Bennett, M. V. L., 1966, Physiology of electrotonic junctions, *Ann. N.Y. Acad. Sci.* **137**:509–539.

Bennett, M. V. L., 1973, Permeability and structure of electrotonic junctions and intercellular movement of tracers, in: *Intracellular Staining Techniques in Neurobiology* (S. D. Kater and C. Nicholson, eds.), Elsevier North-Holland, New York, pp. 115–133.

Bennett, M. V. L., 1978, Junctional permeability, in: *Intercellular Junctions and Synapses* (J. Feldman, N. B. Gilula, and J. D. Pitts, eds.), John Wiley and Sons, New York, pp. 25–36.

Bennett, M. V. L., and Dunham, P. B., 1970, Sucrose permeability of junctional membrane at an electrotonic synapse, *Biophys. J.* **10:**114a.

Bennett, M. V. L., Dunham, P. B., and Pappas, G. D., 1967, Ion fluxes through a "tight junction," *J. Gen. Physiol.* **50:**1094a.

Bennett, M. V. L., Spira, M. E., and Pappas, G. D., 1972, Properties of electrotonic junctions between embryonic cells of Fundulus, *Dev. Biol.* **29:**419–435.

Bennett, M. V. L., Spira, M., and Spray, D. C., 1977, Permeability of electrotonic junctions between embryonic cells, a reevaluation, *J. Cell Biol.* **75:**64a.

Berdan, R. C., and Caveney, S., 1981, Gap junction ultrastructure in three states of conduction, *J. Cell Biol.* **91:**99a.

Bernardini, G., and Peracchia, C., 1981, Gap junction crystallization in lens fibers after an increase in cell calcium, *Invest. Ophthalmol. Vis. Sci.* **21:**291–299.

Bernardini, G., Peracchia, C., and Venosa, R. A., 1981, Healing over in rat crystalline lens, *J. Physiol. (London)* **320:**187–192.

Bernardini, G., Peracchia, C., and Peracchia, L. L., 1982, Reversible gap junction crystallization and electrical uncoupling by heptanol, *Biophys. J.* **37:**285a.

Blalock, J. E., and Baron, S., 1977, Interferon-induced transfer of viral resistance between animal cells, *Nature (London)* **269:**422–425.

Blalock, J. E., and Stanton, J. D., 1980, Common pathways of interferon and hormonal action, *Nature (London)* **283:**406–408.

Bloemendal, H., Vermorken, A. J. M., Kibbelaar, M., Dunia, I., and Benedetti, E. L., 1977, Nomenclature for the polypeptide chains of lens plasma membranes, *Exp. Eye Res.* **24:**413–415.

Bok, D., Dockstader, J., and Horwitz, J., 1982, Immuno-cytochemical localization of the lens main intrinsic polypeptide (MIP26) in communicating junctions, *J. Cell Biol.* **92:**213–220.

Borek, C., Higashino, S., and Loewenstein, W. R., 1969, Intercellular communication and tissue growth. IV. Conductance of membrane junctions of normal and cancerous cells in culture, *J. Membr. Biol.* **222:**274–293.

Brink, P. R., and Barr, L., 1977, The resistance of the septum of the median giant axon of the earthworm, *J. Gen. Physiol.* **69:**517–536.

Brink, P. R., and Dewey, M. M., 1978, Nexal membrane permeability to anions, *J. Gen. Physiol.* **72:**67–86.

Brink, P. R., and Dewey, M. M., 1980, Evidence for fixed charge in the nexus, *Nature (London)* **285:**101–102.

Broekhuyse, R. M., Kuhlmann, E. D., and Stols, A. L. H., 1976, Lens membranes. II. Isolation and characterization of the main intrinsic polypeptide (MIP) of bovine lens fiber membranes, *Exp. Eye Res.* **23:**365–371.

Broekhuyse, R. M., Kuhlmann, E. D., and Winkens, H. J., 1979, Lens membranes. VII. MIP is an immunologically specific component of lens fiber membranes and is identical with 26K band protein, *Exp. Eye Res.* **29:**303–313.

Caspar, D. L. D., Goodenough, D. A., Makowski, L., and Phillips, W. C., 1977, Gap junction structures. I. Correlated electron microscopy and X-ray diffraction, *J. Cell Biol.* **74:**605–628.

Caveney, S., and Blennerhassett, M. G., 1980, Elevation of ionic conductance between insect epidermal cells by β-ecdysone in vitro, *J. Insect Physiol.* **26:**13–25.

Caveney, S., Berdan, R. C., and McLean, S., 1980, Cell-to-cell ionic communication stimulated by 20-hydroxyecdysone occurs in the absence of protein synthesis and gap junction growth, *J. Insect Physiol.* **26:**557–567.

Colaco, C. A. L. S., and Evans, W. H. A., 1981, A biochemical dissection of the cardiac intercalated disk. Isolation of subcellular fractions containing fasciae adherentes and gap junctions, *J. Cell Sci.* **52:**313–325.

Conti, F., De Felice, L. J., and Wanke, E., 1975, Potassium and sodium ion current noise in the membrane of the squid giant axon, *J. Physiol. (London)* **248:**45–82.

Conti, F., Hille, B., Neumeke, B., Nonner, W., and Stampfli, R., 1976, Conductance of the sodium channel in myelinated nerve fibers with modified sodium inactivation, *J. Physiol.* **262:**729–742.

Corsaro, C. M., and Migeon, B. R., 1977, Contact-mediated communication of ouabain resistance in mammalian cells in culture, *Nature (London)* **268:**737–739.

Cox, R. P., Krauss, M. R., Balis, M. E., and Dancis, J., 1970, Evidence for transfer of enzyme product as the basis of metabolic cooperation between tissue culture fibroblasts of Lesch-Nyhan disease and normal cells, *Proc. Natl. Acad. Sci. USA* **67:**1573–1579.

Dahl, G., and Berger, W., 1978, Nexus formation in the myometrium during parturition and induced by estrogen, *Cell Biol. Int. Rep.* **2:**381–387.

Dahl, G., and Isenberg, G., 1980, Decoupling of heart muscle cells: Correlation with increased cytoplasmic calcium and with changes of nexus ultrastructure, *J. Membr. Biol.* **53:**63–75.

Dahl, G., Azarnia, R., and Werner, R., 1981, Induction of cell–cell channel formation by mRNA, *Nature (London)* **289:**683–685.

Decker, R. S., and Friend, D. S., 1974, Assembly of gap junctions during amphibian neurulation, *J. Cell Biol.* **62:**32–47.

De Haan, R. L., 1982, *In vitro* models of entrainment of cardiac cells, in: *Cardiac Rate and Rhythm* (L. N. Bouman and H. J. Jongsma, eds.), Martinus Nijhoff Publishers, The Hague, pp. 323–361.

De Haan, R. L., and Hirakow, R., 1972, Synchronization of pulsation rates in isolated cardiac myocytes, *Exp. Cell Res.* **70:**214–220.

Délèze, J., 1964, Calcium ions and the healing-over of heart fibers, in: *Electrophysiology of the Heart* (B. Taccardi and C. Marchetti, eds.), Pergamon Press, Oxford, p. 147.

Délèze, J., and Hervé, J. C., 1983, Effects of several uncouplers of cell-to-cell communication on gap junction morphology in mammalian heart, *J. Membr. Biol.* **74:**203–215.

Délèze, J., and Loewenstein, W. R., 1976, Permeability of a cell junction during intracellular injection of divalent cations, *J. Membr. Biol.* **28:**71–86.

De Mello, W. C., 1979, Effect of intracellular injection of Ca^{++} and Mn^{++} on electrical coupling of heart cells, *Cell Biol. Int. Rep.* **3:**113–119.

De Mello, W. C., 1980a, Influence of intracellular injection of H^+ on the electrical coupling in cardiac Purkinje fibres, *Cell Biol. Int. Rep.* **4:**51–58.

De Mello, W. C., 1980b, Intercellular communication and junctional permeability, in: *Membrane Structure and Function,* Vol. 3 (E. E. Bittar, ed.), John Wiley and Sons, New York, pp. 128–170.

Dreifuss, J. J., Girardier, L., and Forssman, W. G., 1966, Étude de la propagation de l' excitation dans le ventricule de rat au moyen de solutions hypertoniques, *Pflügers Arch. Ges. Physiol.* **292:**13–33.

Duguid, J. R., and Revel, J.-P., 1975, The protein composition of the gap junction, *Cold Spring Harbor Symp. Quant. Biol.* **40:**45–47.

Eisenberg, R. S., and Johnson, E. A., 1970, Three-dimensional electrical field problems in physiology, in: *Progress in Biophysics and Molecular Biology,* Vol. 20 (J. A. Butler and D. Noble, eds.), Pergamon Press, New York, pp. 1–65.

Engelmann, T. W., 1877, Vergleichende Untersuchungen zur Lehre von der Muskel und Nervenelektrizität, *Pflügers Arch. Ges. Physiol.* **15:**116–148.

Evans, W. H., and Gurd, J. W., 1972, Preparation and properties of nexuses and lipid-enriched vesicles from mouse liver plasma membranes, *Biochem. J.* **128:**691–700.

Fallon, R. F., and Goodenough, D. A., 1981, 5-Hour half-life of mouse liver gap junction protein, *J. Cell Biol.* **90:**521–526.

Finbow, M., Yancey, S. B., Johnson, R., and Revel, J.-P., 1980, Independent lines of evidence suggesting a major gap junction protein with a molecular weight of 26,000, *Proc. Natl. Acad. Sci. USA* **77:**970–974.

Flagg-Newton, J. L., and Loewenstein, W. R., 1979, Experimental depression of junctional membrane permeability in mammalian cell culture. A study with tracer molecules in the 300 to 400 dalton range, *J. Membr. Biol.* **50:**65–100.

Flagg-Newton, J. L., and Loewenstein, W. R., 1980, Asymmetrically permeable membrane channels in cell junction, *Science* **207:**771–773.

Flagg-Newton, J. L., and Loewenstein, W. R., 1981, Cell junction and cyclic-AMP. 2. Modulations of junctional membrane permeability, dependent on serum and cell density, *J. Membr. Biol.* **63:**123–131.

Flagg-Newton, J. L., Simpson, I., and Loewenstein, W. R., 1979, Permeability of cell-to-cell membrane channels in mammalian cell junction, *Science* **205**:404–407.

Flagg-Newton, J. L., Dahl, G., and Loewenstein, W. R., 1981, Cell junction and cyclic-AMP. 1. Upregulation of junctional membrane permeability and junctional membrane particles by administration of cyclic nucleotide or phosphodiesterase inhibitor, *J. Membr. Biol.* **63**:105–121.

Friedlander, M., 1980, Immunological approaches to the study of myogenesis and lens fiber junction formation, *Curr. Top. Dev. Biol.* **14**:321–358.

Furshpan, E. J., and Potter, D. D., 1959, Transmission at the giant synapse of the crayfish, *J. Physiol. (London)* **145**:289–325.

Garfield, R. E., Sims, S., and Daniel, E. E., 1977, Gap junctions: Their presence and necessity in myometrium during parturition, *Science* **198**:958–960.

Garfield, R. E., Sims, S. M., Kannan, M. S., and Daniel, E. E., 1978, Possible role of gap junctions in activation of myometrium during parturition, *Am. J. Physiol.* **235**:C168–C179.

Giaume, C., Spira, M. E., and Korn, H., 1980, Uncoupling of invertebrate electrotonic synapses by carbon dioxide, *Neurosci. Lett.* **17**:197–202.

Gilula, N. B., 1978, Structure of intercellular junctions, in: *Intercellular Junctions and Synapses* (J. Feldman, N. B. Gilula, and J. D., Pitts, eds.), John Wiley and Sons, New York, pp. 1–22.

Gilula, N. B., Reeves, O. R., and Steinbach, A., 1972, Metabolic coupling, ionic coupling and cell contacts, *Nature (London)* **235**:262–265.

Girsch, S. J., and Peracchia, C., 1982, Calcium-activated calmodulin induces conformational changes in lens gap junction protein, *J. Cell Biol.* **95**:103a.

Girsch, S. J., and Peracchia, C., 1983a, Lens junction protein changes conformation in the presence of calcium-activated calmodulin, *Invest. Ophthalmol. Vis. Sci.* (Suppl. 24):27.

Girsch, S. J., and Peracchia, C., 1983b, Lens junction protein (MIP 26) self-assembles in liposomes forming large channels regulated by calmodulin (CaM), *J. Cell Biol.* **97**:83a.

Goodenough, D. A., 1974, Bulk isolation of mouse hepatocyte gap junctions. Characterization of the principal protein, connexin, *J. Cell Biol.* **61**:557–563.

Goodenough, D. A., 1976, *In vitro* formation of gap junction vesicles, *J. Cell Biol.* **68**:220–231.

Goodenough, D. A., 1979, Lens gap junctions: A structural hypothesis for non-regulated low resistance intercellular pathways, *Invest. Opthamol. Vis. Sci.* **18**:1104–1122.

Goodenough, D. A., and Gilula, N. B., 1974, The splitting of hepatocyte gap junctions and zonulae occludentes with hypertonic disaccharides, *J. Cell Biol.* **61**:575–590.

Goodenough, D. A., and Stoeckenius, W., 1972, The isolation of mouse hepatocyte gap junctions. Preliminary chemical characterization and X-ray diffraction, *J. Cell Biol.* **54**:646–656.

Goodenough, D. A., Caspar, D. L. D., Phillips, W. C., and Makowski, L., 1978, Preparation and characterization of highly ordered gap junction structure, *J. Cell Biol.* **79**:223a.

Goodenough, D. A., Dick, J. S. B., II, and Lyons, J. E., 1980, Lens metabolic cooperation: A study of mouse lens transport and permeability visualized with freeze-substitution, autoradiography and electron microscopy, *J. Cell Biol.* **86**:576–589.

Goodman, C. S., and Spitzer, N. C., 1979, Embryonic development of identified neurones: Differentiation from neuroblast, *Nature (London)* **280**:208–214.

Goshima, K., 1969, Synchronized beating of and electrotonic transmission between myocardial cells, mediated by heterotypic strain cells in monolayer culture, *Exp. Cell Res.* **58**:420–426.

Goshima, K., 1970, Formation of nexuses and electrotonic transmission between myocardial and FL cells in monolayer culture, *Exp. Cell Res.* **63**:124–130.

Hama, K., and Saito, K., 1977, Gap junctions between the supporting cells in some acoustico-vestibular receptors, *J. Neurocytol.* **6**:1–12.

Hanna, R. B., Keeter, J. S., and Pappas, G. D., 1978, Fine structure of a rectifying electrotonic synapse, *J. Cell Biol.* **79**:764–773.

Hanna, R. B., Reese, T. S., Ornberg, R. L., Spray, D. C., and Bennett, M. V. L., 1981, Fresh frozen gap junctions: Resolution of structural details in the coupled and uncoupled states, *J. Cell Biol.* **91**:125a.

Harris, A. L., Spray, D. C., and Bennett, M. V. L., 1981, Kinetic properties of a voltage-dependent junctional conductance, *J. Gen. Physiol.* **77**:95–117.

Hauswirth, O., 1968, Influence of halothane on electrical properties of cardiac Purkinje fibers, *J. Physiol. (London)* **201:**42P–43P.

Henderson, D., Eibl, H., and Weber, K., 1979, Structure and biochemistry of mouse hepatic gap junctions, *J. Mol. Biol.* **132:**193–218.

Hermsmeyer, K., 1980, Angiotensin-II increases electrical coupling in mammalian ventricular myocardium, *Circ. Res.* **47:**524–539.

Hertzberg, E. L., 1980, Biochemical and immunological approach to the study of gap junctions and cell-to-cell communication, *In Vitro (Rockville)* **16:**1057–1067.

Hertzberg, E. L., and Gilula, N. B., 1979, Isolation and characterization of gap junctions from rat liver, *J. Biol. Chem.* **254:**2138–2147.

Hertzberg, E. L., and Gilula, N. B., 1981, Liver gap junctions and lens fiber junctions: Comparative analysis and calmodulin interaction, *Cold Spring Harbor Symp. Quant. Biol.* **46:**639–645.

Hertzberg, E. L., Anderson, D. J., Friedlander, M., and Gilula, N. B., 1982, Comparative analysis of the major polypeptides from liver gap junctions and lens gap junctions, *J. Cell Biol.* **92:**53–59.

Hess, P., and Weingart, R., 1980, Intracellular free calcium modified by pH_i in sheep cardiac Purkinje fibers, *J. Physiol. (London)* **307:**60P–61P.

Heuser, J. E., Reese, T. S., and Landis, D. M. D., 1975, Preservation of synaptic structure by rapid freezing, *Cold Spring Harbor Symp. Quant. Biol.* **40:**17–24.

Higashino, S., Suzuki, K., and Okazaki, H., 1980, Protective effects of glucocorticoids on gland stiffness and cell coupling in salivary gland of Chironomus plumosus larva, *Exp. Cell Res.* **125:**63–71.

Hirokawa, N., and Heuser, J., 1982, The inside and the outside of gap junction membranes visualized by deep etching, *Cell* **30:**395–406.

Hooper, M. L., and Subak-Sharpe, J. H., 1981, Metabolic cooperation between cells, *Int. Rev. Cytol.* **69:**45–104.

Hyde, A., Blondel, B., Matter, A., Cheneval, J. P., Filloux, B., and Girardier, L., 1969, Homo and heterocellular junctions in cell cultures: An electrophysiological and morphological study, *Prog. Brain Res.* **31:**283–311.

Hyman, L. H., 1940, *The Invertebrates: Protozoa through Ctenophora,* Vol. 1, McGraw-Hill, New York.

Ito, S., Sato, E., and Loewenstein, W. R., 1974, Studies on formation of a permeable cell membrane junction. 2. Evolving junctional conductance and junctional insulation, *J. Membr. Biol.* **19:**339–355.

Iwatsuki, N., and Petersen, O. H., 1979, Pancreatic acinar cells: The effect of carbon dioxide, ammonium chloride and acetylcholine on intercellular communication, *J. Physiol. (London)* **291:**317–326.

Johnson, R., Hammer, M., Sheridan, J., and Revel, J.-P., 1974, Gap junction formation between reaggregated Novikoff hepatoma cells, *Proc. Natl. Acad. Sci. USA* **71:**4536–4540.

Johnston, M. F., and Ramón, F., 1981, Electrotonic coupling in internally perfused crayfish segmented axons, *J. Physiol. (London)* **317:**509–518.

Johnston, M. F., Simon, S. A., and Ramón, F., 1980, Interaction of anesthetics with electrical synapses, *Nature (London)* **286:**498–500.

Jongsma, H. J., and Van Rijn, H. E., 1972, Electrotonic spread of current in monolayer cultures of neonatal rat heart cells, *J. Membr. Biol.* **9:**341–360.

Kakiuchi, S., Yamazaki, R., Teshima, Y., Uenishi, K., Yasuda, S., Kashiba, A., Sobue, K., Ohshima, M., and Nakajima, T., 1978, Membrane-bound protein modulator and phosphodiesterase, *Adv. Cycl. Nucleotide Res.* **9:**253–263.

Katz, B., and Miledi, R., 1972, Stastical nature of acetylcholine potential and its molecular components, *J. Physiol. (London)* **224:**665–699.

Kawamura, K., and Konishi, T., 1967, Ultrastructure of the cell junction of heart muscle with special reference to its functional significance in excitation conduction and in the concept of "disease of intercalated disc," *Jpn. Circ. J.* **31:**1533–1543.

Keeters, J. S., Deschenes, M., Pappas, G. D., and Bennett, M. V. L., 1974, Fine structure and permeability studies of a rectifying electrotonic synapse, *Biol. Bull. (Woods Hole)* **147:**485–501.

Kensler, R. W., and Goodenough, D. A., 1980, Isolation of mouse myocardial gap junctions, *J. Cell Biol.* **86:**755–764.

Kistler, J., and Bullivant, S., 1980a, Lens gap junctions and orthogonal arrays are unrelated, *FEBS Lett.* **111:**73–78.

Kistler, J., and Bullivant, S., 1980b, The connexon order in isolated lens gap junctions, *J. Ultrastruct. Res.* **72**:27–38.

Kriebel, M. E., 1968, Electrical characteristics of tunicate heart cell membranes and nexuses, *J. Gen. Physiol.* **52**:46–59.

Kuffler, S. W., and Potter, D. D., 1964, Glia in the leach central nervous system: Physiological properties and neuro-glial relationship, *J. Neurophysiol.* **27**:290–320.

Kuszak, J. R., Alcala, J., and Maisel, H., 1981, Biochemical and structural features of chick lens gap junctions, *Exp. Eye Res.* **33**:157–166.

Lane, N. J., and Skaer, H. le B., 1980, Intercellular junctions in insect tissues, *Adv. Insect Physiol.* **15**:35–213.

Larsen, W. J., 1977, Structural diversity of gap junctions. A review, *Tissue Cell* **9**:373–394.

Larsen, W. J., Tung, H. N., Murray, S. A., and Swenson, C. A., 1979, Evidence for the participation of actin microfilaments and bristle coats in the internalization of gap junction membrane, *J. Cell Biol.* **83**:576–587.

Lawrence, P. A., and Green, S., 1975, Anatomy of a compartment border-intersegmental boundary in Oncopelitus, *J. Cell Biol.* **65**:373–382.

Lawrence, T. S., Beers, W. H., and Gilula, N. B., 1978, Transmission of the hormonal stimulation by cell-to-cell communication, *Nature (London)* **272**:501–506.

Lea, T. J., and Ashley, C. C., 1978, Increase in free Ca^{++} in muscle after exposure to CO_2, *Nature (London)* **275**:236–238.

Ledbetter, M. L., and Lubin, M., 1979, Transfer of potassium. New measure of cell–cell coupling, *J. Cell Biol.* **80**:150–165.

Lo, C. W., and Gilula, N. B., 1979, Gap junctional communication in the preimplantation mouse embryo, *Cell* **18**:399–409.

Loewenstein, W. R., 1966, Permeability of membrane junctions, *Ann. N.Y. Acad. Sci.* **137**:441–472.

Loewenstein, W. R., 1967, On the genesis of intercellular communication, *Dev. Biol.* **15**:503–520.

Loewenstein, W. R., 1975, Permeable junctions, *Cold Spring Harbor Symp. Quant. Biol.* **40**:49–63.

Loewenstein, W. R., 1979, Junctional intercellular communication and the control of growth, *Biochim. Biophys. Acta* **560**:1–65.

Loewenstein, W. R., and Kanno, Y., 1964, Studies on an epithelial (gland) cell junction. I. Modifications of surface membrane permeability, *J. Cell Biol.* **22**:565–586.

Loewenstein, W. R., and Rose, B., 1978, Calcium in (junctional) intercellular communication and a thought on its behavior in intercellular communication, *Ann. N.Y. Acad. Sci.* **307**:285–307.

Loewenstein, W. R., Nakas, M., and Socolar, S. J., 1967, Junctional membrane uncoupling: Permeability transformation at a cell membrane junction, *J. Gen. Physiol.* **50**:1865–1891.

Loewenstein, W. R., Kanno, Y., and Socolar, S. J., 1978, Quantum jump of conductance during formation of membrane channels of cell–cell junction, *Nature (London)* **274**:133–136.

Maisel, H., Alcala, J., and Lieska, N., 1976, The protein structure of chick lens fiber cell membranes and intercellular matrix, *Doc. Ophthalmol.* **8**:121–133.

Makowski, L., Caspar, D. L. D., Phillips, W. C., and Goodenough, D. A., 1977, Gap junction structures. II. Analysis of the X-ray diffraction data, *J. Cell Biol.* **74**:629–645.

Makowski, L., Caspar, D. L. D., Goodenough, D. A., and Phillips, W. C., 1982, Gap junction structures. III. The effects of variations in the isolation procedure, *Biophys. J.* **37**:189–191.

McNutt, N. S., and Weinstein, R. S., 1970, The ultrastructure of the nexus. A correlated thin-section and freeze-cleave study, *J. Cell Biol.* **47**:666–688.

Meda, P., Perrelet, A., and Orci, L., 1980, Gap junctions and β-cell function, *Horm. Metabol. Res.* **10(Suppl.)**:157–162.

Meech, R., and Thomas, R. C., 1977, The effect of calcium injection on the intracellular sodium and pH of snail neurones, *J. Physiol. (London)* **265**:867–879.

Nakas, M., Higashino, S., and Loewenstein, W. R., 1966, Uncoupling of an epithelial cell membrane junction by calcium-ion removal, *Science* **151**:89–91.

Nicholls, J., and Purves, D., 1970, Monosynaptic chemical and electrical connexions between sensory and motor cells in the central nervous system of the leech, *J. Physiol. (London)* **209**:647–667.

Nicholson, B. J., Hunkapiller, M. W., Grim, L. B., Hood, L. E., and Revel, J.-P., 1981, Rat liver gap junction protein: Properties and partial sequence, *Proc. Natl. Acad. Sci. USA* **78**:7594–7598.

Nicholson, B. J., Takemoto, L. J., Hunkapiller, M. W., Hood, L. E., and Revel, J.-P., 1983. Differences between liver gap junction protein and lens MIP26 from rat: Implications for tissue specificity of gap junctions, *Cell* **32**:967–978.

Oliveira-Castro, G. M., and Loewenstein, W. R., 1971, Junctional membrane permeability. Effects of divalent cations, *J. Membr. Biol.* **5**:51–77.

Pappas, G. D., Asada, Y., and Bennett, M. V. L., 1971, Morphological correlates of increased coupling resistance at an electrotonic synapse, *J. Cell Biol.* **49**:173–188.

Payton, B. W., Bennett, M. V. L., and Pappas, G. D., 1969a, Temperature-dependence of resistance at an electrotonic synapse, *Science* **165**:594–597.

Payton, B. W., Bennett, M. V. L., and Pappas, G. D., 1969b, Permeability and structure of junctional membranes at an electrotonic synapse, *Science* **166**:1641–1643.

Peracchia, C., 1973a, Low resistance junctions in crayfish. I. Two arrays of globules in junctional membranes, *J. Cell Biol.* **57**:54–65.

Peracchia, C., 1973b, Low resistance junctions in crayfish. II. Structural details and further evidence for intercellular channels by freeze-fracture and negative staining, *J. Cell Biol.* **57**:66–76.

Peracchia, C., 1974, A structure-function correlation in gap junctions of crayfish, *Proc. 8th. Int. Congr. Electron Microsc.*, Vol. II (J. V. Sanders and D. J. Goodchild, eds.), The Australian Academy of Sciences, Canberra, pp. 226–227.

Peracchia, C., 1977, Gap junctions—structural changes after uncoupling procedures, *J. Cell Biol.* **72**:628–641.

Peracchia, C., 1978, Calcium effects on gap junction structure and cell coupling, *Nature (London)* **271**:669–671.

Peracchia, C., 1980, Structural correlates of gap junction permeation, *Int. Rev. Cytol.* **66**:81–146.

Peracchia, C., 1982a, Gap junction transport regulation, in: *Membranes and Transport,* Vol. II (A. Martonosi, ed.), Plenum Press, New York, pp. 271–278.

Peracchia, C., 1982b, Calmodulin and gap junctions: Reversible inhibition of electrical uncoupling by calmidazolium, *J. Cell Biol.* **95**:102a.

Peracchia, C., and Dulhunty, A. F., 1974, Gap junctions: Structural changes associated with changes in permeability, *J. Cell Biol.* **63**:263a.

Peracchia, C., and Dulhunty, A. F., 1976, Low resistance junctions in crayfish. Structural changes with functional uncoupling, *J. Cell Biol.* **70**:419–439.

Peracchia, C., and Girsch, S. J., 1983, Asymmetric incorporation of lens junction particles into phospholipid vesicles, *Invest. Ophthalmol. Vis. Sci.* (Suppl. 24):31.

Peracchia, C., and Peracchia, L. L., 1978, Orthogonal and rhombic arrays in gap junctions exposed to low pH, *J. Cell Biol.* **79**:217a.

Peracchia, C., and Peracchia, L. L., 1980a, Gap junction dynamics: Reversible effects of divalent cations, *J. Cell Biol.* **87**:708–718.

Peracchia, C., and Peracchia, L. L., 1980b, Gap junction dynamics: Reversible effects of hydrogen ions, *J. Cell Biol.* **87**:719–727.

Peracchia, C., and Peracchia, L. L., 1983, Gap junction particles are interlinked by extracellular filamentous bridges, *J. Cell Biol.* **97**:82a.

Peracchia, C., Bernardini, G., and Peracchia, L. L., 1981, A calmodulin inhibitor prevents gap junction crystallization and electrical uncoupling, *J. Cell Biol.* **91**:124a.

Peracchia, C., Bernardini, G., and Peracchia, L. L., 1983, Is calmodulin involved in the regulation of gap junction permeability? *Pflügers Arch.* **399**:152–154.

Pitts, J. D., 1971, Molecular exchange and growth control in tissue culture, in: *Growth Control in Cell Cultures* (G. E. W. Wolstenholme and J. Knight, eds.), Churchill Livingston, London, pp. 89–105.

Pitts, J. D., 1976, Junctions as channels of direct communication between cells, in: *Developmental Biology of Plants and Animals* (C. F. Graham and P. E. Waring, eds.), Blackwell Scientific Publications, Oxford, pp. 96–110.

Pitts, J. D., 1977, Direct communication between animal cells, in: *International Cell Biology* (B. R. Brinkley and K. R. Porter, eds.), The Rockefeller University Press, New York, pp. 43–49.

Pitts, J. D., and Finbow, M. E., 1977, Junctional permeability and its consequences, in: *Intercellular Communication* (W. C. De Mello, ed.), Plenum Publishing Corp., New York, pp. 61–86.

Pitts, J. D., and Simms, J. W., 1977, Permeability of junctions between animal cells: Intercellular transfer of nucleotides but not macromolecules, *Exp. Cell Res.* **104:**153–163.

Politoff, A., and Pappas, G. D., 1972, Mechanism of increase in coupling resistance at an electrotonic synapse of crayfish septate axon, *Anat. Rec.* **172:**384–385.

Politoff, A., Pappas, G. D., and Bennett, M. V. L., 1974, Cobalt ions cross an electrotonic synapse if cytoplasmic concentration is low, *Brain Res.* **76:**343–346.

Rae, J. L., 1974, The movement of procion dye in the crystalline lens, *Invest. Ophthalmol.* **13:**147–150.

Rae, J. L., 1979, The electrophysiology of the crystalline lens, *Curr. Top. Eye Res.* **1:**37–90.

Rae, J. L., Thompson, R. D., and Eisenberg, R. S., 1982, The effect of 2-4 dinitrophenol on cell-to-cell communication in the frog lens, *Exp. Eye Res.*, **35:**598–609.

Raviola, E., Goodenough, D. A., and Raviola, G., 1978, The native structure of gap junctions rapidly frozen at 4°K, *J. Cell Biol.* **79**(2, Pt. 2):229a.

Raviola, E., Goodenough, D. A., and Raviola, G., 1980, Structure of rapidly frozen gap junctions, *J. Cell Biol.* **87:**273–279.

Reber, W. R., and Weingart, R., 1982, Ungulate cardiac Purkinje fibers: The influence of intracellular pH on the electrical cell-to-cell coupling, *J. Physiol. (London)* **328:**87–104.

Revel, J.-P., 1968, Studies on the fine structure of intercellular junctions, *26th Proc. Electron Microsc. Soc. Am.*, Claitor's Publishing Division, Baton Rouge, Louisiana, p. 40.

Revel, J.-P., and Karnovsky, M. J., 1967, Hexagonal array of subunits in intercellular junctions of the mouse heart and liver, *J. Cell Biol.* **33:**C7–C12.

Rink, T. J., Tsien, R.-Y., and Warner, A. E., 1980, Free calcium in Xenopus embryos measured with ion-selective microelectrodes, *Nature (London)* **283:**658–660.

Rose, B., and Loewenstein, W. R., 1975, Permeability of cell junction depends on local cytoplasmic calcium activity, *Nature (London)* **254:**250–252.

Rose, B., and Loewenstein, W. R., 1976, Permeability of a cell junction and the local cytoplasmic free ionized calcium concentration: A study with aequorin, *J. Membr. Biol.* **28:**87–119.

Rose, B., and Rick, R., 1978, Intracellular pH, intracellular free Ca, and junctional cell–cell coupling, *J. Membr. Biol.* **44:**377–415.

Rose, B., Simpson, I., and Loewenstein, W. R., 1977, Calcium ion produces graded changes in permeability of membrane channels in cell junction, *Nature (London)* **267:**625–627.

Satterlie, R. A., and Case, J. F., 1978, Gap junctions suggest epithelial conduction within the comb plates of the ctenophore Pleurobrachia bachei, *Cell Tissue Res.* **193:**87–91.

Schwarzmann, G., Wiegand, H., Rose, B., Zimmermann, A., Ben-Haim, D., and Loewenstein, W. R., 1981, Diameter of the cell-to-cell junctional channels as probed with neutral molecules, *Science* **213:**551–553.

Sheridan, J. D., Hammer-Wilson, M., Preus, D., and Johnson, R. G., 1978, Quantitative analysis of low-resistance junctions between cultured cells and correlation with gap junctional areas, *J. Cell Biol.* **76:**532–544.

Shibata, Y., and Page, E., 1981, Gap junctional structure in intact and cut sheep cardiac Purkinje fibers. A freeze-fracture study of Ca^{++}-induced resealing, *J. Ultrastruct. Res.* **75:**195–204.

Siegenbeck Van Henkelom, J., Van Der Gon, J. J. D., and Prop, F. J. A., 1972, Model approaches for evaluation of cell coupling on monolayers, *J. Membr. Biol.* **7:**88–110.

Sikerwar, S., and Malhotra, S., 1981, Structural correlates of glutaraldehyde-induced uncoupling in mouse liver gap junctions, *Eur. J. Cell Biol.* **25:**319–323.

Sikerwar, S. S., Tewari, J. P., and Malhotra, S. K., 1981, Subunit structure of the connexons in hepatocyte gap junctions, *Eur. J. Cell Biol.* **24:**211–215.

Simms, J. W., 1973, Interactions between animal cells in tissue culture, Ph.D. thesis, University of Glasgow, Scotland.

Simpson, I., Rose, B., and Loewenstein, W. R., 1977, Size limit of molecules permeating the junctional membrane channels, *Science* **195:**294–296.

Slack, C., and Palmer, J. F., 1969, The permeability of intercellular junctions in the early embryo of Xenopus laevis studied with a fluorescent tracer, *Exp. Cell Res.* **55:**416–419.

Slack, C., and Warner, A. E., 1975, Properties of surface and junctional membranes of embryonic cells isolated from blastula stages of Xenopus laevis, *J. Physiol. (London)* **248:**97–120.

Smith, T. G., and Bauman, B. T., 1969, The functional organization within the omnatidium of the lateral eye of Limulus, *Prog. Brain Res.* **31:**313–349.

Socolar, S. J., and Loewenstein, W. R., 1979, Methods for studying transmission through permeable cell-to-cell junctions, in: *Methods in Membrane Biology,* Vol. 10 (E. D. Korn, ed.), Plenum Publishing Corp., New York, pp. 123–179.

Spira, A. W., 1971, Nexus in intercalated disk of canine heart—quantitative data for estimation of its resistance, *J. Ultrastruct. Res.* **34:**409–425.

Spray, D. C., Harris, A. L., and Bennett, M. V. L., 1981a, Gap junctional conductance is a simple and sensitive function of intracellular pH, *Science* **211:**712–715.

Spray, D. C., Harris, A. L., and Bennett, M. V. L., 1981b, Equilibrium properties of a voltage-dependent junctional conductance, *J. Gen. Physiol.* **77:**77–93.

Spray, D. C., Harris, A. L., and Bennett, M. V. L., 1981c, Glutaraldehyde differentially affects gap junctional conductance and its pH and voltage dependence, *Biophys. J.* **33:**108a.

Spray, D. C., Stern, J. H., Harris, A. L., and Bennett, M. V. L., 1982, Gap junctional conductance: Comparison of sensitivities to H and Ca ions, *Proc. Natl. Acad. Sci. USA* **79:**441–445.

Staehelin, L. A., 1974, Structure and function of intercellular junctions, *Int. Rev. Cytol.* **39:**191–283.

Subak-Sharpe, H., Burk, R. R., and Pitts, J. D., 1966, Metabolic cooperation by cell-to-cell transfer between genetically different mammalian cells in tissue culture, *Heredity* **21:**342–343.

Subak-Sharpe, H., Burk, R. R., and Pitts, J. D., 1969, Metabolic cooperation between biochemically marked mammalian cells in tissue culture, *J. Cell Sci.* **4:**353–367.

Suzuki, K., and Higashino, S., 1977, Protective effect of corticosteroids on intercellular junctions, *Exp. Cell Res.* **109:**263–268.

Tamm, S. L., 1973, Mechanisms of ciliary coordination in ctenophores, *J. Exp. Biol.* **59:**231–245.

Tsien, R. W., and Weingart, R., 1974, Cyclic-AMP: Cell-to-cell movement and inotropic effect in ventricular muscle, studied by a cut-end method, *J. Physiol. (London)* **242:**95P–96P.

Tupper, J. T., and Sanders, J. W., Jr., 1972, Intercellular permeability in the early Asterias embryo, *Dev. Biol.* **27:**546–554.

Turin, L., 1977, New probes for studying intercellular permeability, *J. Physiol (London)* **269:**6P–7P.

Turin, L., and Warner, A. E., 1977, Carbon dioxide reversibly abolishes ionic communication between cells of early amphibian embryo, *Nature (London)* **270:**56–57.

Turin, L., and Warner, A. E., 1980, Intracellular pH in early Xenopus embryos: Its effect on current flow between blastomeres, *J. Physiol. (London)* **300:**489–504.

Unwin, P. N. T., and Zampighi, G., 1980, Structure of the junction between communicating cells, *Nature (London)* **283:**545–549.

Van Belle, H., 1981, R 24571: A potent inhibitor of calmodulin-activated enzymes, *Cell Calcium* **2:**483–494.

Van Harreveld, A., Crowell, J., and Malhotra, S. K., 1965, A study of extracellular space in central nervous tissue by freeze-substitution, *J. Cell Biol.* **25:**117–137.

Waggoner, P. R., and Maisel, H., 1978, Immunofluorescent study of a chick lens fiber cell membrane polypeptide, *Exp. Eye Res.* **27:**151–157.

Warner, A. E., and Lawrence, P. A., 1982, Permeability of gap junctions at the segmented border in insect epidermis, *Cell* **28:**243–252.

Watanabe, A., and Grundfest, H., 1961, Impulse propagation at the septal and commissural junctions of crayfish lateral giant axons, *J. Gen. Physiol.* **45:**267–308.

Weidman, S., 1952, The electrical constants of Purkinje fibres, *J. Physiol. (London)* **118:**348–360.

Weidman, S., 1966, Diffusion of radiopotassium across intercalated disks of mammalian cardiac muscle, *J. Physiol. (London)* **187:**323–342.

Weingart, R., 1974, Permeability of tetraethylammonium ions of surface membrane and intercalated disks of sheep and calf myocardium, *J. Physiol.* **240:**741–762.

Weingart, R., 1977, Action of ouabain on intercellular coupling and conduction-velocity in mammalian ventricular muscle, *J. Physiol. (London)* **264:**341–365.

Weingart, R., and Reber, W., 1979, Influence of internal pH on r_i of Purkinje fibers from mammalian heart, *Experientia* **35:**17a.

Welsh, M. J., Aster, J., Ireland, M., Alcala, J., and Maisel, H., 1981, Calmodulin and gap junctions: Localization of calmodulin and calmodulin binding sites in chick lens cells, *J. Cell Biol.* **91:**123a.

Welsh, M. J., Aster, J. C., Ireland, M., Alcala, J., and Maisel, H., 1982, Calmodulin binds to chick lens gap junction protein in a calcium-independent manner, *Science* **216:**642–644.

Wiener, J., Spiro, D., and Loewenstein, W. R., 1964, Studies on an epithelial (gland) cell junction. II. Surface structure, *J. Cell Biol.* **22:**587–598.

Wojtczak, J., 1979, Contractures and increase in internal longitudinal resistance of cow ventricular muscle induced by hypoxia, *Circ. Res.* **44:**88–95.

Woodbury, J. W., and Crill, W. E., 1961, On the problem of impulse conduction in the atrium, in: *Nervous Inhibition* (E. Florey, ed.), Pergamon Press, Oxford, pp. 124–135.

Yancey, S. B., Nicholson, B. H., and Revel, J.-P., 1981, The dynamic state of liver gap junctions, *J. Supramol. Str. Cell. Biochem.* **16:**221–232.

Zampighi, G., and Robertson, J. D., 1977, Macula communicans structure, *Biophys. J.* **17:**31a.

Zampighi, G., and Unwin, P. N. T., 1979, Two forms of isolated gap junctions, *J. Mol. Biol.* **135:**451–464.

Zampighi, G., Corless, J. M., and Robertson, J. D., 1980, On gap junction structure, *J. Cell Biol.* **86:**190–198.

Zampighi, G., Simon, S. A., Robertson, J. D., McIntosh, T. J., and Costello, M. J., 1982, On the structural organization of isolated bovine lens fiber junctions, *J. Cell Biol.* **93:**175–189.

Zigler, J. S., Jr., and Horwitz, J., 1981, Immunochemical studies on the major intrinsic polypeptides from human lens membranes, *Invest. Ophthalmol. Vis. Sci.* **21:**46–51.

4

Lipid Polymorphism and Membrane Function

B. de Kruijff, P. R. Cullis, A. J. Verkleij, M. J. Hope, C. J. A. Van Echteld, and T. F. Taraschi

I. INTRODUCTION

One of the fundamental problems in membrane biology is that of lipid diversity. The number of chemically different membrane lipids is much larger than other key biological building blocks such as nucleotides, amino acids, and carbohydrates. For instance, a relatively simple biomembrane such as that of the red blood cell contains well over a hundred different lipid species.

Except for some metabolic and receptor-type functions, current models of biological membranes do not take into account this lipid diversity. In the fluid mosaic model (Singer and Nicolson, 1972), the lipids are thought to form a fluid semipermeable bilayer which acts as a matrix for the functional membrane proteins. On the basis of our present understanding, a single unsaturated phosphatidylcholine (PC) species could easily satisfy such demands.

The incomplete nature of such membrane models is reinforced by two basic membrane phenomena. First, biological membranes are highly dynamic structures which are continuously involved in a variety of biochemical processes during which transient departures from bilayer structure must occur. Examples include the transbilayer transport of lipids and proteins as well as membrane fusion, among others. Second, although the ability of lipids to adopt a variety of phases has been recognized

B. de Kruijff and A. J. Verkleij • Department of Molecular Biology, State University of Utrecht, 3584 CH Utrecht, The Netherlands. *C.J.A. Van Echteld* • Department of Biochemistry, State University of Utrecht, 3584 CH Utrecht, The Netherlands. *P.R. Cullis and M.J. Hope* • Department of Biochemistry, University of British Columbia, Vancouver, British Columbia, V6T 1W5, Canada. *T.F. Taraschi* • Hahnemann Medical College, Department of Pathology, Philadelphia, Pennsylvania 19102.

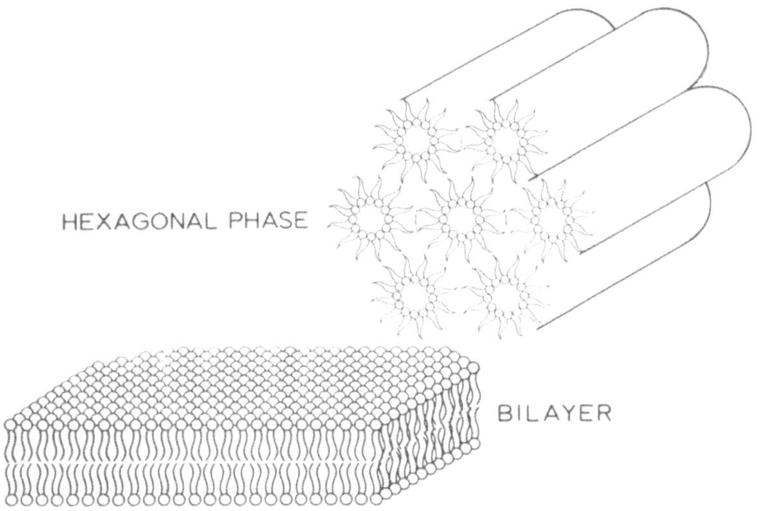

Figure 1. Schematic representation of the bilayer and hexagonal H_{II} phase.

for quite some time (Lucy, 1964; Luzzati *et al.*, 1968; Reiss-Husson, 1967; Shipley, 1973), only in recent years has it become apparent that several major membrane lipids assume the nonlamellar hexagonal, type II (H_{II}) phase (see Figure 1) when dispersed in isolated form in excess buffer under physiological conditions (pH, ionic strength, and temperature). This phase consists of hexagonally organized lipid cylinders in which the polar head groups of the lipid molecules surround a narrow aqueous channel. It is obvious that the presence of such lipids in membranes poses important questions concerning their structural and functional roles.

In this chapter, we present an inclusive review of the various aspects of lipid polymorphism. It will be shown that the ability of lipids to adopt different structures and the rationalization of this behavior via the "molecular shape" concept offers attractive possibilities for understanding the reasons for lipid diversity and provides new perspectives on structure–function relationships in biomembranes.

II. MEMBRANE LIPID POLYMORPHISM: TECHNICAL ASPECTS

Small angle X-ray diffraction is the classical technique to elucidate the macroscopic organization of (phospho) lipid aggregates (for reviews see Luzzati, 1968; Shipley, 1973). In lipid systems that possess some form of long-range order, analysis of the relationship between first and higher order reflections can provide an unambiguous means for lipid phase determination.

A number of authors (Luzzati *et al.*, 1968; Shipley, 1973; Reiss-Husson, 1967; Luzzati, 1968; Rand *et al.*, 1971; Janiak *et al.*, 1976; Harlos and Eibl, 1980) have shown that, in agreement with theory, in multibilayer systems, the spacings of the first and higher order reflections relate as 1 : 1/2 : 1/3. Depending on the sample particulars, sometimes fourth, fifth, or even higher order reflections can be seen. The intensity of the various reflections is highly variable and depends also on the nature of the sample. In fact, in some multilamellar systems, only the first order reflection is observed (Harlos and Eibl, 1980; Marsh and Seddon, 1982), which makes a conclusive assignment of the phase rather difficult. The largest first-order spacing corresponds with the lamellar repeat distance, which is comprised of the bilayer thickness and the thickness of the water layer in between the bilayers.

Alternatively, sharp reflections in the ratio $1 : 1/\sqrt{3} : 1/2 : 1\sqrt{7}$ are also commonly encountered in lipid preparations (Luzzati *et al.*, 1968; Shipley, 1973; Reiss-Husson, 1967; Luzzati, 1968; Rand *et al.*, 1971; Janiak *et al.*, 1976; Harlos and Eibl, 1980; Marsh and Seddon, 1982) and found to be typical for the two-dimensional periodicity of hexagonally packed cylinders. Experimentally, the $1/\sqrt{7}$ and higher order reflections are seldom found. For naturally occurring lipids, the hexagonal H_{II} phase is found most often, rather than the hexagonal, type I (H_I) phase, where the polar head groups are at the periphery of the cylinders. The largest spacing of the hexagonal H_{II} phase (d) relates to the diameter of the lipidic cylinders (a) as $d = (\sqrt{3}/2).a$.

When a three-dimensional periodicity is present in a lipid preparation, often many reflections can be found (Luzzati *et al.*, 1968; Shipley, 1973; Rivas and Luzzati, 1969; Larsson *et al.*, 1980; Lindblom *et al.*, 1979). Provided that enough reflections are observed, an attempt can be made to characterize the lattice type and space group to which the structure belongs. Frequently, a cubic symmetry is found for lipid structures. As a possible model for some cubic lipid phases, an interesting structure based upon deformed bilayer units that are attached to create two different continuous aqueous networks has been proposed (Larsson *et al.*, 1980; Lindblom *et al.*, 1979).

When the long-range order is distorted or not present at all, the sharp Bragg reflections are replaced by continuous scattering profiles that show far less detail and are much more difficult to interpret in an unambiguous way. This particularly complicates the analysis of the phase state in samples that contain different phases in which only one of the phases is ordered. In this case, X-ray diffraction can totally overlook the presence of the disordered phase.

The application of X-ray techniques to biological membranes is hampered by the fact that the exposure times are often long compared to the biological stability of the preparations. Furthermore, only some specialized membrane systems such as nerve myelin, retinal rod outer segments, and chloroplasts are composed of ordered arrays of membranes, facilitating interpretation of the data. In general, biological membranes lack this order and only the scattering profile can be used to evaluate the properties of the membrane (for review see Blaurock, 1982).

Among the electron microscopic techniques to study lipid polymorphism, freeze-fracturing has been shown to be the most valuable method. This can be attributed to (1) the reliable fixation procedure by fast freezing, (2) the fact that beam damage in the microscope is excluded by the replication method, (3) the understanding of the fracture mechanism, (4) the resolution power of about 30 Å (Zingsheim, 1972; Verkleij and Ververgaert, 1978), and (5) like any electron microscopic method, it provides direct visualization of the macromolecular structure in contrast to the spectroscopic and diffraction methods, which give averaged information originating from many sites simultaneously. This feature is also evident in the study of lipid polymorphism in that it has revealed unique structural details such as lipidic particles, transitional intermediates, and overall morphology and heterogeneity of the various systems.

Pure lipid phases characterized by X-ray diffraction have been visualized by freeze-fracturing (Deamer *et al.*, 1970). Liposomes consisting of concentric bilayers show alternating smooth fracture faces (Figure 2). The hexagonal H_{II} phases show distinct fracture faces composed of long parallel lines (giving a ribbed appearance) which occur along at least two fracture planes at angles of approximately 120° to each other (see Figure 2). The repeat distances of the bilayer and the different diameters of hexagonal H_{II} phase agreed well with X-ray repeat distances, which clearly demonstrates that the structure is well preserved during the freezing method. However, it is clear from other studies that because of temperature-induced disorder–order or hexagonal H_{II}-lamellar phase transitions, the high-temperature phase may not be preserved with the current quenching procedures. However, with the development of ultrarapid freezing devices, one is able to prevent alterations in the lipid organization although phase transitions occurring at temperatures higher than 20–30°C may not be prevented. A second argument to use fast-freezing devices is that one does not need to use cryoprotectants which could affect lipid structures (Boni *et al.*, 1981). Therefore, it is highly recommended to use ultrarapid quenching methods, e.g., spray-freezing

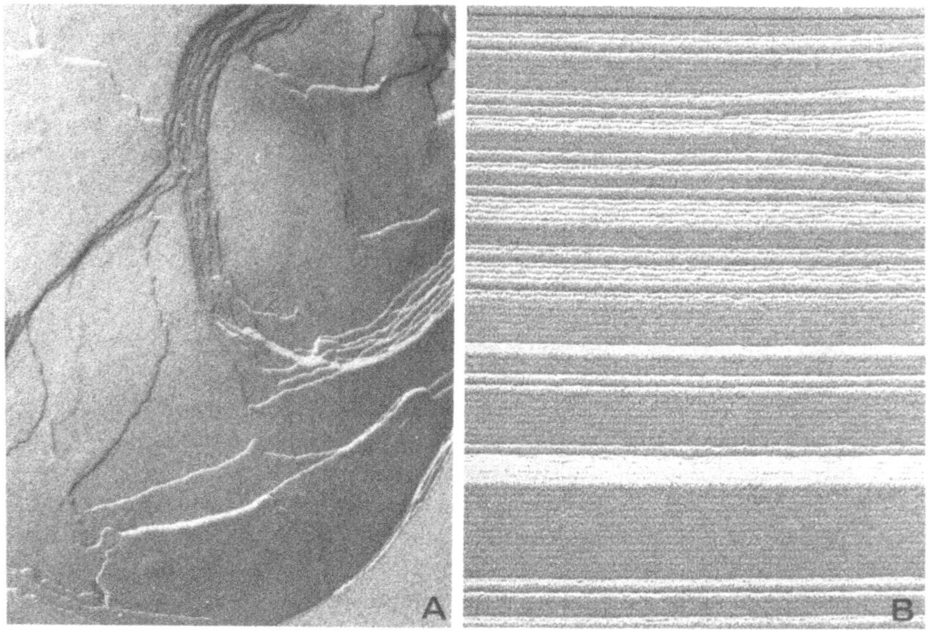

Figure 2. Freeze-fracture electron microscopy of (A) a multilamellar liposome (egg-PC) and (B) a hexagonal H_{II} phase (18 : 1_c/18 : 1_c-PE). Final magnification ×100,000.

(Bachmann and Schmitt, 1971), the copper black method (Heuser *et al.*, 1979), or the jet-freezing method (Moor *et al.*, 1976) to study the various lipid phases.

The introduction of nuclear magnetic resonance (NMR) techniques, especially ^{31}P NMR, has been greatly advantageous in studying lipid polymorphism. They allow for a convenient and quantitative discrimination between the most important phases found for hydrated membrane lipids. Samples can be heterogeneous, long-range order is not required, and these techniques can be equally well applied to model and biological membranes. Since NMR signals contain contributions both from structural and motional origin, the use of NMR to obtain structural information is extrapolative. Therefore, we will first relate the different types of ^{31}P NMR lineshapes to their corresponding phospholipid structures and give some insight in the theories describing the various lineshapes (for a detailed review see Seelig, 1978).

The lineshape of the ^{31}P NMR signal of phospholipids is determined both by the chemical shielding (or shift) anisotropy of the lipid phosphate moiety and ^{1}H–^{31}P dipolar interactions. The dipolar interactions complicate interpretation of the various lineshapes and therefore are most often reduced by applying strong proton-decoupling irradiation.

Dry lipid powders represent a completely immobilized system and are used as a reference state. Figure 3 shows a dry powder proton-decoupled ^{31}P NMR spectrum of 16 : 0/16 : 0-PC · H_2O. Since the electron density around the phosphorus nucleus is not isotropic but depends on the bonding pattern, different orientations of the phosphate segment in a magnetic field will result in different shielding and hence give rise to resonances at different frequencies. This phenomenon is known as chemical shift anisotropy (CSA). In a dry lipid powder, all possible orientations of the phosphate segment occur. Therefore, the spectrum in Figure 3 is a superposition of resonance fréquencies corresponding with all those orientations and weighed by the distribution function of all microcrystalline regions. The resonance frequencies of the edges and the peak of the dry powder spectrum correspond with the principal elements of a static axially asymmetric tensor, which describes the CSA. Inspection of these principal tensor elements of various phospholipids from natural and synthetic origin, including PCs (Griffin, 1976; Kohler and Klein, 1977; Herzfeld *et al.*, 1978; Van Echteld *et al.*, 1981b), phosphatidylethanolamines (PEs) (Kohler and Klein, 1977; Herzfeld *et al.*, 1978), phosphatidylserines (PSs) (Kohler and Klein, 1977; Hope and Cullis, 1980), phosphatidylglycerols (PGs) (Farren and Cullis, 1980), phosphatidylinositol (PI) (Nayar *et al.*, 1982), sphingomyelin (Cullis and Hope, 1980), and lyso-PC (Van Echteld *et*

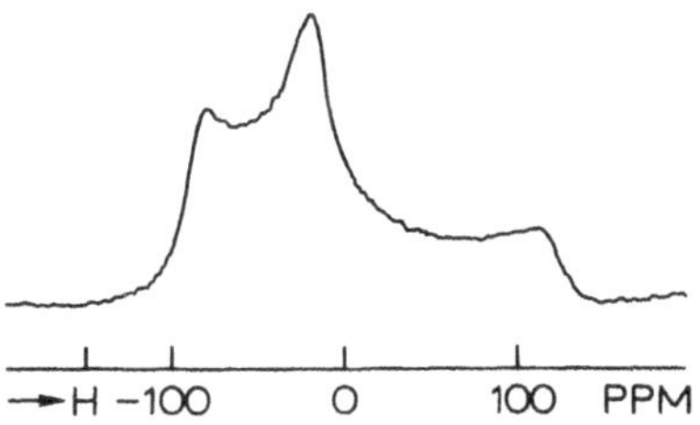

Figure 3. Dry powder proton-decoupled 81.0 MHz ^{31}P NMR spectrum of 16 : 0/16 : 0-PC · H_2O at 25°C. For experimental details see Van Echteld *et al.* (1981a,b). In this and other spectra of 0 ppm position corresponds to the chemical shift of the ^{31}P NMR resonance of sonicated egg-PC vesicles.

al., 1981b), reveals a great similarity with the exception of the phosphomonoester phosphatidic acid (PA) (Kohler and Klein, 1977). This similarity indicates very similar local conformation in the phosphate region of all these lipids. Also, other immobilized systems such as the calcium salts of certain PGs (Farren and Cullis, 1980) and PSs (Hope and Cullis, 1980) next to hydrated PCs and PEs at very low temperatures (Herzfeld *et al.*, 1978) give rise to powder-type ^{31}P NMR spectra with principal values of the CSA tensor similar to those observed for the dry phospholipid powders.

The hydration of phospholipids leads to the onset of motion, which partially averages the CSA. Hydrated phospholipids organized in extended bilayers give rise to a ^{31}P NMR spectrum which is characteristic for a shielding tensor that is axially symmetric around a director axis. Figure 4 shows an example of such a spectrum with a high field peak and a low field shoulder. An insight in the nature of this lineshape can be obtained by orienting the phospholipid bilayers between glass plates such that the long axes of the molecules are perpendicular to the glass plates. Stepwise rotation of these oriented bilayers in the magnetic field gives rise to individual narrow resonances with an angular-dependent chemical shift (McLaughlin *et al.*, 1975, 1981; Seelig and Gally, 1976; Hemminga and Cullis, 1982). The high field peak of the unoriented sample has been found to correspond with phospholipid molecules, which have their long axes perpendicular to the magnetic field ($\sigma_\perp$), whereas the shoulder resonance position corresponds with a parallel orientation ($\sigma_\parallel$). In general, the resonance frequency $\Delta\nu$ of phospholipid molecules having director axes at an angle α with the magnetic field is given by (Seelig, 1978; Cullis and De Kruijff, 1978a)

$$\Delta\upsilon(\alpha) = -\,2/3\;\Delta\sigma\left(\frac{3\cos^2\alpha - 1}{2}\right)\;\text{ppm} \qquad (1)$$

where the residual chemical shift anisotropy is:

$$-\Delta\sigma = \Delta\upsilon(0°) - \Delta\upsilon(90°) = \sigma_\parallel - \sigma_\perp$$

When these resonance frequencies $\Delta\upsilon(\alpha)$ are plotted vs. $(3\cos^2\alpha - 1)/2$ and taking α, the angle between the bilayer normal and the magnetic field, a straight line results for several phospholipids studied (McLaughlin *et al.*, 1981; Hemminga and Cullis, 1982; Cullis and De Kruijff, 1978a). This proves that the director axis coincides with

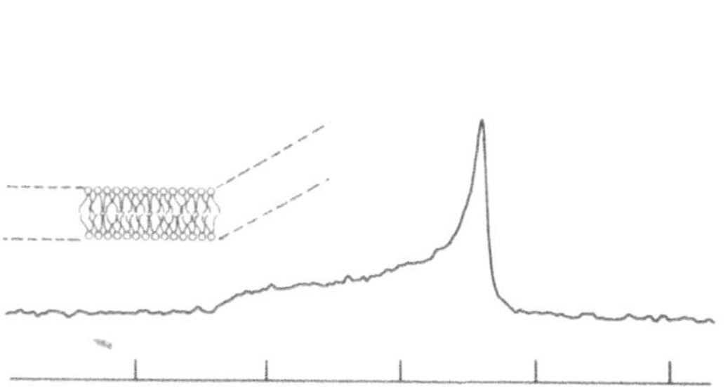

Figure 4. Proton-decoupled 81.0 MHz ^{31}P NMR spectrum of an aqueous dispersion of dierucoylphosphatidylcholine at 25°C. For experimental details see Van Echteld *et al.* (1982).

the bilayer normal and that the phosphate segment (and likely the whole molecule) rapidly rotates ($\tau_c \leq 10^{-6}$ sec) around the bilayer normal.

The lineshape of the spectrum in Figure 4 can be understood when realizing that, in liposomal lipid dispersions, molecules are found in every direction and therefore the spectrum will be a superposition of individual resonance lines from Eq. (1), weighed by the probability of finding lipid molecules with their long axes at angles α. This can be visualized by regarding the low number of lipid molecules in a large spherical bilayer with their long axes parallel to a given magnetic field as opposed to the high amount of lipid molecules with their long axes perpendicular to the field direction.

Up until now, for all naturally occurring membrane phospholipids organized in liquid-crystalline lamellar phases, ^{31}P NMR lineshapes have been observed such as shown in Figure 4. The $\Delta\sigma$-values of the ^{31}P NMR spectra of various phospholipids are all of the same magnitude and range from approximately 55 to 30 ppm (Seelig, 1978; Cullis and De Kruijff, 1978a). Comparable values of $\Delta\sigma$ are found for the ^{31}P NMR spectra of some biological membranes (McLaughlin *et al.*, 1981; Cullis, 1976a; De Kruijff *et al.*, 1976a; Seelig *et al.*, 1981). These data together with the observations that the ^{31}P NMR lineshape and $\Delta\sigma$ of bilayer organized phospholipids are hardly affected by the presence of membrane proteins, cations, cholesterol, fatty acid composition, and temperature demonstrate that the structure of the phosphate region is highly conserved (Seelig and Seelig, 1980). The structural stability of that part of the lipid molecule is also born out by ^{2}H NMR (Seelig and Seelig, 1980), neutron (Büldt *et al.*, 1978), and X-ray (Hauser *et al.*, 1981) diffraction studies. Suggestions (Thayer and Kohler, 1981) that changes in local structure can give rise to spectral changes resembling those noticed during polymorphic phase transitions (which will be discussed later) are theoretically valid, but in the absence of any evidence for the occurrence of such changes, rather speculative.

However, factors that do influence the ^{31}P NMR lineshape include the inter- and intramolecular ^{1}H–^{31}P dipolar interactions. Experimentally, these dipolar contributions are suppressed by applying proton decoupling. Unfortunately, this decoupling is imperfect in situations where dipolar couplings are relatively strong as for gel-state lipids. This results in substantial signal intensity at the edges of the spectrum beyond the resonance positions $\sigma_{\parallel}$ and $\sigma_{\perp}$ and a less well-defined lineshape.

Another important parameter which may determine the ^{31}P NMR lineshape of lamellar organized phospholipids is the reorientation rate of the molecules. For isotropic rotational motion of a spherical vesicle, this reorientation rate can be described by a correlation time τ_c that is related to the vesicle size by (Cullis, 1976b; Burnell *et al.*, 1980a)

$$\frac{1}{\tau_c} = \frac{6}{r^2}(D_t + D_{\text{diff}}) \tag{2}$$

where $D_t = kT/8\pi r\eta$ (η = medium viscosity) is the vesicle-tumbling-dependent part describing Brownian rotational diffusion and D_{diff} is the rate of lateral diffusion of the lipid molecules in the bilayer. When the reorientation rate of the phospholipid molecules is sufficiently fast, i.e., $\tau_c < 1/\Delta\sigma$, motional narrowing of the ^{31}P NMR spectrum will occur, until a final situation is reached where isotropic motional averaging will lead to a narrow symmetrical "high resolution" resonance line. From Eq. (2) it can be seen that τ_c is dependent on the vesicle radius. To gain insight in the influence of the vesicle size on the ^{31}P NMR lineshape, spectra were simulated for different radii, which are shown in Figure 5. It can be seen that the asymmetrical lineshape with a high field peak and a low field shoulder only appears for vesicles with radii greater

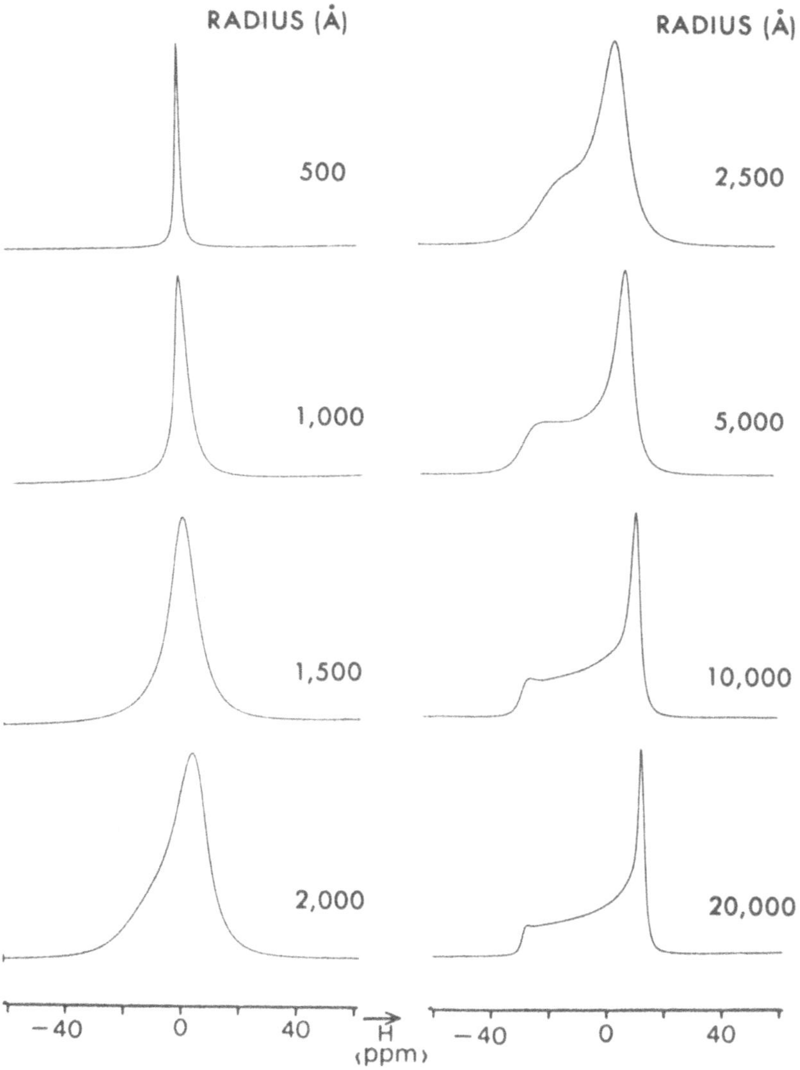

Figure 5. Simulated ^{31}P NMR spectra of 18 : 1$_c$/18 : 1$_c$-PC vesicles of different sizes at 30°C. The spectra were simulated using $\Delta\sigma$ = 3550 Hz, η = 0.008 P, orientationally independent Lorentzian full line width at half height = 60 Hz, D_{diff} = 6.2 · 10^{-7} cm^2/sec and r as indicated. Reproduced with permission from Burnell *et al.* (1982a).

than 2000 Å (assuming lateral diffusion rates in the order of 5×10^{-7} cm^2/sec) (Burnell *et al.*, 1980a).

The narrow symmetrical lineshape seen in Figure 6, resulting from isotropic motional narrowing, is not restricted to lipid bilayer vesicles with small radii, but such ^{31}P NMR spectra will result in all those situations where the phospholipid molecules can reorient themselves fast ($\tau_c < 10^{-5}$ sec) such as in micelles, inverted micelles, and cubic phases. Examples of these "isotropic" structures are presented in Figure 6.

For phospholipids organized in the hexagonal H_{II} phase, lateral diffusion of the lipid molecules around the cylinders which are the constituents of this phase will result in an additional averaging of the CSA. This can be understood in the following way: when a hexagonal phase cylinder is oriented with the cylinder axis parallel to the magnetic field, all the long axes of the phospholipid molecules are oriented perpendicular to the magnetic field and hence a resonance will be observed at $\sigma_{\perp}$ ppm. However, when the cylinder axis is perpendicular to the magnetic field, a completely different situation arises, since now a cylindrical distribution of phospholipid molecules is seen by the spectrometer. In general, it can be shown (Seelig, 1978; Cullis and De Kruijff, 1978a) that the resonance frequency $\Delta\nu(\beta)$ of phospholipid molecules in a particular cylinder having its axis at an angle β with the magnetic field direction is given by

$$\Delta\nu(\beta) = 1/3\ \Delta\sigma \left(\frac{3\cos^2\beta - 1}{2}\right) \text{ ppm}$$

The difference between the two extreme orientations is then

$$\Delta\nu(\beta = 0^\circ) - \Delta\nu(\beta = 90^\circ) = 1/2\ \Delta\sigma \qquad (3)$$

For a random distribution of hexagonal cylinders, the probability of finding cylinder axes perpendicular to the magnetic field is higher than the probability of finding cylinder

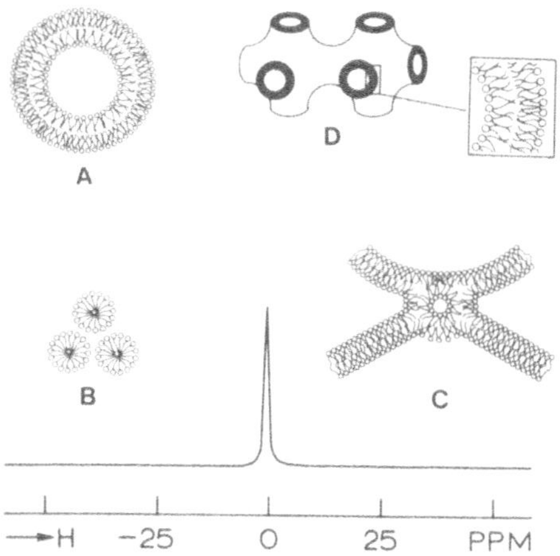

Figure 6. Various phospholipid structures that can give rise to a narrow symmetrical ^{31}P NMR lineshape at 0 ppm. (A) Sonicated vesicle, (B) Micelles, (C) inverted micelle, and (D) example of a cubic phase (Larsson *et al.*, 1980).

axes parallel to the magnetic field. Therefore, it can be shown in an analogous way as for the bilayer situation that for the hexagonal phase in the superposition of the individual resonance lines at all angles β a peak will arise at $\Delta\nu(= 90°)$ and a shoulder at $\Delta\nu(\beta = 0°)$, resulting in a reversed asymmetry, compared to the bilayer spectra, and a reduction in the separation of the spectral edges by a factor of two as shown in Figure 7. Experimentally, this separation between peak and shoulder has indeed been found to be approximately equal to 1/2 $\Delta\sigma$ (Cullis and De Kruijff, 1978a; Cullis *et al.*, 1980a). These observations indicate that there are few (if any) conformational and motional differences between polar head groups in bilayer and hexagonal H_{II} phases at least as far as the phosphate segment is concerned.

^2H NMR has been widely used in membrane research (see for reviews Seelig and Seelig, 1980; Seelig, 1977). Chemical or biochemical techniques allow for a selective replacement of ^1H by ^2H in lipid molecules, which is not expected to affect the molecular organization in the membrane. Since virtually all parts of a lipid molecule can be labeled, ^2H NMR has provided detailed insight on the conformational and motional properties of polar region, glycerol backbone, and acyl chains of membrane lipid molecules.

Briefly, the deuterium quadrupole interactions in a sample of hydrated specifically ^2H-labeled lipids, organized in large bilayer membranes, give rise to a characteristic ^2H NMR spectrum, which is dominated by two distinct peaks as shown in Figure 8. The separation of the two peaks is the ^2H quadrupole splitting $\Delta\nu_Q$ which can be used to calculate the ^2H order parameter S_{CD} according to

$$\Delta\nu_Q = (3/4)\left(\frac{e^2qQ}{h}\right)S_{CD} \tag{4}$$

where e_q^2Q/h is the static quadrupole coupling constant which has been found to be 170 kHz for aliphatic C-D bonds (Burnett and Muller, 1971) and 175 kHz for olefinic C-D bonds (Achlama and Zur, 1979). S_{CD} is defined as $(3\ \overline{\cos^2\theta} - 1/2)$, where θ is the instantaneous angle between the C-D bond and the bilayer normal and the bar denotes the time average.

Analogous to what is found with ^{31}P NMR, the reorientation rate of the lipid molecules influences the ^2H NMR spectrum and $\Delta\nu_Q$ is dependent on the size of the lipid structures. A final situation is reached when isotropic motional averaging results in a collapse of the quadrupole splitting and the spectrum only shows a single resonance.

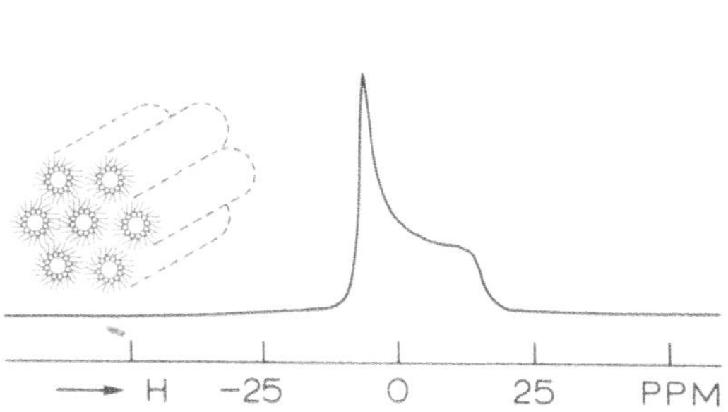

Figure 7. Proton decoupled 81.0 MHz ^{31}P NMR spectrum of an aqueous dispersion of 18 : 1_c/18 : 1_c-PE at 25°C. For experimental details see Van Echteld *et al.* (1981a,b).

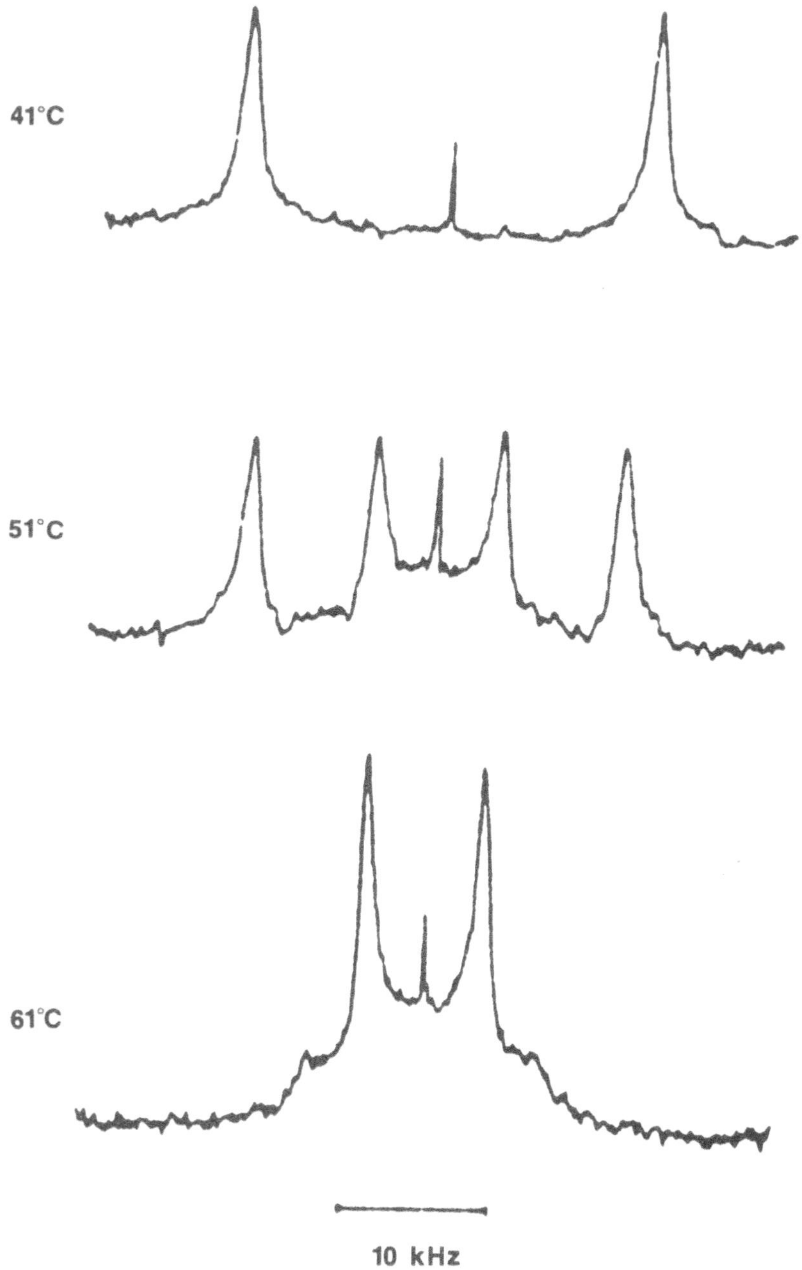

Figure 8. ^2H NMR of [9,10-^2H]-18 : 1_t/18 : 1_t-PE in 0.01 M Na_3PO_4, pH 7.0, 0.1 M NaCl, 0.001 M EDTA, 41°C bilayer, 51°C coexistence of bilayer and H_{II} phase, and 61°C H_{II} phase. Reproduced with permission from Gally *et al.* (1980).

Such isotropic motion for part of the lipid molecules has been observed for mixtures of acyl chain labeled 18 : 1_c/18 : 1_c-PE and 18 : 1_c/18 : 1_c-PC and has been found to closely correlate with results obtained with ^{31}P NMR (Akutsu and Seelig, 1981). However, since minor conformational changes may have large influences on ^{2}H NMR spectra of lipids, the observations of a singlet without the concomitant observation of an "isotropic" peak in the ^{31}P NMR spectrum (Akutsu and Seelig, 1981) indicates a local conformational change, rather than isotropic motion of the whole molecule.

Lipids in a hexagonal H_{II} phase can also be readily detected by ^{2}H NMR, since it can be shown that in such a situation the quadrupolar splitting will be described by

$$\Delta\nu_Q = (3/8)(e^2qQ/h) \cdot S_{CD} \tag{5}$$

so that in a transition from a lamellar to a hexagonal phase the splitting will be reduced by a factor of two, provided that there is no change in S_{CD}. This is illustrated in Figure 8 for ^{2}H labeled 18 : 1_t/18 : 1_t-PE.

The use of ^{2}H NMR in studying lipid polymorphism has been relatively limited. However, it may be very useful to surmount some of the shortcomings of ^{31}P NMR in this respect. For instance, not all membrane lipids are phospholipids and thus are not detected by ^{31}P NMR, whereas those lipids are in principle accessible by ^{2}H NMR. Also, specific labeling allows for the determination of individual phase behavior in (phospho)lipid mixtures (Tilcock *et al.*, 1982). Finally, in those situations where ^{31}P NMR data are not straightforward to interpret, such as for PA (Cullis and De Kruijff, 1976), ^{2}H NMR may give more unambiguous results.

III. PHASE PREFERENCES OF MEMBRANE LIPIDS

In this section, we will classify the major membrane lipids according to their structural preferences. Before doing so, it is important to realize two aspects of lipid polymorphism. First, early studies indicated that virtually every lipid can adopt a large variety of phases dependent upon water content and temperature (Luzzati, 1968). As nonlamellar structures were often observed at low (nonphysiological) water contents and high temperatures, the physiological relevance of lipid polymorphism was not obvious. Here, we only consider those structures adopted by isolated membrane lipids in excess (>50% by weight) water, at "physiological" (0–40°C) temperatures. Second, as will be discussed in detail in the following sections, the phase preferences of membrane lipids depend on a variety of factors such as fatty acid composition, pH, and the presence of divalent cations. Therefore, any given classification is somewhat arbitrary as a particular lipid can adopt different structures under different conditions.

It is remarkable that the large majority of the membrane lipids will adopt only two types of structure in isolation, the lamellar or the hexagonal H_{II} phase. Micellar phospholipid organizations have only been observed for minority membrane lipids such as gangliosides and lysophospholipids.

Table 1 summarizes the phase preferences of various membrane lipid classes. It

Table 1. Phase Preferences of Membrane Lipid Classes

	Lamellar	H_{II}	Micellar
Zwitterionic phospholipids			
Phosphatidylcholine	+	−	−
Sphingomyelin	+	−	−
Phosphatidylethanolamine	+	+	−
Negatively charged phospholipids			
Phosphatidylserine	+	+	−
Phosphatidylglycerol	+	+	−
Phosphatidylinositol	+	−	−
Phosphatidic acid	+	+	−
Cardiolipin	+	+	−
Glycolipids			
Monoglucosyldiglyceride	−	+	−
Monogalactosyldiglyceride	−	+	−
Diglucosyldiglyceride	+	−	−
Digalactosyldiglyceride	+	−	−
Cerebroside	+	−	−
Cerebroside sulfate	+	−	−
Gangliosides	−	−	+
Lysophospholipids	−	−	+

should be noted that this table is incomplete in that information is given only for commonly occurring major membrane lipid species. When a lipid can adopt more than one structure depending on the experimental conditions, it is classified as such.

The quantitatively most important type of H_{II} lipids are PE (Cullis and De Kruijff, 1979) and monoglucosyl- (Wieslander *et al.*, 1981b) and monogalactosyldiglycerides (Shipley, 1973; Sen *et al.*, 1982). It is intriguing that in bacterial membranes there appears to be an interrelationship between the amount of PE and the proportions of these glycolipids (Minnikin *et al.*, 1971), despite large differences in their chemical structures, suggesting a regulated requirement for H_{II}-favoring lipids in these membranes.

As will be shown in subsequent sections, the observation that the bilayer-preferring diglucosyl- and digalactosyldiglycerides (Shipley, 1973; Sen *et al.*, 1982; Wieslander *et al.*, 1981a) are metabolically directly coupled to the monoglucosyl and galactosyl derivatives offers fascinating regulatory possibilities for the net structure preferences of certain membrane systems.

Both lamellar and hexagonal H_{II} phases have been observed in the total lipid extracts of biological membranes. The total lipid extract of the human erythrocyte has a ^{31}P NMR spectrum which is typical of extended bilayers (Cullis and Grathwohl, 1977), despite the presence of PE (20% of phospholipid), which prefers the H_{II} phase in isolation (Cullis and De Kruijff, 1978b). In contrast, X-ray (Huynk, 1973), freeze-fracture, and ^{31}P NMR (De Grip *et al.*, 1979) studies demonstrate the presence of H_{II} phase structure for an aqueous dispersion of the total rod outer segment lipids at 37°C.

In addition, ^{31}P NMR studies often reveal the presence of an "isotropic" component such as observed for total *E. coli* lipids (Burnell *et al.*, 1980b). As will be shown in later sections, in these cases other inverted lipid structures are also present.

IV. THE HEXAGONAL H_{II} PHASE

Of the two major structures adopted by hydrated membrane lipids, the lamellar phase is the most familiar and its properties have been reviewed in great detail and will not be explicitly dealt with here. However, since many of the ideas developed over the last years concerning the structural and functional aspects of nonbilayer structures rely on the fact that many membrane lipids prefer the hexagonal H_{II} phase, a closer look at the properties of this phase is warranted.

A distinction can often be made visually between lipids in lamellar and hexagonal H_{II} organizations. Whereas bilayer-forming lipids in excess buffer usually form homogeneous milky dispersions readily, hexagonal H_{II} type lipids often do not disperse at all or form large aggregates. This phase separation is likely related to the low hydration capacity of these lipids (Hauser *et al.*, 1981). With regard to the nature of the interface between the bulk water and the H_{II} phase-two questions come into mind. First, are the aqueous channels present in the hexagonal H_{II} phase in open contact with the surrounding aqueous phase? Second, what is the lipid organization at the interface between the hydrophobic cylinder and the water?

For unsaturated PEs in the H_{II} organization, it has been found that the addition of Mn^{2+} broadened the entire ^{31}P NMR signal of the phospholipids beyond detection (C. J. A. Van Echteld, unpublished observations) demonstrating that Mn^{2+} can interact with all lipid head groups. This suggests that the aqueous tubes are open to the bulk aqueous phase. Freeze-fracture electron microscopy on similar systems indicates that the outermost tubes of the hexagonal H_{II} phase in excess water are covered with a monolayer of lipid (Van Venetië and Verkleij, 1981), thereby shielding the hydrophobic acyl chains from the aqueous phase. Another interesting feature revealed by these studies is that in some cases the tubes (often 10^4–10^5 Å long) are straight and in other cases highly curved (Van Venetië and Verkleij, 1981; Verkleij *et al.*, 1980). There is as yet no explanation for this phenomenon. The intertube distance (tube diameter) of hexagonal H_{II} phase lipids have been determined from both freeze-fracture electron microscopy and X-ray. Values obtained from some selected systems are presented in Table 2.

Typically, the tube diameters for PEs range from 70 to 74 Å and appear to be rather independent of the fatty acid composition. For complexes of cardiolipin (CL) and $18:1_c/18:1_c$-PA with various cations, the tube diameter ranges from 52 to 75 Å, depending on the type of cation. A unique feature of the hexagonal H_{II} phase is that the phospholipid molecules reside in very curved monolayers. For instance, in the most highly curved lamellar system (sonicated vesicles) the inner diameter is still approximately 170 Å. This property, together with the notion that the molecules in the hexagonal H_{II} phase have an inverted orientation, forms the basis for the potential ionophoric properties of these types of phospholipids (see Section VIII).

Table 2. Repeat Distances of Selected Hexagonal H_{II} Phase Lipid Systems

	Repeat distance (nm)	
	Freeze-fracture	X-ray[h]
18 : 1_c/18 : 1_c-PE (20°C)	7.4[a]	
18 : 1_t/18 : 1_t-PE (60°C)		7.0[b]
18 : 3_c/18 : 3_c-PE (20°C)	7.7[c]	
20 : 4/20 : 4-PE (20°C)	4.4[c]	
22 : 6/22 : 6-PE (20°C)	4.3[c]	
Tetrahymena PE (10°C)		7.4[d]
Soya PE (5°C)		7.3[e]
CL, Ca^{2+}-salt (20°C)	5.2[a]	5.3[f]
CL, Mg^{2+}-salt (20°C)	6.5[a]	
CL, Mn^{2+}-salt (20°C)	7.5[a]	
18 : 1_c/18 : 1_c-PA, Ca^{2+}-salt (20°C)	5.2[g]	
18 : 1_c/18 : 1_c-PA, Mg^{2+}-salt (20°C)	5.7[g]	
18 : 1_c/18 : 1_c-PA, Mn^{2+}-salt (20°C)	7.4[g]	

[a] Van Venetië and Verkleij (1981).
[b] De Kruijff, (unpublished observations).
[c] Dekker *et al.* (1983).
[d] Ferguson *et al.* (1982).
[e] Hui *et al.* (1981).
[f] Rand and Sengupta (1972).
[g] Verkleij *et al.* (1982).
[h] To obtain the tube diameter multiply by $\sqrt{3}/2$.

The acyl chains in hexagonal H_{II} phases adopted by hydrated membrane lipids are in a liquid-crystalline state. From ^{31}P NMR (Seelig, 1978; Cullis and De Kruijff, 1979) and ^{2}H NMR (Seelig, 1977; Burnell *et al.*, 1980b) studies, it can be inferred that the lipid molecules undergo lateral diffusion around the aqueous cylinders at rates comparable to that in the liquid-crystalline lamellar phase. These studies further indicate that the molecular order in the polar head group region is very similar for lipids organized in the hexagonal H_{II} and the liquid-crystalline lamellar phase (Seelig, 1977; Gally *et al.*, 1980; Taylor and Smith, 1981). In contrast, ^{2}H NMR (Tilcock *et al.*, 1982; Gally *et al.*, 1980), electron spin resonance (ESR) (Hardman, 1982), and Fourier transform infrared spectroscopy (FT-IR) (Mantsch *et al.*, 1981) studies indicate that the acyl chains are progressively more disordered towards the terminal methyls, consistent with the strong curvature of the lipid monolayers surrounding the aqueous channels.

V. MODULATION OF MEMBRANE LIPID POLYMORPHISM

An important aspect of lipid polymorphism is that the macroscopic structure adopted by lipids depends very much on the experimental conditions. As regulation of lipid structure is of crucial importance for potential functional roles of nonlamellar lipid structures in membranes, we will review in this section the various ways membrane lipid polymorphism can be modulated.

A. One-Lipid Systems

1. Temperature and Fatty Acid Composition

Temperature is an important experimental parameter which determines the macroscopic structure of hydrated membrane lipids. This is illustrated in Figure 9, which shows the ^{31}P NMR spectra of $18:1_t/18:1_t$-PE dispersed in excess aqueous buffer at pH 7.0. At 30°C, this phospholipid is organized in a gel state lamellar phase as is shown by the characteristic lineshape and the large line width, which results from incomplete removal of the strong dipolar ^{1}H–^{31}P coupling (Seelig and Gally, 1976). At 40°C, the acyl chains melt and a ^{31}P NMR lineshape typical of liquid-crystalline lamellar phospholipids is observed. Above 50°C, a second spectral component appears which gradually increases in intensity with increasing temperature such that at 70°C, the entire spectrum consists of this spectral component. From the chemical shift position of the dominant high-temperature spectral feature, the reversal of the asymmetry in the spectrum, and the reduced width, it can be concluded that the phospholipid molecules are organized in cylindrical structures around which rapid diffusion of the molecules occurs. In conjunction with freeze-fracture and X-ray data, these results demonstrate that $18:1_t/18:1_t$-PE adopts the hexagonal H_{II} phase above 50°C. When the temperature is increased through the 50–70°C temperature range, these transitions are also manifested in differential scanning calorimetry (DSC) scans of the same

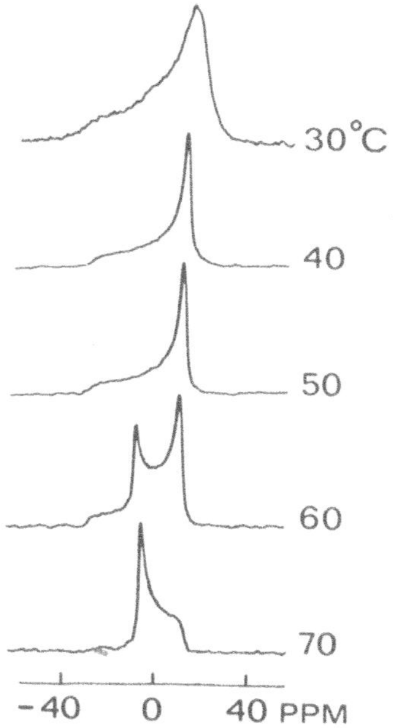

Figure 9. 81.0 MHz ^{31}P NMR spectra of $18:1_t/18:1_t$-PE in 100 mM NaCl, 10 mM Tris/HCl, and pH 7.0. For experimental details see Van Echteld *et al.* (1981a,b).

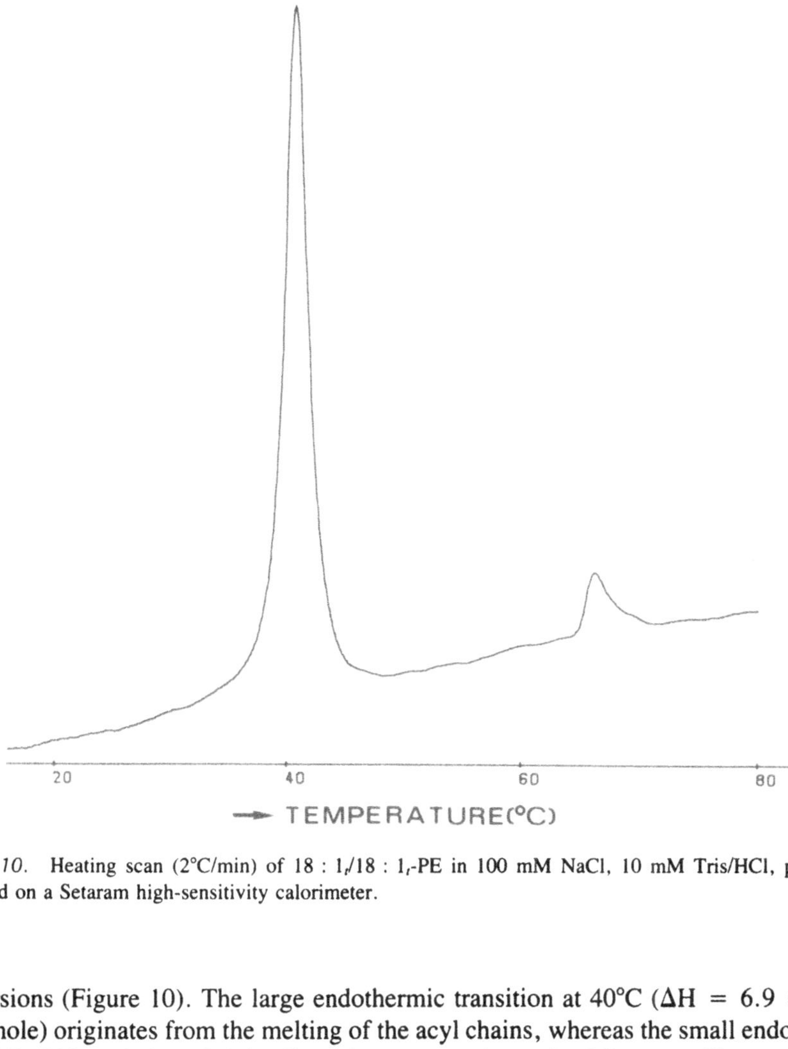

Figure 10. Heating scan (2°C/min) of 18 : 1_t/18 : 1_t-PE in 100 mM NaCl, 10 mM Tris/HCl, pH 7.0, recorded on a Setaram high-sensitivity calorimeter.

dispersions (Figure 10). The large endothermic transition at 40°C ($\Delta H = 6.9 \pm 0.1$ kcal/mole) originates from the melting of the acyl chains, whereas the small endotherm at 66°C ($\Delta H = 0.4 \pm 0.1$ kcal/mole) corresponds to the bilayer → H_{II} transition. The low heat content of the latter transition must be related to the fact that in both phases, the acyl chains are in the liquid-crystalline state. From the functional point of view, it is intriguing that there is such a low energy barrier between these macroscopically very different structures.

Temperature-dependent bilayer → H_{II} transitions have been observed in a large variety of both synthetic and natural PEs (Table 3). The bilayer → H_{II} transition depends strongly on the fatty acid composition of the lipid. Increasing unsaturation results in decreased bilayer → H_{II} transition temperatures. However, it should be realized that there is no direct correlation between membrane "fluidity" and H_{II} formation. For instance, whereas for 18 : 1_t/18 : 1_t- and 18 : 1_c/18 : 1_c-PE there is approximately a 15–20°C interval between the gel → liquid-crystalline and the bi-

Table 3. *Bilayer–Hexagonal* H_{II} *Transition Temperatures of PEs*

Product	Temperature (0°C)	Remarks
Synthetic		
Saturated (diester)		
18 : 0/18 : 0-PE	105[a]	1 M NaCl
Saturated (diether)		
1,2-Dihexadecyl-*sn*-glycero-3-phosphoethanolamine	87[a], 88[b]	1 M NaCl
1,2-Ditetradecyl-*rac*-glycero-3-phosphoethanolamine	93.5[a]	1 M NaCl
Saturated (ether-ester)		
1-Hexadecyl-2-palmitoyl-*sn*-glycero-3-phosphoethanolamine	102[b]	
Unsaturated		
$16:1_c/16:1_c$-PE	~0[c]	
$18:1_c/18:1_c$-PE	10[d]	
$18:1_t/18:1_t$-PE	55[d], 60–63[e], 65[f]	
$18:2_{cc}/18:2_{cc}$-PE	0–25[g], < −15[e]	
$18:3_{ccc}/18:3_{ccc}$-PE	0–30[g]	
20 : 4/20 : 4-PE	< −30[g]	
22 : 6/22 : 6-PE	< −30[g]	
Mixed species		
$16:0/18:1_c$-PE	75[g]	
Natural species		
Egg	25–30[d], 32–45[b], 28[h], 28[i]	
Egg (from egg-PC via phospholipase D)	50[i], 63[b]	
E. coli	55–60[d]	Wild-type
Endoplasmic reticulum	7[j]	Rat liver
Sarcoplasmic reticulum	−10[k]	Rabbit
Inner mitochondrial membrane	10[l]	Muscle
Erythrocyte membrane	8[d]	Rat liver
Soya	−10[m]	Human
Soya (from soya-PC, via phospholipase D)	0 to −20[n]	
Bovine white matter ethanol phospholipid	18[b]	Containing plasmalogens

[a] Harlos and Eibl (1981).	[h] Hardman (1982).
[b] Boggs *et al.* (1981).	[i] Mantsch *et al.* (1981).
[c] Van Dyck *et al.* (1976).	[j] De Kruijff *et al.* (1980b).
[d] Cullis and De Kruijff (1978b).	[k] Cullis *et al.* (1982).
[e] Tilcock and Cullis (1982).	[l] Cullis *et al.* (1980b).
[f] Ghosh and Seelig (1982).	[m] Cullis and De Kruijff (1978a).
[g] Dekker *et al.* (1983).	[n] Cullis and Hope (1980).

layer → H_{II} transitions, no H_{II} phase has been observed up to 70°C above the gel → liquid-crystalline transition of 14 : 0/14 : 0-PE. Furthermore, in a series of disaturated PEs, the gel → liquid-crystalline transition temperature increases with increasing chain length, whereas the bilayer → H_{II} phase transition decreases in temperature with increasing chain length (Boggs *et al.*, 1981). Hydrated PEs isolated from biological membranes are typically organized in the H_{II} phase at physiological temperatures. Very similar temperature- and fatty-acid-composition-dependent bilayer → H_{II} transitions have been observed for monoglucosyldiglyceride isolated from *A. laidlawii* (Wieslander *et al.*, 1981b). Although no information is available on the exact kinetics of the bilayer → H_{II} transition in PEs, freeze-fracture experiments have shown that it is very difficult to preserve the hexagonal H_{II} phase when the bilayer → H_{II} transition occurs above 30°C. In these cases, only lamellar phases are observed. From the freezing rates, it can be estimated that the hexagonal → lamellar transition is completed in the order of milliseconds (Van Venetië *et al.*, 1981). Studies employing ^{31}P NMR demonstrate that there is no detectable hysteresis in this transition, and also indicate (using oriented PE multilayers) that the tubes of the hexagonal H_{II} phase run parallel to the bilayers from which they are formed, suggesting that the transition is a highly cooperative interbilayer fusion process (Cullis *et al.*, 1980a). The bilayer → H_{II} transition of 18 : 1_c/18 : 1_t-PE is accompanied by an increase in repeat distance from 55 to 70 Å (De Kruijff, unpublished observations). Assuming that the length of a lipid molecule is 20 Å in each phase, then the interbilayer aqueous space is 15 Å thick, whereas the aqueous cylinder in the H_{II} phase has a diameter of 30 Å. This corresponds to very similar water contents of 27 and 25% (vol./vol.) for both these phases. Thus, no massive, long-range water movements have to occur during the transition. That the water structure at the lipid–water interface probably is important for the transition is indicated by the decrease in transition temperature induced by glycerol in 18 : 1_t/18 : 1_t-PE dispersions (Van Echteld, unpublished observations). Therefore, the use of this cryoprotectant in freeze-fracture studies on such systems is questionable.

2. *Electrostatic Interactions*

The molecular packing of negatively charged phospholipids in particular is extremely sensitive to the ionic composition of the aqueous phase. It is therefore not surprising that lipid polymorphism is dependent on the magnitude and nature of electrostatic lipid head group interactions.

At neutral pH, when PEs are uncharged, the unsaturated species prefer the H_{II} phase. Upon increasing the pH towards the pK of the amine, thereby increasing the net negative charge and corresponding interheadgroup repulsion, the bilayer phase is preferred (Cullis and De Kruijff, 1978b; Hardman, 1982). Similarly, lipids such as PA and PS are negatively charged at neutral pH and form bilayers. Lowering the pH below the respective pKs reduces the charge density resulting in H_{II} phase formation (Hope and Cullis, 1980; Farren *et al.*, 1983). Increasing NaCl concentrations will decrease the lamellar to hexagonal H_{II} phase transition temperature of PEs (Harlos and Eibl, 1981).

Divalent cation–lipid interactions are of special interest for a number of reasons.

In the case of negatively charged lipids, these interactions are very strong and can result in the formation of divalent cation–lipid salts with unique structural properties. Moreover, many membrane functions appear to be highly dependent on the presence of Ca^{2+} in particular suggesting that such interactions might play important structural and functional roles.

Aqueous dispersions of PS, PG or PI are organized in a lamellar phase at neutral pH. The addition of divalent cations such as Ca^{2+} will result in strong Ca^{2+} binding and, at high concentrations, in lipid precipitation. However, in these precipitates, the phospholipid-Ca^{2+} complexes remain organized in a lamellar arrangement (Hope and Cullis, 1980; Farren and Cullis, 1980; Papahadjopoulos *et al.*, 1975a; Verkleij *et al.*, 1974). In some cases, such as the more saturated species of PS and PG, the Ca^{2+} salt forms anhydrous cochleate or cylindrical structures in which the acyl chains are in the gel state (Hope and Cullis, 1980; Farren and Cullis, 1980; Papahadjopoulos *et al.*, 1975a; Verkleij *et al.*, 1974).

It is important to note that although bilayer structure is maintained in all these cases, such interactions can play important roles in lamellar → nonlamellar transitions in mixed lipid systems.

Divalent cations can also directly influence the macroscopic organization of two other negatively charged lipids, CL and PA, and we will discuss these effects in more detail. CL is in many respects a unique membrane lipid (for review see Ioannou and Golding, 1979). Chemically, it can be defined as a "double" phospholipid containing four fatty acids and at neutral pH, two charged phosphates. In eukaryotic cells, it is found exclusively in the inner mitochondrial membrane where it is the only major (20 mole% of the total lipids) negatively charged phospholipid. In addition, CL is found in several bacterial membranes. The mitochondrial species is highly unsaturated and specifically enriched in linoleic acid, whereas the bacterial CL is much more saturated. Beef heart CL, when dispersed in salt solutions at neutral pH, is organized in a lamellar configuration (Rand and Sengupta, 1972; Cullis *et al.*, 1978a; De Kruijff *et al.*, 1982b). Upon gradually increasing the Ca^{2+} concentration, via dialysis methods such that the liposomes are never exposed to high local Ca^{2+} concentrations, the following events occur. The lamellar phase is maintained up to a free Ca^{2+} concentration of 1 mM, whereas in the 1–3-mM free Ca^{2+} concentration range the lipids precipitate and a transition to the hexagonal H_{II} phase is observed (De Kruijff *et al.*, 1982b; see Figure 11). This structural transition is endothermic (ΔH = 1.8 kcal/mole) and is accompanied by a marked increase in Ca^{2+}-binding from a maximum of 0.35 Ca^{2+}/CL in the bilayer phase to the stoichiometric value of 1.0 Ca^{2+}/CL in the H_{II} phase (De Kruijff *et al.*, 1982b). Addition of Ca^{2+} from a concentrated stock solution, thereby transiently exposing the liposomes to local high Ca^{2+} concentrations, results in the formation of a so-called "isotropic" intermediate structure, which will be discussed in detail in Section VI.

Removal of Ca^{2+} from the hexagonally organized Ca^{2+}–CL (1 : 1) complex by addition of ethylenediaminetetra-acetic acid (EDTA) or by dialysis results in the formation of large unilamellar vesicles (De Kruijff *et al.*, 1982b) and an "isotropic" structure (De Kruijff *et al.*, 1982b), respectively. The Ca^{2+}-induced H_{II} phase formation in beef heart CL can be effectively blocked by the cardiotoxic anticancer drug

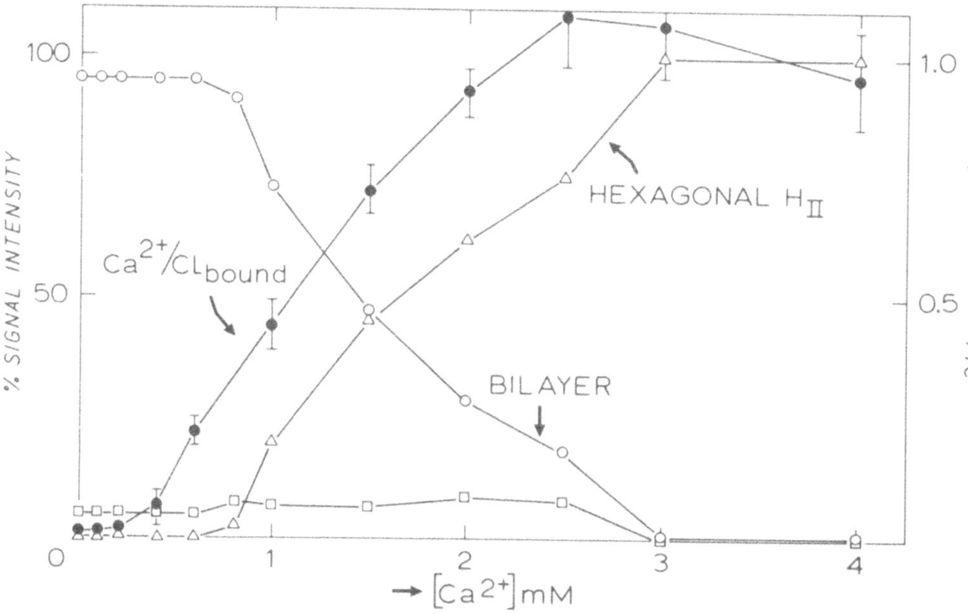

Figure 11. Ca^{2+} binding to and structure of CL liposomes. CL liposomes in 100 mM NaCl, 10 mM Tris/HCl, and pH 7.0 were dialyzed against excess Ca^{2+}-containing buffer, then the Ca^{2+} binding and the structure of the lipids were determined. For experimental details see De Kruijff *et al.* (1982b).

adriamycin (Goormaghtigh *et al.*, 1982) and the specific inner mitochondrial Ca^{2+} transport inhibitor ruthenium red (Cullis *et al.*, 1980a), suggesting that nonlamellar structures formed by CL might play functional roles in the inner mitochondrial membrane (see also Section IX-C).

Other divalent cations such as Mg^{2+}, Mn^{2+}, and Ba^{2+} also interact strongly with CL, leading to the formation of the respective salts (Van Venetië and Verkleij, 1981; Rand and Sengupta, 1972; Vasilenko *et al.*, 1982a) which show temperature-dependent lamellar → H_{II} phase transitions (Figure 12, see Table 2 for H_{II} tube dimensions). The transition temperatures increase in the order $Mg^{2+} < Ca^{2+} < Ba^{2+}$, which is also the order of the size of the dehydrated cations, suggesting, together with other data (Shaw and Schulman, 1965), that the H_{II}-promoting ability of divalent cations is related to the formation of an intramolecular phosphate–metal–phosphate complex in which the two phosphates are pulled together resulting in a decreased head group size (De Kruijff *et al.*, 1982b).

For the more saturated bacterial CL, the bilayer → H_{II} transition of the divalent cation–CL salts occurs at a much higher temperature (Vasilenko *et al.*, 1982a), a result which is consistent with the phase behavior of PEs (see Figure 12). In côntrast to beef heart CL at physiological temperatures the lamellar phase is always preferred.

PA, a key intermediate in phospholipid metabolism, occurs in small amounts in many membranes and undergoes a rapid turnover which in some cases is related to

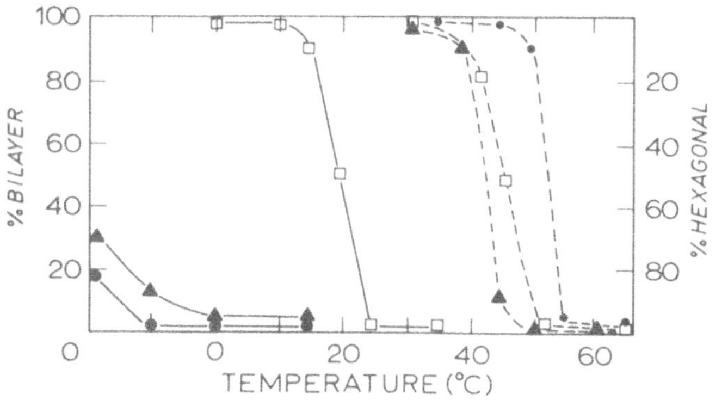

Figure 12. Temperature dependency of the bilayer → H_{II} transition of various CL salts. Beef heart CL: ▲———▲, Ca^{2+}-salt; ●———●, Mg^{2+}-salt; and □———□, Ba^{2+}-salt. *B. subtilis* CL: ▲- - - -▲, Ca^{2+}-salt; ●- - - -●, Mg^{2+}-salt; □- - - -□, Ba^{2+}-salt. Reproduced with permission from Vasilenko *et al.* (1982a). For further details see this reference.

hormonal stimulation (Van den Bosch, 1974; Salmon and Honeyman, 1980; Putney *et al.*, 1980). As a phosphomonoester, it has two ionizable groups with pKs of 3.5 and 8.0, respectively (Koter *et al.*, 1978). 18 : 1_c/18 : 1_c-PA forms bilayers when the molecule is singly or doubly charged (Verkleij *et al.*, 1982; Farren *et al.*, 1983). A lamellar organization is also found for the Ca^{2+} and Mg^{2+} salts above pH 8.0 and below pH 4.0 (Verkleij *et al.*, 1982; Farren *et al.*, 1983). However, at neutral pH, when the molecule only bears one negative charge, Ca^{2+} addition results in H_{II} phase formation. Similar effects have been reported for egg-PA (Papahadjopoulos *et al.*, 1976). This finding offers intriguing regulatory possibilities as the stimulation of nonlamellar structures by divalent cations is dependent on the charge of the lipid molecules which in turn will be dependent on environmental conditions.

3. *Anaesthetics*

Anaesthetics are a broad class of molecules which greatly differ in chemical structure but which have as common properties their affinity for membranes and their inhibitory action on the Na^+ channel in excitable membranes (Seeman, 1972). The correlation between solubility in an apolar environment and anaesthetic potency (Seeman, 1972) suggests a nonaspecific type of membrane interaction. Several models have been put forward for their mode of action, most of which evoke some pertubation in fluidity at a protein–lipid interface (Lee, 1976).

The observation that lipid polymorphism is greatly affected by the presence of these molecules has led to the formulation of alternative molecular mechanisms (Cullis *et al.*, 1980c). Several effects can be observed dependent on the type of lipid or anaesthetic. In the case of beef heart CL (Verkleij *et al.*, 1982; Cullis *et al.*, 1978a) or 18 : 1_c/18 : 1_c-PA (Verkleij *et al.*, 1982a), under conditions where divalent cations would promote H_{II} formation, the local anaesthetic dibucaine and chlorpromazine produce similar effects. This is shown in Figure 13 for 18 : 1_c/18 : 1_c-PA. At pH 6.0, in the absence of the drug, freeze-fracturing reveals bilayer structure. Addition of stoichiometric amounts of chlorpromazine results in lipid precipitation and the formation of the hexagonal H_{II} phase in conjunction with lipidic particles (the structural features of this latter component will be discussed in Section VI). It should be emphasized that these effects are specific for these types of lipids and are not observed for other negatively charged lipids. Furthermore, other positively charged amphipatic molecules such as adriamycin militate against hexagonal H_{II} phase formation (Goormaghtigh *et al.*, 1982).

In PE systems, the local anaesthetic chlorpromazine, dibucaine, tetracaine, and procaine stabilize bilayer structure (Hornby and Cullis, 1981), their relative anaesthetic potency being mirrored by their relative ability to stabilize bilayer structure. Of the neutral anaesthetics such as alkanes and alcohols, differential effects have been observed. Whereas ethanol and butanol stabilize bilayer structure for egg-PE, the longer chain ($n > 6$) normal alcohols and alkanes strongly promote hexagonal H_{II} phase formation (Hornby and Cullis, 1981). It is suggested that the ability of these molecules

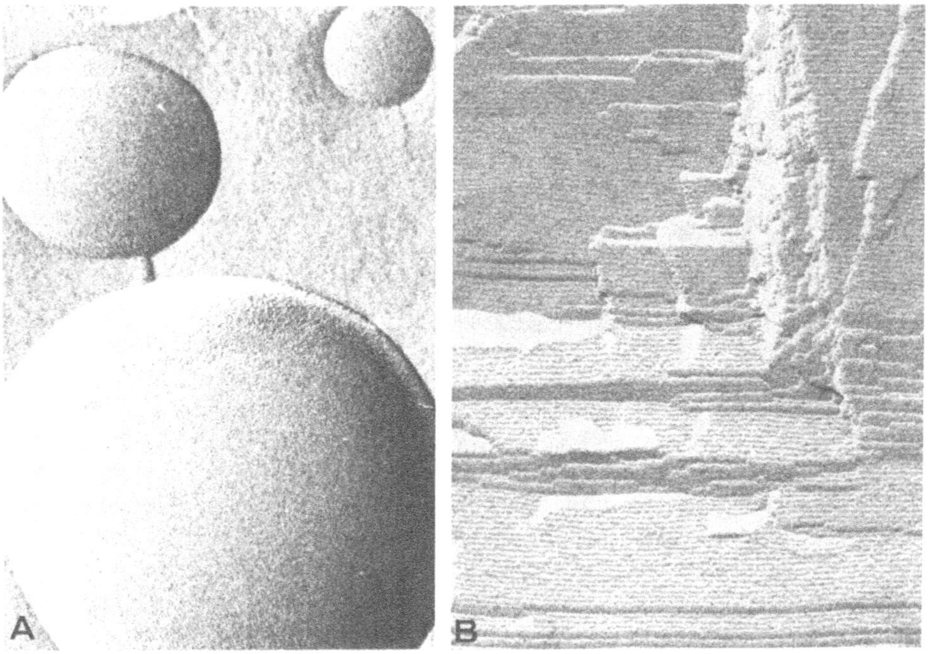

Figure 13. Freeze-fracture micrographs of 18 : 1_c/18 : 1_c-PA dispersed in 10 mM Mes, 100 mM NaCl at pH 6.0 before (A) and after (B) the addition of chlorpromazine at a ratio chlorpromazine/18 : 1_c/18 : 1_c-PA of 1 : 1. Final magnification $\times$ 100,000. Besides the H_{II} phase formation, addition of the anaesthetic also induces the formation of lipidic particles. For further details see Verkleij *et al.* (1982).

to affect the phase structure of PEs is related to their dynamic shape (Hornby and Cullis, 1981).

B. Mixed Lipid Systems

As biological membranes contain both bilayer-and H_{II}-forming lipids, it is essential to obtain insight into the phase properties of mixtures of these lipids under various physiologically relevant conditions. Most experimental work has been directed towards obtaining an understanding of the phase behavior of PE in mixtures with PCs, sterols, and negatively-charged lipids which we will review in that order.

1. PE–PC Mixtures

Incorporation of PC in unsaturated PE dispersions stabilizes the bilayer configuration of the PE. The degree of stabilization depends on the acyl chain composition, the phase state of the PC (gel vs. liquid-crystalline), and the temperature (Cullis and De Kruijff, 1979). In general, decreasing unsaturation or temperature results in a stronger preference for bilayer organization. Figure 14 illustrates this for soya egg-PE–PC mixtures at 30°C. In the absence of PC, the PE is hexagonally organized, whereas in the equimolar mixture, all lipids adopt the bilayer organization (Cullis and De Kruijff, 1978a). At intermediate concentrations, an isotropic ^{31}P NMR signal is observed. Such a signal is commonly observed in lipid systems intermediate between lamellar and H_{II} configurations and indicates the presence of structures in which isotropic motional averaging occurs (see Section VI). In the related 18 : 1_c/18 : 1_c-PE–18 : 1_c/18 : 1_c-PC system, as little as 25 mole% PC is sufficient to convert the PE into the bilayer phase (Cullis *et al.*, 1978b). In equimolar 18 : 1_c/18 : 1_c-PE–16 : 0/16 : 0-PC mixtures, the magnitude of the stabilization is strongly temperature-dependent (Cullis *et al.*, 1978b; Vasilenko *et al.*, 1982b). For example, the use of the thion analogue of PC (P=S instead of P=O), for which the ^{31}P NMR signal is completely separated from that of the normal species, allows the observation of the phase state of each lipid in these mixed systems (Vasilenko *et al.*, 1982b). It was observed that at low temperatures, where the PC is organized in the gel state, phase separation occurs between gel state 16 : 0/16 : 0-PC bilayers and a liquid-crystalline 18 : 1_c/18 : 1_c-PE H_{II} phase. Melting of the palmitic acid chains results in mixing of the two lipids and the formation of a lamellar phase. At higher temperatures, both molecules are incorporated into an "isotropic" structure. In mixtures of two PE species, the gel → liquid-crystalline phase transition can also induce similar structural transitions (Dekker *et al.*, 1983; Tilcock and Cullis, 1982). This is shown in Figure 15 for 16 : 0/18 : 1_c-PE–22 : 6/22 : 6-PE mixtures for which a gel state lamellar and a hexagonal H_{II} phase is observed below the gel → liquid-crystalline transitions of 16 : 0/18 : 1_c-PE. Upon melting of the 16 : 0/18 : 1_c-PE, mixing occurs in the liquid-crystalline bilayer producing a net bilayer structure whereas at higher temperatures, the lipids convert to the H_{II} phase. It may be noted that the other major zwitterionic membrane lipid with a choline head group, sphingomyelin, exerts similar strong bilayer stabilization in model membranes containing unsaturated PE (Cullis and Hope, 1980).

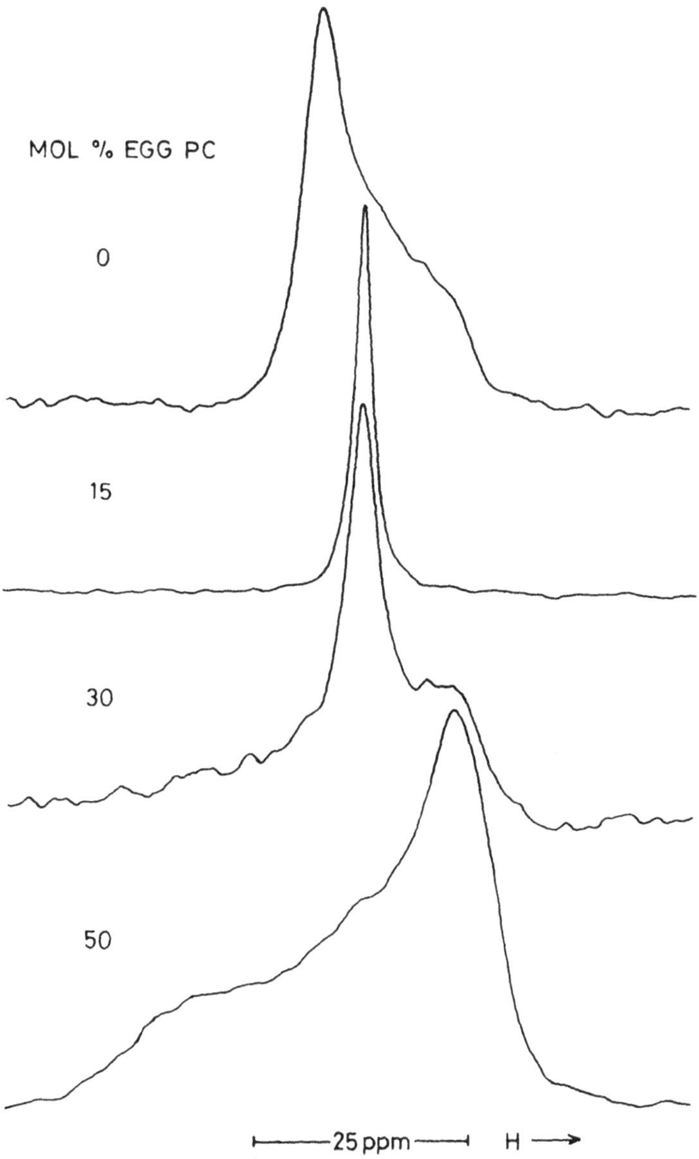

Figure 14. 36.4 MHz ^{31}P NMR spectra at 30°C of aqueous dispersions of soya-PE containing increasing amounts of egg-PC. Reproduced with permission from Cullis and De Kruijff (1976).

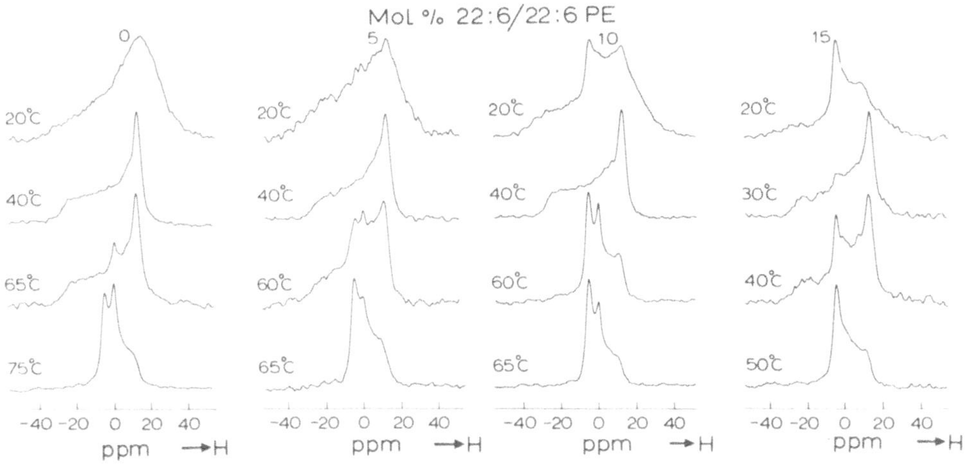

Figure 15. 36.4 MHz ^{31}P NMR spectra of aqueous dispersions of various mixtures of 16 : 0/18 : 1_c-PE and 22 : 6/22 : 6-PE. For further details see Dekker *et al.* (1983).

2. Mixtures with Sterols

Sterols are main constitutents of many biological membranes. Despite extensive effort, no clear picture has emerged of their functional role(s). Cholesterol, the most abundant mammalian sterol, can solidify or "condense" liquid-crystalline lipid systems and liquidify or expand gel state phospholipids leading to an intermediate state of "fluidity" (Demel and De Kruijff, 1976). For both effects, a 3β-OH group, a planar ring system, and an aliphatic side chain at C_{17} are required. *A priori,* there are several reasons to suppose that cholesterol–PE interactions might be special and could affect membrane structure. In the first place, there appears to be an inverse relationship between the occurrence of PE and sterols in membranes (Demel and De Kruijff, 1976). PE-rich membranes, such as occur in bacteria, have a very low sterol content whereas in mammalian plasma membranes, such as that of the human erythrocyte, the opposite is true. Second, in PE–PC mixtures under conditions of lateral phase separation, cholesterol preferentially interacts with PC (Van Dyck *et al.*, 1976). Third, from the preferential localization of cholesterol in the inner monolayer of sonicated vesicles (De Kruijff *et al.*, 1976b), it can be inferred that cholesterol will have a cone shape with the hydroxyl group at the smaller end of the cone. As cone-shaped lipids favor H_{II} phase formation (see Section VII), it can be expected that the incorporation of sterols in membranes will destabilize bilayer structure. Indeed, in mixed PE–PC systems, strong bilayer destabilization effects have been observed on incorporation of cholesterol. For instance, whereas 25 mole% 18 : 1_c/18 : 1_c-PC stabilizes bilayer structure at 40°C for 18 : 1_c/18 : 1_c-PE, the additional incorporation of cholesterol (equimolar with respect to PC) induces the H_{II} phase for all the phospholipids (Cullis *et al.*, 1978b). Using lower cholesterol concentrations and selective ^{2}H labeling (Tilcock *et al.*, 1982), it could be demonstrated (employing ^{31}P and ^{2}H NMR) that when both lamellar and H_{II} phases were present, the phospholipid composition of both phases was very similar (Figure 16). This is a rather remarkable result, as it could have been expected that the H_{II} phase would be enriched in PE and the bilayer phase in PC. It has also been shown for other lipid mixtures that substantial amounts of the bilayer-preferring lipid can be incorporated into nonlamellar structures (Vasilenko *et al.*, 1982b; De Kruijff *et al.*, 1979).

The magnitude of the bilayer destabilization by sterols depends very much on their chemical structure (Gallay and De Kruijff, 1982). Figure 17 summarizes the effect of the incorporation of different sterols on the bilayer → H_{II} transition temperature of 18 : 1_t/18 : 1_t-PE. The bilayer destabilization potential is found to be directly related to the mean molecular area occupied by these sterols in pure and mixed monolayers at the air–water interface (Figure 17). Interestingly, those sterols which do not show condensing or liquifying effects in PC bilayers cause the strongest bilayer destabilization. Thus, observations such as the increased fragility of erythrocytes caused by replacing cholesterol with 3-ketosterols (Bruckdorfer *et al.*, 1969) could well result from the strong bilayer-destabilizing action of the latter molecule.

The effects observed upon incorporation of cholesterol in PE-containing lipid mixtures are paralleled by similar effects observed upon incorporation of cholesterol in monoglucosyl- and diglucosyl-containing lipid systems (Khan *et al.*, 1981). The strong bilayer-destabilization ability of cholesterol is illustrated by the observation that

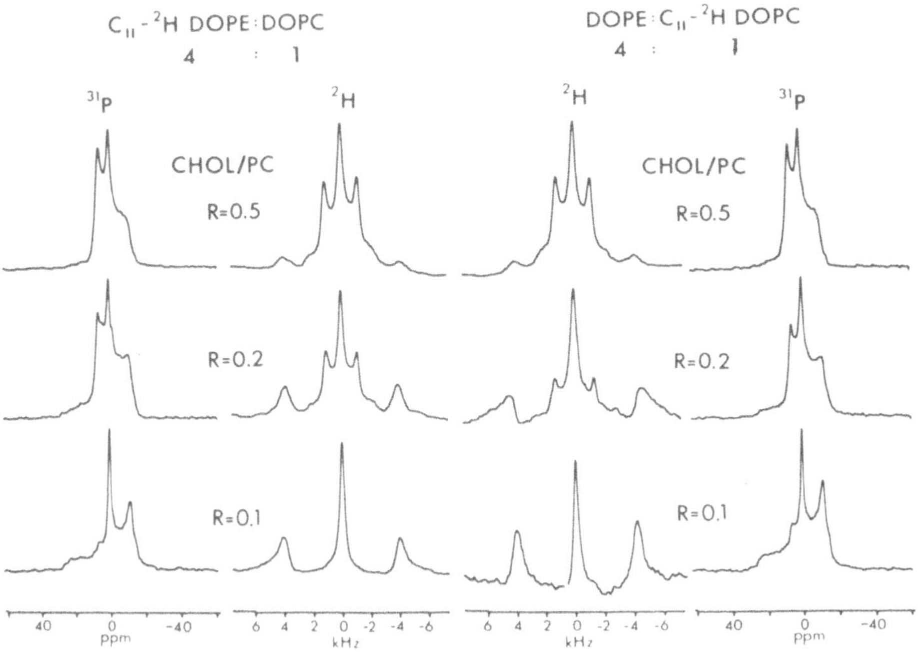

Figure 16. 81.0 MHz ^{31}P NMR and 30.7 MHz ^{2}H NMR spectra at 40°C arising from aqueous dispersions of mixtures of 18 : 1_c/18 : 1_c-PE, 18 : 1_c/18 : 1_c-PC and cholesterol. Reproduced with permission from Tilcock *et al.* (1982).

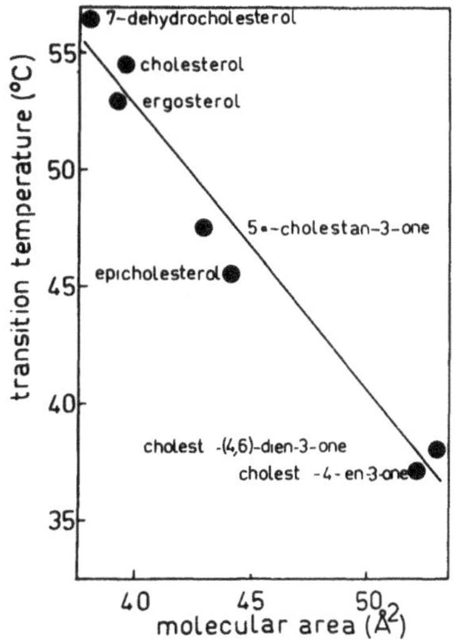

Figure 17. Relationship between the effect of equimolar sterols on the bilayer → H_{II} transition of 18 : 1_t/18 : 1_t-PE and the molecular area of the pure sterols in monolayers at the air–water interface. Reproduced with permission from Gallay and De Kruijff (1982).

incorporation of cholesterol in polyunsaturated PC species such as 18 : 3/18 : 3-PC, 20 : 4/20 : 4-PC, and 22 : 6/22 : 6-PC will induce H_{II} phase formation (Dekker *et al.*, 1983).

3. *Negatively Charged Lipid-Containing Mixtures*

The ability to isothermally modulate lipid structure is a prerequisite for possible functional roles of nonlamellar lipid structures in biomembranes. The results described in Section V-A.2, indicating that divalent cations can influence the molecular packing of negatively charged lipids, together with the observation that bilayer-forming lipids stabilize bilayer structure in PE-containing model membranes, suggest that in mixed PE negatively charged phospholipid systems the lipid structure will be very sensitive to the presence of divalent cations and the nature of the charged lipid.

This is certainly the case as illustrated in Figure 18, which shows the ^{31}P NMR characteristics of mixtures of unsaturated PE with unsaturated PS (Tilcock and Cullis, 1980), PG (Farren and Cullis, 1980), PI (Nayar *et al.*, 1982), and CL (De Kruijff and Cullis, 1980b) as a function of the molar ratio of Ca^{2+} to negatively charged lipid. The presence of 30 mole% or more of the charged lipid stabilizes bilayer structure of the PE. In the case of PS-containing systems, Ca^{2+} addition results in the formation of the gel state (lamellar) Ca^{2+}-PS salt. Phase separation occurs and the PE reverts to the H_{II} phase. In contrast, in the case of PG both lipids are incorporated into the H_{II} phase upon the Ca^{2+}–PG interaction despite the fact that the Ca^{2+}-PG salt favors a lamellar organization. On the other hand, Ca^{2+}–PI interactions appear to result in lateral phase segregation of PI into liquid-crystalline lamellar domains leaving the PE to adopt the H_{II} phase. Since CL itself favors the H_{II} phase in the presence of Ca^{2+} in mixtures with PE, the addition of Ca^{2+} results, as may be expected, in the formation of the H_{II} phase for all the lipids. Correspondingly, when the PE in the mixture with

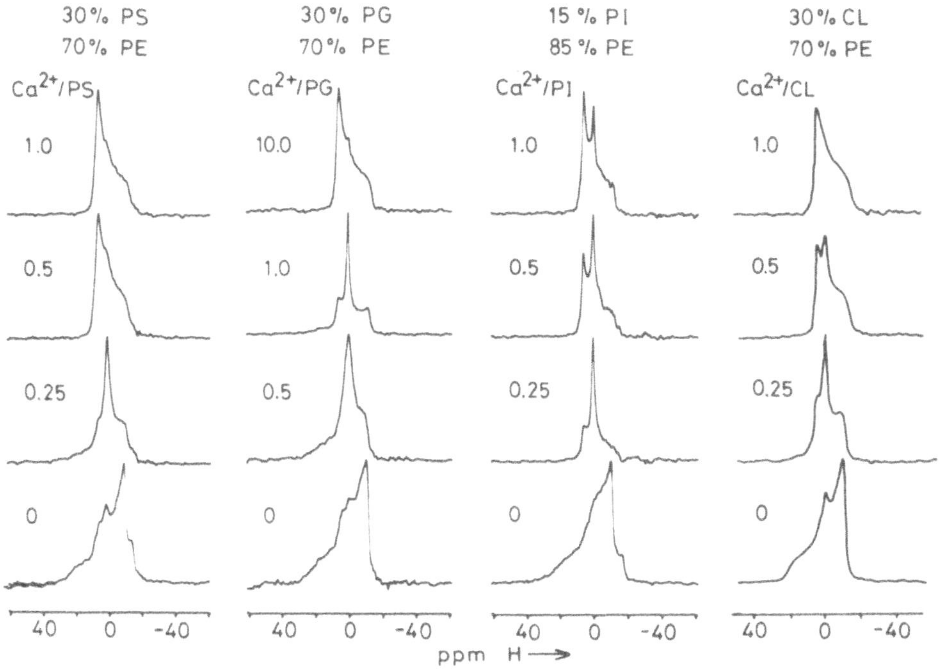

Figure 18. 81.0 MHz ^{31}P NMR spectra arising from mixtures of acidic phospholipids with soya-PE in the presence of various molar ratios of Ca^{2+}. For experimental details for PS-PE see Tilcock and Cullis (1980), for PG-PE see Farren and Cullis (1980), for PI-PE see Nayar *et al.* (1982), and for CL-PE see De Kruijff and Cullis (1980b). All previously published spectra reproduced with permission.

CL is replaced by PC, the H_{II} phase-stimulating capacity of Ca^{2+} is much reduced. In this latter case, next to the H_{II} phase in the PE-containing systems, a large isotropic ^{31}P NMR signal is observed with associated lipidic particles as detected by freeze-fracture (see Section VI). The detailed morphology of the hexagonal H_{II} phase in the divalent cation-PC/CL (1 : 1) system depends on the nature of the cation. Whereas in the presence of Ca^{2+}, one type of H_{II} phase cylinder is observed by freeze-fracturing (with an increased diameter as compared to the pure CL-Ca^{2+} H_{II} phase), addition of Mg^{2+} results in the occurrence of two types of H_{II} cylinders. One of these has the size observed for the H_{II} phase of the Mg^{2+}-CL salt, whereas the other exhibits an increased diameter, suggesting phase separation (Van Venetië and Verkleij, 1981).

It should be noted that not only the absolute amounts of Ca^{2+} (Figure 18), but also the Ca^{2+} concentrations needed to induce these structural changes, are different for the various negatively charged phospholipids. In several biological membrane lipid extracts, Ca^{2+} addition causes similar structural changes. For instance, an appreciable fraction of the total rat liver inner mitochondrial membrane lipids (composition 40% PC, 40% PE, and 20% CL) converts to the H_{II} phase upon addition of Ca^{2+} (Cullis *et al.*, 1980b). Similar effects have been noted for an equimolar mixture of the phospholipids derived from the inner leaflet of the human erythrocyte membrane and cholesterol (Hope and Cullis, 1979).

C. Lipid–Protein and Lipid–Peptide Interactions

1. Extrinsic Proteins and Peptides

Poly-L-lysines have been commonly used as model peptides to gain insight regarding electrostatic peptide–lipid interactions (Hartmann and Galla, 1978; Papahadjopoulos *et al.*, 1975b). The high affinity of the basic poly-L-lysine for negatively charged lipids is also apparent in studies on the effect of this compound on lipid polymorphism (De Kruijff and Cullis, 1980a).

Addition of poly-L-lysine to CL liposomes results in strong peptide–lipid binding and immediate precipitation of the lipids. In this precipitate, the lamellar phase is maintained (Figure 19). Further, whereas Ca^{2+} addition in the absence of poly-L-lysine triggers a bilayer → H_{II} phase transition, the presence of the peptide completely blocks this transition (Figure 19), revealing a strong bilayer-stabilization effect of the poly-L-lysine in this system. Alternatively, if poly-L-lysine is added to mixed PE-CL

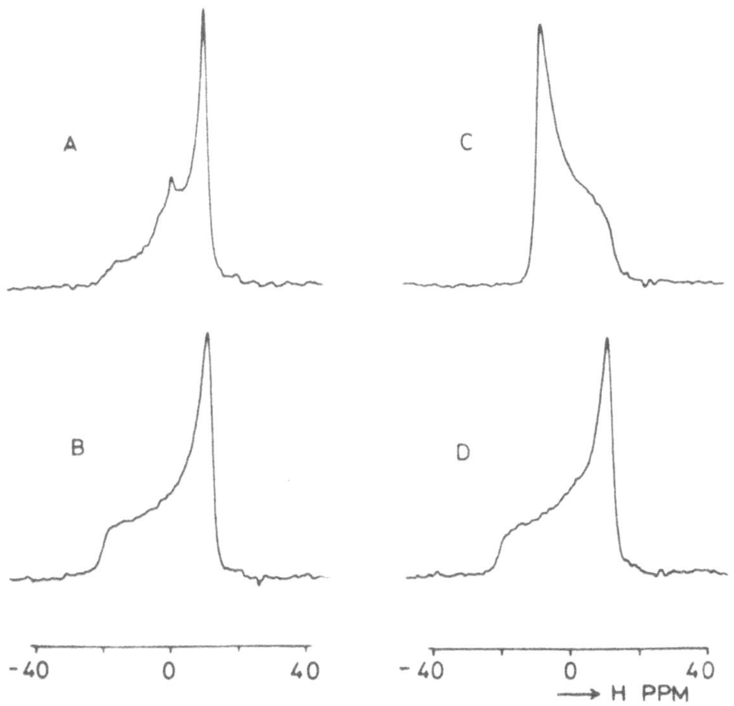

Figure 19. 81.0 MHz ^{31}P NMR spectra at 30°C of (A) 50 μmoles of CL in 1.0 ml buffer; (B) 50 μmoles of CL in 1.0 ml buffer, 5 min after the addition of 40 mg poly-L-lysine; and (C) as in (A) after addition of 100 μl 1 M $CaCl_2$. Reproduced with permission from De Kruijff and Cullis (1980a).

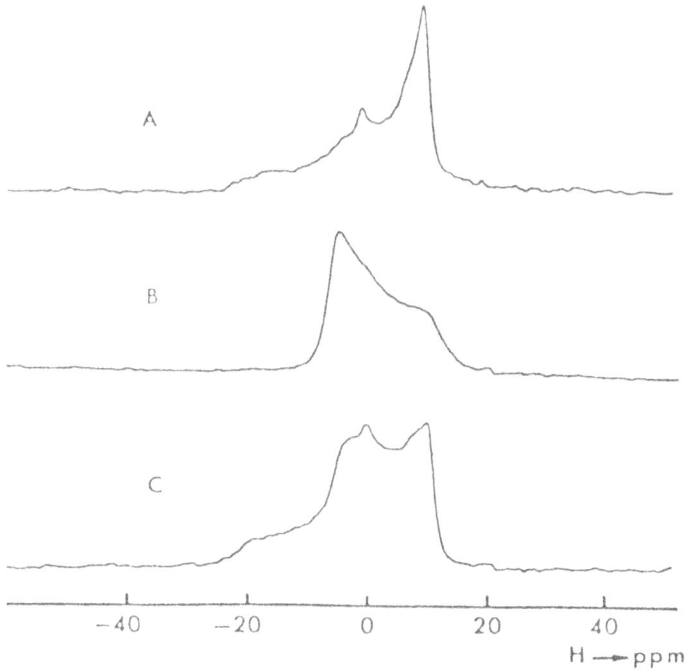

Figure 20. 81.0 MHz ^{31}P NMR spectra at 30°C of an aqueous dispersion of CL (A) in the presence of Ca^{2+} (B) and cytochrome *c* (C). Beef heart CL (50 μmoles) was dispersed in 1.0 ml 100 mM NaCl, 10 mM Tris/HCl, 0.2 mM EDTA, and pH 7.0. In (B) 0.1 ml 1 M $CaCl_2$ and in (C) 0.2 ml buffer containing 36 mg oxidized cytochrome *c* was added. For further details see De Kruijff and Cullis (1980b).

liposomes in which CL stabilizes bilayer structure of the PE component, a phase separation of the CL/poly-L-lysine complex occurs, leading to a bilayer → H_{II} transition of the PE (De Kruijff and Cullis, 1980a).

The inner mitochondrial membrane protein cytochrome *c* is another example of a highly basic protein which experiences strong interactions with a variety of negatively charged membrane phospholipids (Nicholls, 1974). It is interesting that only in the case of the inner mitochondrial CL does this interaction result in changes in macroscopic organization of the lipids (De Kruijff and Cullis, 1980b). ^{31}P NMR (De Kruijff and Cullis, 1980b), electron microscopy (De Kruijff and Cullis, 1980b; Borovjagin and Moshkov, 1973) and X-ray (Gulik-Krzywicky *et al.*, 1969) have shown that the protein can induce nonlamellar lipid structures in this case. This is shown in Figure 20 by ^{31}P NMR. The addition of cytochrome *c* to CL liposomes results in the formation of an isotropic component together with a spectral component indicative of the H_{II} phase.

2. *Gramicidin–Lipid Interactions*

The helical membrane spanning dimer of the hydrophobic pentadecapeptide gramicidin (Urry, 1971; Weinstein *et al.*, 1980) may serve as a model for the hydrophobic segments of intrinsic membrane proteins (Chapman *et al.*, 1977). Due to its hydrophobicity, gramicidin can be easily incorporated in phospholipids. When gramicidin is incorporated in 18 : 1_t/18 : 1_t-PE it can be seen from Figure 21 that the onset of the formation of the hexagonal H_{II} phase is shifted to much lower temperatures (Van Echteld *et al.*, 1981a). At a gramicidin/18 : 1_t/18 : 1_t-PE ratio of 1 : 25 the transition is already completed at 50°C. This hexagonal H_{II} phase-promoting ability of gramicidin, which can be seen to be clearly concentration-dependent, has also been found for 18 : 1_c/18 : 1_c-PE (Van Echteld *et al.*, 1981a). A straightforward implication of these findings is the association of gramicidin with phospholipid molecules within the hexagonal H_{II} phase.

Most surprisingly, however, this strong influence of gramicidin on the structural organization of lipids is not restricted to H_{II} type of lipids that undergo a lamellar to hexagonal H_{II} phase transition by themselves, but has also been found with lipids that are known to pre-eminently organize themselves in bilayers such as 18 : 1_c/18 : 1_c-PC (Van Echteld *et al.*, 1981a). In Figure 22, it is shown that incorporation of a small amount of gramicidin in 18 : 1_c/18 : 1_c-PC gives rise to a small isotropic component in the ^{31}P NMR spectrum and a reduction of $\Delta\sigma$. These spectral changes are compatible with a reduction in size of the gramicidin-containing structures compared to the pure

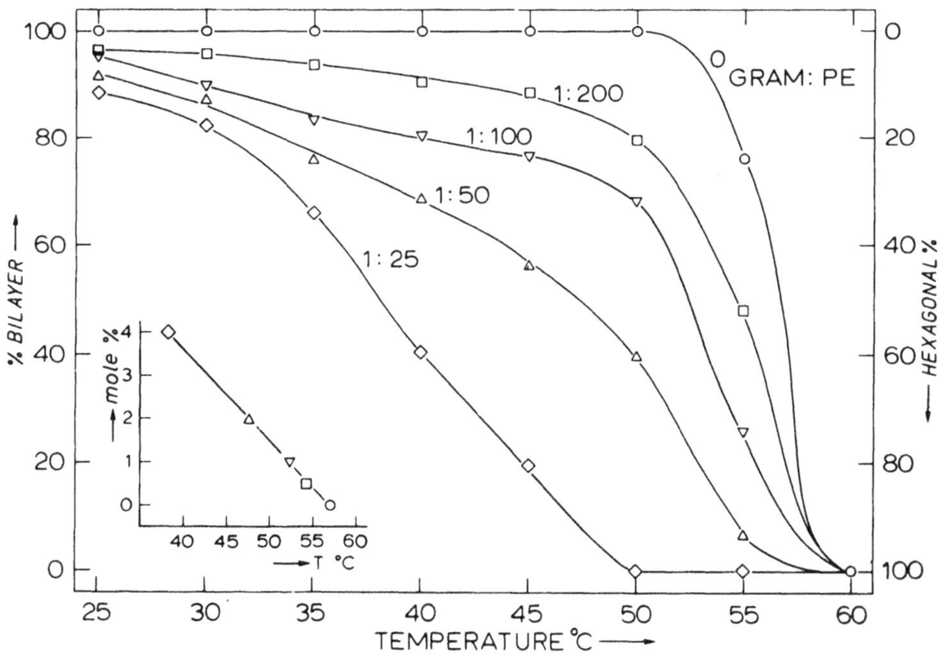

Figure 21. Effect of gramicidin on the bilayer → H_{II} transition of 18 : 1_t/18 : 1_t-PE. The gramicidin/18 : 1_t/18 : 1_t-PE molar ratios are indicated in the figure. The insert shows the relationship between gramicidin concentration (mole%) and the temperature at which 50% of the lipid is organized in the H_{II} phase. Reproduced with permission from Van Echteld *et al.* (1981a). See this reference for further details.

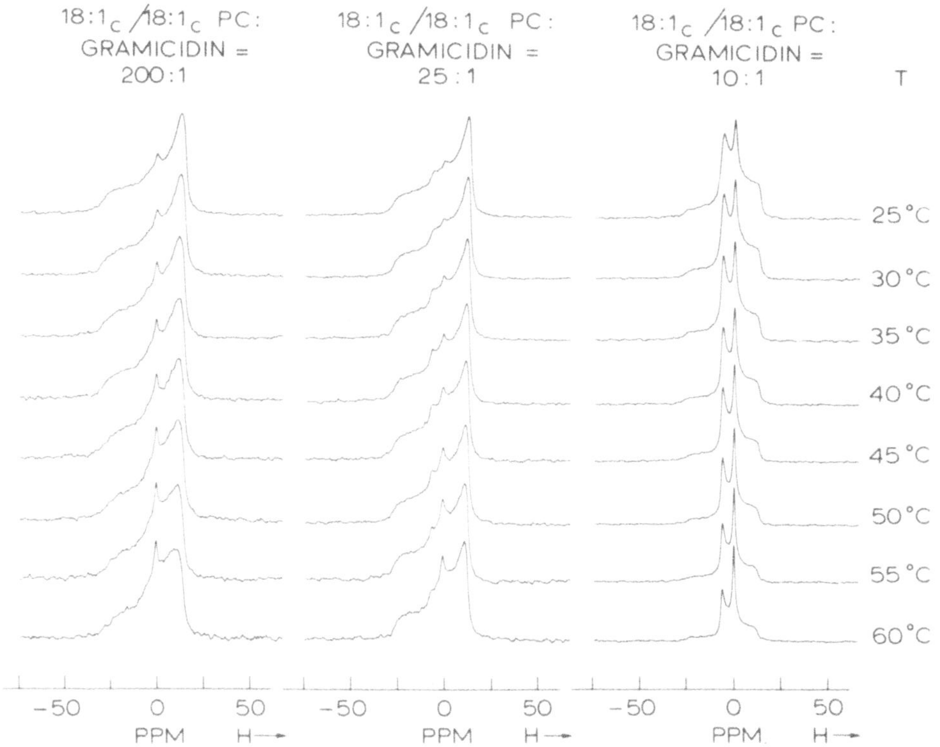

Figure 22. 81.0 MHz ^{31}P NMR spectra of 18 : 1_c/18 : 1_c-PC-gramicidin mixtures at various temperatures. Reproduced with permission from Van Echteld *et al.* (1981a). See this reference for further details.

lipid, which also was found with freeze-fracture electron microscopy. However, increasing the gramicidin content induces an additional spectral component indicating PC organized in a hexagonal H_{II} phase, which can be seen most clearly in Figure 22 as the resonance intensity at approximately -6 ppm. Freeze-fracturing of the same sample shows large areas with a typical striated fracture pattern (Van Echteld *et al.*, 1981a).

The hexagonal H_{II}-promoting ability of gramicidin could possibly be the result of a mismatch of the length of the gramicidin dimer with the thickness of the hydrophobic part of the bilayer. To test this latter hypothesis, the influence of gramicidin on PCs with varying fatty acid chain lengths has been investigated (Van Echteld *et al.*, 1982). As can be seen from Figure 23, the onset of the induction of the hexagonal H_{II} phase occurs when the fatty acid chain exceeds 16 carbon atoms. The isotropic peaks seen in this figure originate from lipid in smaller structures. The chain length-dependent formation of the hexagonal H_{II} phase by gramicidin is not restricted to unsaturated PCs. As shown in Figure 24, the onset of the induction of the hexagonal phase in liquid-crystalline-saturated PCs again coincides with a fatty acid chain exceeding 16 carbon atoms. Also, with PC from natural sources, e.g., egg-PC and soya-

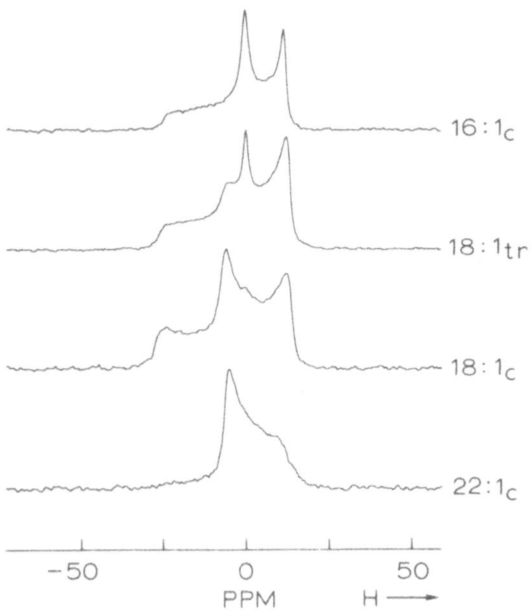

Figure 23. 81.0 MHz ^{31}P NMR spectra of aqueous dispersions of mixtures of gramicidin with various unsaturated PCs in a 1 : 10 molar ratio at 25°C. The fatty acids present in the various PC species are indicated in the figure. Reproduced with permission from Van Echteld *et al.* (1982). See this reference for further details.

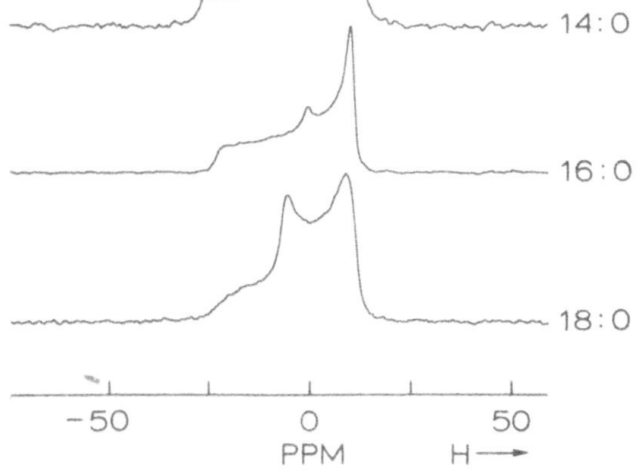

Figure 24. 81.0 MHz ^{31}P NMR spectra of aqueous dispersions of mixtures of gramicidin with various saturated liquid-crystalline PC species in a 1 : 10 molar ratio. Fatty acid composition as indicated in the figure. Recording temperatures: 12 : 0/12 : 0-PC at 30°C, 14 : 0/14 : 0-PC at 40°C, 16 : 0/16 : 0-PC at 50°C, and 18 : 0/18 : 0-PC at 60°C. Reproduced with permission from Van Echteld *et al.* (1982). See this reference for further details.

PC, induction of a hexagonal phase for part of the lipid molecules was found (Van Echteld *et al.*, 1982). From neutron diffraction studies (Büldt *et al.*, 1978), the thickness of the hydrophobic part of liquid-crystalline 16 : 0/16 : 0-PC may be estimated to be 30–31 Å, similar to the length of the gramicidin dimer of approximately 30 Å (Wallace *et al.*, 1981). When the length of the hydrophobic part of the phospholipid molecules exceeds this length, apparently an instable situation arises which results in the formation of a hexagonal H_{II} phase for part of the molecules. The longer the fatty acid chains, the more lipid molecules enter the hexagonal phase, probably as a result of a larger imbalance. To accommodate the gramicidin in the hexagonal H_{II} phase, a model has been proposed (Van Echteld *et al.*, 1982) in which the gramicidin dimer spans adjacent cylinders, thereby maintaining a similar orientation as in the bilayer situation.

Alternatively, the H_{II}-promoting ability could be related to the shape of gramicidin (or the dimer). The bulky tryptophan residues located at the C-terminal renders the molecule cone-shaped, thus favoring H_{II} phase formation (see Section VII). In agreement with this suggestion is the observation that in mixtures with lyso-PC gramicidin is organized in a lamellar phase (Killian *et al.*, 1983).

3. *Glycophorin–Phospholipid Interactions*

Glycophorin, the major integral sialoglycoprotein of the human erythrocyte membrane, is well characterized biochemically (Marchesi *et al.*, 1976), has been the subject of many physicochemical studies in reconstituted phospholipid systems, and has been found to cause extensive perturbation of membrane phospholipid hydrocarbon chain conformation and packing (Taraschi and Mendelsohn, 1980; Van Zoelen *et al.*, 1978; Rüppel *et al.*, 1982). Furthermore, it is known to carry receptors for several sugar-specific lectins such as wheat germ agglutinin (WGA) (Verpoorte, 1975), MN blood group substances (Marchesi *et al.*, 1972), influenza virus (Marchesi *et al.*, 1972), and malaria (Pasvol *et al.*, 1982) and therefore when incorporated into membranes offers systems amenable to effector–receptor studies.

The introduction of glycophorin into liposomes of 18 : 1_c/18 : 1_c-PC gives rise to unilamellar vesicles (1000–5000 Å diameter) with intramembranous protein particles as observed by freeze-fracture electron microscopy (Taraschi *et al.*, 1982a). When examined by ^{31}P NMR, these vesicles (300/1; moles lipid/mole protein) exhibit bilayer spectra over a wide range of temperatures (0–60°C) (Taraschi *et al.*, 1982a).

Reconstitution of glycophorin with 18 : 1_c/18 : 1_c-PE, which prefers the hexagonal H_{II} phase above 10°C, results in the formation of small, unilamellar bilayer vesicles (300–1500 Å diameter) with ill-defined intramembranous protein particles (Taraschi *et al.*, 1982a). In contrast to the 18 : 1_c/18 : 1_c-PC–glycophorin system where the lipid–protein ratio could be varied over a wide range, the 18 : 1_c/18 : 1_c-PE–glycophorin recombinants exclusively form a 25/1 (moles/mole PE : glycophorin) complex. The protein particles observed by freeze-fracture electron microscopy are much larger in the 18 : 1_c/18 : 1_c-PE–glycophorin vesicles than in the 18 : 1_c/18 : 1_c-PC–glycophorin system suggesting that the protein was more highly aggregated in this former system.

The temperature dependence of the ^{31}P NMR lineshapes arising from unilamellar vesicles of 18 : 1_c/18 : 1_c-PE and glycophorin is shown in Figure 25. At 0°C, a

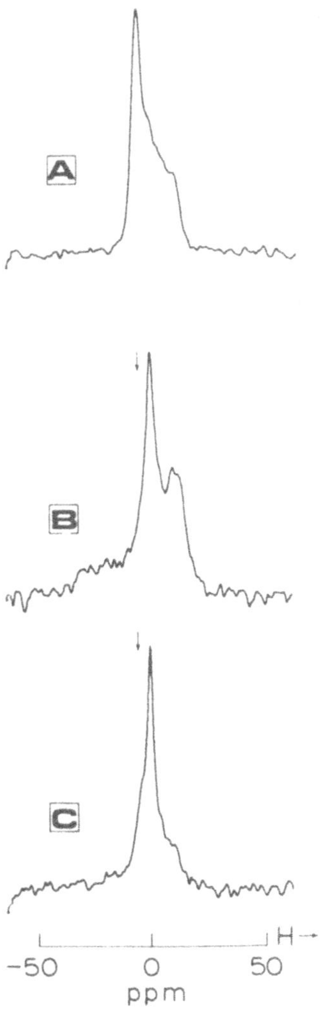

Figure 25. 36.4 MHz ^{31}P NMR spectra of 18 : 1_c/18 : 1_c-PE in the absence (A) 0°C and presence of glycerophorin (B) 0°C and (C) 25°C. Arrow indicates the position of the main spectral component of 18 : 1_c/18 : 1_c-PE in the H_{II} phase. Reproduced with permission from Taraschi *et al.* (1982a). See this reference for further details.

considerable amount of the phospholipid (50%) is in a structure allowing isotropic motion and the remaining lipid is arranged in a bilayer organization. The isotropic signal arises from the smaller glycophorin-containing vesicles present in the preparation. In comparison, 18 : 1_c/18 : 1_c-PE in isolation assumes a hexagonal H_{II} phase at this temperature. The spectra of the glycophorin–PE system become increasingly "isotropic" at higher temperatures (Taraschi *et al.*, 1982a); however, the signals remain devoid of hexagonal H_{II} phase characteristics. In summary, the combination of ^{31}P NMR and freeze-fracture electron microscopy demonstrates a strong bilayer-stabilizing capacity of glycophorin.

It may be noted that treatment of the vesicles with trypsin, which removes the large, bulky hydrophilic sugar residues [which are oriented (95%) towards the exterior of the vesicles] but leaves the hydrophobic segment of the protein incorporated in the membrane (Tomita and Marchesi, 1975), does not cause any appreciable changes in vesicle structure. Thus, the intrinsic hydrophobic segment of glycophorin appears to be mainly responsible for the strong bilayer-stabilization effects (Taraschi *et al.*, 1982a).

Glycophorin has also been reported to stabilize bilayer organization in reconstituted vesicle systems containing CL, a lipid which undergoes an isothermal bilayer to hexagonal H_{II} phase transition in the presence of Ca^{2+} (Taraschi *et al.*, 1983). Trypsin treatment failed again to alter the bilayer-stabilizing capacity of this protein.

During the past 10 years, considerable attention has been focused on the complex contact events that occur at the eukaryotic cell surfaces. These phenomena are thought to be specifically mediated by complex carbohydrates which coat the surface of cells and are covalently linked to membrane lipids and proteins. Following the initial recognition of the effector, e.g., hormone, toxin, antibody, lectin, or virus, a chain of events transpire that eventually evoke a metabolic response by the cell (Nicolson, 1976). The mechanistic details of the events which occur following the recognition of the effector by the receptor and the resulting response remain largely a mystery. Although a number of theories have appeared (Nicolson, 1976; Edelman, 1976), none have directly addressed the possibility that the binding event and subsequent receptor rearrangement ("patching") may lead to an alteration in the molecular organization of the lipids in that region of the membrane in which the receptor glycolipids and glycoproteins are located. This hypothesis is appealing in that major membrane lipids can form nonbilayer structures of potential functional importance. In order to investigate these possibilities, glycophorin–WGA interactions have been characterized in model membrane systems (Taraschi *et al.*, 1982b).

The addition of WGA to cosonicated vesicles of 18 : 1_c/18 : 1_c-PC and glycophorin (200/1; moles lipid/mole protein), in which 100% of the receptor-bearing carbohydrates are externally oriented, induces aggregation but not fusion of the vesicles as indicated by both ^{31}P NMR and freeze-fracture electron microscopy. This aggregation can be completely reversed by *N*-acetyl-D-glucosamine, which is a (sugar hapten) inhibitor of WGA agglutination. It was therefore concluded that the binding and subsequent vesicle aggregation caused by WGA proceeds without a structural reorganization of the membrane phospholipid bilayer.

The effect of WGA on two 18 : 1_c/18 : 1_c-PE–glycophorin vesicle preparations (moles/mole PE : glycophorin; 200/1; 25/1) has been obtained from ^{31}P NMR and freeze-fracture electron microscopy (Figure 26). The 25 : 1 cosonicated recombinant gives rise to a narrow, symmetric ($\nu_{1/2}$ = 100 Hz), "high resolution" ^{31}P NMR spectrum which arises from phospholipids present in small bilayer vesicles as indicated by freeze-fracture electron microscopy (Taraschi *et al.*, 1982b). The addition of increasing amounts of WGA to the sample (up to WGA/glycophorin mole ratio = 1.8) induces a substantial broadening in the spectrum although the lipids maintain a bilayer organization (Figure 26). Some membrane reorganization does occur, most likely as a result of vesicle fusion; however, a gross morphological change from a bilayer to nonbilayer (hexagonal H_{II} phase) is not observed.

Cosonication of a glycophorin–18 : 1_c/18 : 1_c-PE (1 : 200 sample) results in a narrow, symmetric ($\nu_{1/2}$ = 130 Hz), ^{31}P NMR lineshape, typical of small bilayer vesicles

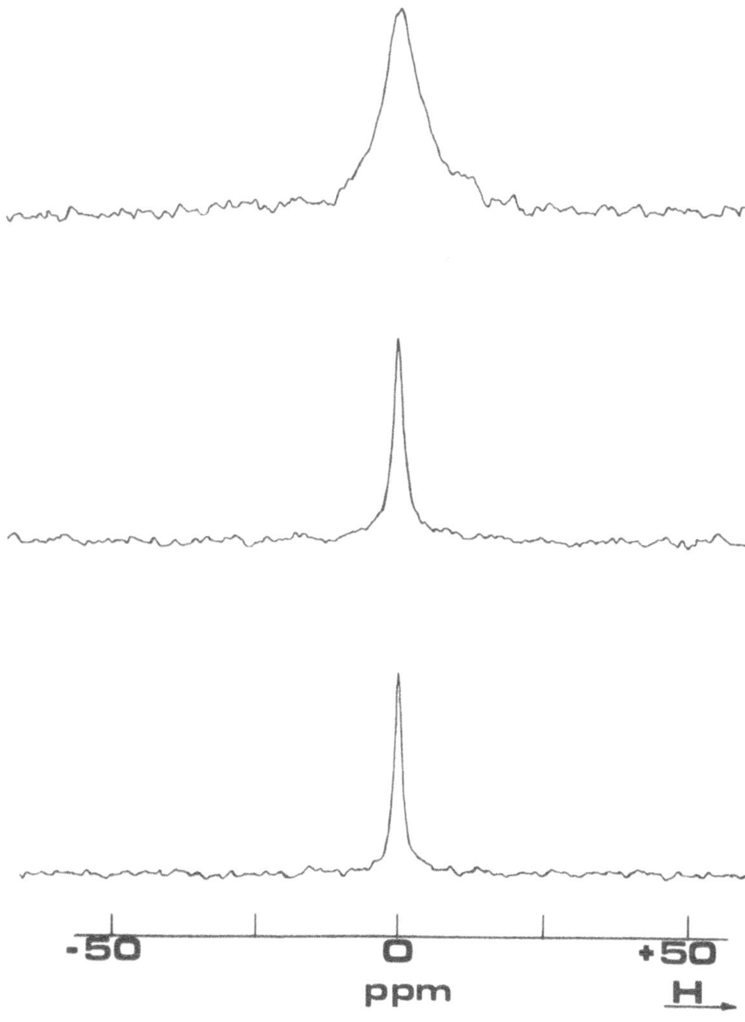

Figure 26. 81.0 MHz ^{31}P NMR spectra of sonicated 18 : 1_c/18 : 1_c-PE–glycophorin vesicles (25 : 1, moles lipid/mole protein; 15 μmoles 18 : 1_c/18 : 1_c-PE, 30.0 mg glycophorin) in the absence (A) and presence of 420 nmoles WGA (B) and 840 nmoles WGA (C) at 30°C. The molar ratios WGA/glycophorin were 0.7 (A) and 1.4 (B). For further details see Taraschi *et al.* (1982b).

(Figure 27). Titration with WGA causes drastic changes in the vesicles as indicated by both ^{31}P NMR and freeze-fracture electron microscopy, as higher levels of WGA (WGA/glycophorin, mole/mole = 1.0) induced a nearly complete (90%) bilayer to hexagonal H_{II} phase transition in the vesicle preparation as evidenced from ^{31}P NMR (Figure 27) and freeze-fracture experiments (Figure 28).

Treatment of the 25/1 and 200/1 18 : 1_c/18 : 1_c-PE–glycophorin samples with trypsin provides insight into the origin of the differing effects that WGA exerts on these membrane recombinants. The ^{31}P NMR spectra obtained for the enzymatically treated vesicles are shown in Figure 29. In the case of the 200/1 sample, trypsin treatment triggers a nearly complete bilayer to hexagonal H_{II} phase transition (with 95% of the ^{31}P NMR signal arising from lipid in the hexagonal H_{II} phase), whereas the 25/1 sample showed only an increase in line width of the isotropic NMR signal. Thus, it was concluded that the large hydrophilic carbohydrate residues which present a steric barrier to fusion and consequently prevent hexagonal H_{II} phase formation were mainly responsible for the bilayer-stabilizing effect of glycophorin in the 200 : 1 vesicles (Taraschi *et al.*, 1982b). The absence of this transition in the 25/1 sample likely results from a direct bilayer stabilization by the intrinsic, hydrophobic portion of the protein.

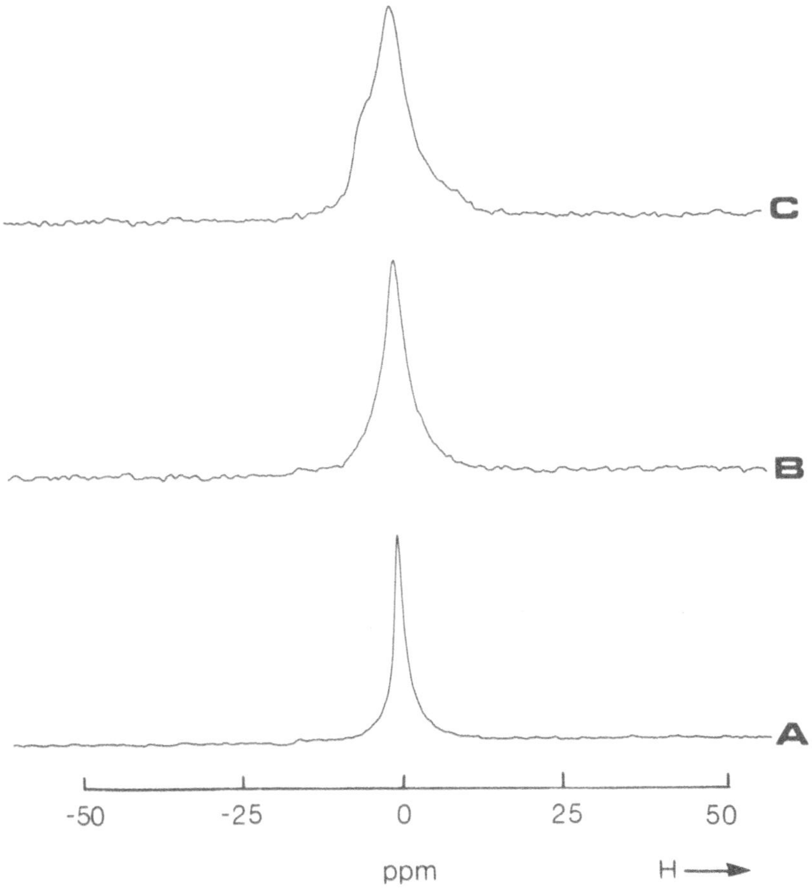

Figure 27. 81.0 MHz ^{31}P NMR spectra of cosonicated 18 : 1_c/18 : 1_c-PE–glycophorin vesicles (200 : 1, moles lipid / mole protein; 37.5 μmoles 18 : 1_c/18 : 1_c-PE, 7.5 mg glycophorin) in the absence (A) and presence of 100 nmoles WGA (B) and 200 nmoles WGA (C) at 30°C. The molar ratios WGA/glycophorin were 0.5 and 1.0 in (B) and (C), respectively. For further details see Taraschi *et al.* (1982b).

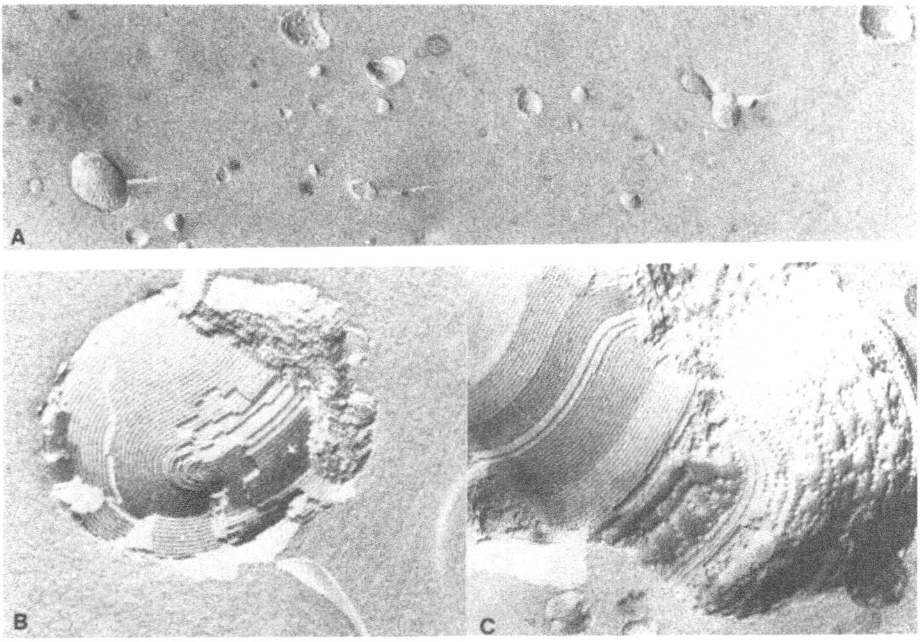

Figure 28. Freeze-fracture electron micrographs of cosonicated 18 : 1_c/18 : 1_c-PE–glycophorin vesicles (200 : 1, moles lipid/mole protein) in the absence (A) and presence (B,C) of 200 nmoles WGA (WGA/glycophorin 1.0). Final magnification × 100,000. Reproduced with permission from Taraschi *et al.* (1982b). See this reference for further details.

It is interesting to discuss these results further. Upon WGA binding to the 18 : 1_c/18 : 1_c-PE–glycophorin vesicles (200 : 1), crosslinking of the receptor head groups of glycophorin within the individual vesicles by the multivalent WGA results in aggregation of the protein into "patches," thus removing the distributed bilayer-stabilizing effect of glycophorin. As a result, large domains of lipid are freed from

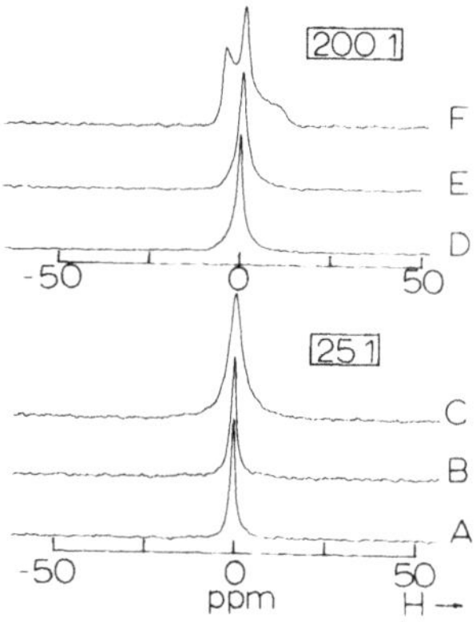

Figure 29. Effect of neuraminidase and trypsin treatment on the 81.0 MHz ^{31}P NMR spectra of cosonicated 18 : 1_c/18 : 1_c-PE–glycophorin vesicles. The molar ratios of 18 : 1_c/18 : 1_c-PE–glycophorin are indicated in the figure. Vesicles were treated with (B,E) neuraminidase (50 units/1 mg of protein, 2 hr, 37°C) followed by (C,F) trypsin treatment (5% wt./wt., 2 hr, 37°C). Reproduced with permission from Taraschi *et al.* (1982b).

the influence (both hydrophobic and steric) of the protein and may form intrabilayer hexagonal H_{II} phase cylinders in the patch or fuse with protein-free lipid areas present in adjacent vesicles to form the hexagonal H_{II} phase. It is also likely that intervesicle receptor crosslinking may act to aggregate the vesicles, allowing fusion (and consequent H_{II} phase formation) to proceed. It is interesting that lipidic particles (see Section VI) are observed by freeze-fracture electron microscopy in these preparations (see also Figure 28). In the light of the dramatic membrane destabilization resultant from WGA binding to membrane glycoprotein, it can be suggested that varying degrees of cellular-mediated protein aggregation in a membrane may trigger local structural reorganizations of the lipid, which may participate in and facilitate certain cellular functional processes.

VI. "ISOTROPIC" LIPID STRUCTURES AND LIPID PARTICLES

From the previous sections, it is clear that the hexagonal H_{II} phase is commonly observed for various hydrated membrane phospholipids. It is, however, hard to imagine that a membrane can contain extensive hexagonal H_{II} phase lipid and still maintain a permeability barrier. The possibility must be considered whether lipids can adopt alternative nonlamellar structures and if so, whether such structures are more compatible with membrane structure and function. As noted previously, many lipid systems at conditions intermediate between lamellar or H_{II} phase reveal a component experiencing "isotropic motion" as indicated by ^{31}P NMR. This is commonly observed in mixtures of bilayer and H_{II} phase-preferring lipids (see for instance Figure 14), including the total lipid extracts of various biological membranes. Although the isotropic signal can, in some cases, be attributed to the presence of small vesicles in the great majority of the systems studied, the phospholipids are organized in macroscopically large structures. From the NMR lineshape alone, it is impossible to derive information on the exact nature of these "isotropic" structures, however. Among various possibilities, lateral diffusion around highly curved bilayer surfaces or tumbling and diffusion in inverted micellar-like structures are the most attractive.

The X-ray data on these "isotropic" structures are limited, which is due to a large extent to the structural heterogeneity in such samples. Often, also, lamellar and/or hexagonal H_{II} structures are present as well as those structures allowing isotropic motion. In the case of an "isotropic" structure observed for glycolipids extracted from *A. laidlawii* cell membranes, evidence has been presented for a cubic type of phase (Wieslander *et al.*, 1981a,b) similar to an interwoven bilayer network system as shown in Figure 6.

A clearer picture on the nature of the "isotropic" structures has emerged from freeze-fracture studies (Sen *et al.*, 1982; De Kruijff *et al.*, 1979; Verkleij *et al.*, 1979a; Rand *et al.*, 1981) in such systems, which often reveal small spherical particles and pits (Figure 30). These "lipidic particles" are in some cases organized in a string or sometimes are randomly dispersed and have uniform sizes in a particular lipid system. The diameter can range from 80 to 130 Å, depending on the type of lipid (Van Venetië and Verkleij, 1981; De Kruijff *et al.*, 1979). When the H_{II} phase is also present, lipidic particles can often be seen in transitional regions where they seem to originate from

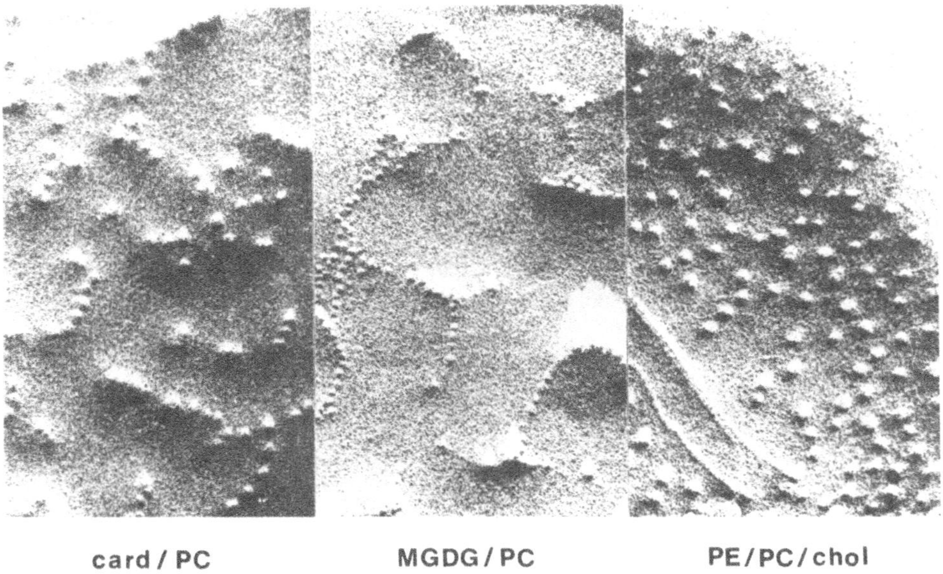

Figure 30. Freeze-fracture micrographs of lipidic particles in CL-(Ca^{2+})-PC (card/PC), monoglucosyldiglyceride-PC (MGDG/PC), and PE/PC/cholesterol (PE/PC/chol) mixtures. Reproduced with permission from De Kruijff *et al.* (1980a). See this reference for further details.

(or form) an H_{II} cylinder (Figure 31; Van Venetië and Verkleij, 1981; Verkleij *et al.*, 1980, 1982). In systems such as mixtures of 18 : 1_c/18 : 1_c-PC–18 : 1_c/18 : 1_c-PA in the presence of chlorpromazine, the particles are organized in a three-dimensional lattice (Verkleij *et al.*, 1982). Also, large irregular cusp-shaped particles have been noticed (Verkleij *et al.*, 1979a; Hui and Stewart, 1981; Verkleij and De Kruijff, 1981; Miller, 1980; De Kruijff *et al.*, 1980a; Borovjagin *et al.*, 1982; Rand *et al.*, 1981). Several models have been proposed for the molecular organization of the lipidic particles. Although conclusive evidence is lacking, the general consensus is that the homogeneously sized lipidic particles represent inverted micelles present in one bilayer (Figure 32b) or at the nexus of intersecting bilayers (Figure 32a). The larger cusp-shaped particles possibly represent intermembrane attachment sites without inverted micelles (Figure 32c). The inverted micelle, which can be considered to be a short H_{II} cylinder with closed-off ends, is a nonlamellar lipid structure which clearly is more compatible with membrane structure and function than the extended areas of H_{II} phase. It is not surprising that the observation of these particles has greatly contributed to our present ideas concerning lipid polymorphism in relation to membrane structure and to a renewed interest in the nature of intramembrane particles.

In model membrane systems containing lipidic particles, the bilayers often appear to form an interwoven network with particles present predominantly at the intersections of bilayers (Figure 33). Such honeycomb (Cullis *et al.*, 1980a) or sponge (Noordam *et al.*, 1980) structures in which membrane continuity exists with compartmentalization of aqueous spaces are reminiscent of structures found in some biological membranes, which will be discussed in Section IX. Two other aspects of lipidic particles are of interest. First, those factors which can modulate bilayer → H_{II} transitions also can regulate the presence of these particles, which is a prerequisite for a possible functional role of these structures. Second, although the lipidic particles observed in model membranes are long-lived structures, it can be expected that similar structures might be (transiently) formed at discrete sites in a membrane dependent on local environmental conditions.

The amount of isotropic ^{31}P NMR signals arising from such systems is often considerably larger than the fraction of the lipids present in lipidic particles (De Kruijff *et al.*, 1979). This might indicate the presence of short-lived lipidic particles which might escape detection by freeze-fracture electron microscopy or might indicate that the isotropic signal originates from both lipids present in inverted micellar structures and from lipids present in strongly curved bilayers present in the honeycomb type of liposomal structure (Figure 33). It is intriguing to note that ^{31}P NMR has demonstrated that both bilayer and hexagonal H_{II} type lipids can be present in these isotropic structures in relative amounts identical to those found for the entire sample (De Grip *et al.*, 1979; Vasilenko *et al.*, 1982b; De Kruijff *et al.*, 1979).

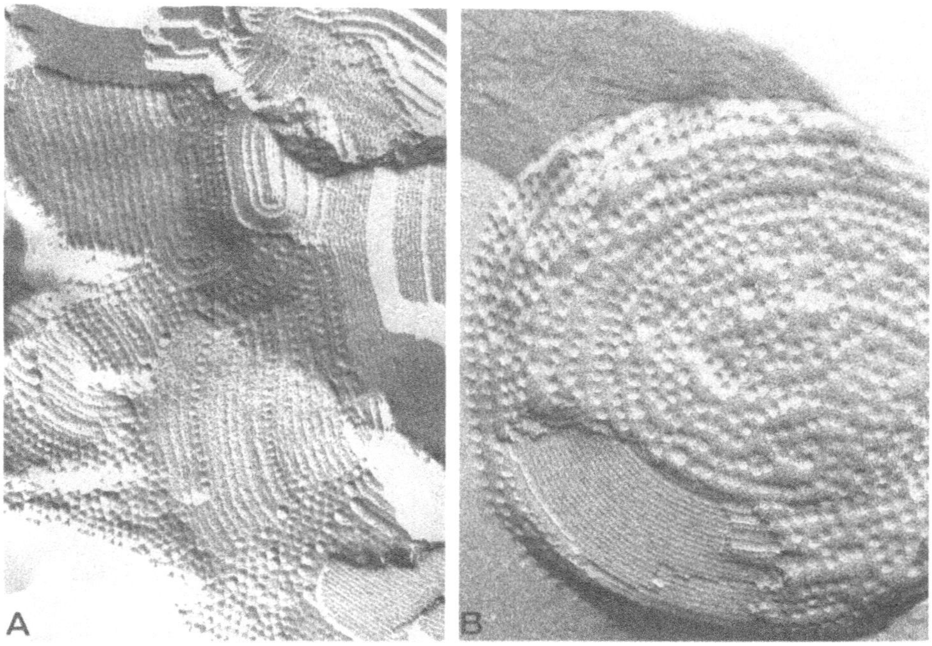

Figure 31. Freeze-fracture morphology of transitional regions between the H_{II} phase and lipidic particles in (A) chlorpromazine–18 : 1_c/18 : 1_c-PA mixtures (Verkleij *et al.*, 1982) and (B) 18 : 1_c/18 : 1_c-PC–CL (1 : 1) vesicles in the presence of 2 mM $MnCl_2$ (Verkleij *et al.*, 1980). Final magnification ×100,000. Reproduced with permission from Verkleij *et al.* (1980, 1982).

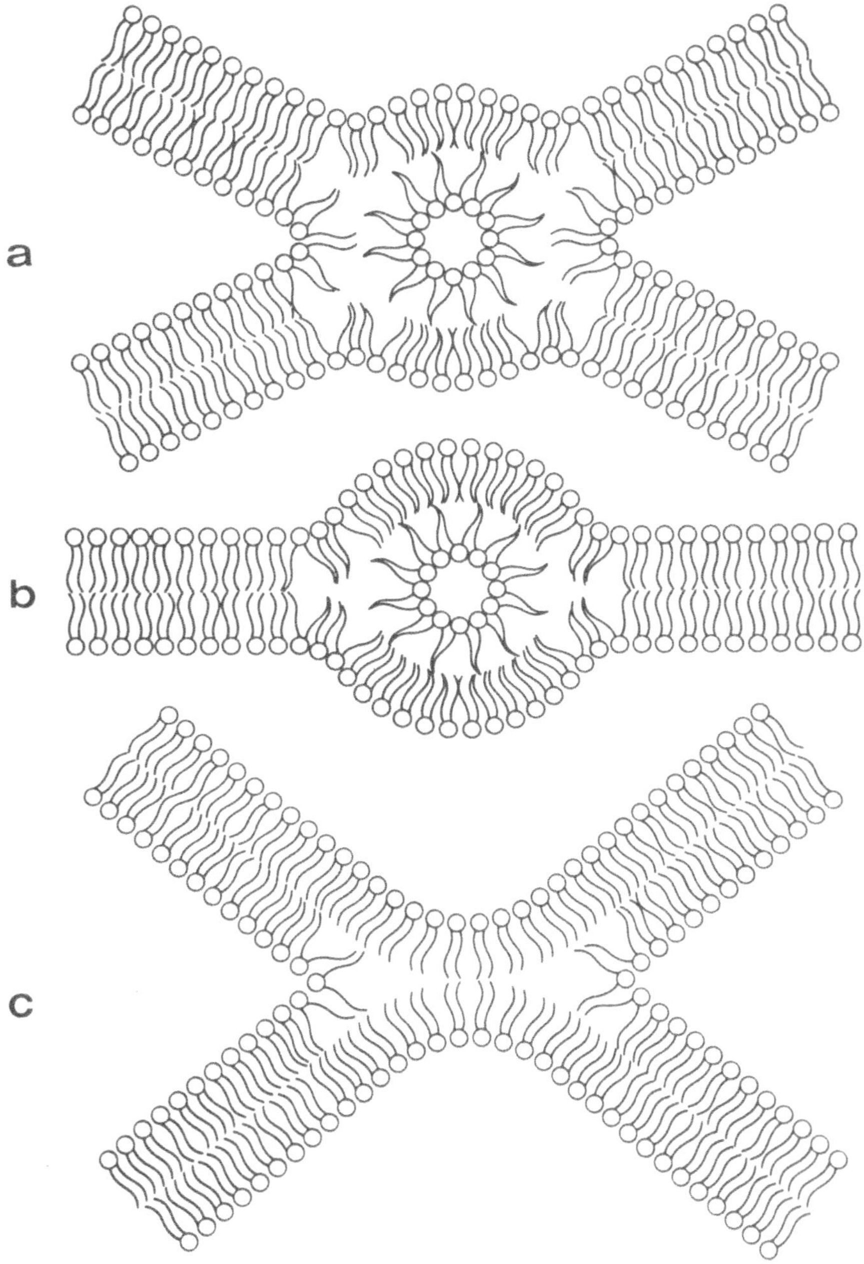

Figure 32. Inverted micelles at the nexus of intersecting bilayers (a) within one bilayer (b) and intermembrane attachment sites without inverted micelles (c).

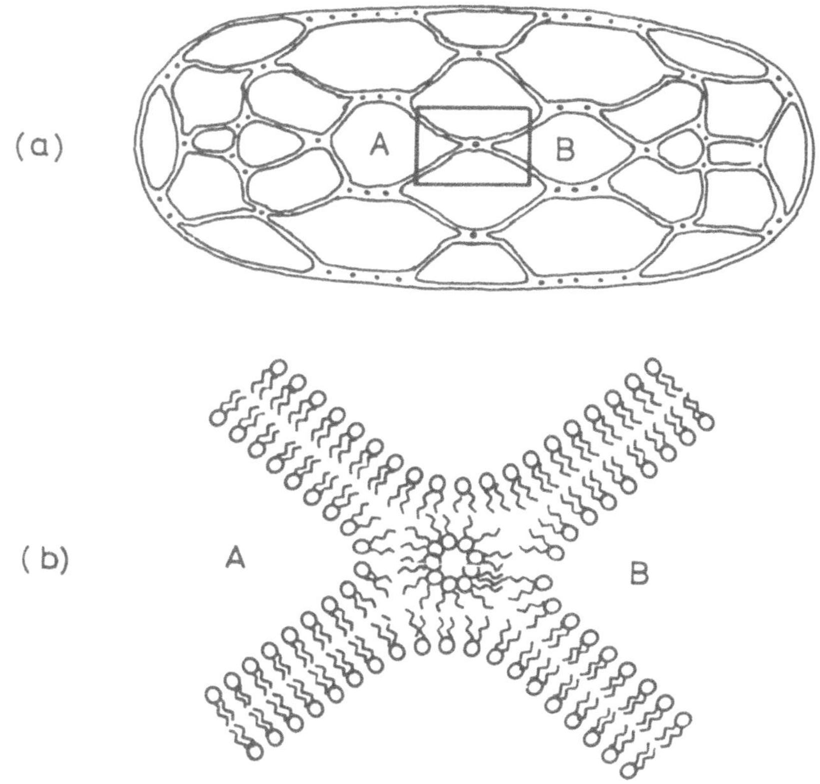

Figure 33. (a) A "honeycomb" structure compatible with ^{31}P NMR, freeze-fracture, and permeability results derived from systems containing mixtures of lipids which assume bilayer and H_{II} phase structure upon isolation. Compartmentalization within a continuous membrane structure is emphasized for compartments A and B in the expanded diagram (b).

VII. THE SHAPE CONCEPT: A RATIONALE FOR LIPID POLYMORPHISM

In order to explain the phase preferences of membrane lipids, a simple concept has been developed which relates the shape of the lipid molecule to the phase preferred by that lipid in aqueous dispersion (Cullis and De Kruijff, 1979). In this "molecular shape"-concept, lipids with a relatively large head group as compared to the cross-sectional area of the hydrocarbon chains are defined as "inverted cone"-shaped molecules. Such lipids include lysophospholipids and detergents in general, which pack optimally in micellar or H_I type configurations (Figure 34). The opposite case of "cone"-shaped lipids includes lipids where the head group is smaller and thus, inverted structures such as the H_{II} phase or inverted micelles are preferred. When the head

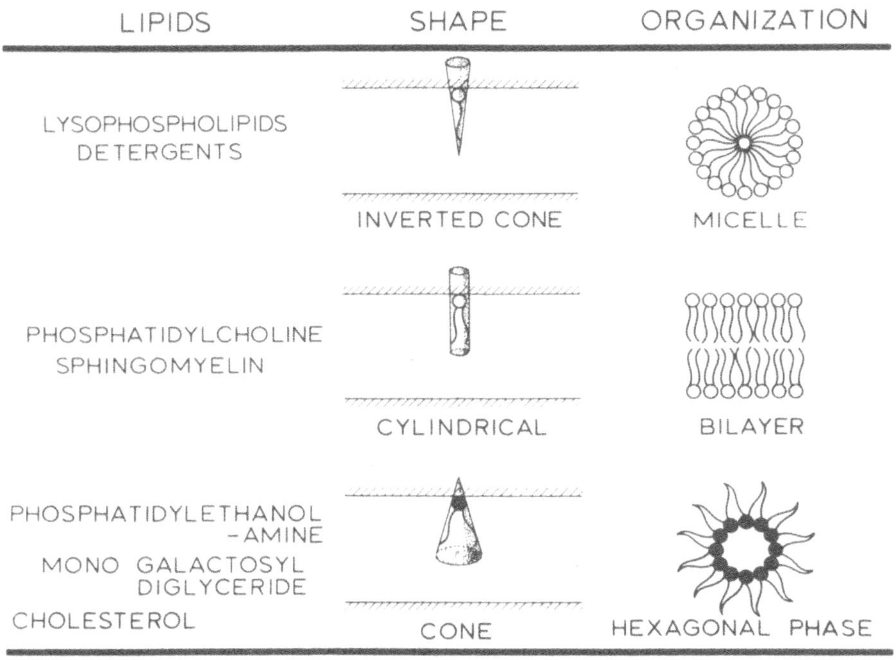

Figure 34. The molecular shape concept of lipid polymorphism.

groups and the acyl chains have comparable cross-sectional areas, the molecules have an overall cylindrical shape and organize themselves in bilayers. In some of the cubic phases which are intermediate between the lamellar and H_{II} phase, the shape of the molecules is also intermediate between that resulting in lamellar or H_{II} phase formation (Larsson *et al.*, 1980).

It should be noted that the shape concept is inclusive, and takes into account the following factors: (1) dynamic properties of the molecules (for instance fast rotation about the long axis), resulting in a time-averaged shape, (2) intra- and intermolecular interactions (in particular the strong intermolecular hydrogen bonding possible for PEs; Hauser *et al.*, 1981), and (3) the hydration properties of the lipid head groups.

Although the shape concept may initially appear naive, it describes surprisingly well many of the phase properties described in the previous sections. The temperature-dependent bilayer $\rightarrow$ H_{II} transitions can be understood in terms of an increase in hydrocarbon area with increasing temperature, resulting in an increased cone shape which eventually triggers the transition. Similarly, the more unsaturated PE species are more cone-shaped and thus have lower transition temperatures. Addition of Ca^{2+} to CL liposomes results in Ca^{2+}-binding which reduces the head group size because of dehydration, decreased electrostatic repulsion, as well as possibly decreasing the distance between the phosphates in one molecule because of intramolecular P-Ca-P salt formation (De Kruijff *et al.*, 1982b). Whereas monoglucosyldiglyceride prefers the H_{II} phase, the additional presence of another glucose in the head group to produce diglucosyldiglyceride increases the head group size and results in a preference for bilayer structure.

The shape hypothesis also allows the structural preferences of certain mixed lipid systems to be understood. In an equimolar mixture of the inverted cone-shaped lyso-PC and the cone-shaped unsaturated PE, bilayer structure is observed (Figure 35; Madden and Cullis, 1982). Also, in mixtures with cholesterol (Van Echteld *et al.*, 1981b; Rand *et al.*, 1975) and fatty acids (Jain *et al.*, 1980), which both can induce H_{II} structures (Marsh and Seddon, 1982; Cullis and De Kruijff, 1979), the lyso-PC can form bilayers.

It is important to note that the shape–structure relationship is discontinuous. Although shapes appear to be continuously variable, structural transitions occur only at discrete steps. Nevertheless, within one type of structure, for instance the lamellar phase, the molecular shape of the components might still be important to particular properties of the system. For instance, the membrane spanning portion of integral proteins most likely will have an irregular contour. It can therefore be expected that in bilayers containing different lipid species, those lipids which partition into the protein–lipid interface are those which match the shape of that part of the protein (Israelachvili, 1977). In the case of a mismatch of shapes, packing defects could exist which might affect the barrier function of the bilayer. Alternatively, protein aggregation could possibly be modulated by matching protein and lipid shapes (Israelachvili, 1977). Some evidence for these hypotheses has accumulated in recent years in studies on reconstituted lipid protein systems. The general observation is that incorporation of integral membrane proteins into artificial bilayers made from one lipid species greatly increases the bilayer permeability towards small solutes. On the other hand, incorporation into mixed lipid systems results in "sealed" vesicles. This is shown in Figure 36 for glycophorin. Whereas the protein induces a large increase in glucose permeability through 18 : 1_c/18 : 1_c-PC bilayers, glycophorin incorporation in vesicles containing the total lipid extract of the erythrocyte (in which lipids with a large variety of dynamic

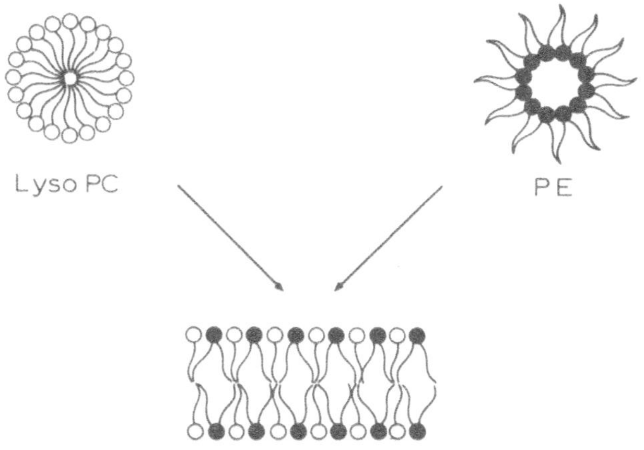

Figure 35. Molecular shape concept in mixed lipid systems.

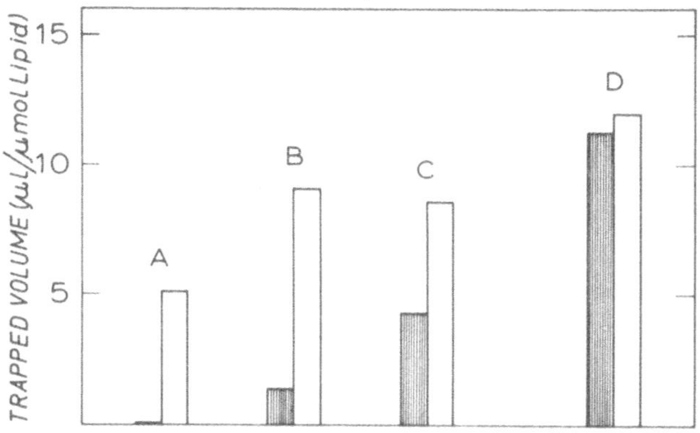

Figure 36. Glucose trap (shaded) and dextran trap (unshaded) in large unilamellar glycophorin-containing vesicles and protein-free vesicles made of various lipid mixtures. (A) Glycophorin-containing vesicles made of 18 : 1_c/18 : 1_c-PC (PC/glycophorin, 500 : 1, moles/mole). (B) Glycophorin-containing vesicles made of cholesterol-free erythrocyte lipids (lipid/glycophorin, 310 : 1, moles/mole). (C) Glycophorin-containing vesicles made of erythrocyte lipids in the presence of 45 mole% cholesterol (lipid/glycophorin, 330 : 1, moles/mole), and (D) protein-free 18 : 1_c/18 : 1_c-PC vesicles. Reproduced with permission from Van der Steen *et al.* (1981). See this reference for further details.

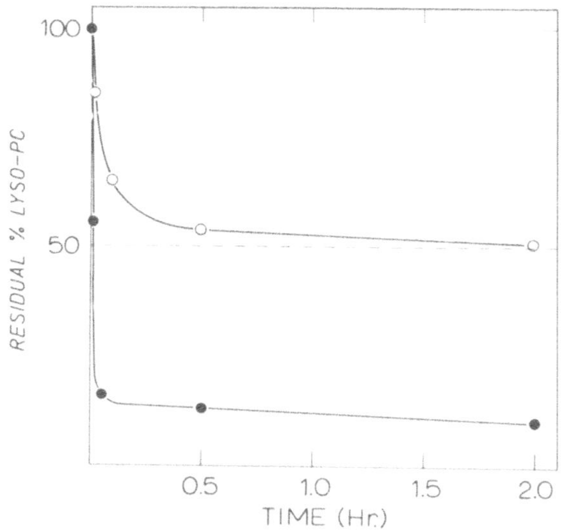

Figure 37. Time course of lyso-PC hydrolysis by lysophospholipase in glycophorin-containing large unilamellar vesicles of 18 : 1_c/18 : 1_c-PC (●—●) and total erythrocyte lipids (○—○). The vesicles contained 5 mole% of lyso-PC. For further details see Van der Steen *et al.* (1981).

shapes are present), are considerably less leaky (Van der Steen *et al.*, 1982). The presence of defects induced by the protein in the 18 : 1_c/18 : 1_c-PC bilayer is also apparent from the fast transbilayer movement of lipids in these vesicles. Lyso-PC incorporated into these vesicles is nearly completely degraded by externally added lysophospholipase, demonstrating that lyso-PC molecules rapidly move from the inner to the outer monolayer of these vesicles (Figure 37). In contrast, for the total lipid extract from the erythrocyte membrane, only 50% of lyso-PC located in the outer monolayer of the vesicles can be degraded (Van der Steen *et al.*, 1981).

Regulation of lipid composition apparently based on a balance of molecular shapes instead of "fluidity" has recently been observed for *A. laidlawii* cell membranes (Wieslander *et al.*, 1980, 1981a; Christiansson *et al.*, 1981). Growing this micro-organism in the presence of cholesterol or anaesthetics greatly influences the monoglucosyl–diglucosyldiglyceride ratio in the membrane. Similar types of adaptation were noticed when the growth temperature was varied. When the perturbation induced an increase in the cone character of the lipids, this ratio was decreased, and vice versa. Although at present these shape-conservation ideas are still speculative, the initial results are most promising and could lead to a better understanding of lipid diversity in membranes.

VIII. FUNCTIONAL ASPECTS OF LIPID POLYMORPHISM

In addition to those functional properties related to the barrier function of the membrane including the proper sealing at lipid–protein interfaces by lipids with appropriate shapes, the observation that nonlamellar lipid structures can occur in model membranes under "physiological" conditions suggests new possibilities for the functional roles of lipids in biological membranes. We discuss some of these possibilities here.

A. Fusion

Membrane fusion is an ubiquitous event in cell biology. Virtually every membrane will fuse with another membrane at some stage, and this process is very fast (in the order of milliseconds) and strictly regulated. Complexities associated with different types of fusion (for instance endocytosis, exocytosis, and cell division) and large variations in membrane composition have hampered the formulation of a universal mechanism for this process. Although in recent years it has become apparent that membrane proteins play essential roles [such as in the case of receptor-mediated endocytosis of viruses (Helenius *et al.*, 1980)], it is obvious that membrane lipids must be actively involved in the actual fusion event. In particular, it is difficult to imagine that bilayer structure of lipids at the fusion interface is continuously preserved during this step. At some stage, the lipids will have to locally adopt a (transiently) nonbilayer lipid configuration. This forms the basis for the hypothesis that H_{II}-preferring lipids, by virtue of their ability to adopt nonlamellar lipid structures, actively participate in the fusion process.

This hypothesis is now supported by a large number of observations made in model systems. Addition of fusogenic lipids to erythrocytes results in membrane fusion, which process is accompanied by the formation of the hexagonal H_{II} phase (Cullis and Hope, 1978; Hope and Cullis, 1981). The fusion process and H_{II} phase formation appear to be related as is shown in Figure 38 for the glyceromonooleate-induced fusion of human erythrocytes. Chemically related compounds such as glyceromonostearate which do not fuse erythrocytes also do not cause the H_{II} phase (Hope and Cullis, 1981). Unilamellar vesicles made of a mixture of bilayer and H_{II}-preferring lipids will fuse when they are brought under conditions in which the tendency to form the H_{II} phase is increased (Verkleij *et al.*, 1979b, 1980; Hope *et al.*, 1983). In the fused vesicles, lipidic particles are observed which often appear to be localized at the side of fusion. Figure 39 shows this for unilamellar PC/CL (1 : 1) vesicles which were fused upon the addition of Ca^{2+}. Recent rapid-freezing experiments performed on systems in which vesicle fusion could be kinetically monitored (via an assay in which mixing of aqueous compartments was measured) reveal that at the earliest moments of vesicle fusion, no lipidic particles can be observed (Verkleij *et al*, 1983; Baerer *et al.*, 1982). They only appear after several rounds of fusion. This suggests that the lipidic particles are intermediates which are stabilized by as yet unknown structural parameters. In view of the very high rate of vesicle fusion (half-times less than 1 sec), the possibility cannot be excluded that transient formation of inverted micelles during vesicle fusion might escape detection by freeze-fracture electron microscopy. It also

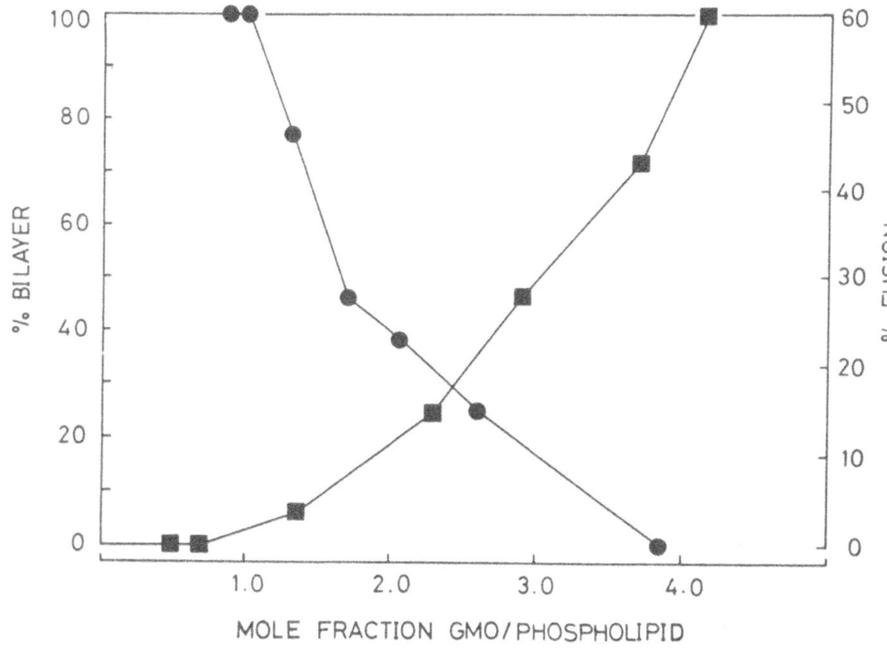

Figure 38. A comparison of the extent of fusion between erythrocytes and the amount of phospholipid remaining in the bilayer phase in erythrocyte membranes at various membrane concentrations of glyceromonooleate (GMO). Reproduced with permission from Hope and Cullis (1981).

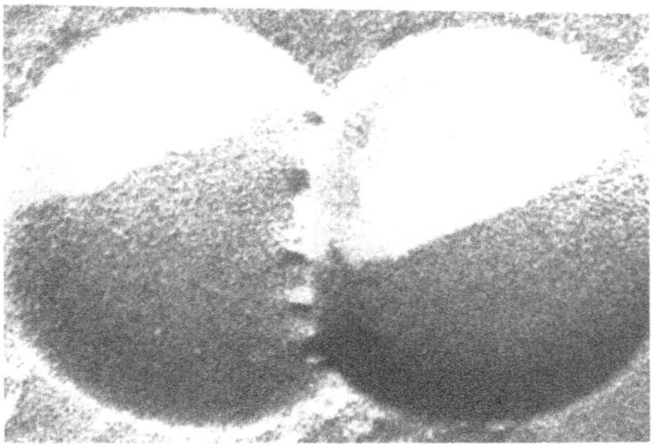

Figure 39. Lipidic particles at the site of fusion of PC-CL vesicles induced by Ca^{2+}. Final magnification ×100,000. Reproduced with permission from Verkleij *et al.* (1979a). See this reference for further details.

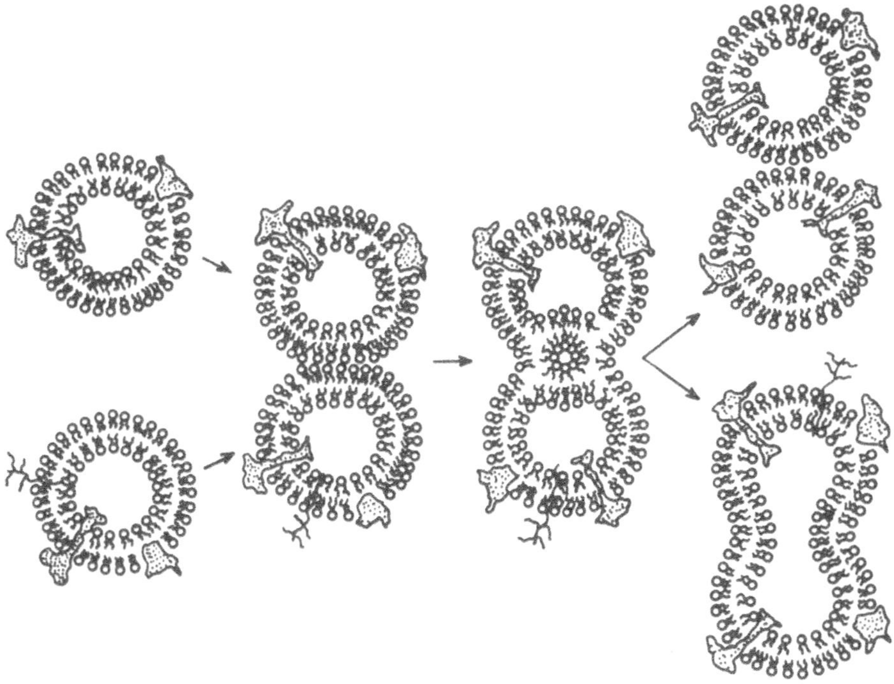

Figure 40. Proposed mechanism of membrane fusion proceeding via an inverted cylinder or inverted micellar intermediate. The process whereby the membranes come into close apposition is possibly protein-mediated, whereas the fusion event itself is proposed to involve formation of an "inverted" fusion intermediate. Reproduced with permission from Cullis *et al.* (1982).

should be noted that the presence of nonbilayer lipids is not a prerequisite for vesicle fusion. For instance, 14 : 0/14 : 0-PC vesicles fuse upon repeatedly passing the gel → liquid-crystalline phase transition (Van Dyck *et al.*, 1978) as well as PS vesicles upon the addition of Ca^{2+} (Papahadjopoulos *et al.*, 1978). However, in view of the fact that in every biological membrane, H_{II} type of lipids occur (which is not the case for PS, for instance), this suggests more general possibilities for the involvement of nonlamellar lipids in fusion events.

Prior to fusion, two membranes must move into close contact, an energetically unfavorable event because of strong intermembrane repulsion forces. The energy barrier for fusion includes head group dehydration. It is interesting that the H_{II} type of lipids have low head group hydration which will facilitate the fusion event. A model taking into account the involvement of inverted structures as fusion intermediates is shown in Figure 40.

B. Transport

Dynamic formation of inverted structures in bilayers can provide an attractive pathway for the transbilayer transport of lipids and polar solutes (Figure 41). Local compositional fluctuations or agents which trigger H_{II} structure could cause bilayer invagination, possibly followed by inverted micelle formation. When the inverted micelle subsequently "dissolves" in the opposite monolayer, transbilayer transport of lipids and polar molecules in the aqueous compartment has occurred.

The transbilayer movement of lipids is indeed increased under conditions in which lipidic particles can form as indicated by several studies (Noordam *et al.*, 1981; Gerritsen *et al.*, 1980). Egg-PE/18 : 1_c/18 : 1_c-PC/cholesterol/18 : 1_c/18 : 1_c-PA (50 : 16 : 30 : 4) dispersions undergo a temperature-dependent transition from extended bilayers to structures characterized by isotropic ^{31}P NMR signals and associated lipidic particles. This transition is accompanied by a threefold increase in the PC pool which can be exchanged by phospholipid exchange protein (Figure 42) suggesting a direct relationship between the occurrence of nonbilayer lipid structures and an increased transbilayer movement of PC (Noordam *et al.*, 1981).

Similar behavior is observed for PC/CL liposomes in which Ca^{2+} addition (which induced lipidic particle formation) is accompanied by a large increase in the exchangeable PC pool (Gerritsen *et al.*, 1980). In general, in order to act as an ionophore for a particular component, the lipid molecules must be able to form a complex with the solute to be transported which is "soluble" in a hydrophobic environment. Using a simple two-phase system it is observed that phospholipids can indeed dissolve polar solutes into an organic phase (Cullis *et al.*, 1980a; Tyson *et al.*, 1976). Table 4 shows that CL and PA effectively sequester Ca^{2+} into a chloroform phase. The neutral lipid PC is ineffective as is another negatively charged phospholipid, PS. As both CL and PA can form the H_{II} phase upon addition of Ca^{2+} this suggests a correlation between the ability of a lipid to dissolve a divalent cation into an organic phase and the formation of the H_{II} phase in aqueous dispersions. This correlation can be extended to the effect of ruthenium red, the classical inhibitor of Ca^{2+} transport in mitochondria. This molecule inhibits the uptake of Ca^{2+} by CL into the chloroform phase (Table 4) as

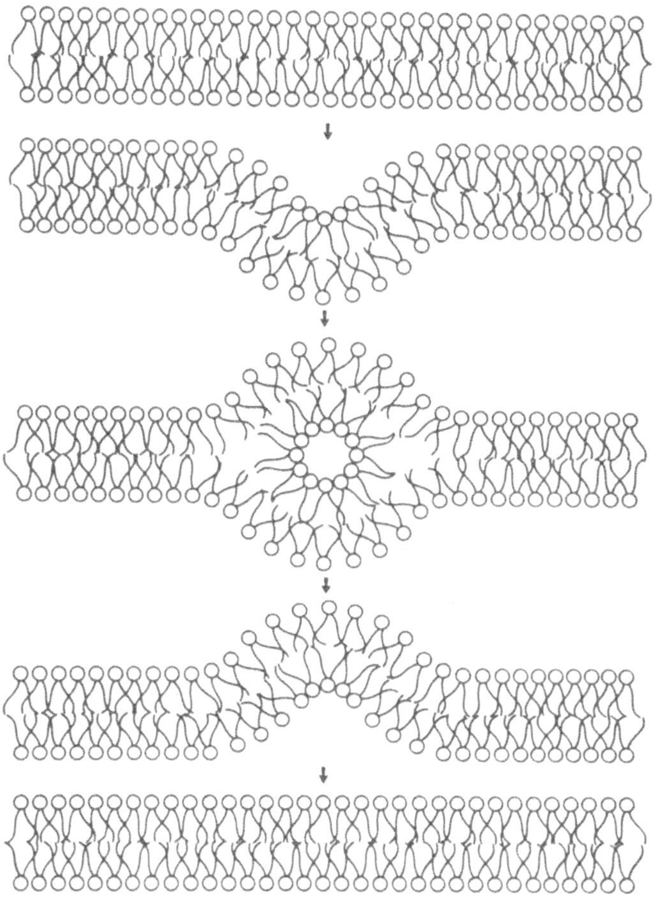

Figure 41. Dynamic formation of inverted micelles provides a permeation pathway for lipids and polar solutes.

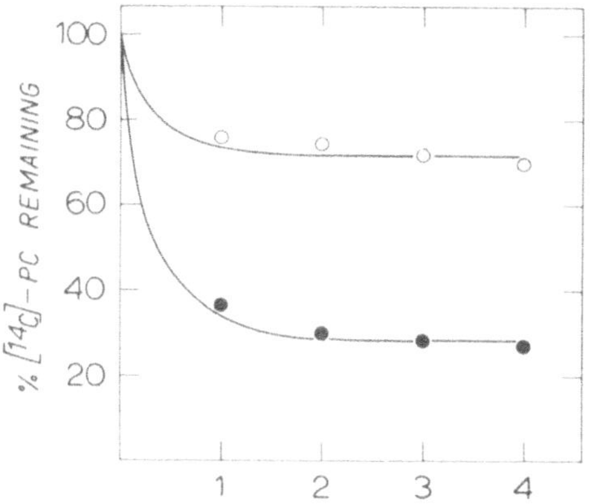

Figure 42. Phospholipid exchange protein catalyzed transfer at 20°C of [^{14}C]-18 : 1_c/18 : 1_c-PC from aqueous dispersions of PE/PC/PA/cholesterol liposomes (○——○). The exchangeable pool greatly increases upon briefly heating the liposomes to 60°C followed by cooling to 20°C (●——●). Under this condition, an isotropic ^{31}P NMR signal and lipidic particles are induced in the liposomes. See Noordam *et al.* (1981) for further details.

Table 4. Amount of Ca^{2+} Taken up into Chloroform in the Presence of Various Phospholipid Species[a]

Lipid species	Amount of Ca^{2+} taken up (μmoles)
CL	4.7
CL + ruthenium red (1 : 1)	0.5
PA	10.7
PS	1.1
PE	0.0
PC	0.4
No lipid	0.0

[a]Amount of phospholipid present in all cases corresponds to 6.0 μmoles phosphorus. For details see Cullis *et al.* (1980a).

well as the formation by Ca^{2+} of the H_{II} phase of CL (Figure 43; Cullis *et al.*, 1980a). Together with Ca^{2+} and Mn^{2+} permeability studies in liposomal systems (De Kruijff *et al.*, 1979; Gerritsen *et al.*, 1980; Serhan *et al.*, 1981), this suggests that CL and PA might act as ionophores for divalent cations. It is tempting to speculate that the receptor-mediated increase in cytosolic Ca^{2+} levels may result from an ionophoric

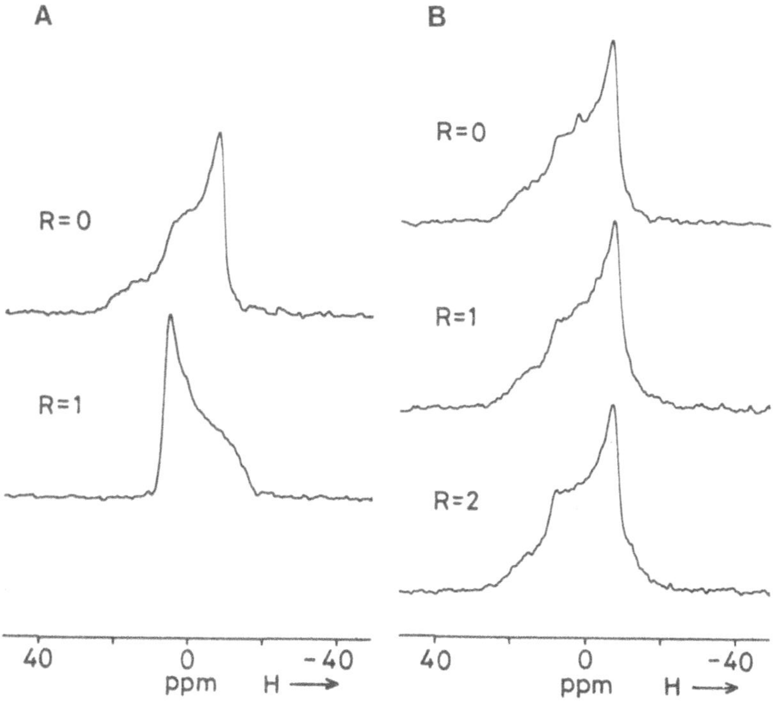

Figure 43. 81.0 MHz ^{31}P NMR spectra at 30°C of CL liposomes in the (A) absence and (B) presence of equimolar ruthenium red. The ratio *R* refers to the molar ratio of Ca^{2+} to CL. Reproduced with permission from Cullis *et al.* (1980a). See this reference for further details.

effect of PA which is formed during the receptor-mediated breakdown of PI (Salmon and Honeyman, 1980; Putney *et al.*, 1980).

C. Protein Insertion and Transport

Many proteins synthesized on free or bound ribosomes have to be inserted or transported across membranes. In recent years, considerable progress has been made in understanding the molecular details of these processes. This applies particularly to recognition and processing of the proteins. However, no universal picture has yet emerged on the exact way in which the polypeptide chain is inserted in and moved across the membrane. Some models postulate protein channels (Blobel and Dobberstein, 1975) as transport pathways, whereas in others, "spontaneous" insertion into the lipid bilayer is proposed (Wickner, 1980; Engelman and Steitz, 1981). For bacterial secretory proteins, the involvement of acidic phospholipids has recently been suggested (Nesmeyanova, 1982). In this model, the transport is thought to be coupled to transbilayer movements of negatively charged lipids linked to metabolic interconversion of CL to PG. In view of the phase properties of the bacterial lipids, hydrophilic channels (inverted short cylinders) were proposed as the structures through which the major hydrophilic part of a secretory protein is linearly translocated. An alternative but related model for protein insertion and translocation is presented in Figure 44. The partially charged leader sequence interacts with negatively charged lipids in the outer monolayer which, by adopting a nonbilayer configuration, are cotranslocated with the leader peptide. When the opposite membrane–water interface is reached, these negatively charged phospholipids are released in the inner monolayer. During translocation of hydrophobic stretches of peptide, no cotranslocation of negatively charged lipids occurs. The nonbilayer intermediate in Figure 44 resembles the semi-inverted micelle proposed recently (Fromherz, 1983).

These models are supported by the observations that proteins and peptides such as cytochrome *c* and gramicidin can induce nonlamellar lipid structures in model membranes. In particular, the notion that a mismatch in length of the hydrophobic part of a peptide and the bilayer thickness can trigger such transitions (van Echteld *et al.*, 1982) opens the possibility that during polypeptide chain synthesis and transport, the lipid structure around the polypeptide chain is responding to the progress of the insertion process.

IX. LIPID STRUCTURE IN BIOLOGICAL MEMBRANES

The model membrane studies described so far have provided a framework relating structural and functional abilities of nonbilayer lipids. We now turn to the biological membrane with two important questions in mind. First, is there any evidence for nonbilayer lipid structure in biological membranes, and second, which processes occurring in biological membranes can be better understood given these new insights regarding membrane lipid behavior? To answer these questions we analyze the structural and functional properties of selected membrane systems.

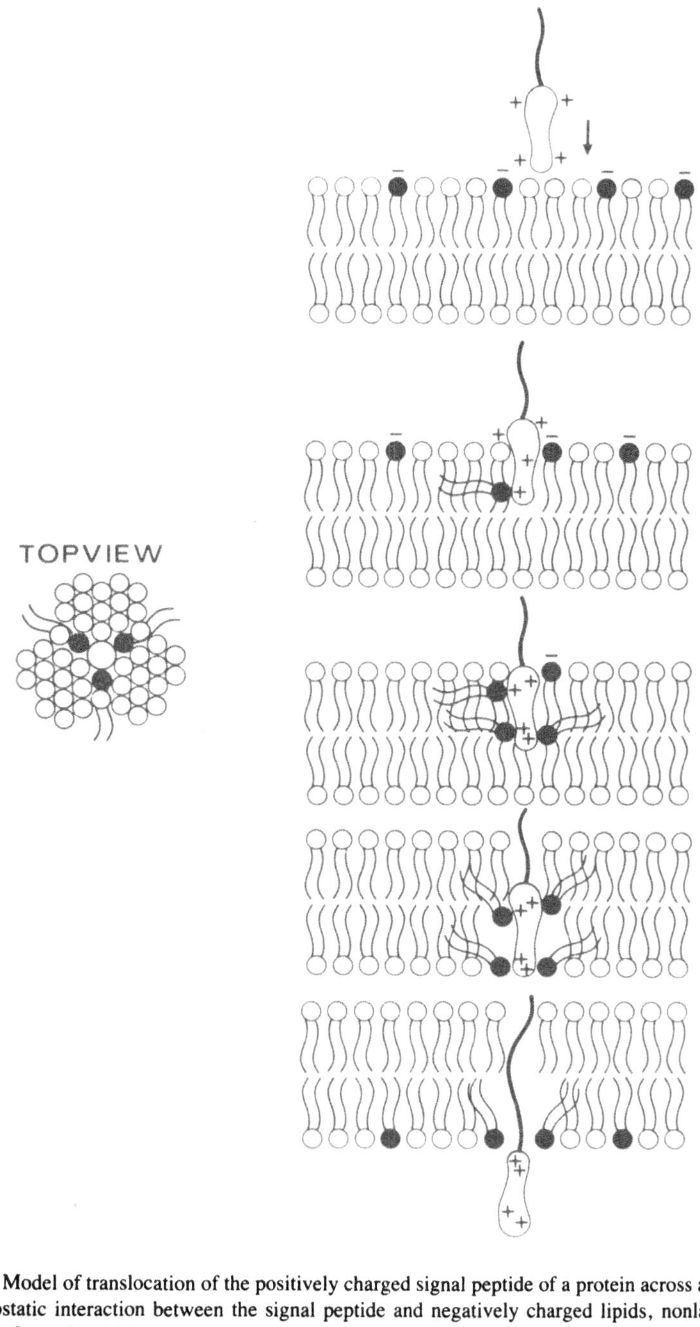

Figure 44. Model of translocation of the positively charged signal peptide of a protein across a membrane. Upon electrostatic interaction between the signal peptide and negatively charged lipids, nonlamellar lipid structures are formed, which enable the signal peptide to cross the membrane. The translocation is driven among other factors by the protein synthesis or the membrane potential. The dark line represents a hydrophobic stretch of the polypeptide chain. In the top view, it is shown that the acyl chains of the phospholipids adopting this nonlamellar configuration are shielded from aqueous medium by the head groups of the lipids in the bilayer (see also Nesmeyanova, 1982; Fromherz, 1983).

A. Erythrocyte Membrane

Within the detection limits (which are in the order of a few percent), it can be concluded that the phospholipids in the erythrocyte membrane are organized in a bilayer configuration (Cullis and Grathwohl, 1977). With the exception of the induction of the H_{II} phase by chemical fusogens (Cullis and Hope, 1978; Hope and Cullis, 1981), no other structural organization of the lipids in the erythrocyte has been observed even after such harsh treatments as extensive phospholipase degradation (Van Meer *et al.*, 1980) or proteolytic digestion of the membrane proteins (Cullis and Grathwohl, 1977). Figure 45 illustrates by ^{31}P NMR that the bilayer structure is an inherent property of the total lipid extract of that membrane. The stability of the bilayer structure might be related to the low metabolic activity of that membrane or to the large mechanical stresses the red cell has to overcome in narrow capillaries during its long lifespan in the circulation.

Special consideration has to be given to the well-known lipid asymmetry (Op den Kamp, 1979) of the erythrocyte, which appears to be typical for plasma membranes. The preferential localization of PS and the H_{II}-preferring PE in the inner monolayer suggests that the structural stability of that layer might be particularly sensitive to electrostatic interaction between PS and other polar molecules. For instance, Ca^{2+} addition to a lipid extract mimicking the inner monolayer will trigger a bilayer → H_{II} phase transition (Hope and Cullis, 1979). Phospholipid flip-flop and vesicle release might be two areas related to a possible inner monolayer instability. Whereas phospholipid flip-flop in erythrocytes normally is very slow, chemical modification of the underlying spectrin network increases the transbilayer movement of PS and PC (for review see Haest, 1982). From model membrane studies it is known that spectrin will bind to PS (Mombers *et al.*, 1979) and this binding inhibits Ca^{2+}–PS interactions. It therefore can be speculated that, upon chemical modification of spectrin, the interaction with PS is altered thereby affecting the structural properties of the inner monolayer,

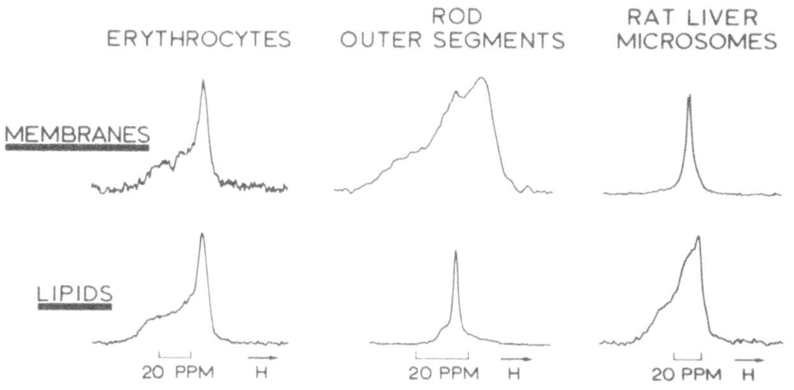

Figure 45. 36.4 MHz ^{31}P NMR spectra of intact biological membranes and derived aqueous extracts of the total membrane lipids at 37°C. For details see Cullis and de Kruiff (1979), De Grip *et al.* (1979), and De Kruijff *et al.* (1978).

possibly resulting in transbilayer transport via transiently formed inverted micellar structures. Protocols which result in an increased intracellular Ca^{2+} concentration lead to blebbing off of small vesicles from the erythrocyte membrane (Lutz *et al.*, 1977). A model for this process which takes into account the structural preferences of the inner monolayer lipids has been proposed (Cullis and De Kruijff, 1979; Cullis *et al.*, 1982). Finally, although the erythrocyte can hardly be considered as a general model for a plasma membrane, the notion that in model membranes WGA–glycophorin interactions can modulate bilayer–nonbilayer transitions (Section V-C.3) could well be relevant to lipid–receptor interactions and structural changes (for instance endocytosis of the clustered receptor–effector complex) occurring in plasma membranes.

B. Endoplasmic Reticulum (Microsomes)

The endoplasmic reticulum of eukaryotic cells is a complex intracellular membrane network and can be considered to be the main metabolic factory of the cell. It is the organelle in which the synthesis and processing of the cellular proteins, lipids, and carbohydrates occurs, and can be isolated after rupturing the cells in the form of small 1000–2000 Å diameter vesicles, which are called microsomes.

^{31}P NMR techniques have revealed that at 37°C a considerable proportion of the endogenous phospholipids of rat (De Kruijff *et al.*, 1978, 1980b), rabbit (Stier *et al.*, 1978), and beef liver (De Kruijff *et al.*, 1978) microsomes undergo rapid isotropic motion (Figure 45). It is interesting that at lower nonphysiological temperatures, ^{31}P NMR spectra typical of the lamellar organization were observed (De Kruijff *et al.*, 1978; 1980b; Stier *et al.*, 1978). The isolated lipids were found to be organized in a bilayer. Similar results were obtained from ^{13}C NMR experiments (De Kruijff *et al.*, 1980c).

In the intact perfused rat liver, a temperature-dependent bilayer → "isotropic" transition of some 15% of the endogenous rat liver membrane phospholipids can also be observed by ^{31}P NMR (De Kruijff *et al.*, 1980b). Unfortunately, NMR techniques cannot give information on the exact nature of the structure giving rise to the isotropic signal. Although lateral diffusion of lipids around highly curved bilayer surfaces could be partially responsible for this signal, the possibility that (transiently forming) inverted micelles arising from the influence of proteins such as cytochrome P450 (Stier *et al.*, 1978) give rise to such a signal cannot be excluded. This is particularly possible given the hexagonal H_{II} phase-preferring character of the PE component (De Kruijff *et al.*, 1980b), the observations that upon dehydration H_{II} phase lipid is found in microsomes (Crowe and Crowe, 1982), the network structure of the endoplasmic reticulum (Morré *et al.*, 1979), and the various metabolic processes occurring in this membrane. For instance, the phospholipid-synthesizing enzymes are asymmetrically oriented into the microsomal membrane. However, the products are found on both sides implying rapid transbilayer movements of the lipids. At 37°C, such rapid flip-flop of PC and PE has been observed (Zilversmit and Hughes, 1977; Van den Besselaar *et al.*, 1978; Hutsen and Higgins, 1982). At lower temperatures, under conditions where ^{31}P NMR indicates bilayer structure, the rate of this process is greatly decreased (Van den Besselaar *et al.*, 1978). As discussed in Section VIII-C, protein insertion and transport, which occurs on a massive scale in the rough endoplasmic reticulum, might involve dynamic structural reorganization of the lipid. For the processing and sorting of newly synthesized proteins in the endoplasmic reticulum and Golgi apparatus, membrane fusion events appear to play vital roles (Rothman, 1981). Although studies on fusion of isolated microsomes are limited (Paiement *et al.*, 1980), the fact that upon mild homogenization of the tissue of the various membranes it is only the endoplasmic reticulum network which readily breaks up into small vesicles demonstrates that this membrane has the ability to vesiculate readily.

C. The Inner Mitochondrial Membrane

For a review of the possible involvement of nonbilayer lipid structures in inner mitochondrial membrane functioning see De Kruijff *et al.* (1981). Freeze-fracturing (Van Venetië and Verkleij, 1982; Hackenbrock *et al.*, 1976) and ^{31}P NMR studies (Hackenbrock *et al.*, 1976) have indicated that the endogenous phospholipids in the inner membrane of functionally intact rat liver mitochondria at 37°C are predominantly (>95%) organized in a liquid-crystalline lamellar phase. This characteristic is maintained during oxidative phosphorylation processes (De Kruijff *et al.*, 1982a). Several observations suggest that the bilayer stability of the inner mitochondrial membrane is limited. ^{31}P NMR of isolated inner mitochondrial membrane ghosts showed a tem-

Figure 46. Freeze-fracture electron microscopic pictures of rat liver mitochondria incubated with (A) 5 mM Mn^{2+} or (B) 5 mM Ca^{2+}. The patchwork-like structure on the fracture face of the mitochondrion is indicative of the contact sites proposed in Ogawa *et al.* (1975). Reproduced with permission from Van Venetië and Verkleij (1982).

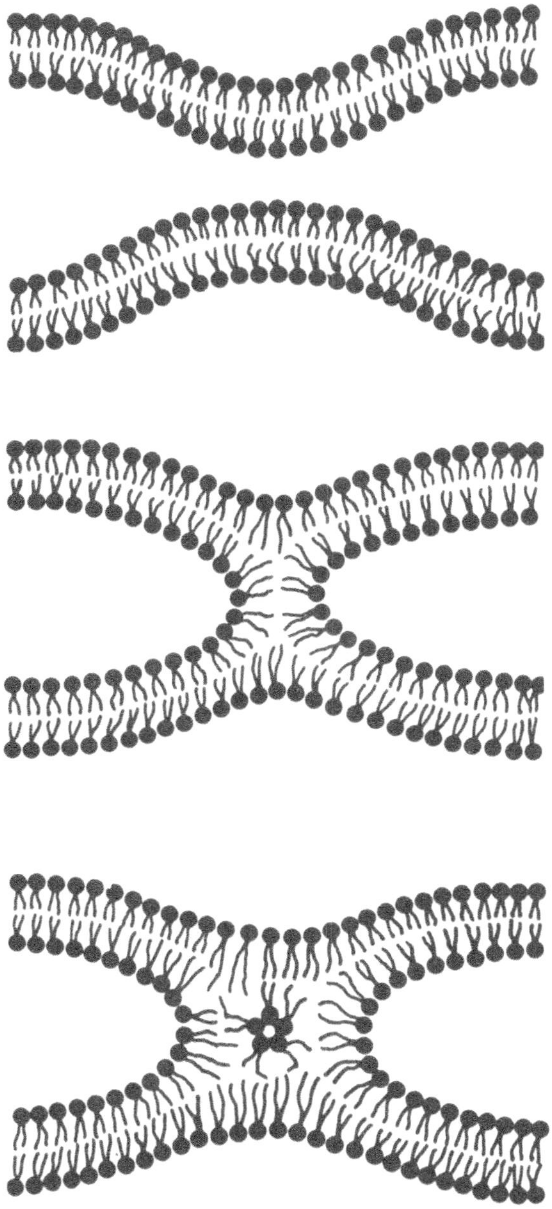

Figure 47. Structural possibilities of the lipid part at the contact site of the outer and inner mitochondrial membrane.

perature-dependent bilayer-isotropic transition such that at 37°C part of the phospholipids undergo isotropic motion (Cullis *et al.*, 1980b). This fraction is greatly increased upon the addition of Ca^{2+}. Incubation of mitochondria in divalent cation solutions has revealed some interesting freeze-fracture results (Van Venetië and Verkleij, 1982). In the presence of low concentrations of Mn^{2+} membrane, lipids segregate and form small areas of H_{II} phase in the mitochondria (Figure 46). This result is of interest because of the relatively high Mn^{2+} concentrations occurring in mitochondria (Ogawa *et al.*, 1975), and is compatible with the finding that this cation has the highest affinity for negatively charged phospholipids and promotes H_{II} phase formation in CL liposomes (Van Venetië and Verkleij, 1981). With other divalent cations, contact sites between the outer and inner membrane are induced (Figure 46). These contact sites between the inner and outer membrane were observed under all conditions which favor H_{II} phase formation of the lipids (Van Venetië and Verkleij, 1982). Figure 47 illustrates the various structural possibilities for these contact sites. Biochemical evidence supporting the existence of such sites comes from studies on the ATP-ADP translocation (Brdiczka and Kolb, 1978) and the transport of PC from the outer to the inner membrane in the presence of Ca^{2+} (Ruygrok *et al.*, 1972). These experiments make model A of Figure 47 unlikely, however. No evidence is available to discriminate between the other models. The suggestion that transitory inverted micellar structures are involved in divalent cation transport is supported by these structural changes induced by relatively high concentrations of the divalent cations. Furthermore, the finding that apocytochrome *c* which is the precursor of cytochrome *c* induces such structures (Van Venetië and Verkleij, 1982), suggests the involvement of contact sites in the import of mitochondrial proteins as has also been suggested from *in vitro* protein translocation studies (Gasser *et al.*, 1982).

D. Bacterial Membranes

In common with other membranes, spectroscopic studies have revealed that the majority of the lipids in bacterial membranes are organized in a lamellar phase. Nevertheless, several studies again indicate that local departures from bilayer structure are required for proper membrane functioning. In analogy with the mitochondrial system, the contact sites (sites of adhesion) between outer and inner membrane of gram-negative bacteria could well be formed via nonbilayer lipid structures (Lugtenberg and Van Alphen, 1983; Bayer, 1979). The fact that the isolated lipids readily adopt nonbilayer lipid structures (Burnell *et al.*, 1980b) and the rapid flip-flop of newly synthesized phospholipids (Rothman and Lenard, 1977) suggest a transport pathway via inverted micellar systems. The evidence favoring a model of translocation of secretory bacterial proteins involving nonbilayer lipid structures (Nesmeyanova, 1982) is discussed in Section VIII-C (see also Figure 44). For *A. laidlawii* it has been shown that the membrane lipid composition is regulated in such a way as to maintain a proper balance of molecular shapes (Wieslander *et al.*, 1980, 1981a; Christiansson *et al.*, 1981) suggesting the importance of both bilayer and nonbilayer lipids for membrane function.

E. Rod Outer Segment (ROS)

Evidence for existence of the hexagonal H_{II} phase in biological systems comes from freeze-fracture studies on disk membranes of frog retinal rod outer segments (Corless and Costello, 1981). Embedded within the lamellar array of ROS disk membranes, a series of H_{II} type lipid inclusions were observed. Although the functional significance of these inclusions is unknown, they may have relevance to the enhanced rates of lipid redistribution and turnover (Corless and Costello, 1981). ^{31}P NMR on bovine ROS disk membranes revealed only lamellar structures, which suggests a profound bilayer-stabilizing role of rhodopsin as, in the total lipid extract at 37°C, nonbilayer lipid structures are preferred (Figure 45; De Grip *et al.*, 1979). That the lipid composition in retinal cells is such that the membrane is very close to a transition to a hexagonal H_{II} type of organization is indicated by experiments in which liposidosis is induced chemically in rats. Under these conditions, hexagonal H_{II} phase lipid inclusions were observed in the adrenal cells and the retinal ganglia cells (Buchheim *et al.*, 1979). Possible functional roles of the H_{II}-preferring PE in this membrane might be its involvement in the sites of high bilayer curvature at the end of the disks or by allowing close proximates of the disks due to the low hydration of the PE head group.

F. Chloroplast and Prolamellar Body

The etioplasts which are found in leaves of plants grown from seed in the dark, represent an arrested stage in the normal development of proplastids into chloroplasts. In the etioplast which transforms into chloroplasts after exposure to light, a highly organized membrane structure, the prolamellar body, can be observed by electron microscopy (Simpson, 1978). Figure 48 shows the fine structure of such a structure. The prolamellar body is considered to consist of interconnected tetrahedrally-branched tubes, which greatly resembles the structure of some of the cubic phases observed in lipid–water systems (Larsson *et al.*, 1980). The functional significance of this structural organization is unknown. Of the various factors which might be responsible for the unique structure of the prolamellar body, the H_{II} phase-preferring monogalactosyldiglyceride (Shipley, 1973; Sen *et al.*, 1982) is the most likely. Since, in the chloroplast, the thykaloid membranes are organized in the form of stacked lamellae, the etioplast → chloroplast transformation resembles the cubic → lamellar transition. For further details on the relationship of nonbilayer lipids and chloroplast structure see Murphy (1982).

Another example of an interwoven bilayer network is the tubular myelin of the developing rat lung (Chi and Lagunoff, 1978). At the contact sites of the bilayers string-wise organized particles and pits have been observed by freeze-fracturing. Whether these structures are related to the lipidic particles is at yet unknown.

G. Tight Junction

Tight junctions are structures by which cells can be joined and which separate the apical and basolateral plasma membrane domains of polarized cells. Since the protein (Boulan and Sabatini, 1978) and lipid (Van Meer and Simons, 1982) com-

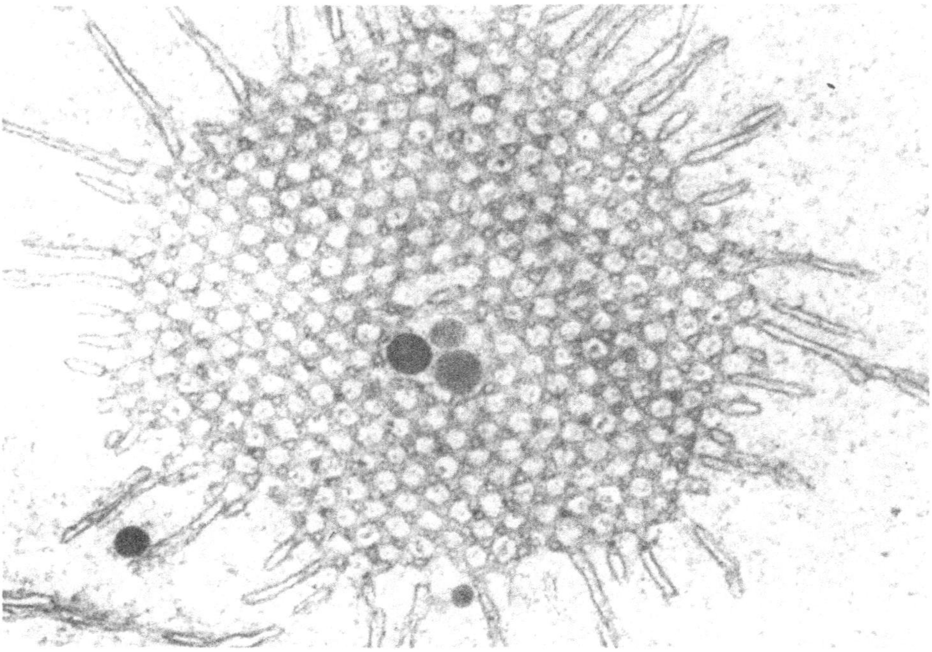

Figure 48. High-magnification micrographs of the prolamellar body of etiolated seedlings of barley. Reproduced with permission from Simpson (1978).

positions of these domains are different, it appears that the tight junction acts as a barrier to lateral diffusion. From experiments using fluorescent-labeled lipids, it could be demonstrated that lipids introduced in the outer layer of the apical part of the cell membrane remains in that layer, whereas lipids which can move to the inner layer of that part of the plasma membrane will subsequently redistribute over the entire cell membrane (Dragsten *et al.*, 1981).

Morphologically, the tight junction can be visualized in the form of tubes. From freeze-fracture results, it has been suggested (Kachar and Reese, 1982; Pinto da Silva and Kachar, 1982; Verkleij, 1980) that the tight junction is formed by lipids which are organized in intrabilayer H_{II} type of tubes (Figure 49). In this model, the cells are in a state of arrested fusion and the tubes can be considered to be stable fusion intermediates, which possibly are stabilized by membrane proteins in conjunction with membrane–cytoskeleton interactions. Evidence for similar arrested fusion intermediates has been obtained for adipose tissue of rats (Blanchette-Mackie and Scow, 1981a,b).

X. CONCLUDING REMARKS

The general picture on the structure of membranes which has emerged over the last decade is that the lipid bilayer forms the backbone of the membranes and maintains

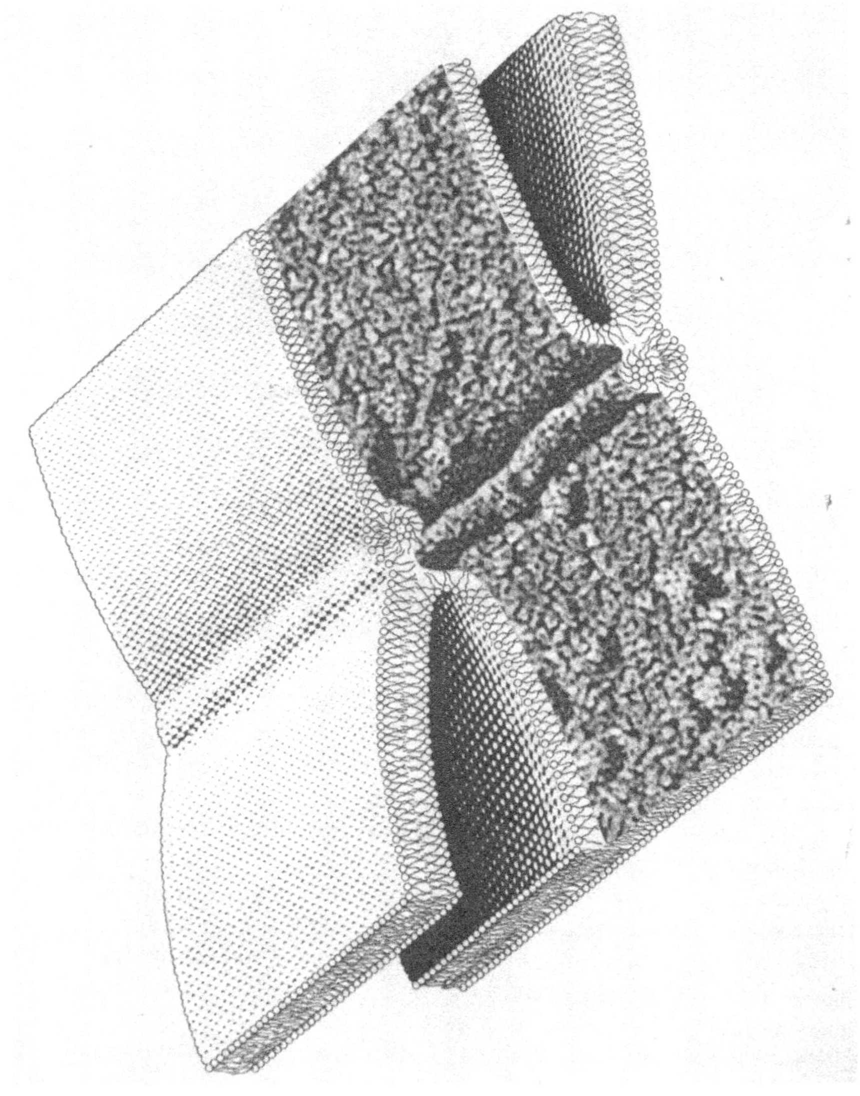

Figure 49. Diagram of a cross section of a tight junction strand combined with freeze-fracture micrographs. Reproduced with permission from Kachar and Reese (1982).

the most important permeability barrier function. However, for specific functions, local and long or shortened departures from bilayer structure must occur. The recognition of the various structural preferences of membrane lipids and the underlying molecular shape concept has opened new avenues toward a molecular understanding of these processes. Observation in model membranes of the stable structural intermediates has given insight regarding the possible molecular structure of nonbilayer lipid structures in membranes.

In the immediate future, these concepts and models will have to be tested. Due to greatly improved techniques and increased sophistication of protein and peptide chemistry, our understanding of lipid polymorphism and lipid–protein interactions should soon result in a better understanding of the vital interplay between these membrane components.

REFERENCES

Achlama, A., and Zur, Y., 1979, The electric field gradient tensor at the olefinic deuterons of potassium hydrogen maleate, *J. Magn. Res.* **36**:249–258.

Akutsu, H., and Seelig, J., 1981, Interaction of metal ions with phosphatidylcholine bilayer membranes, *Biochemistry* **20**:7366–7373.

Bachmann, L., and Schmitt, W. W., 1971, Weniger Artefakte in der Gefrieratzung durch erhohte Einfriergeschwindigkeit, *Naturwissenschaften* **58**:217–218.

Baerer, E. L., Düzgünes, N., Friend, D. S., and Papahadjopoulos, D., 1982, Fusion of phospholipid vesicles arrested by quick-freezing. The question of lipidic particles as intermediates in membrane fusion, *Biochim. Biophys. Acta* **693**:93–98.

Bayer, M. E., 1979, The fusion sites between outer membrane and cytoplasmic membrane of bacteria. Their role in membrane assembly and virus infection, in: *Bacterial Outer Membranes, Biogenesis and Functions* (M. Inoye, ed.), Wiley Interscience, New York, pp. 167–202.

Blanchette-Mackie, E. J., and Scow, R. O., 1981a, Membrane continuities within cells and intercellular contacts in white adipose tissue of young rats, *J. Ultrastruct. Res.* **77**:277–294.

Blanchette-Mackie, E. J., and Scow, R. O., 1981b, Lipolysis and lamellar structures in white adipose tissue of young rats: Lipid movement in membranes, *J. Ultrastruct. Res.* **77**:295–318.

Blaurock, A. E., 1982, Evidence of bilayer structure and of membrane interactions from X-ray diffraction analysis, *Biochim. Biophys. Acta* **650**:167–207.

Blobel, S., and Dobberstein, B., 1975, Transfer of proteins across membranes. I. Presence of proteolytically processed and unprocessed nascent immunoglobulin light chains on membrane-bound ribosomes of murine myeloma, *J. Cell Biol.* **67**:835–851.

Boggs, J. M., Stamp, D., Hughes, D. W., and Deber, C. M., 1981, Influence of ether linkage on the lamellar to hexagonal phase transition of ethanolamine phospholipids, *Biochemistry* **20**:5728–5735.

Boni, L. T., Stewart, T. P., Aldorfer, J. L., and Hui, S. W., 1981, Lipid-polyethylene glycol interactions. II. Formation of defects in bilayers, *J. Membr. Biol.* **62**:71–77.

Borovjagin, V. L., and Moshkov, D. A., 1973, A study of the ultrastructural organization of cytochrome-c-phospholipid membranes as revealed by various experimental treatments, *J. Membr. Biol.* **13**:245–262.

Borovjagin, V. L., Vergara, J. A., and McIntosh, T. J., 1982, Morphology of the intermediate stages in the lamellar to hexagonal lipid phase transition, *J. Membr. Biol.* **69**:199–212.

Boulan, E., and Sabatini, D. D., 1978, Asymmetric budding of viruses in epithelial monolayers: A model system for study of epithelial polarity, *Proc. Natl. Acad. Sci. USA* **75**:5071–5075.

Brdiczka, D., and Kolb, V., 1978, Reduction of ADP/ATP exchange rates after dissociation of the contact sites between the two boundary membranes in rat liver mitochondria, *Hoppe-Seyler's Z. Physiol. Chem.* **359**:1063–1068.

Bruckdorfer, K. R., Demel, R. A., de Gier, J., and Van Deenen, L. L. M., 1969, The effects of partial replacements of membrane cholesterol by other steroids on the osmotic fragility and glycerol permeability of erythrocytes, *Biochim. Biophys. Acta* **183**:334–345.

Buchheim, W., Drenckhahn, D., and Lüllmann-Rauch, R., 1979, Freeze-fracture studies of cytoplasmic inclusions occurring in experimental lipidosis as induced by amphiphilic cationic drugs. *Biochim. Biophys. Acta* **575**:71–80.

Büldt, G., Gally, H. U., Seelig, A., Seelig, J., and Zaccai, G., 1978, Neutron diffraction studies on selectively deuterated phospholipid bilayers, *Nature* **271**:182–184.

Burnell, E. E., Cullis, P. R., and De Kruijff, B., 1980a, Effects of tumbling and lateral diffusion on phosphatidylcholine model membrane ^{31}P-NMR lineshapes, *Biochim. Biophys. Acta* **603**:63–69.

Burnell, E., Van Alphen, L., Verkleij, A. J., De Kruijff, B., and Lugtenberg, B., 1980b, ^{31}P nuclear magnetic resonance and freeze-fracture electron microscopy studies on Escherichia coli. III. The outer membrane, *Biochim. Biophys. Acta* **597**:518–532.

Burnett, L. J., and Muller, B. H., 1971, Deuteron quadrupole coupling constants in three solid deuterated paraffin hydrocarbons: C2D6, C4D10, C6D14, *J. Chem. Phys.* **55**:5829–5831.

Chapman, D., Cornell, B. A., Eliasz, A. W., and Perry, A., 1977, Interactions of helical polypeptide segments which span the hydrocarbon region of lipid bilayers. Studies of the gramicidin A lipid-water system, *J. Mol. Biol.* **113**:517–538.

Chi, E. Y., and Lagunoff, D., 1978, Linear arrays of intramembranous particles in pulmonary tubular myelin, *Proc. Natl. Acad. Sci. USA* **75**:6225–6229.

Christiansson, A., Gutman, H., Wieslander, Å, and Lindblom, G., 1981, Effects of anaesthetics on water permeability and lipid metabolism in Acholeplasma laidlawaii membranes, *Biochim. Biophys. Acta* **645**:24–32.

Corless, J. M., and Costello, M. J., 1981, Paracrystalline inclusions associated with the disk membranes of frog retinal rod outer segments, *Exp. Eye Res.* **32**:217–228.

Crowe, L. M., and Crowe, J. H., 1982, Hydration dependent hexagonal phase lipid in a biological membrane, *Arch. Biochem. Biophys.* **217**:582–587.

Cullis, P. R., 1976a, Hydrocarbon phase transitions, heterogenous lipid distributions and lipid-protein interactions in erythrocyte membranes, *FEBS Lett.* **68**:173–176.

Cullis, P. R., 1976b, Lateral diffusion rates of phosphatidylcholine in vesicle membranes: Effects of cholesterol and hydrocarbon phase transitions, *FEBS Lett.* **70**:233–228.

Cullis, P. R., and De Kruijff, B., 1976, ^{31}P-NMR studies of unsonicated aqueous dispersions of neutral and acidic phospholipids. Effects of phase transitions p^{2}H and divalent cations on the motion in the phosphate region of the polar headgroup, *Biochim. Biophys. Acta* **436**:523–540.

Cullis, P. R., and De Kruijff, B., 1978a, Polymorphic phase behaviour of lipid mixtures as detected by ^{31}P-NMR, *Biochim. Biophys. Acta* **507**:207–218.

Cullis, P. R., and De Kruijff, B., 1978b, The polymorphic phase behaviour of phosphatidylethanolamines of natural and synthetic origin. A ^{31}P-NMR study, *Biochim. Biophys. Acta* **513**:31–42.

Cullis, P. R., and De Kruijff, B., 1979, Lipid polymorphism and the functional role of lipids in biological membranes, *Biochim. Biophys. Acta* **559**:399–420.

Cullis, P. R., and Grathwohl, Ch., 1977, Hydrocarbon phase transitions and lipid-protein interactions in the erythrocyte membrane, *Biochim. Biophys. Acta* **471**:213–226.

Cullis, P. R., and Hope, M. J., 1978, Effects of fusogenic agent on membrane structure of erythrocyte ghosts and the mechanism of membrane fusion, *Nature* **271**:672–675.

Cullis, P. R., and Hope, M. J., 1980, The bilayer stabilizing role of sphingomyelin in the presence of cholesterol. A ^{31}P-NMR study, *Biochim. Biophys. Acta* **597**:533–542.

Cullis, P. R., Verkleij, A. J., and Ververgaert, P. H. J. Th., 1978a, Polymorphic phase behaviour of cardiolipin as detected by ^{31}P-NMR and freeze fracture techniques. Effects of calcium, dibucaine and chlorpromazine, *Biochim. Biophys. Acta* **513**:*11–20*.

Cullis, P. R., Van Dijck, P. W. M., De Kruijff, B., and De Gier, J., 1978b, Effects of cholesterol on the properties of equimolar mixtures of synthetic phosphatidylethanolamine and phosphatidylcholine. A ^{31}P-NMR and differential scanning calorimetry study, *Biochim. Biophys. Acta* **513**:*21–30*.

Cullis, P. R., De Kruijff, B., Hope, M. J., Nayar, R., and Schmid, S. L., 1980a, Phospholipids and membrane transport, *Can. J. Biochem.* **58**:1091–1100.

Cullis, P. R., De Kruijff, B., Hope, M. J., Nayar, R., Rietveld, A., and Verkleij, A. J., 1980b, Structural properties of phospholipids in the rat liver inner mitochondrial membrane. A ^{31}P-NMR study, *Biochim. Biophys. Acta* **600**:625–635.

Cullis, P. R., Hornby, A. P., and Hope, M. J., 1980c, Lipid polymorphism and the molecular of anaesthesia, in: *Molecular Mechanisms of Anaesthesia, Progress in Anaesthesia,* Vol. 25 (B. R. Fisk, ed.), Raven Press, New York.

Cullis, P. R., De Kruijff, B., Hope, M. J., Verkleij, A. J., Nayar, R., Farren, S. B., Tilcock, C., Madden, T. D., and Bally, M. B., 1982, Structural properties of lipids and their functional roles in biological membranes. In: *Membrane Fluidity,* Vol. 2 (R. C. Aloia, ed.), Academic Press, New York, pp. 40–79.

Deamer, D. W., Leonard, R., Tardieu, A., and Branton, D., 1970, Lamellar and hexagonal lipid phases visualized by freeze etching, *Biochim. Biophys. Acta* **219**:47–60.

De Grip, W. J., Drenthe, E. H. S., Van Echteld, C. J. A., De Kruijff, B., and Verkleij, A. J., 1979, A possible role of rhodopsin in maintaining bilayer structure in the photoreceptor membrane, *Biochim. Biophys. Acta* **558**:330–337.

Dekker, C. J., Geurts van Kessel, W. S. M., Klomp, J. P. G., Pieters, J., and De Kruijff, B., 1983, Synthesis and polymorphic phase behaviour of polyunsaturated phosphatidylcholines and phosphatidylethanolamines, *Chem. Phys. Lipids,* in press.

De Kruijff, B., and Cullis, P. R., 1980a, The influence of poly(L-lysine) on phospholipid polymorphism. Evidence that electrostatic polypeptide-phospholipid interactions can modulate bilayer-non-bilayer transitions, *Biochim. Biophys. Acta* **601**:235–240.

De Kruijff, B., and Cullis, P. R., 1980b, Cytochrome c specifically induces non-bilayer structures in cardiolipin-containing model membranes, *Biochim. Biophys. Acta* **602**:477–490.

De Kruijff, B., Cullis, P. R., Radda, G. K., and Richards, R. E., 1976a, Phosphorus nuclear magnetic resonance of *Acholeplasma laidlawaii* cell membranes and derived liposomes, *Biochim. Biophys. Acta* **419**:411–424.

De Kruijff, B., Cullis, P. R., and Radda, G. K., 1976b, Outside–inside distributions and sizes of mixed phosphatidylcholine cholesterol vesicles, *Biochim. Biophys. Acta* **436**:729–740.

De Kruijff, B., Van den Besselaar, A. M. H. P., Cullis, P. R., Van den Bosch, H., and Van Deenen, L. L. M., 1978, Evidence for isotropic motion of phospholipids in liver microsomal membranes, A ^{31}P-NMR study, *Biochim. Biophys. Acta* **514**:1–8.

De Kruijff, B., Verkleij, A. J., Van Echteld, C. J. A., Gerritsen, W. J., Mombers, C., Noordam, P. C., and De Gier, J., 1979, The occurrence of lipid particles in lipid bilayers as seen by ^{31}P-NMR and freeze fracture electron microscopy, *Biochim. Biophys. Acta* **555**:200–209.

De Kruijff, B., Cullis, P. R., and Verkleij, A. J., 1980a, Nonbilayer lipid structures in model and biological membranes, *Trends Biochem. Sci.* **5**:79–81.

De Kruijff, B., Rietveld, A., and Cullis, P. R., 1980b, ^{31}P-NMR studies on membrane phospholipids in microsomes, rat liver slices and intact perfused rat liver, *Biochim. Biophys. Acta* **600**:343–357.

De Kruijff, B., Rietveld, A., and Van Echteld, C. J. A., 1980c, ^{13}C-NMR detection of lipid polymorphism in model and biological membranes, *Biochim. Biophys. Acta* **600**:597–606.

De Kruijff, B., Verkleij, A. J., Van Echteld, C. J. A., Gerritsen, W. J., Noordam, P. C., Mombers, C., Rietveld, A., De Gier, J., Cullis, P. R., Hope, M. J., and Nayar, R., 1981, Non-bilayer lipids and the inner mitochondrial membrane, in: *International Cell Biology, 1980–1981* (H. G. Schweig, ed.), Springer Verlag, Berlin, pp. 559–571.

De Kruijff, B., Nayar, R., and Cullis, P. R., 1982a, ^{31}P-NMR studies on phospholipid structure in membranes of intact, functionally-active, rat liver mitochondria, *Biochim. Biophys. Acta* **684**: 47–52.

De Kruijff, B., Verkleij, A. J., Leunissen-Bijvelt, J., Van Echteld, C. J. A., Hille, J., and Rijnbout, H., 1982b, Further aspects of the Ca^{2+}-dependent polymorphism of bovine heart cardiolipin, *Biochim. Biophys. Acta* **693**:1–12.

Demel, R. A., and De Kruijff, B., 1976, The function of sterols in membranes, *Biochim. Biophys. Acta* **457**:109–132.

Dragsten, P. R., Blumenthal, R., and Handler, J. S., 1981, Membrane asymmetry in epithelia: Is the tight junction a barrier to diffusion in the plasma membrane? *Nature* **294**:718–722.

Edelman, G. M., 1976, Surface modulation in cell recognition and cell growth, *Science* **192**:218–226.

Engelman, D. M., and Steitz, T. A., 1981, The spontaneous insertion of proteins into and across membranes: The helical hairpin hypothesis, *Cell* **23:**411–422.
Farren, S. B., and Cullis, P. R., 1980, Polymorphism of phosphatidylglycerol-phosphatidylethanolamine model membrane systems: A ^{31}P-NMR study, *Biochem. Biophys. Res. Commun.* **97:**182–191.
Farren, S. B., Hope, M. J., and Cullis, P. R., 1983, Polymorphic phase preferences of phosphatidic acid: A ^{31}P and ^{2}H-NMR study, *Biochem. Biophys. Res. Commun.*, in press.
Ferguson, K. A., Hui, S. W., Stewart, T. P., and Yeagle, P. I., 1982, Phase behavior of the major lipids of tetrahymena ciliary membranes, *Biochim. Biophys. Acta* **684:**179–186.
Fromherz, P., 1983, The assembly of lipids and surfactants: Molecular and phenomenological concepts, in: *Biological and Technological Relevance of Reversed Micelles and Other Amphiphilic Structures in Apolar Media* (P. L. Luisi, ed.), Plenum Press, New York, in press.
Gallay, J., and De Kruijff, B., 1982, Correlation between molecular shape and hexagonal H_{II} phase promoting ability of sterols, *FEBS Lett.* **143:**133–136.
Gally, H. U., Pluschke, G., Overath, P., and Seelig, J., 1980, Structure of *Escherichia coli* membranes. Fatty acyl chain order parameters of inner and outer membranes and derived liposomes, *Biochemistry* **19:**1638–1643.
Gasser, S. M., Ohashi, A., Daum, G., Bohni, P. C., Gibson, J., Reid, G. A., Yonetami, T., and Schatz, G., 1982, Imported mitochondrial proteins cytochrome-b2 and cytochrome-c1 are processed in two steps, *Proc. Natl. Acad. Sci. USA* **79:**267–271.
Gerritsen, W. J., De Kruijff, B., Verkleij, A. J., De Gier, J., and Van Deenen, L. L. M., 1980, Ca^{2+}-induced isotropic motion and phosphatidylcholine Flip-Flop in phosphatidylcholine-cardiolipin bilayers, *Biochim. Biophys. Acta* **598:**554–560.
Ghosh, R., and Seelig, J., 1982, The interaction of cholesterol with bilayers of phosphatidylethanolamine, *Biochim. Biophys. Acta* **691:**151–160.
Goormaghtigh, E., Van den Branden, M., Ruysschaert, J. M., and De Kruijff, B., 1982, Adriamycin inhibits the formation of non-bilayer lipid structures in cardiolipin-containing model membranes, *Biochim. Biophys. Acta* **685:**137–143.
Griffin, R. G., 1976, Observation of the effect of water on the ^{31}P nuclear magnetic resonance spectra of dipalmitoyllecithin, *J. Am. Chem. Soc.* **98:**851–853.
Gulik-Krzywicky, T., Shechter, E., Luzzatti, V., and Foure, M., 1969, Interactions of proteins and lipids: Structure and polymorphism of protein-lipid-water phases, *Nature* **223:**1116–1117.
Hackenbrock, C. R., Höchli, M., and Chau, R. M., 1976, Calorimetric and freeze fracture analysis of lipid phase transitions and lateral translational motion of intramembrane particles in mitochondrial membranes, *Biochim. Biophys. Acta* **455:**466–484.
Haest, C. W. M., 1982, Interactions between membrane skeleton proteins and the intrinsic domain of the erythrocyte membrane, *Biochim. Biophys. Acta* **694:**331–353.
Hardman, P. D., 1982, Spin-label characterization of the lamellar-to-hexagonal (H_{II}) phase transition in egg phosphatidylethanolamine aqueous dispersions, *Eur. J. Biochem.* **124:**95–101.
Harlos, K., and Eibl, H., 1980, Influence of calcium on phosphatidylglycerol. Two separate lamellar structures, *Biochemistry* **19:**896–899.
Harlos, K., and Eibl, H., 1981, Hexagonal phases in phospholipids with saturated chains: Phosphatidylethanolamines and phosphatidic acids, *Biochemistry* **20:**2888–2892.
Hartmann, W., and Galla, H. J., 1978, Binding of poly-lysine to charged bilayer membranes: Molecular organization of a lipid-peptide complex, *Biochim. Biophys. Acta* **509:**474–490.
Hauser, H., Pascher, L., Pearson, R. H., and Sunbell, S., 1981, Preferred conformation and molecular packing of phosphatidylethanolamine and phosphatidylcholine, *Biochim. Biophys. Acta* **650:**21–51.
Helenius, A., Marsh, M., and White, J., 1980, The entry of viruses into animal cells, *Trends Biochem. Sci.* **5:**104–106.
Hemminga, M. A., and Cullis, P. R., 1982, Phosphorus-31 NMR studies of orientated phospholipid multilayers, *J. Magn. Res.* **47:**307–323.
Herzfeld, J., Griffin, R. G., and Haberkorn, R. A., 1978, Phosphorus-31 chemical shift tensors in bariumdiethylphosphate and urea-phosphoric acid: Model compounds for phospholipid head-group studies, *Biochemistry* **17:**2711–2718.

Heuser, J. E., Reese, T. S., Dennis, M. J., Ian, V., Ian, L., and Evans, L., 1979, Synaptic vesicle exocytosis captured by quick freezing and correlated with quantal transmitter release, *J. Cell Biol.* **81**:275–300.

Hope, M. J., and Cullis, P. R., 1979, The bilayer stability of inner monolayer lipids from the human erythrocytes, *FEBS Lett.* **107**:323–326.

Hope, M. J., and Cullis, P. R., 1980, Effects of divalent cations and pH on phosphatidylserine model membranes: A ^{31}P-NMR study, *Biochem. Biophys. Res. Commun.* **92**:846–852.

Hope, M. J., and Cullis, P. R., 1981, The role of non-bilayer lipid structures in the fusion of human erythrocytes induced by lipid fusogens, *Biochim. Biophys. Acta* **640**:82–90.

Hope, M. J., Walker, D. C., and Cullis, P. R., 1983, Ca^{2+} and pH induced fusion of small lamellar vesicles consisting of phosphatidylethanolamine and negatively charged phospholipids: A freeze fracture study. *Biochem. Biophys. Res. Commun.* **110**:15–23.

Hornby, A. P., and Cullis, P. R., 1981, Influence of local and neutral anaesthetics on the polymorphic phase preferences of egg yolk phosphatidylethanolamine, *Biochim. Biophys. Acta* **647**:285–292.

Hui, S. W., and Stewart, T. P., 1981, "Lipidic particles" are intermembrane attachment sites, *Nature* **290**:427.

Hui, S. W., Stewart, T. P., Yeagle, P. L., and Albert, A. D., 1981, Bilayer to non-bilayer transition in mixtures of phosphatidylethanolamine and phosphatidylcholine: Implications for membrane properties, *Arch. Biochem. Biophys.* **207**:227–240.

Hutsen, J. L., and Higgins, J. A ., 1982, Asymmetric synthesis followed by transmembrane movement of phosphatidylethanolamine in rat liver endoplasmic reticulum, *Biochim. Biophys. Acta* **687**:247–256.

Huynk, S., 1973, Etude par diffraction des rayons X du systeme lipides totaux de batonnets retiniens-eau, *Biochimie* **55**:431–434.

Ioannou, P. V., and Golding, B. T., 1979, Cardiolipins: Their chemistry and biochemistry, *Progr. Lipid Res.* **17**:279–318.

Israelachvili, J. N., 1977, Refinement of the fluid-mosaic model of membrane structure, *Biochim. Biophys. Acta* **469**:221–225.

Jain, M. K., Van Echteld, C. J. A., Raminez, F., De Gier, J., De Haas, G. H., and Van Deenen, L. L. M., 1980, Association of lypophosphatidylcholine with fatty acids in aqueous phase to form bilayers, *Nature* **284**:486–487.

Janiak, M. S., Small, D. M., and Shipley, G. G., 1976, Nature of the thermal pretransition of synthetic phospholipids: Dimyristoyl- and dipalmitoyllecithin, *Biochemistry* **25**:4575–4580.

Kachar, B., and Reese, T. S., 1982, Evidence for the lipidic nature of tight junction strands, *Nature* **296**:464–466.

Khan, A., Rilfors, L., Wieslander, A., and Lindblom, B., 1981, The effect of cholesterol on the phase structure of glucolipids from *Acholeplasma laidlawaii* membranes, *Eur. J. Biochem.* **116**:215–220.

Killian, J. A., De Kruijff, B., Van Echteld, C. J. A., Verkleij, A. J., Leunissen-Bijvelt, J., and De Gier, J., 1983, Mixtures of gramicidin and lysophosphatidylcholine from lamellar structures, *Biochim. Biophys. Acta* **728**:141–144.

Kohler, S. J., and Klein, M. P., 1977, Orientation and dynamics of phospholipid head groups in bilayers and membranes determined from ^{31}P nuclear magnetic resonance chemical shielding tensors, *Biochemistry* **16**:519–526.

Koter, M., De Kruijff, B., and Van Deenen, L. L. M., 1978, Calcium-induced aggregation and fusion of mixed phosphatidylcholine-phosphatidic acid vesicles as studied by ^{31}P-NMR, *Biochim. Biophys. Acta* **514**:255–263.

Larsson, K., Fontell, K., and Krog, N., 1980, Structural relationships between lamellar, cubic and hexagonal phases in monoglyceride–water systems. Possibility of cubic structures in biological systems, *Chem. Phys. Lipids* **27**:321–328.

Lee, A. G., 1976, Interactions between phospholipids and barbiturates, *Biochim. Biophys. Acta* **455**:102–108.

Lindblom, G., Larsson, K., Johansson, L., Fontell, K., and Forsen, S., 1979, The cubic phase of monoglyceride–water systems. Arguments for a structure based upon lamellar bilayer unit, *J. Am. Chem. Soc.* **101**:5465–5470.

Lucy, J. A., 1964, Globular lipid micelles and cell membranes, *J. Theoret. Biol.* **7**:360–375.

Lugtenberg, B., and Van Alphen, L., 1983, Molecular architecture and functioning of the outer membrane of *E. coli* and other gram-negative bacteria, *Biochim. Biophys. Acta* **737**:51–115.

Lutz, H. A., Shin-Chun, L., and Palek, J., 1977, Release of spectrin-free vesicles from human erythrocytes during ATP depletion. I. Characterization of spectrin-free vesicles, *J. Cell Biol.* **73**:548–560.

Luzzati, V., 1968, X-ray diffraction studies of lipid–water systems, in: *Biological Membranes* (D. Chapman, ed.), Academic Press, New York, pp. 71–123.

Luzzati, V., Gulik-Krzywicki, T., and Tardieu, A., 1968, Polymorphism of lecithins, *Nature* **218**:1031–1034.

Madden, T. D., and Cullis, P. R., 1982, Stabilization of bilayer structure for unsaturated phosphatidylethanolamines by detergens, *Biochim. Biophys. Acta* **684**:149–153.

Mantsch, H. H., Martin, A., and Cameron, D. G., 1981, Characterization by infra-red spectroscopy of the bilayer to non-bilayer phase transition of phosphatidylethanolamines, *Biochemistry* **20**:3138–3145.

Marchesi, V. T., Tillack, T. W., Jackson, R. L., Segrest, J. P., and Scott, R. E., 1972, Chemical characterization and surface orientation of the major glycoprotein of the human erythrocyte membrane, *Proc. Natl. Acad. Sci. USA* **69**:1445–1449.

Marchesi, V. T., Furthermayr. H., and Tomita, M., 1976, The red cell membrane, *Annu. Rev. Biochem.* **45**:667–698.

Marsh, D., and Seddon, J. M., 1982, Gel-to-inverted hexagonal (L_β-H_{II}) phase transitions in phosphatidylethanolamines and fatty acid-phosphatidylcholine mixtures, demonstrated by ^{31}P-NMR spectroscopy and X-ray diffraction, *Biochim. Biophys. Acta* **690**:117–123.

McLaughlin, A. C., Cullis, P. R., Hemminga, M. A., Hoult, D. I., Radda, G. K., Ritchie, G. A., Seeley, P. J., and Richards, R. E., 1975, Application of ^{31}P-NMR to model and biological membrane systems, *FEBS Lett.* **57**:213–218.

McLaughlin, A. C., Herbette, L., Blasie, J. K., Wang, C. T., Hymel, L., and Fleischer, S., 1981, ^{31}P-NMR studies of oriented multilayers formed from isolated sarcoplasmic reticulum and reconstituted sarcoplasmic reticulum. Evidence that "boundary-layer" phospholipid is not immobilized, *Biochim. Biophys. Acta* **643**:1–16.

Miller, R. G., 1980, Do "lipidic particles" represent intermembrane attachment sites? *Nature* **287**:166–167.

Minnikin, D. E., Abdolrahimzadeh, H., and Baddiley, J., 1971, The interrelation of phosphatidylethanolamine and glycosyldiglycerides in bacterial membranes, *Biochem. J.* **124**:447–448.

Mombers, C., Verkleij, A. J., De Gier, J., and Van Deenen, L. L. M., 1979, The interaction of spectrin-actin and synthetic phospholipids. II. The interaction with phosphatidylserine. *Biochim. Biophys. Acta* **551**:271–281.

Moor, H., Kistler, J., and Muller, M., 1976, Freezing in a propane jet, *Experientia* **32**:805–815.

Morré, D. J., Kartenbeck, J., and Franke, W. W., 1979, Membrane flow and interconversions among endomembranes, *Biochim. Biophys. Acta* **559**:71–152.

Murphy, D. J., 1982, The importance of non-bilayer regions in photosynthetic membranes and their stabilization by galactolipids, *FEBS Lett.* **150**:19–27.

Nayar, R., Schmid, S. L., Hope, M. J., and Cullis, P. R., 1982, Structured preferences of phosphatidylinositol and phosphatidylinositol–phosphatidylethanolamine model membranes. Influence of Ca^{2+} and Mg^{2+}, *Biochim. Biophys. Acta* **688**:169–176.

Nesmeyanova, M. A., 1982, On the possible participation of acid phospholipids in the translocation of secreted proteins through the bacterial cytoplasmic membrane, *FEBS Lett.* **142**:189–193.

Nicholls, P., 1974, Cytochrome c binding to enzymes and membranes, *Biochim. Biophys. Acta* **346**:261–310.

Nicolson, G. L., 1976, Transmembrane control of the receptors on normal and tumor cells. I. Cytoplasmic influence over cell surface components, *Biochim. Biophys. Acta* **457**:57–108.

Noordam, P. C., Van Echteld, C. J. A., De Kruijff, B., Verkleij, A. J., and De Gier, J., 1980, Barrier characteristics of membrane model systems containing unsaturated phosphatidylethanolamines, *Chem. Phys. Lipids* **27**:221–232.

Noordam, P. C., Van Echteld, C. J. A., De Kruijff, B., and De Gier, J., 1981, Rapid transbilayer movement of phosphatidylcholine in unsaturated phosphatidylethanolamine containing model systems, *Biochim. Biophys. Acta* **646**:483–487.

Ogawa, S., Rottenberg, H., Brown, T. R., Schulman, R. G., Costello, C. L., and Glynn, P., 1975, High-resolution ^{31}P nuclear magnetic resonance study of rat liver mitochondria, *Proc. Natl. Acad. Sci. USA* **75**:1796–1800.

Op Den Kamp, J. A. F., 1979, Lipid asymmetry in membranes, *Annu. Rev. Biochem.* **48:**47–71.

Paiement, J., Beaufay, H., and Godelaine, D., 1980, Coalescence of microsomal vesicles from rat liver: A phenomenon occurring in parellel with enhancement of the glycosylation activity during incubation of stripped rough microsomes with GTP, *J. Cell Biol.* **86:**29–37.

Papahadjopoulos, D., Vail, W. J., Jacobson, K., and Poste, G., 1975a, Cochleate lipid cylinders: Formation by fusion of unilamellar lipid vesicles, *Biochim. Biophys. Acta* **394:**483–491.

Papahadjopoulos, D., Moscarello, M., Eylar, E. H., and Isac, T., 1975b, Effects of proteins on thermotropic phase transitions of phospholipid membranes, *Biochim. Biophys. Acta* **401:**317–335.

Papahadjopoulos, D., Vail, W. J., Pangborn, W. A., and Poste, G., 1976, Induction of fusion in pure phospholipid membranes by calcium ions and other divalent metals, *Biochim. Biophys. Acta* **448:** 265–283.

Papahadjopoulos, D., Portis, A., and Pangborn, W., 1978, Calcium-induced lipid phase transitions and membrane fusion, *Ann. N.Y. Acad. Sci.* **308:**50–66.

Pasvol, G., Wainscoat, J. S., and Weatherall, D. J., 1982, Erythrocytes deficient in glycophorin resist invasion by the material parasite Plasmodium falcipaurus, *Nature* **297:**64–66.

Pinto da Silva, P., and Kachar, B., 1982, On tight-junction structure, *Cell* **28:**441–450.

Putney, J. W., Weiss, S. J., Van Der Walle, C. M., and Haddas, R. A., 1980, Is phosphatidic acid a calcium ionophore under neurohumoral control? *Nature* **284:**345–347.

Rand, R. P., and Sengupta, S., 1972, Cardiolipin forms hexagonal structures with divalent cations, *Biochim. Biophys. Acta* **255:**484–492.

Rand, R. P., Tinker, D. A., and Fast, P. G., 1971, Polymorphism of phosphatidylethanolamines from two natural sources, *Chem. Phys. Lipids* **6:**333–342.

Rand, R., Pangborn, W. A., Purdas, A. D., and Tinker, D. O., 1975, Lysolecithin and cholesterol interact stoichiometrically forming bimolecular lamellar structures in the presence of excess water, or lysolecithin or cholesterol, *Can. J. Biochem.* **53:**189–195.

Rand, P. R., Reese, T. S., and Miller, R. G., 1981, Phospholipid bilayer deformations associated with interbilayer contact and fusion, *Nature* **293:**237–238.

Reiss-Husson, F., 1967, Structure des phases liquides-cristallines de differentes phospholipides, monoglycerides, sphingolipides, anhydres ou en presence d'eau, *J. Mol. Biol.* **25:**363–382.

Rivas, E., and Luzzatti, V., 1969, Polymorphisme des lipides polaires et des galacto-lipides de choroplastes de mais, en presence d'eau, *J. Mol. Biol.* **41:**261–281.

Rothman, J. E., 1981, The Golgi apparatus: Two organelles in tandem, *Science* **213:**1212–1219.

Rothman, J. E., and Lenard, J., 1977, Membrane asymmetry: The nature of membrane asymmetry provides clues to the puzzle of how membranes are assembled, *Science* **195:**743–753.

Rüppel, D., Kapitza, H.-G., Galla, H. J., Sixl, F., and Sachmann, E., 1982, On the microstructure and phase diagram of dimyristoylphosphatidylcholine–glycophorin bilayers. The role of defects and the hydrophilic lipid–protein interactions, *Biochim. Biophys. Acta* **692:**1–17.

Ruygrok, T. J. C., Van Zaane, D., Wirtz, K. W. A., and Scherphof, G. L., 1972, The effects of calcium acetate on mitochondria in the perfused rat liver. II. Enhanced transfer of phosphatidylcholine from outer to inner mitochondrial membranes, *Cytobiologie* **5:**412–421.

Salmon, D. M., and Honeyman, T. W., 1980, Proposed mechanism of cholinergic action in smooth muscle. *Nature* **284:**344–345.

Seelig, J., 1977, Deuterium magnetic resonance: Theory and application to lipid membranes, *Q. Rev. Biophys.* **10:**353–418.

Seelig, J., 1978, ^{31}P nuclear magnetic resonance and the head group structure of phospholipids in membranes, *Biochim. Biophys. Acta* **515:**105–140.

Seelig, J., and Gally, H. U., 1976, Investigation of phosphatidylethanolamine bilayers by deuterium and phosphorus-31 nuclear magnetic resonance, *Biochemistry* **15:**5199–5204.

Seelig, J., and Seelig, A., 1980, Lipid conformation in model membranes and biological membranes, *Q. Rev. Biophys.* **13:**19–61.

Seelig, J., Tamm, L., Hymel, L., and Fleischer, S., 1981, Deuterium and phosphorus nuclear magnetic resonance and fluorescence depolarization studies of functional reconstituted sarcoplasmic reticulum membrane vesicles, *Biochemistry* **20:**3922–3932.

Seeman, P., 1972, The molecular mechanism of anaesthesia, *Pharmacol. Rev.* **24:**583–655.

Sen, A., Brain, A. P. R., Quinn, P. J., and Williams, W. P., 1982, Formation of inverted lipid micelles in aqueous dispersions of mixed *sn*-3-galactosyldiacylglycerols induced by heat and ethylene glycol, *Biochim. Biophys. Acta* **686**:215–224.

Serhan, C., Anderson, P., Goodman, E., Durham, P., and Weissman, G., 1981, Phosphatidate and oxidized fatty acids are calcium ionophores. Studies employing arsenazo III in liposomes, *J. Biol. Chem.* **256**:2736–2741.

Shaw, D. O., and Schulman, J. H., 1965, Binding of metal ions to monolayers of lecithins, plasmalogen, cardiolipin, and dicetyl phosphate, *J. Lipid Res.* **6**:341–349.

Shipley, G. G., 1973, Recent X-ray diffraction studies of biological membranes and membrane components, in: *Biological Membranes*, Vol. 2 (D. Chapman and D. F. H., Wallach, eds.), Academic Press, London and New York, pp. 1–89.

Simpson, D. J., 1978, Freeze-fracture studies on barley plastid membranes. I. Wild-type etioplast, *Carlsberg Res. Commun.* **43**:145–170.

Singer, S. J., and Nicolson, G. L., 1972, The fluid mosaic model of the structure of cell membranes, *Science* **175**:720–731.

Stier, A., Finch, S. A. E., and Bösterling, B., 1978, Non-lamellar structure in rabbit liver microsomal membranes, *FEBS Lett.* **91**:109–112.

Taraschi, T. F., and Mendelsohn, R., 1980, Lipid–protein interaction in the glycophorin–dipalmitoylphosphatidylcholine system: Raman spectroscopic investigation, *Proc. Natl. Acad. Sci. USA* **77**:2362–2366.

Taraschi, T. F., De Kruijff, B., Verkleij, A. J., and Van Echteld, C. J. A., 1982a, Effect of glycophorin on lipid polymorphism. A ^{31}P-NMR study, *Biochim. Biophys. Acta* **685**:153–161.

Taraschi, T. F., Van der Steen, A. T. M., De Kruijff, B., Tellier, C., and Verkleij, A. J., 1982b, Lectin-receptor interactions in liposomes: Evidence that binding of wheat germ agglutinin to glycoprotein-phosphatidylethanolamine vesicles induces nonbilayer structures, *Biochemistry* **21**:5756–5764.

Taraschi, T. F., De Kruijff, B., and Verkleij, A. J., 1983, The effect of an integral membrane protein on lipid polymorphism on the cardiolipin-Ca^{2+} system, *Eur. J. Biochem.* **129**:621–625.

Taylor, M. G., and Smith, I. C. P., 1981, A comparison of spin probe ESR, ^{2}H- and ^{31}P-nuclear magnetic resonance for the study of hexagonal phase lipids, *Chem. Phys. Lipids* **28**:119–136.

Thayer, A. M., and Kohler, S. J., 1981, Phosphorus-31 nuclear magnetic resonance spectra characteristic of phosphatidylethanolamine in the bilayer phase, *Biochemistry* **20**:6831–6834.

Tilcock, C. P. S., and Cullis, P. R., 1980, The polymorphic phase behaviour of mixed phosphatidylserine–phosphatidylethanolamine model systems as detected by ^{31}P-NMR. Effects of divalent cations and pH, *Biochim. Biophys. Acta* **641**:189–201.

Tilcock, C. P. S., and Cullis, P. R., 1982, The polymorphic phase behaviour and miscibility properties of synthetic phosphatidylethanolamines, *Biochim. Biophys. Acta* **684**:212–218.

Tilcock, C. P. S., Bally, M. B., Farren, S. B., and Cullis, P. R., 1982, Influence of cholesterol on the structural preferences of dioleoyl-phosphatidylethanolamine–dioleoylphosphatidylcholine systems: A phosphorus-31 and deuterium nuclear magnetic resonance study, *Biochemistry* **21**:4596–4601.

Tomita, M., and Marchesi, V. T., 1975, Amino-acid sequence and oligosaccharide attachment sites of human erythrocyte glycophorin, *Proc. Natl. Acad. Sci. USA* **72**:2964–2968.

Tyson, C. A., Zande, H. V., and Green, D. E., 1976, Phospholipids as ionophores, *J. Biol. Chem.* **251**:1326–1332.

Urry, D. W., 1971, The gramicidin A transmembrane channel: A proposed (L.D) Helix, *Proc. Natl. Acad. Sci. USA* **68**:672–676.

Van den Besselaar, A. M. H. P., De Kruijff, B., Van den Bosch, H., and Van Deenen, L. L. M., 1978, Phosphatidylcholine mobility in liver microsomal membranes, *Biochim. Biophys. Acta* **510**:242–255.

Van den Bosch, H., 1974, Phosphoglyceride metabolism, *Annu. Rev. Biochem.* **43**:243–277.

Van der Steen, A. T. M., De Jong, W. A. C., De Kruyff, B., and Van Deenen, L. L. M., 1981, Lipid dependence of glycophorin-induced transbilayer movement of lysophosphatidylcholine in large unilamellar vesicles, *Biochim. Biophys. Acta* **647**:63–72.

Van der Steen, A. T. M., de Kruijff, B., and De Gier, J., 1982, Glycophorin incorporation increases the bilayer permeability of large unilamellar vesicles in a lipid-dependent manner, *Biochim. Biophys. Acta* **691**:13–23.

Van Dyck, P. W. M., De Kruijff, B., Van Deenen, L. L. M., De Gier, J., and Demel, R. A., 1976, The preference of cholesterol for phosphatidylcholine in mixed phosphatidylcholine– phosphatidylethanolamine bilayers, *Biochim. Biophys. Acta* **455:**576–587.

Van Dyck, P. W. M., De Kruijff, B., Aarts, P. A. M. M., Verkleij, A. J., and De Gier, J., 1978, Transitions in phospholipid model membranes of different curvature, *Biochim. Biophys. Acta* **506:**183–191.

Van Echteld, C. J. A., Van Stigt, R., De Kruijff, B., Leunissen-Bijvelt, J., Verkleij, A. J., and De Gier, J., 1981a, Gramicidin promotes formation of the hexagonal H_{II} phase in aqueous dispersions of phosphatidylethanolamine and phosphatidylcholine, *Biochim. Biophys. Acta* **648:**287–291.

Van Echteld, C. J. A., De Kruijff, B., Mandersloot, J. G., and De Gier, J., 1981b, Effects of lysophosphatidylcholines on phosphatidylcholine and phosphatidylcholine/cholesterol liposome systems as revealed by ^{31}P-NMR, electron microscopy and permeability studies, *Biochim. Biophys. Acta* **649:** 211–220.

Van Echteld, C. J. A., De Kruijff, B., Verkleij, A. J., Leunissen-Bijvelt, J., and De Gier, J., 1982, Gramicidin induces the formation of non-bilayer structures in phosphatidylcholine dispersions in a fatty acid chain length depended way, *Biochim. Biophys. Acta* **692:**126–138.

Van Meer, G., and Simons, K., 1982, Viruses budding from either the apical or the basolateral membrane domain of MDCK cells have unique phospholipid compositions, *EMBO J.* **1:**847–852.

Van Meer, G., De Kruijff, B., Op Den Kamp, J. A. F., and Van Deenen, L. L. M., 1980, Preservation of bilayer structure in human erythrocytes and erythrocyte ghosts after phospholipase treatment, *Biochim. Biophys. Acta* **596:**1–9.

Van Venetië, R., and Verkleij, A. J., 1981, Analysis of hexagonal II phase and its relations to lipidic particles and the lamellar phase. A freeze-fracture study, *Biochim. Biophys. Acta* **645:**262–269.

Van Venetië, R., and Verkleij, A. J., 1982, Possible role of non-bilayer lipids in the structure of mitochondria, *Biochim. Biophys. Acta* **692:**397–405.

Van Venetië, R., Hage, W. J., Bluemink, J. G., and Verkleij, A. J., 1981, Propane jet-freezing: A valid ultra-rapid freezing method for preservation of temperature-dependent lipid phases, *J. Microsc.* **123:**287–292.

Van Zoelen, E. J. J., Van Dyck, P. W. M., De Kruijff, B., Verkleij, A. J., and Van Deenen, L. L. M., 1978, Effect of glycophorin incorporation on the physico-chemical properties of phospholipid bilayers, *Biochim. Biophys. Acta* **514:**9–24.

Vasilenko, I., De Kruijff, B., and Verkleij, A. J., 1982a, Polymorphic phase behaviour of cardiolipin from bovine heart and from Bacillus subtilis as detected by ^{31}P-NMR and freeze-fracture. Effects of Ca^{2+}, Mg^{2+}, Ba^{2+} and temperature, *Biochim. Biophys. Acta* **684:**282–286.

Vasilenko, I., De Kruijff, B., and Verkleij, A. J., 1982b, The synthesis and use of thionphospholipids in ^{31}P-NMR studies of lipid polymorphism, *Biochim. Biophys. Acta* **685:**144–152.

Verkleij, A. J., 1980, The nature of the intramembrane particle, *Proc. Electr. Microsc. Soc. Am.* **38:**688–691.

Verkleij, A. J., and De Kruijff, B., 1981, Reply to: "Lipidic particles" are intermembrane attachment sites, *Nature* **290:**427–428.

Verkleij, A. J., and Ververgaert, P. H. J. Th., 1978, Freeze-fracture morphology of biological membranes, *Biochim. Biophys. Acta* **515:**303–327.

Verkleij, A. J., De Kruijff, B., Ververgaert, P. H. J. Th., Tocanne, J. F., and Van Deenen, L. L. M., 1974, The influence of pH, Ca^{2+} and protein on the thermotropic behaviour of the negatively charged phospholipid phosphatidylglycerol, *Biochim. Biophys. Acta* **339:**432–437.

Verkleij, A. J., Mombers, C., Leunissen-Bijvelt, J., and Ververgaert, P. H. J. Th., 1979a, Lipidic intramembranous particles, *Nature* **279:**162–163.

Verkleij, A. J., Mombers, C., Gerritsen, W. J., Leunissen-Bijvelt, J., and Cullis, P. R., 1979b, Fusion of phospholipid vesicles in association with the appearance of lipidic particles as visualized by freeze fracturing, *Biochim. Biophys. Acta* **555:**358–362.

Verkleij, A. J., Van Echteld, C. J. A., Gerritsen, W. J., Cullis, P. R., and De Kruijff, B., 1980, The lipidic particle as an intermediate structure in membrane fusion processes and bilayer to hexagonal H_{II} transitions, *Biochim. Biophys. Acta* **600:**620–624.

Verkleij, A. J., De Maagd, R., Leunissen-Bijvelt, J., and De Kruijff, B., 1982, Divalent cations and chlorpromazine can induce non-bilayer structures in phosphatidic acid-containing model membranes, *Biochim. Biophys. Acta* **684:**255–262.

Verkleij, A. J., Van Venetie, R., Leunissen-Bijlevelt, J., De Kruijff, B., Hope, M. J., and Cullis, P. R., 1983, Membrane fusion and polymorphism, in: *Physical Methods on Biological Membranes and Their Model Systems* (F. Conti, ed.), Plenum Press, New York, in press.

Verpoorte, J. A., 1975, Purification and characterization of glycoprotein from human erythrocyte membranes, *Int. J. Biochem.* **6**:855–862.

Wallace, B. A., Veatch, W. R., and Blout, E. R., 1981, Conformation of gramicidin A in phospholipid vesicles: Circular dichroism studies of effects of ion binding, chemical modification, and lipid structure, *Biochemistry* **20**:5754–5760.

Weinstein, S., Wallace, B. A., Morrow, J. S., and Veatch, W. R., 1980, Conformation of the gramicidin A transmembrane channel: A ^{13}C nuclear magnetic resonance study of ^{13}C-enriched gramicidin in phosphatidylcholine vesicles, *J. Mol. Biol.* **143**:1–19.

Wickner, W., 1980, Assembly of proteins into membranes, *Science* **210**:861–863.

Wieslander, Å, Christiansson, A., Rilfors, L., and Lindblom, G., 1980, Lipid bilayer stability in membranes. Regulation of lipid composition in Acholeplasma laidlawii as governed by molecular shape, *Biochemistry* **19**:3650–3655.

Wieslander, Å, Christiansson, A., Rilfors, L., Khan, A., Johansson, L. B. A., and Lindblom, G., 1981a, Lipid phase structure governs the regulation of lipid composition in membranes of Acholeplasma laidlawii, *FEBS Lett.* **124**:273–278.

Wieslander, Å, Rilfors, L., Johansson, L. B. A., and Lindblom, G., 1981b, Reversed cubic phase with membrane glucolipids from Acholeplasma laidlawii, H, H and diffusion nuclear magnetic resonance measurements. *Biochemistry* **20**:730–735.

Zilversmit, D. B., and Hughes, M. E., 1977, Extensive exchange of rat liver microsomal phospholipids, *Biochim. Biophys. Acta* **469**:99–110.

Zingsheim, H. P., 1972, Membrane structure and electron microscopy. The significance of physical problems and technics (freeze etching), *Biochim. Biophys. Acta* **265**:339–366.

5

Intrinsic Protein–Lipid Interactions in Biomembranes

Jeff Leaver and Dennis Chapman

I. INTRODUCTION

The nature of the interface between membrane lipids and intrinsic protein molecules has been the subject of a good deal of research in recent years. An understanding of these interactions are important in understanding the functioning of the membrane as a whole. A number of increasingly sophisticated physical techniques have been applied to the study of protein–lipid interactions, but due to the complex interplay which can occur in even model membranes, interpretation of the data is often difficult since different techniques can yield apparently contradictory results.

In this chapter, we shall consider the various techniques which have been applied to the study of protein–lipid interactions and the results obtained. Although certain aspects are still not completely resolved, a clearer picture is beginning to emerge.

II. PROPERTIES OF BIOMEMBRANE COMPONENTS

A. Lipids

Biomembranes in general contain a broad spectrum of lipids, variations occurring in both the structure of the polar head group and the degree of unsaturation and branching of the acyl chains. Whereas the basic arrangement of phospholipids in biomembranes is a bilayer, other, nonbilayer structures, particularly inverted micelles, have been reported to occur in some biomembranes. Such structures may be important in cell fusion and the post-translational insertion of protein into membranes (see review

Jeff Leaver and Dennis Chapman • Department of Biochemistry and Chemistry, Royal Free Hospital School of Medicine, University of London, London NW3 2PF, England.

by De Kruijff *et al.*, 1982). The existence of such structures in biomembranes is still controversial and we shall confine our attentions to lipid bilayers.

When hydrated phospholipids and certain biomembranes containing little or no cholesterol are heated, changes occur in the packing of the lipids at a temperature which is specific for the lipid composition of the system. Below this "phase-transition temperature," the molecules are arranged in an ordered quasicrystalline or gel state. As the temperature is raised, the mobility of the lipid acyl chains and polar head groups increases as does the number of trans–gauche isomerizations along the chains. This less ordered arrangement is the fluid state. Lipid phase transitions have been detected by a number of techniques including differential scanning calorimetry (DSC), differential thermal analysis (DTA), nuclear magnetic resonance (NMR), electron paramagnetic resonance (EPR), X-ray diffraction, and fluorescence measurements. The structural and functional aspects of the lipid phase transition and the methods used to measure it have been reviewed by Overath and Thilo (1978). Calorimetric methods (DSC and DTA) are the techniques most frequently used to determine phase transitions by measuring changes in the enthalpy of the system. Calorimetric studies of biomembranes have recently been reviewed by McElhaney (1982).

The physical state of the lipid component of the membrane could potentially have significant effects on the properties of the proteins associated with the membrane. This will be discussed later.

B. Proteins

The proteins associated with biological membranes may be divided into two main classes: those which are associated with the hydrophilic, polar region of the membrane (the extrinsic proteins), and those associated with the hydrophobic core (the intrinsic proteins). Extrinsic proteins are unaffected by changes in the physical properties and composition of the membrane lipids and are characterized by their ease of separation from the membrane by the use of relatively mild conditions, e.g., high-ionic-strength buffers and chelating agents. In contrast, the intrinsic proteins are much more intimately associated with the membrane, which must be physically disrupted by the use of detergents and organic solvents in order to solubilize them. In their turn, the intrinsic proteins can be classified into those which completely span the membrane and those which only partially span it. Certain proteins, e.g., the F_0-F_1-ATPase, consist of a hydrophobic intrinsic region covalently coupled to a hydrophilic extrinsic portion. It is the intrinsic proteins with which we shall be concerned in this chapter.

The lipids of most biomembranes appear to exist mainly in the fluid state under normal physiological conditions, and there is considerable evidence to show that the vast majority of the membrane proteins are associated with this fluid lipid. There are exceptions to this. The purple patches of the membrane of the bacterium *Halobacterium halobium*, for example, contain a membrane-spanning protein, bacteriorhodopsin. This protein is normally arranged in a crystalline hexagonal lattice structure at very low lipid–protein ratios.

A group of extrinsic membrane-bound proteins which are important in some cells, are those which form a cytoskeleton. These proteins are located on the cytoplasmic face of the membrane, and the cytoskeleton itself is important in the control of cell shape and deformation. As will be discussed later, the lateral and rotational diffusion of certain intrinsic proteins has been found to be affected by the presence of cytoskeletal components, e.g., actin and spectrin.

III. LIPID COMPOSITION AND ENZYME ACTIVITY

In order to study the effect of membrane lipid composition on enzyme activity, various methods have been used to modify biomembranes. Model membrane systems in which specific proteins are reconstituted with lipids of a defined composition have also been used.

By using the endogenous membrane-bound lysolecithin acyltransferase to incorporate specific fatty acids into the phosphatidylcholines of isolated lymphocyte plasma membranes, Szamel and Resch (1981) produced significant changes in the activity of various membrane-bound enzymes. Whereas the incorporation of saturated fatty acids had no effect and unsaturated fatty acids had only a marginal effect on enzyme activity, incorporation of polyunsaturated fatty acids strongly modulated enzyme activities. Some enzymes were strongly activated and others were inhibited.

Similarly, Poon *et al.* (1981) modified the plasma membrane phospholipids of the murine T lymphocyte tumor EL4 and studied the effect on the activity of adenylate cyclase and a Na^+, K^+-ATPase. Both were inhibited by saturated fatty acids, whereas unsaturated fatty acids increased activity.

Certain strains of bacteria, e.g., the Mycoplasmas and mutant strains of *Escherichia coli,* lack the ability to synthesize or degrade particular fatty acids. The fatty acid composition of their membrane lipids can therefore be readily modified and the effet on enzymatic activity studied. De Kruijff *et al.* (1973) noted the effect of temperature on the activity of various membrane-bound enzymes in *Acholeplasma laidlawii* cells having different fatty acid compositions. Whereas the Arrhenius plots (specific activity vs. 1/temperature) of several of these enzymes showed no discontinuities even though the membranes underwent a phase transition in the temperature range studied, that of the Na^+, Mg^{2+}-ATPase activity showed distinct breaks. These discontinuities were found to be dependent upon the fatty acid composition of the membrane lipids and occurred at the lower ends of the lipid phase transitions. Silvius and McElhaney (1980) analyzed the Arrhenius plots of this enzyme in more detail. They suggest that the ATPase is inactivated when its "boundary lipids" undergo a phase transition which, although it is driven by the bulk lipid transition, is less cooperative.

A number of membrane functions of *E. coli* have been shown to exhibit breaks in their Arrhenius plots. Thilo *et al.* (1977) studied the correlation between the activity of β-glucoside and β-galactoside transport and the phase transition as measured using fluorescent probes in unsaturated fatty acid-requiring strains. In addition to a break in the Arrhenius plot which occurred near the middle of the phase transition temperature range, they also observed a second break in the region of the low temperature end of the transition. This was interpreted in terms of proteins partitioning between the fluid and ordered lipid domains in the membrane, and the proteins showed a 2- to 20-fold preference for the fluid domains.

A new method for modifying the viscosity of membranes has recently been reported (Leaver *et al.*, 1983). Fatty acids containing diacetylene groups in the center of the molecule have been biosynthetically incorporated into the plasma membrane of *A. laidlawii*. When cooled to below the phase transition temperature and irradiated with ultraviolet light, these groupings crosslink to form polymers (see Figure 1). The activity of NADH oxidase, an intrinsic membrane-bound protein, showed a large decrease in the polymerized membrane, whereas the activity of ribonuclease, an extrinsic protein, was unaffected. The exact mode of inhibition is as yet unclear but may reflect the increased viscosity of the membrane.

Lipid mutants of *Saccharomyces cerevisiae* have been studied. Mitochondria containing less than 20% unsaturated fatty acids lack the ability to couple phosphorylation to oxidation (Haslam *et al.*, 1971). Furthermore, mitochondrial ATPase from cells containing high and low levels of sterols showed breaks in the Arrhenius plots which varied with the sterol content. The oligomycin sensitivity of the enzyme was also affected by the sterol content of the membrane which may indicate that sterol facilitates the interaction of the ATPase complex with the inhibitor (see Haslam and Al Mahdawi, 1980, for a review).

The Na^+,K^+-ATPase from rabbit kidney outer medulla has been incorporated into phospholipid vesicles (Kimbelberg and Paphadjopoulous, 1974). The temperature at which the discontinuity in the Arrhenius plot occurred varied depending upon the lipid. Enzymatic activity and membrane fluidity as measured by calorimetry were found to be linked, although the break in the Arrhenius plots and the phase transition

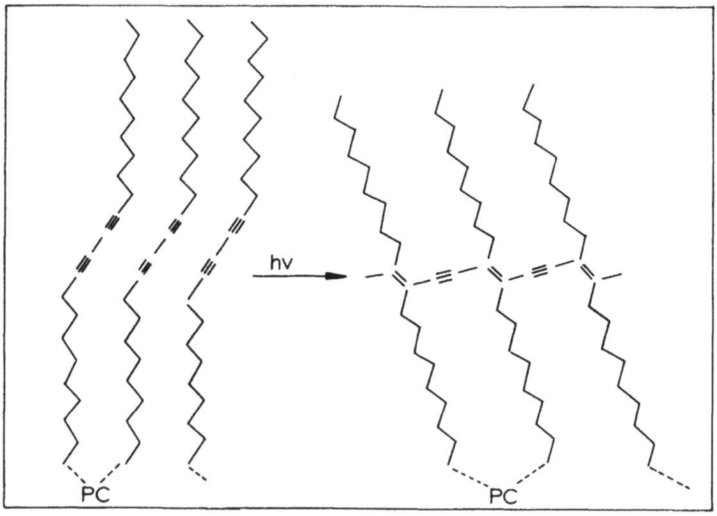

Figure 1. The formation of a phospholipid polymer by the action of light. The crosslinking forms a conjugated path so that the polymer is a colored material. From Johnston *et al.* (1980).

of the bulk of the lipid did not coincide. Cholesterol, which also decreased the fluidity of the phospholipid acyl chains, inhibited the activity of the enzyme.

Although breaks in the Arrhenius plots of membrane-bound enzymes have often been correlated with changes in the physical state of the membrane phospholipids, such breaks also occur in some soluble enzymes where they probably reflect changes in the protein structure. Similarly, Sondergaard (1979) studied the membrane-bound succinate-cytochrome and reductase from mitochondria obtained from Drosophila. The enzyme from a mutant strain exhibits a break in the Arrhenius plot at 18°C which is not observed with the enzyme from the wild type. After delipidation and reactivation with various lipids, only the enzyme from the mutant strain showed a break in the Arrhenius plot indicating that this was due to a change in the protein rather than in the lipid structure.

IV. SPECIFICITY OF PROTEIN–LIPID INTERACTIONS

Are proteins associated with specific lipid classes in biomembranes? A variety of techniques have been used in an attempt to answer this question.

Phospholipid-converting enzymes, particularly the phospholipases, have been used to induce changes in the polar head-group composition of membranes. A number of studies (see review by Bonting and De Pont, 1980) with adenylate cyclase, Na^+, K^+-ATPase, and rhodopsin have shown that, provided these proteins are embedded in a phospholipid bilayer, there is no apparent specific phospholipid requirement.

Johansson *et al.* (1981) used detergents to selectively solubilize specific lipid classes from the membranes of *A. laidlawii* and determined the identity of the proteins released. Their results are consistent with the hypothesis that the extractability of membrane proteins by neutral detergents and their interaction with lipids in the membrane are partially dependent upon the molecular shapes of the lipid molecules. Other workers have reported that the treatment of membranes with phospholipase C causes the release of some proteins from the membrane suggesting that protein–phosphatidylinositol interactions may occur in some systems (see Finean and Shukla, 1980 for a short review).

Phosphatidylserine has been reported to be the only phospholipid effective in activating the Ca^{2+}-activated, phospholipid-dependent protein kinase from rat brain in the presence of micromolar concentrations of calcium (Kaibuchi *et al.*, 1981). Other phospholipids displayed positive or negative cooperativity in the activation of the kinase.

Cytochrome oxidase and complex I and III of the mitochondrial respiratory chain have an absolute requirement for cardiolipin for enzymatic activity (Fry and Green, 1980, 1981). The cytochrome oxidase has two tightly bound cardiolipin molecules per molecule of protein and these are essential for enzymatic activity. These lipid molecules appear to be buried within the protein complex, fulfilling some kind of catalytic function.

The relative binding constants of various lipids to Ca^{2+}-ATPase have been measured by the technique of fluorescence quenching which involves determining the quenching of the fluorescence from tryptophan and tyrosine residues of the protein by contact with a nitroxide spin-labeled lipid (London and Feigenson 1981). At neutral pH, there is no selective binding, but negative phospholipids bind somewhat more tightly as the protein becomes positively charged.

Differences in the binding affinity of various spin-labeled derivatives of 14-proxylstearic acid to Na^+, K^+-ATPase have been reported (Brotherus *et al.*, 1980). The binding affinity decreased in the order negative > neutral > positive charge. High salt concentrations reversibly abolished any selectivity indicating the importance of change in the binding equilibria.

V. DISTRIBUTION OF PROTEINS IN MEMBRANES

Freeze-fracture electron microscopy of various cell types and of protein–lipid recombinants has revealed the presence of numerous granular particles in the hydrophobic fracture faces of the bilayers. Above the membrane transition temperature, these particles are randomly distributed in the plane of the membrane. Below the transition temperature, they were aggregated. Since such particles have not been observed in pure lipid systems, they were believed to be due to the presence of the membrane proteins. Phase separation at temperatures below the phase transition apparently produced protein-rich and lipid-rich patches.

In order to confirm that these particles were proteinaceous, Tourtellote and Zuprik (1973) modified the protein–lipid content of *A. laidlawii* cells. Growing cells were transferred to medium containing puromycin or lacking nitrogen. Under these conditions, protein biosynthesis is blocked, whereas lipid synthesis continues. Freeze-fracture electron microscopy showed that the density of the particles was greatly decreased in the absence of protein biosynthesis confirming their identity as proteins.

This temperature-dependent phase separation is important in the interpretation of a number of the results obtained from other studies of protein–lipid interactions. This will be discussed in more detail in later sections.

VI. PERTURBATION OF LIPID DYNAMICS BY INTRINSIC PROTEINS

The incorporation of large, irregularly shaped protein molecules into a lipid bilayer would be expected to affect the packing and physical properties of the lipid molecules in the neighborhood of the protein (Figure 2). Information regarding this aspect of membrane dynamics has been obtained by a number of techniques. Indications of the effect of protein on membrane fluidity have been obtained from the fluorescence polarization of lipophilic probes. The number of lipid molecules influenced by the addition of protein can be determined from DSC, and the effect of proteins on lipid mobility can be measured by EPR and NMR.

A. NMR and EPR Spectroscopy

Both NMR and EPR have been used to study the effect of protein on the ordering of lipid molecules in membranes. While NMR measures the properties of nuclei such

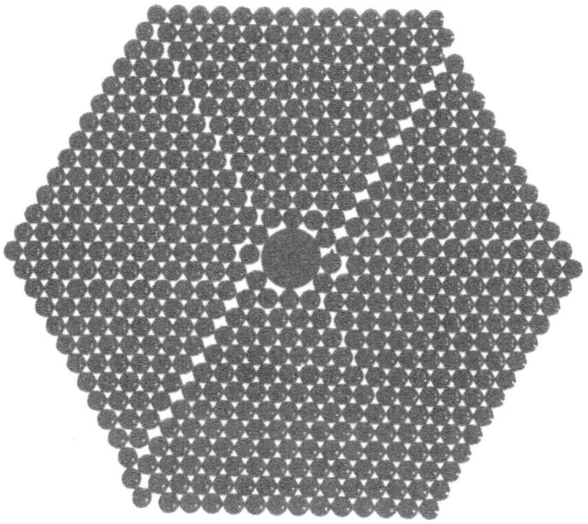

Figure 2. The packing of a large (polypeptide) and small (lipid chains) circles in a hexagonal array. The large circle diameter corresponds to a size of 1.18 nm and the small circle diameter to 0.48 nm. The ratios below each arrangement correspond to the ratio of small to large particles in a continuous two-dimensional array of that particular packing configuration. After Chapman *et al.* (1977).

as ^1H, ^2H, ^{13}C, and ^{31}P which occur naturally or which can be incorporated by isotope enrichment and, therefore, do not perturb the natural system, EPR requires the addition of spin labels either to the lipid phase or covalently bound to protein molecules. The interpretation of the results obtained by these two techniques has been the subject of some controversy.

Early EPR studies on cytochrome oxidase in lipid bilayers (Jost *et al.*, 1973) showed the presence of two distinct spectral components: one which they attributed to immobilized lipid at the protein–lipid boundary, and one due to the mobile bulk phase. Calculations of the size of the immobilized component suggested that it corresponded to the first layer of lipid surrounding the protein. Similarly, an immobilized boundary layer of lipid which excluded cholesterol was reported to surround the sarcoplasmic reticulum ATPase (Hesketh *et al.*, 1976; Warren *et al.*, 1975). From such studies, the concept of the lipid annulus was developed. This annulus was believed to be a long-lived, stoichiometric lipid shell bound to the protein and capable of excluding molecules such as cholesterol. The lifetime of the annulus was believed to be several minutes in some cases.

NMR studies have failed to confirm the EPR findings. Only one type of lipid environment is observed and no signal corresponding to immobilized lipid can be detected. A closer examination of the results has shown that much of the difference can be explained by consideration of the dynamic time scales of the two techniques. For EPR measurements, this is 10^{-8} sec, whereas, for NMR, it is 10^{-4} sec. A molecule can therefore appear to be immobile on the short time scale yet mobile on the longer one. Hence, the rate of exchange into and out of the boundary layer may explain the difference in the spectra obtained by the two techniques (see Figure 3 for a comparison). It is clear that the presence of an intrinsic protein does have some influence on the surrounding phospholipids. So, are the differences between EPR and ^2H-NMR experiments merely a time-scale effect? (Jost and Griffith, 1980). This conclusion clashes with the studies of Davoust *et al.* (1980). In this case, spin-label experiments have been reported in which spin-labeled fatty acids have been covalently attached to a membrane protein, rhodopsin, and no appreciable perturbation of the fatty acyl chains occurs. (There are, however, alternative interpretations of these data.)

If we accept the idea that a time-scale effect is basically the reason for the discord, then do we need to accept the concept that there is a stoichiometric complex in all intrinsic protein–lipid systems? The answer appears to be no.

B. Fluorescence Depolarization

Fluorescence depolarization studies have been performed with biomembranes and also model systems of defined composition. The technique involves incorporating a fluorescent lipophilic probe such as 1,6-diphenyl-1,3,5-hexatriene (DPH) into the membrane. In the fluorescence of such probes, the excitation and emission dipoles lie in the plane of the molecule and are displaced by a small angle. Depolarization of the fluorescence can be measured by exciting the probe with vertically polarized monochromatic light and determining the intensity of the emission through analyzers orientated parallel or perpendicular to the direction of the excitation beam. Rotational depolarization of the fluorescence will occur if the excited molecules are displaced from their original orientation during the lifetime of the excited state. From the polarization value, information regarding the tumbling motion of the probe and, hence, the microviscosity of the region around it can be obtained.

Increasing the protein–lipid ratio in model systems causes a steady decrease in the mobility of the probe suggesting that the addition of protein results in an increase in the ordering of the lipid chains or in the viscosity of the lipid. However, as noted by Hoffman *et al.* (1981), the probe itself can perturb its environment and, furthermore, the DPH molecule is larger and more rigid than the C–C bonds of the acyl chain. Hence, it may not be sensitive to small changes in the order of the $-CH_2$ groups. Furthermore, in the presence of protein, the motion of the probe molecules may be affected mainly by collisions with protein molecules. Fluorescence depolarization measurements may, therefore, only give a good indication of the microenvironment of the probe molecules rather than of the fluidity of the membrane as a whole.

Discrepancies regarding membrane fluidity as measured in biomembranes by this technique have also been reported. Kleinfield *et al.* (1981) compared fluorescence depolarization in biomembranes with the lateral diffusion of lipid and protein probes as measured by fluorescence recovery after photobleaching (see Section IX-A). They reported a distinct lack of correlation between DPH polarization and macroscopic viscosity and concluded that, although fluorescence polarization measurements may be sensitive indicators of local lipid packing, they do not necessarily give information regarding the rates of lateral mobility of lipids and proteins. This may be due to limitations on the mobility of membrane components due to interactions with the cell cytoskeleton. A number of other studies have also cast doubt on the simple interpretation of fluorescence depolarization measurements.

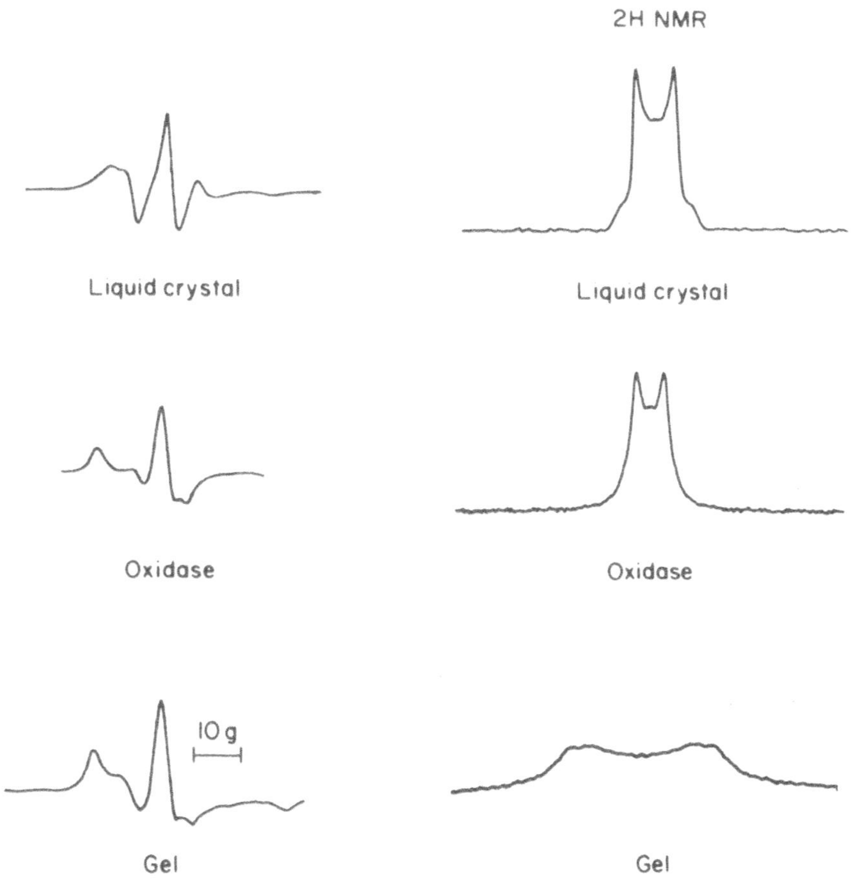

Figure 3. A comparison between ESR and a [^{2}H]-NMR view of hydrocarbon chain order in gel, fluid, and cytochrome oxidase lipid (boundary lipid). From Oldfield *et al.* (1978).

Recent fluorescent probe studies of reconstituted systems are interesting with regard to this situation (Hoffman *et al.*, 1981). A study of the polarization of the probe (in reconstituted systems of an intrinsic polypeptide, gramicidin A, or of various intrinsic proteins as the concentration of the intrinsic molecule increases) shows that the value of the polarization P reaches a limited value. Empirically, each of the curves has been observed to fit a simple exponential equation. The value of the exponent is, in each equation, related to the number of probe molecules or lipids which surround the intrinsic molecule. Polarization data for various mixtures and recombinants are shown in Figure 4.

A careful theoretical analysis using probability theory shows that the occurrence of such exponential curves is consistent with the occurrence of a random arrangement of intrinsic molecules.

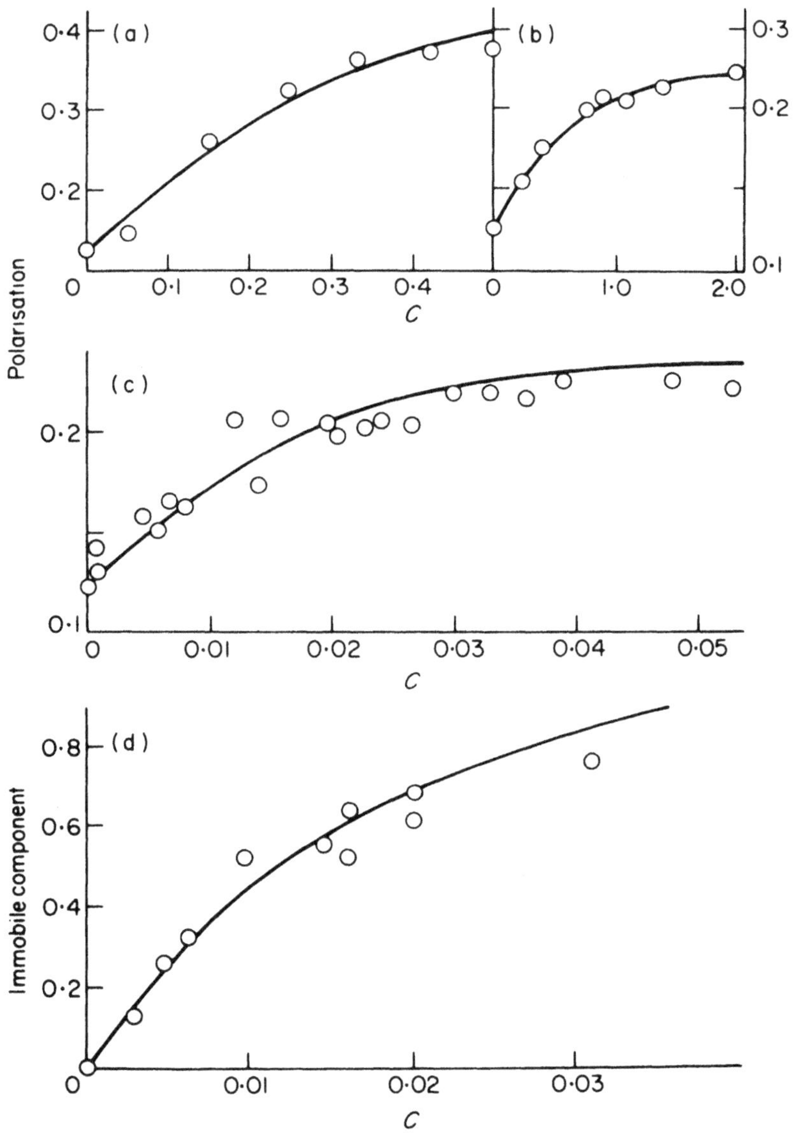

Figure 4. Polarization of 1,6-diphenyl-1,3,5-hexatriene vs. concentration (C) of intrinsic molecules in DMPC vesicles at 36°C. The solid lines are calculated values, and circles are data. (a) DMPC + cholesterol, (b) DMPC + gramicidin A, (c) DMPC + cytochrome oxidase. After Hoffman *et al.* (1981). (d) The immobile component from ESR in dimyristoylphosphatidylcholine + cytochrome oxidase using a nitroxide probe. The solid lines are the calculations. $1 - p_1(M,x)$ for $M = 60$. From Hoffman *et al.* (1981).

The probe molecule is markedly affected by the presence of the intrinsic molecule $P_1 = e^{-Mx}$ where M is the number of DPH molecules which can be accommodated around the intrinsic molecule in half of the lipid bilayer, and x is related to the concentration of intrinsic molecules in the lipid bilayer matrix. The dependence of *P(M,x)* upon x reflects the fact that, as the concentration of polypeptide or protein increases, so does the probability of (say) protein–protein contacts increase and the number of lipid molecules which can contact the protein decrease (Figure 4).

Spin-label data of Knowles *et al.* (1979) also fit an exponential equation similar to that deduced from the fluorescence data (Figure 4d). This indicates that the number of spin probes which can contact the protein varies with concentration of the protein in the reconstituted system; i.e., as the spin probe is intended to mimic a lipid molecule, the number of lipid molecules which contacts the protein varies with protein concentration. Thus, the fluorescent-probe and ESR-probe data are useful for providing an estimate of the number of lipid chains which can contact the protein as the protein concentration changes. The maximum number of lipid chains which can contact the hydrophobic core of an intrinsic protein does not necessarily correspond to tightly bound lipid or specific lipid.

When reconstituted protein–lipid systems are cooled to temperatures below the main lipid T_c transition temperature, upon addition of protein, the transition gradually broadens and the transition enthalpy decreases. At high protein concentrations, the phase transition may not be detectable at all (Chapman *et al.*, 1977; Curatolo *et al.*, 1977; Van Zoelen *et al.*, 1978; Mombers *et al.*, 1979). The broadening of the phase transition has also been confirmed by other methods, for example, fluorescence spectroscopy (Gomez-Fernandez *et al.*, 1980; Heyn, 1979).

C. Range of the Perturbation

An estimate of the extent of the perturbing effect of protein molecules on lipid dynamics can be obtained from DSC measurements of the phase transition in certain biomembranes and protein–lipid recombinants. Generally, the effect of incorporating increasing amounts of intrinsic proteins into lipid systems is to cause a progressive broadening of the phase transition and a decrease in its enthalpy (see review by McElhaney, 1982). As the lipid crystallizes below the phase transition temperature, the proteins form aggregates and take with them solvating lipid molecules which are thus removed from the bulk lipid. The remaining enthalpy arises from the residual pure lipid regions. As more protein is incorporated into the bilayer, the number of lipid molecules free to take part in the cooperative transition decreases causing a corresponding decrease in the enthalpy.

The number of lipid molecules which are withdrawn from the cooperative chain melting can be calculated from graphs of enthalpy as a function of the protein–lipid ratio. Knowing the approximate size and shape of the particular protein and, hence, its circumference, an estimate can be made of the number of concentric shells of lipid whose behavior is influenced by the presence of the protein. Polypeptides such as gramicidin and melittin influence six and ten lipid molecules, which is equivalent to one lipid layer. Lyophilin, a small protein, and bacteriorhodopsin influence around 15 and 22 lipid molecules, respectively, which is also equivalent to one layer (see Alonso *et al.*, 1982). The larger protein glycophorin and the bacteriophage M-13 coat protein influence around 100 molecules, equivalent to three layers of lipid molecules.

VII. THE EFFECT OF PROTEIN ON LIPID CONFORMATION

Having reviewed the effect of protein on the dynamics of the lipid molecules, let us now consider the effect it has on the conformation of lipid molecules.

A. Acyl Chain Region

The effect of intrinsic proteins on the conformation of the acyl chains of the lipids has been studied by both NMR and infrared spectroscopy. The results obtained by both techniques are similar.

By measuring the degree of quadrupole splitting in natural and artificial membranes containing specifically deuterated lipids, deuterium NMR has shown that intrinsic proteins cause either no effect or a small increase in the disorder of the acyl region of the boundary lipids compared to pure lipid bilayers. This is particularly noticeable in the region of the terminal methyl group (Kang *et al.*, 1979a,b; Rice and Oldfield, 1979).

Infrared studies of membranes of *A. laidlawii* enriched biosynthetically with perdeuteriopalmitoyl chains (Casal *et al.*, 1980) have shown that, in the liquid–crystalline phase, proteins cause a disordering of the fatty acyl chains as measured by the number of *gauche* bonds. Infrared studies of gramicidin–lipid complexes (Cortijo and Chapman, 1981) showed a more complex situation. At low protein to lipid ratios, the polypeptide decreases the number of *gauche* isomers above the transition temperature but increases the number below this temperature. At high protein to lipid ratios,

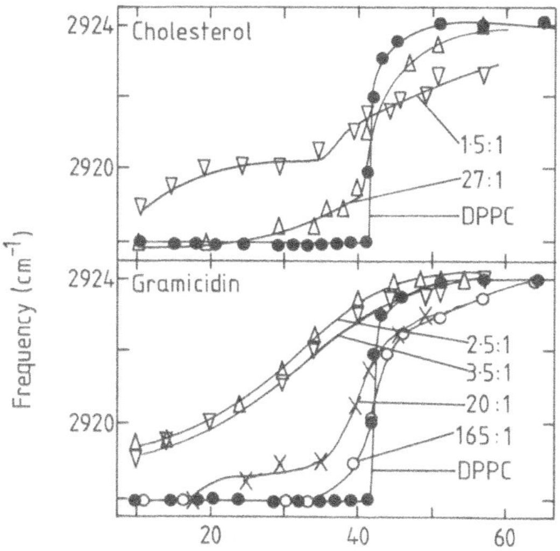

Figure 5. Temperature dependence of the maximum frequency of the CH_2 asymmetric stretching vibration in liposomes of dipalmitoylphosphatidylcholine (DPPC), DPPC–cholesterol, and DPPC–gramicidin mixtures. From Cortijo and Chapman (1981).

the number of *gauche* conformers increased at temperatures above the transition temperature reaching the same level as in pure liposomes. See Figure 5 for a comparison of the effects of cholesterol and gramicidin A.

Pink *et al.* (1981) have presented a model to explain these results. Three populations of lipids are assumed to occur in the membrane: those not adjacent to a polypeptide, those adjacent to a polypeptide, and those trapped between two or three polypeptides. The proportion of each population depends upon the protein–lipid ratio. Lipid chains adjacent to one polypeptide are thought to be more ordered while lipid chains trapped between polypeptides are more disordered than in pure lipid systems.

B. Glycerol Backbone Region

Glycerol deuterated at various positions has been incorporated into the biomembranes of a strain of *E. coli* which is defective in glycerol synthesis (Gally *et al.*, 1981). Using ^{2}H-NMR, the molecular conformation of the glycerol backbone was found to be independent of the phospholipid species and was not changed by the presence of large amounts of membrane proteins.

C. Polar Region

Nitrogen-14-NMR spectroscopy has been used to study the effect of various intrinsic membrane proteins on lipid head-group structure (Rothgeb and Oldfield, 1981). Low concentrations of the polypeptide gramicidin A caused a slight reduction in the quadrupole splitting while high concentrations resulted in the formation of a broad isotropic spectral component, such as is also seen with ^{2}H and ^{31}P-NMR. Incorporation of cytochrome oxidase and Ca^{2+},Mg^{2+}-ATPase, which are much larger molecules, did not result in the same line broadening and seems to have only a small perturbing effect on the polar region of the bilayer. The authors attributed this to the decreased surface area–volume ratio of these large proteins, compared to gramicidin.

VIII. THE INFLUENCE OF LIPIDS ON PROTEIN CONFORMATION

Although the effect of protein on the dynamics of lipids has been studied in detail, little is known about the effect of lipid on the dynamics of the various amino acid side chains of membrane proteins. Recently, some preliminary results of ^{2}H-NMR studies of specifically labeled amino acis in *A. laidlawii* and *H. halobium* membranes have been published (Kinsey *et al.*, 1981a,b) These establish that, in principle, this method should yield data regarding amino acid dynamics but, as yet, the method is still at an early stage of development.

IX. DIFFUSION OF MEMBRANE COMPONENTS

Biomembranes are often dynamic, fluid systems, and considerable diffusion of the various compounds can occur. Two types of diffusion have been measured.

A. Lateral Diffusion

The binding of antibodies and lectins to cell membranes has been shown to result in the clustering of their receptors in the membrane. Measurements of the rate of this lateral diffusion should yield information regarding the fluidity of the membrane and the freedom of the various molecules to diffuse.

Various methods have been used to measure lateral diffusion, the most useful and unambiguous being fluorescence recovery after photobleaching (FRAP). Fluorescent lipid analogues can be incorporated into the membrane, or receptor sites on the membrane surface can be saturated with ligands coupled to fluorescent dyes. Exposing the membrane to a focused laser beam of a suitable wavelength destroys the fluorescence leaving a small bleached patch a few microns in diameter. The rate at which fluorescent molecules from the surrounding membrane diffuse into the bleached region is a relatively simple method of determining the lateral diffusion constant of the labeled complex. A modification of the technique involves bleaching a two-dimensional grid pattern on the membrane which allows the anisotropy of diffusion to be determined. Dye-sensitized photodamage has been shown not to affect the rate of diffusion, and FRAP therefore gives a true measure of the diffusion rate (Koppel and Sheetz, 1981).

Proteins in biomembranes have diffusion constants significantly lower than the lipids with which they are associated. Although some difference is expected due to the different sizes of the lipid and protein molecules, this is not the only explanation. The diffuison rate of receptors in bilayers reconstituted from membrane lipids has been found to be faster by two or more orders of magnitude than that of the same receptors in intact membranes. Two populations of receptors have been measured in cells, one with a mobility less than that of the lipids and one which is immobile. The immobile fraction may result from the receptor being linked to the cytoskeleton of the cell. Koppel *et al.* (1981) measured the lateral diffusion coefficients of fluorescent-labeled lipids and integral proteins in the membranes of normal and spectrin-deficient spherocytic mouse erythrocytes. In the spectrin-deficient cells, the diffusion coefficients of the lipid and protein were $1.5 \pm 0.5 \times 10^{-8}$ cm^2 sec^{-1} and $2.5 \pm 0.6 \times 10^{-9}$ cm^2 sec^{-1}, i.e., a six-fold difference, close to that predicted on the basis of the difference in size of the diffusing molecules. In the normal cells, however, the lipid diffusion coefficient ($1.4 \pm 0.5 \times 10^{-8}$ cm^2 sec^{-1}) was 300-fold greater than that of the membrane proteins ($4.5 \pm 0.7 \times 10^{-11}$ cm^2 sec^{-1}) demonstrating the effect of the cytoskeleton on protein lateral mobility.

The diffusion of the succinylated concanavalin A receptor complex in mouse embryo fibroblasts is ten times greater in directions parallel to the stress fibers than in the perpendicular direction (Austin *et al.*, 1979). Whether this restriction in diffusion is due to direct linkage between the proteins and the cytoskeleton is not yet known.

Other FRAP measurements have shown differences in the lateral mobility of membrane proteins depending upon the physiological state of the cell. The lateral mobility of proteins varies considerably throughout the cell cycle of neuroblastoma cells (De Latt *et al.*, 1980). Similarly, a large proportion of the surface proteins of *Tetrahymena thermophila* are immobile under normal conditions but starvation increases the fraction of mobile components, (Hill, 1981). Changes in lateral mobility are also seen in this organism during the cell cycle.

Table 1. Rotational Relaxation Times of Various Membrane Proteins

Protein	System	Relaxation time (temperature)	References
Rhodopsin	Natural membrane	20 μsec (20°C)	Cone (1972)
Bacteriorhodopsin	Purple membrane	Immobile	Razi-Naqvi *et al.* (1973)
Cytochrome oxidase	Inner mitochondrial membrane	350 μsec, 60% Immobile, 40% (37°C)	Kawato *et al.* (1980)
	PE–PC cardiolipin		
	6000 : 1 Lipid : protein	500 μsec, 100%	Kawato *et al.* (1981)
	1000 : 1 Mole : mole	500 μsec, 83%; Immobile, 17% (20°C)	
Cytochrome P-450	Endoplasmic reticulum	270 ± 90 μsec (21°C)	Richter *et al.* (1979)
	PC–PE–DPPA (2 : 1 : 0.06 wt./wt.), protein : lipid (5 : 1 wt./wt.)	111 μsec (25°C)	Greinert *et al.* (1979)
Cytochrome b_5	DPPC : protein (100 : 1 mole/mole)	0.4 μsec (35°C) above T_c, 9.0 μsec (10°C) below T_c	Vaz *et al.* (1979)
Band 3	Erythrocyte membranes	500 μsec	Cherry *et al.* (1976)
Ca^{2+}, Mg^{2+}-ATPase	Sarcoplasmic reticulum	70–200 μsec (15°C)	Hoffman *et al.* (1979), Moore *et al.* (1979), Burkli and Cherry (1979)

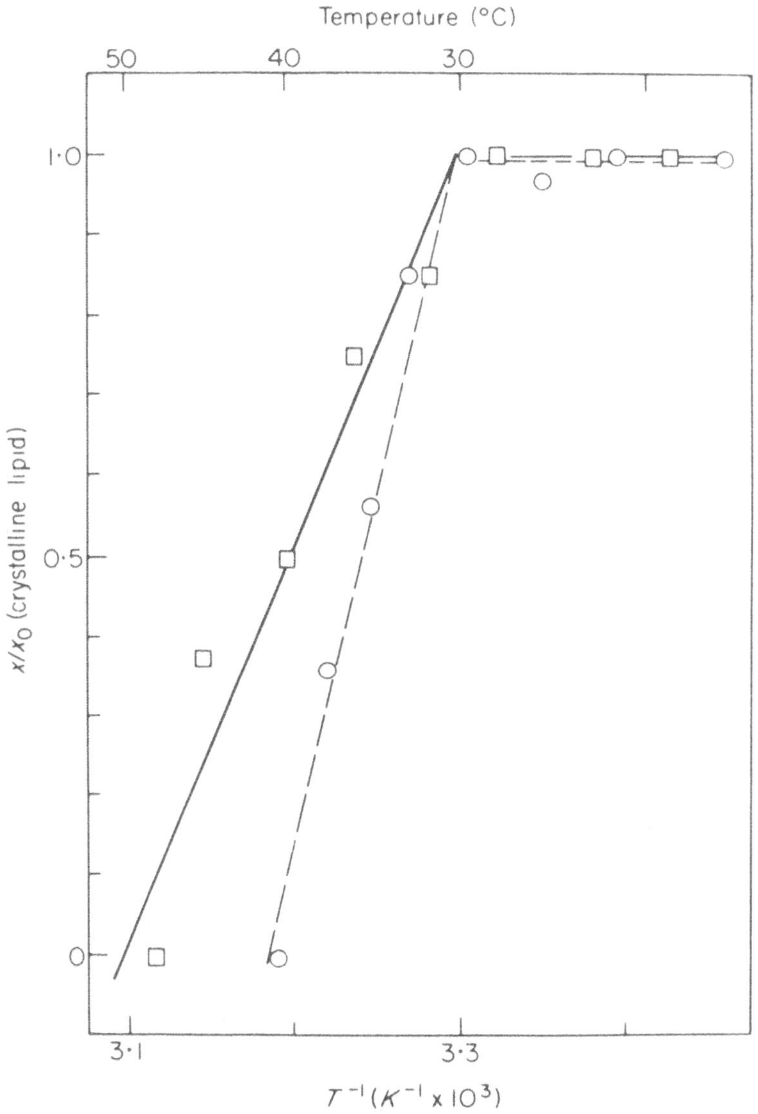

Figure 6. Temperature dependence of the fraction of crystalline lipid x/x_0 for a 93 : 1 (○ - - ○) and 53 : 1 (□—□) lipid : protein molar ratio sample. x is the area corresponding to the 4- to 2-Å intensity and x_0 is the maximum x value for the respective sample. From Hoffman *et al.* (1980).

B. Rotational Diffusion of Proteins

In 1971, Wahl and co-workers, having failed to detect any depolarization of fluorescent-labeled proteins in membranes from the electric organ of the eel, concluded that the proteins were strongly immobilized. However, the technique of fluorescence polarization which has been used to measure the rotational diffusion of proteins in solution proved to be unsuitable for the determination of the slower moving membrane-bound proteins since very little depolarization of the fluorescence occurs during the fluorescence lifetime (about 10^{-9} sec).

Some membrane-bound proteins, e.g., rhodopsin and bacteriorhodopsin, contain chromophores. When excited by a short pulse of plane-polarized light, those molecules whose moment for absorption is parallel or nearly parallel to the plane of the light become excited, and changes in their absorption spectra occur. Information regarding the rotational motion can then be obtained from the polarization of the absorption signal. The triplet probe eosin has been covalently coupled to proteins lacking a suitable chromophore, and the absorption of the dye molecules measured in a similar way (see Cherry, 1978). The depolarization of the phosphorescence of erythrosin-labeled proteins has also been measured using a similar technique enabling rotational correlation times of more than 1 msec to be measured (Moore *et al.*, 1979). Since 1972, when Cone showed that rhodopsin molecules rotated with a relaxation time of 20 μsec, the rotational diffusion of a number of proteins has been determined. Some of these are listed in Table 1.

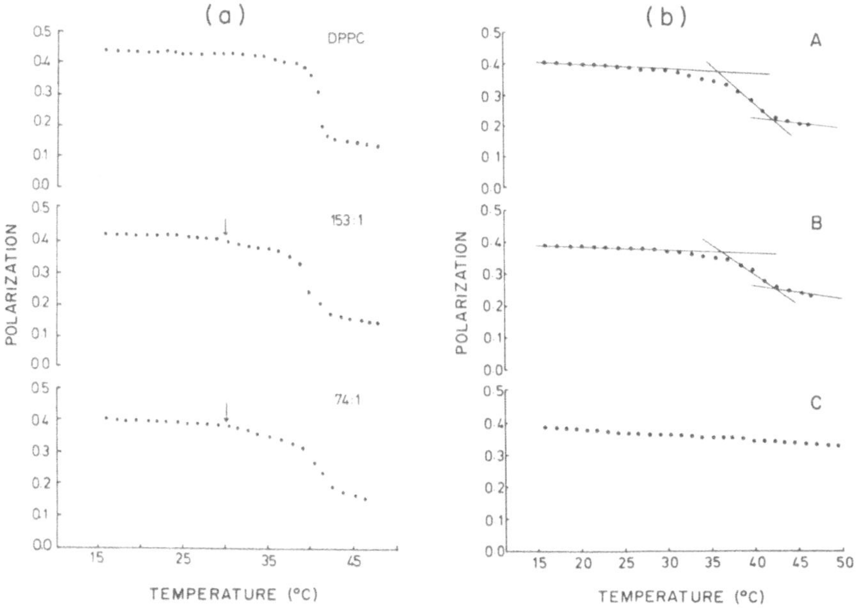

Figure 7. (a) Polarization (*P*) values of a fluorescent probe (1,6-diphenyl-1,3,5-hexatriene) in lipid (dipalmitoylphosphatidylcholine; DPPC) and recombinants with Ca-ATPase as a function of temperature (the arrows mark 30°C). The corresponding lipid/protein molar ratios are indicated. (b) A comparison of *P* as a function of temperature for (A) 37 : 1 dipalmitoylphosphatidylcholine-ATPase recombinant, (B) 19 : 1 recombinant, and (C) a 1 : 1 dipalmitoylphosphatidylcholine–cholesterol mixture. After Gomez-Fernandez *et al.* (1980).

Not all membrane proteins undergo rotational diffusion, and the freedom of some proteins appears to be limited by aggregation or by association with other membrane components. Austin *et al.* (1979) measured the rotational diffusion of the concanavalin A receptors of Friend erythroleukemia cells and the band 3 anion transparent protein of human erythrocytes. Whereas the rotational diffusion of the band 3 protein was strongly temperature-dependent, being faster at higher temperatures, the lectin receptors were immobile at both 4°C and 37°C. Nigg and Cherry (1980) found that the rotation of up to 40% of the band 3 protein was restricted, possibly as a result of interactions with the cytoskeleton. The temperature dependency of band 3 rotation and its limitation by cholesterol (Mühlebach and Cherry, 1982) is probably a result of self-

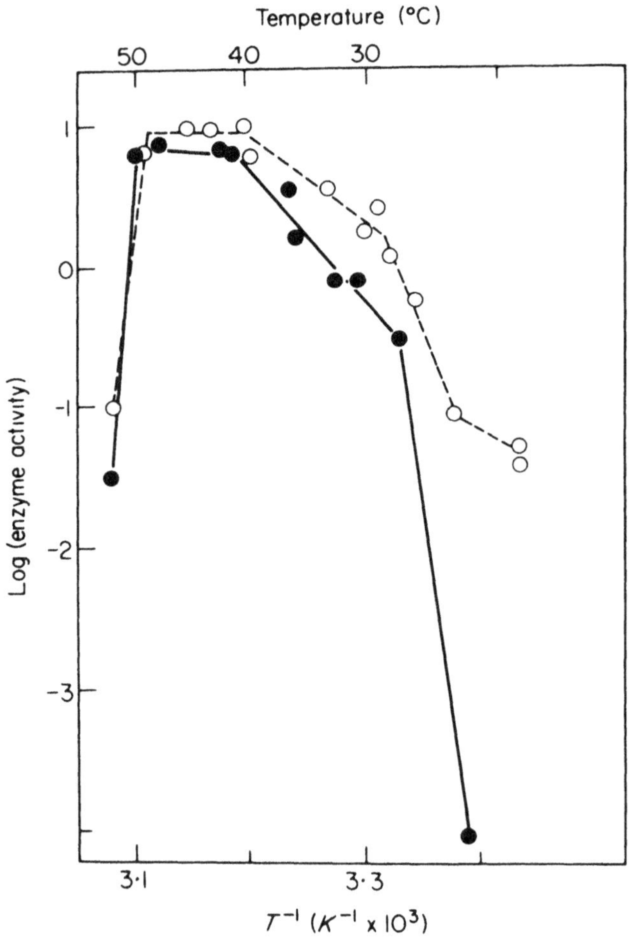

Figure 8. Arrhenius plot of the temperature dependence of Ca^{2+}-ATPase activity for a recombinant having a lipid/protein molar ratio of 60 : 1. ○—○, the sample without eosin-NCS; ●—●, the same sample with eosin-NCS attached (1.8 moles eosin per 10^{-5} g Ca^{2+}-ATPase). From Hoffman *et al.* (1980).

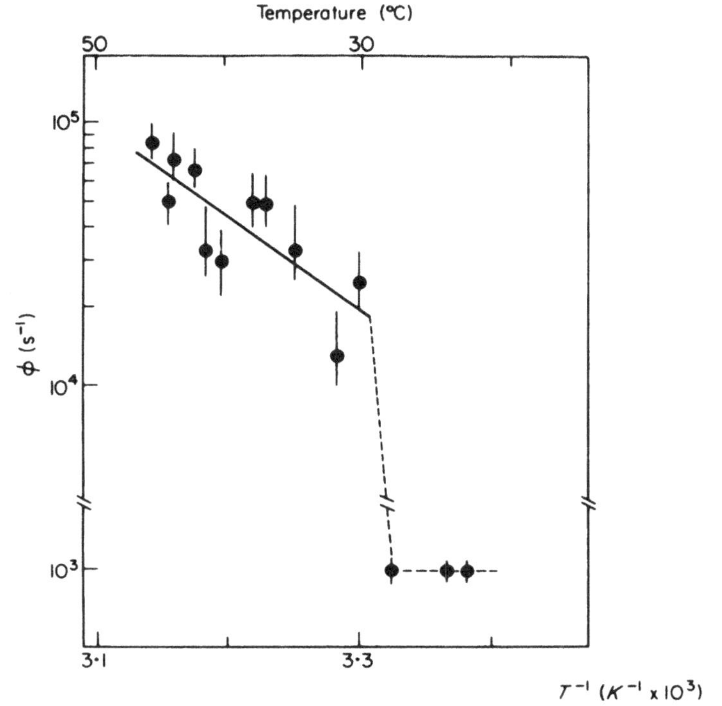

Figure 9. Arrhenius plot of rotational motion parameter for a Ca^{2+}-ATPase recombinant having a lipid/protein ratio of 86 : 1. The line is drawn by linear regression; from the slope, an activation energy of 16 kcal $mole^{-1}$ is calculated. From Hoffman *et al.* (1980).

aggregation. The rotational diffusion of glycophorin, a membrane-spanning sialoglycoprotein from erythrocyte membranes, is also dependent upon the physical form of the protein. The protein reconstituted into dimyristoylphosphatidylcholine shows fast (1–2 μsec) and slow (10–20 μsec) components (Jovin *et al.*, 1981). The fast component appears to be due to the rotation of monomeric protein and the slow due to aggregated protein, only being seen below the transition temperature.

Hoffman *et al.* (1980) investigated the temperature-dependent rotation of Ca^{2+}, Mg^{2+}-ATPase in dipalmitoylphosphatidylcholine vesicles and correlated changes in rotation with enzymatic activity and the physical state of the membrane as measured by X-ray diffraction (Figure 6), fluorescent polarization (Figure 7), and freeze-fracture electron microscopy. The Arrhenius plots of both the enzymatic activity (Figure 8) and the protein rotational motion (Figure 9) showed definite breaks at 28–30°C, which is considerably below the phase transition temperature of the pure lipid (41°C). Below this temperature, no protein rotation is detected. X-ray diffraction measurements showed that at lipid–protein ratios greater than 50 : 1, a portion of the crystalline lipid begins to melt at 28–30°C. In addition, below 25°C, protein-rich aggregates are observed by freeze-fracture electron microscopy. They proposed that the melting of these protein-rich aggregates is responsible for the changes observed at 28–30°C and that rotational motion or lipid fluidity or both are necessary for enzymatic activity. The same enzyme has also been incorporated into vesicles formed from diacetylene-containing phospholipids (Restall, Leaver, Durrani, and Chapman, unpublished results). When irradiated with ultraviolet light below the transition temperature, the diacetylene groups crosslink, forming polymers. Laser flash photolysis and enzymatic activity measurements show that polymerization causes irreversible protein aggregation and loss of enzymatic activity.

X. SUMMARY

The physical properties, particularly the enzymatic activity, of a number of membrane-bound proteins have been shown to be influenced by the lipid composition of the membrane. Some proteins may be specifically associated with particular lipids within the membrane, but this does not appear to be a general phenomenon. While proteins appear to influence the packing and dynamics of the lipid molecules in their neighborhood, the idea of a long-lived stoichiometric annulus of lipid surrounding the protein molecule appears to be erroneous. Other studies have shown how variable is the diffusion of the various membrane components. Although lateral diffusion appears to be important in the response of the cell to physiological changes, the importance of rotational diffusion is not so obvious. Whether certain proteins need to be able to rotate in order to function or whether the measured rotation simply reflects motion of the protein in the fluid bilayer has not yet been determined.

REFERENCES

Alonso, A., Restall, C. J., Turner, M., Gomez-Fernandez, J. C., Goni, F. M., and Chapman, D., Protein–lipid interactions and differential scanning calorimetric studies of bacteriorhodopsin reconstituted lipid–water systems, *Biochim. Biophys. Acta* **689:**283–289.

Austin, R. H., Chan, S. S., and Jovin, T. M., 1979, Rotational diffusion of cell surface components by time-resolved phosphorecence anisotropy, *Proc. Natl. Acad. Sci. USA* **76:** 5650–5654.

Bonting, S. L., and De Pont, J. J. H. H. M., 1980, Use of phospholipid-converting enzymes for the study of membrane-bound enzymes, *Biochem. Soc. Trans.* **8**(1):40–42.

Brotherus, J. R., Jost, P. C., Griffiths, O. H., Keana, J. F. W., and Hokin, L.E., 1980, Charge selectivity at the lipid–protein interface of membranous Na,K-ATPase, *Proc. Natl. Acad. Sci. USA* **77:**272–276.

Burkli, A., and Cherry, R., 1981, Rotational motion and flexibility of Ca^{2+}-Mg^{2+}-dependant adenosine 5-triphosphatase, *Biochemistry* **20:**138–146.

Casal, H., Cameron, D., Smith, I., and Mantsch, H., 1980, *Acholeplasma laidlawii* membranes: A Fourier transform infrared study of the influence of protein on lipid organisation and dynamics, *Biochemistry* **19:**444–451.

Chapman, D., Cornell, B. A., Eliasz, A. W., and Perry, A., 1977, The interactions of helical polypeptide segments which span lipid bilayers. The gramicidin A-lipid water system, *J. Mol. Biol.* **113:**517–538.

Cherry, R. J., 1978, Measurement of protein rotational diffusion in membranes by flash photolysis, in: *Methods in Enzymology*, Vol. LIV (S. Fleischer and L. Packer, eds.), Academic Press, New York.

Cone, R. A., 1972, Rotational diffusion of rhodopsin in the visual receptor membrane, *Nature New Biol.* **236:**39–43.

Cortijo, M., and Chapman, D., 1981, A comparison of the interactions of cholesterol and gramicidin A with lipid bilayers using an infrared data station, *FEBS Lett.* **131**:245–248.

Curatolo, W., Sabura, J. D., Small, D. M., and Shipley, G. G., 1977, Protein-lipid interactions: recombinants of the proteolipid apoprotein of myelin with dimyristoyllecithin, *Biochemistry* **16**:2313–2319.

Davoust, J., Bienvenue, A., Fellman, P., and Devaux, P. F., 1980, Boundary lipids and protein mobility in rhodopsin–phosphatidylcholine vesicles. Effect of lipid phase transition, *Biochim. Biophys. Acta* **596**:28–42.

De Kruijff, B., Van Dijk, P. W. M., Goldbach, R. W., Damel, R. A., and Van Deenen, L. L. M., 1973, Influence of fatty acid and sterol composition on the lipid phase transition and activity of membrane-bound enzymes in *Acholeplasma laidlawii, Biochim. Biophys Acta* **330**:269–282.

De Kruijff, B., Cullis, P. R., and Verkleij, A. J., 1982, Structural and functional aspects of nonbilayer lipids, in: *Membranes and Transport,* Vol. 1 (A. N. Martinosi, ed.), Plenum Press, New York, pp. 43–49.

De Latt, S. W., van der Saag, P. T., Elson, E. L., and Schessinger, J., 1980, Lateral diffusion of membrane lipids and proteins during the cell cycle of neuroblastoma cells, *Proc. Natl. Acad. Sci. USA* **77**:1526–1528.

Devaux, P. F., Davoust, J., and Kousselet, A. 1981, Electron spin resonance studies of lipid protein interactions in membranes, *Biochem. Soc. Symp.* **46**:207–222.

Finean, J. B., and Shukla, S. D., 1980, Enzymes linked to phosphatidylinositol in plasma membranes, *Biochem. Soc. Trans.* **8**(1):43.

Fry, M., and Green, D. E., 1980, Cardiolipin requirement by cytochrome oxidase and the catalytic role of phospholipid, *Biochem. Biophys. Res. Commun.* **93**:1238–1246.

Fry, M., and Green, D. E., 1981, Cardiolipin requirement for electron transfer in Complex I and III of the mitochondrial respiratory chain, *J. Biol. Chem.* **256**:1874–1880.

Gally, H. U., Pluschke, G., Overath, P., and Seelig, J., 1981, Structure of *Escherichia coli* membranes. Glycerol auxotrophs as a tool for the analysis of the phospholipid head-group region by deuterium NMR, *Biochemistry* **20**:1826–1831.

Gomez-Fernandez, J. C., Goni, F. M., Bach, D., Restall, C., and Chapman, D., 1980, Protein–lipid interactions. Biophysical studies of (Ca^{2+}-Mg^{2+})-ATPase reconstituted systems, *Biochim. Biophys. Acta* **598**:502–516.

Greinert, R., Steark, H., Stier, A., and Weller, A., 1979, E-type delayed fluorescence depolarization, a technique to probe rotational motion in the microsecond range, *J. Biochem. Biophys. Meth.* **1**:77–83.

Haslam, J. M., and Al Mahdawi, S. A. H., 1980, The use of lipid mutants of *Saccharomyces cerevisiae* to investigate the role of fatty acids and sterols in membrane functions, *Biochem. Soc. Trans.* **8**:34–37.

Haslam, J. M., Proudlock, J. W., and Linnane, A. W., 1971, Biogenesis of mitochondria 20. The effects of altered membrane lipid composition on mitochondrial oxidative phosphorylation in *Saccharomyces cerevisiae, Bioenergetics* **2**:351–370.

Hesketh, T. R., Smith, G. A., Houslay, M. D., McGill, K. A., Birdsall, N. J. M., Metcalfe, J. C., and Warren, G. B., 1976, Annular lipids determine the ATPase activity of a calcium transport protein complexed with dipalmitoyllecithin, *Biochemistry* **15**:4145–4151.

Heyn, M. D., 1979, Determination of lipid order parameters and rotational correlation times from fluorescence depolarisation experiments, *FEBS Lett.* **108**:359–364.

Hill, R. J., 1981, Restricted lateral diffusion of surface membrane components in *Tetrahymena thermophila, FEBS Lett.* **128**:278–279.

Hoffman, W., Sarzala, M. G., Gomez-Fernandez, J. C., Goñi, F. M., Restall, C. J., Chapman, D., Heppler, G., and Kreutz, W., 1980, Protein rotational diffusion and lipid structure of reconstituted systems of Ca^{2+}-activated adenosine triphosphatase, *J. Mol. Biol.* **141**:119–132.

Hoffman, W., Pink, D. A., Restall, C. J., and Chapman, D., 1981, Intrinsic molecules in fluid phospholipid bilayers. Fluorescence probe studies, *Eur. J. Biochem.* **114**:585–589.

Johansson, K-E., Jägersten, C., Christiansson, A., and Wieslander, A., 1981, Protein composition and extractability of lipid modified membranes from *Acholeplasma laidlawii, Biochemistry* **20**:6073–6079.

Johnston, D. S., Sanghera, S., Pons, M., and Chapman, D., 1980, Phospholipid polymers: Synthesis and spectral characteristics, *Biochim. Biophys. Acta* **602**:57–69.

Jost, P. C., and Griffith, O. H., 1980, The lipid-protein interface in biological membranes, *Ass. New York Acad. Sci.* **348**:391–407.

Jost, P. C., Griffith, O. H., Capaldi, R. A., and Vanderkooi, G., 1973, Evidence for boundary lipid in membranes, *Proc. Natl. Acad. Sci. USA* **70:**480–484.

Jovin, T. M., Bartholdi, M., Vaz, W. L. C., and Austin, R. H., 1981, Rotational diffusion of biological macromolecules by time-resolved delayed luminescence (phosphorescence, fluorescence) anisotropy, *Ann. N.Y. Acad. Sci.* **366:**176–196.

Kaibuchi, K., Takai, Y., and Nishizuka, Y., 1981, Cooperative roles of various membrane phospholipids in the activation of Ca^{2+}-activated phospholipid-dependent protein kinase, *J. Biol. Chem.* **256:**7146–7149.

Kang, S-Y, Gutousky, H. S., Hsung, J. C., Jacobs, R., King, T. E., Rice, D., and Oldfield, E., 1979a, Nuclear magnetic resonance investigation of the cytochrome oxidase–phospholipid interaction: A new model for boundary lipid, *Biochemistry* **18:**3257–3267.

Kang, S-Y., Gutowsky, H. S., and Oldfield, E., 1979b, Spectroscopic studies of specifically deuterium labelled membrane systems. Nuclear magnetic resonance investigation of protein–lipid interactions in *Escherichia coli* membranes, *Biochemistry* **18:**3268–3278.

Kang, S-Y., Kinsey, R. A., Rajan, S., Gutowsky, H. S., Gadebridge, M. G., and Oldfield, E., 1981, Protein–lipid interactions in biological and model membrane systems. Deuterium NMR of *A. laidlawii B.*, *E. coli* and cytochrome systems containing specifically deuterated lipids, *J. Biol. Chem.* **256:**1155–1159.

Kawato, S., Sigel, E., Carafoli, E., and Cherry, R. J., 1980, Cytochrome oxidase rotates in the inner membrane of intact mitochondrial and submitochondrial particles, *J. Biol. Chem.* **255:**5508–5510.

Kawato, S., Sigel, E., Carafoli, E., and Cherry, R., 1981, Rotation of cytochrome oxidase in phospholipid vesicles, *J. Biol. Chem.* **256:**7518.

Kimbelberg, H. K., and Paphadjopoulous, D., 1974, Effects of phospholipid acyl chain fluidity and cholesterol on (Na^{+} + K^{+})-stimulated adenosine triphosphate, *J. Biol. Chem.* **249:**1071–1080.

Kinsey, R. A., Kintaner, A., Tsai, M-D., Smith, R. L., James, N., and Oldfield, E., 1981a, First observation of amino-acid side chain dynamics in membrane proteins using high field deuterium nuclear magnetic resonance spectroscopy, *J. Biol. Chem.* **256:**4146–4149.

Kinsey, R. A., Kintaner, A., and Oldfield, E., 1981b, Dynamics of amino-acid side chains in membrane proteins by high field solid state deuterium NMR. Phenylalanine, tyrosine and tryptophan, *J. Biol. Chem.* **256:**9028–9036.

Kleinfield, A. M., Dragsten, P., Klausner, R. D., Pjura, W. J., and Matayoshi, E. D., 1981, The lack of relationship between fluorescence polarization and lateral diffusion in biological membranes, *Biochim. Biophys. Acta* **649:**471–480.

Knowles, P. F., Watts, A., and Marsh, D., 1979, Spin label studies of lipid immobilisation in dimyristoylphosphatidylcholine-substituted cytochrome oxidase, *Biochemistry* **18:**4480–4487.

Koppel, D. E., and Sheetz, M. P., 1981, Fluorescence photobleaching does not alter the lateral mobility of erythrocyte membrane glycoproteins, *Nature* **293:**159–161.

Koppel, D. E., Sheetz, M. P., and Schindler, M., 1981, Matrix control of protein diffusion in biological membranes, *Proc. Natl. Acad. Sci. USA* **78:**3576–3580.

Leaver, J., Alonso, A., Durrani, A. A., and Chapman, D., 1984, The biosynthetic incorporation of diacetylenic fatty acids into the biomembranes of *Acholeplasma laidlawii* A cells and the polymerization of the biomembranes by UV irradiation, *Biochim. Biophys. Acta,* in press.

London, E., and Fiegenson, G. W., 1981, Fluorescence quenching in model membranes. 2. Determination of local lipid environment of Ca^{2+}-ATPase from sarcoplasmic reticulum, *Biochemistry* **20:**1939–1948.

McElhaney, R. N., 1982, The use of differential scanning calorimetry and differential thermal analysis in studies of model and biological membranes, *Chem. Phys. Lipids* **30:**229–259.

Mombers, C., Verkleij, A. J., De Grier, J., and Van Deenen, L. L. M., 1979, The interaction of spectrin-actin and synthetic phospholipids. II. The interaction with phosphatidylserine, *Biochim. Biophys. Acta* **551:**271–281.

Moore, C., Boxer, D., and Garland, P., 1979, Phosphorescence depolarization and the measurement of rotational motion of proteins in membranes, *FEBS Lett.* **108:**161–166.

Mühlebach, T., and Cherry, R. J., 1982, Influence of cholesterol on the rotation and self-association of band 3 in the human erythrocyte membrane, *Biochemistry* **21:**4225–4228.

Nigg, E. A., and Cherry, R. J., 1980, Anchorage of a band 3 population at the erythrocyte cytoplasmic membrane surface: Protein rotational diffusion measurements, *Proc. Natl. Acad. Sci. USA* **77:**4702–4706.

Oldfield, E., Gilmore, R., Glasser, M., Gutowsky, H. S., Hshung, J. C., Kang, Y-S., King, T. E., Meadows, M., and Rice, D., 1978, Deuterium nuclear magnetic resonance investigation of the effects of proteins and polypeptides on hydrocarbon chain order in model membrane systems, *Proc. Natl. Acad. Sci. USA* **75**:4657–4660.

Overath, P., and Thilo, L., 1978, Structural and functional aspects of biological membranes revealed by lipid phase transitions, in: *International Review of Biochemistry*, Vol. 19 (J. C. Metcalfe, ed.), University Park Press, Baltimore, pp. 1–44.

Pink, D. A., Georgallas, A., and Chapman, D., 1981, Intrinsic proteins and their effect upon lipid hydrocarbon chain order, *Biochemistry* **20**:7152–7157.

Poon, R., Richards, J. M., and Clark, W. R., 1981, The relationship between plasma membrane lipid composition and physical–chemical properties. II. Effect of phospholipid fatty acid modulation of plasma membrane physical properties and enzymic activities, *Biochim Biophys. Acta* **649**:58–66.

Razi-Naqvi, K., Gonzalez-Rodriguez, J., Cherry, R. J., and Chapman, D., 1973, Spectroscopic technique for studying protein rotation in membranes, *Nature New Biol.* **245–251.**

Rice, D., and Oldfield, E., 1979, Deuterium nuclear magnetic resonance studies of the interaction between dimyristoylphosphatidylcholine and gramicidin A, *Biochemistry* **18**:3272–3279.

Richter, C., Winterhalter, K. H., and Cherry, R. J., 1979, Rotational diffusion of cytochrome P-450 in rat liver microsomes, *FEBS Lett.* **102**:151–154.

Rothgeb, T. M., and Oldfield, E., 1981, Nitrogen-14 nuclear magnetic resonance spectroscopy as a probe of lipid bilayer headgroup structure, *J. Biol. Chem.* **256**:6004–6009.

Silvius, J. R., and McElhaney, R. N., 1980, Membrane lipid physical state and modulation of Na^+, Mg^{2+}-ATPase activity in *Acholeplasma laidlawii* B., *Proc. Natl. Acad. Sci. USA* **77**: 1255–1259.

Sondergaard, L., 1979, Role of proteins and lipids in non-linear Arrhenius plots of Drosophila mitochondrial succinate-cyochrome *c* reductase studied by rebinding experiments, *Biochim. Biophys. Acta* **557**:208–216.

Szamel, M., and Resch, K., 1981, Modulation of enzyme activities in isolated lymphocyte plasma membranes by enzymatic modification of phospholipid fatty acids, *J. Biol. Chem.* **256**:11618–11623.

Thilo, L., Trauble, H., and Overath, P., 1977, Mechanistic interpretation of the influence of lipid phase transitions on transport functions, *Biochemistry* **16**:1283–1290.

Tourtellotte, M. E., and Zuprik, J. S., 1973, Freeze-fractured *Acholeplasma laidlawii* membranes: Nature of particles observed, *Science* **179**:84–86.

Van Zoelen, E. J. J., Van Dijck, P. W. M., de Kruijff, B., Verklsij, A. N., and Van Deenen, L. L. M., 1978, Effect of glycophorin incorporation on the physico-chemical properties of phospholipid bilayers, *Biochim. Biophys. Acta* **514**:9–24.

Vaz, W. L. C., Austin, R. H., and Vogel, H., 1979, The rotational diffusion of cytochrome b_5 in lipid bilayer membranes. Influence of lipid physical state, *Biophys. J.* **26**:415–426.

Wahl, P., Kasai, M., and Changeux, J-P., 1971, A study of the motion of proteins in excitable membrane fragments by nanosecond fluorescence polarization spectroscopy, *Eur. J. Biochem.* **18**:332–341.

Warren, G. B., Houslay, M. D., Metcalfe, J. C., and Birdsall, N. J. M., 1975, Cholesterol is excluded from the phospholipid annulus surrounding an active calcium transport protein, *Nature* **255**: 684–687.

Watts, A., Davoust, J., Marsh, D., and Devaux, P. F., 1981, Distinct states of lipid mobility in bovine rod outer segment membranes: Resolution of spin label results, *Biochim. Biophys. Acta* **643**:673–676.

6

On the Molecular Structure of the Gramicidin Transmembrane Channel

Dan W. Urry

I. INTRODUCTION

A. Primary Structure

The primary structure of Gramicidin A (GA) was determined by Sarges and Witkop (1964, 1965a,b) to be HCO-L · Val^1-Gly^2-L · Ala^3-D · Leu^4-L · Ala^5-D · Val^6-L · Val^7-D · Val^8-L · Trp^9-D · Leu^{10}-L · Trp^{11}-D · Leu^{12}-L · Trp^{13}-D · Leu^{14}-L · Trp^{15}NH-CH_2 CH_2OH. The common analogs are Gramicidin B (GB), which is L-Phe^{11} GA, and Gramicidin C (GC), which is L · Tyr^{11} GA. Also, there is a variation in residue-1 which can occur as Ile^1 in GA, GB, and GC. These structures were further confirmed by the solid phase syntheses of Fontana and Gross (1972) and of Noda and Gross (1972). The amino terminus (N) will also be referred to as the formyl end or head and the carboxyl terminus (C) will also be referred to as the ethanolamine end or tail. In single-stranded helical structures of Gramicidin, possible end-to-end associations will be referred to as head-to-head, head-to-tail, and tail-to-tail. In double-stranded structures, parallel and antiparallel associations will be indicated by the proximity of the formyl (HCO) and ethanolamine termini.

Two classes of helical states have been proposed for the Gramicidins. The first class was the single-stranded π_{LD}-helices (Urry, 1971; Urry *et al.*, 1971), also proposed independently by Ramachandran and Chandrasekharan (1972a,b) as LD-helices. The

Dan W. Urry • Laboratory of Molecular Biophysics, University of Alabama in Birmingham School of Medicine, Birmingham, Alabama 35294.

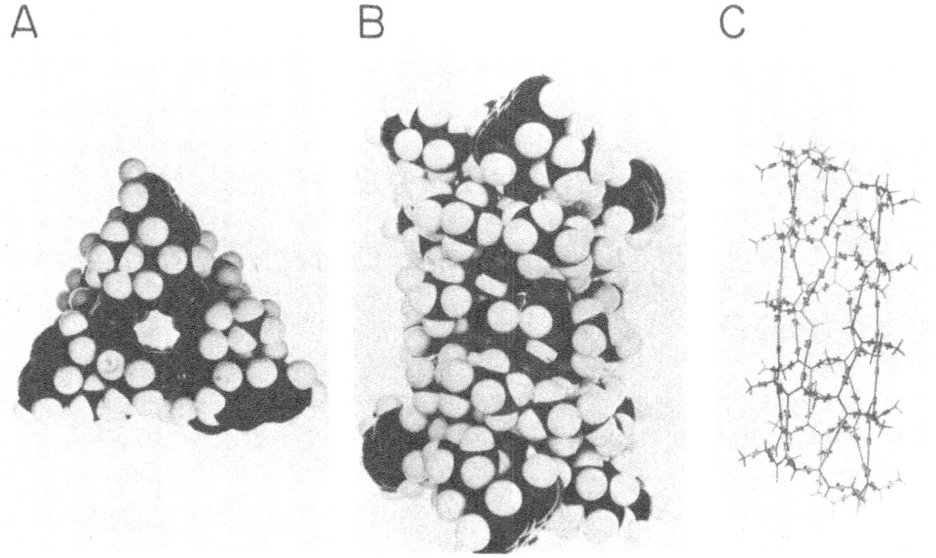

Figure 1. Proposed structure of the Gramicidin A transmembrane channel. This is the head-to-head dimerized, left-handed, single-stranded $\beta^{6.3}_{3.3}$-helix (Urry *et al.*, 1971). (A) Channel view with a 4 Å maximal diameter. (B) Side view showing the two formyl hydrogens at the center of the structure. (C) Wire model of side view oriented as in (B). Reproduced with permission from Urry *et al.* (1975).

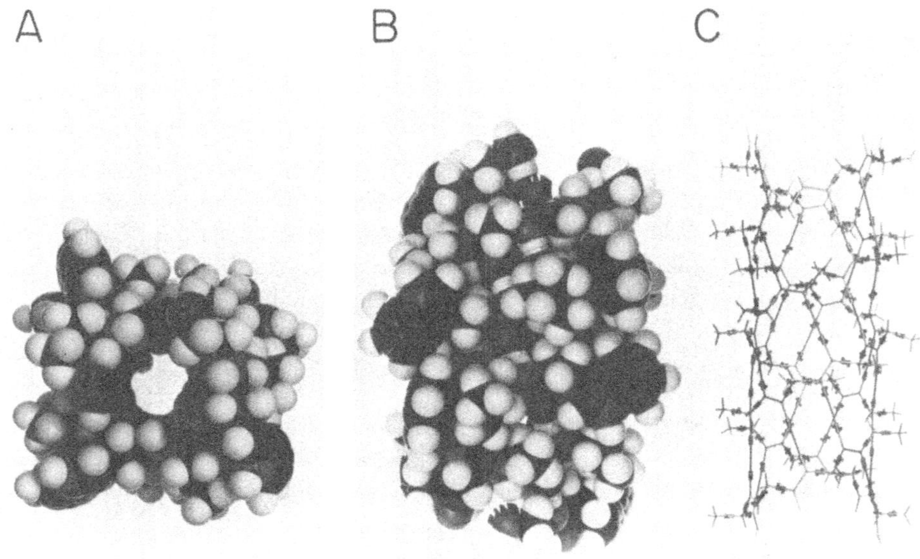

Figure 2. Space-filling model of left-handed, antiparallel, double-stranded $\beta^{7.2}$-helix proposed by Veatch *et al.* (1974). (A) Channel view. (B) Side view. (C) Wire model of side view. Reproduced with permission from Urry *et al.* (1975).

name of this class of helices was subsequently changed to β-helix (Urry, 1972, 1973) because the hydrogen bonding pattern between turns of the helix was the same as between the chains in a parallel β-pleated sheet. The term β-helix will be used throughout this review. The channel structure was proposed to be formed by two left-handed π_{LD}^{6} ($\beta_{3.3}^{6.3}$)-helices associated head-to-head (Urry, 1971, Urry *et al.*, 1971). This structure is given in Figure 1. The second class of Gramicidin conformations that were considered to be candidates for the channel structure were the double-stranded β-helices described by Veatch and Blout (1974). A representative member of this class of conformations is given in Figure 2.

It was early appreciated that the channel may be formed by two Gramicidins

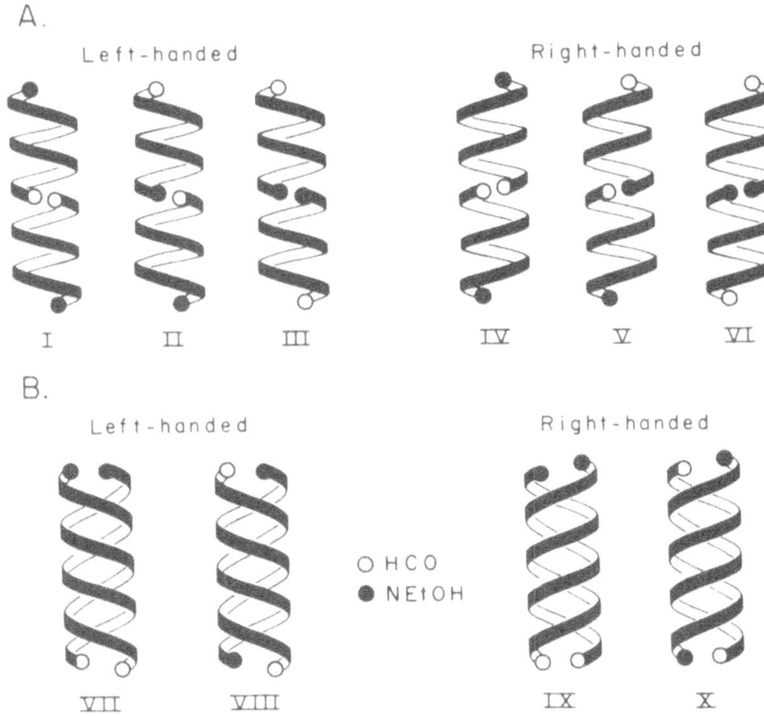

Figure 3. Possible helical dimeric states of Gramicidin which are potential transmembrane channels. (A) Single-stranded β-helices. Structures I, II, and III are left-handed, head-to-head dimerized (as in Figure 1), head-to-tail dimerized, and tail-to-tail dimerized, respectively. Structures IV, V, and VI are right-handed head-to-head, head-to-tail, and tail-to-tail dimerized, respectively. (B) Double-stranded β-helices. Structures VII and VIII are left-handed parallel and antiparallel dimers, respectively. Structures IX and X are right-handed parallel and antiparallel dimers, respectively. In addition, each of these ten structures may occur with a different number of residues per turn; the single-stranded with approximately 4.4, 6.3, 8.1, and 10 residues per turn (Urry *et al.*, 1971) and the double-stranded with approximately 5.6, 7.2, 9.0, and 10.8 residues per turn (Lotz *et al.*, 1976). These possible structures are considered both from the standpoint of the electrical transport studies characterizing the channel state of Gramicidins and in terms of spectroscopic characterization of Gramicidins in suspensions of lipid bilayers.

(Tosteson *et al.*, 1968; Goodall, 1970a,b; Hladky and Haydon, 1971; Urry *et al.*, 1971). This was confirmed in lipid bilayer studies of the Läuger group (Apell *et al.*, 1977; Bamberg and Läuger, 1973; Bamberg and Benz, 1976; Kolb *et al.*, 1975) and of Zingsheim and Neher (1974), and in lipid vesicles by fluorescence studies (Veatch *et al.*, 1975). The set of possible dimeric channel structures arising from the above two classes of helical states is given in Figure 3. It should also be appreciated that, in addition to the ten structures given in Figure 3, each structure can exist with a different number of residues per turn of helix. In the case of the single-stranded helices, there can be approximately four, six, eight or ten residues per turn (Urry *et al.*, 1971). In the case of the double-stranded helices, there can be 5.6, 7.2, 9.0, or 10.8 residues per turn (Lotz *et al.*, 1976). Furthermore, the double-stranded structures can have different hydrogen bond shifts between the two chains. Thus, a complete description of possible dimeric Gramicidin structures involves consideration of approximately 50 different structures. Fortunately, studies on the transport properties of GA and of analogs and derivatives of GA provide a straightforward elimination of these many structures as possibilities and limit the possibilities to two. By entirely independent means, spectroscopic studies of phospholipid packaged channels, the structural argument is carried all the way to a single structure which is one of the two deduced from the planar bilayer studies and which was the original proposed structure (Urry *et al.*, 1971) shown in Figure 1.

II. PLANAR LIPID BILAYER TRANSPORT STUDIES

A. Phenomenology of Channel Transport

The first demonstration that Gramicidin formed transmembrane channels is due to Hladky and Haydon (1970, 1972). Using the approach of black lipid membranes of Mueller and Rudin (1967), they observed step current or conductance changes in membranes doped with very small amounts of Gramicidin. Each step increase in membrane conductance is due to the formation of a single transmembrane channel capable of conducting more than 10^7 ions/sec at 25°C, 100 mV applied potential, and 1 M NaCl (Bamberg and Läuger, 1974). Representative conductance steps, called single-channel conductances, are shown in Figure 4 (Bradley *et al.*, 1981a) for Gramicidin A and for the analog, D · Leu^2 Gramicidin A. There is actually a distribution of single-channel conductances. As shown in Figure 5, both GA and D · Leu^2 GA have very similar distributions of single-channel conductances with the most probable value being the same. As is apparent in Figure 4, however, there is a difference in duration of the channels, i.e., in channel lifetime. The mean value of channel lifetime can be obtained manually by examining records of single channels turning on and off or instrumentally when there are many channels by means of fluctuation analysis using autocorrelation methods (Zingsheim and Neher, 1974; Kolb *et al.*, 1975) or spectral density methods (Bradley *et al.*, 1981a,b; DeFelice, 1977). The power spectra for both GA and D · Leu^2 GA are in Figure 6. The cut-off frequencies, f_c, indicated by the arrows, are used to calculate the mean channel lifetimes, τ ($= 1/2\pi f_c$). The

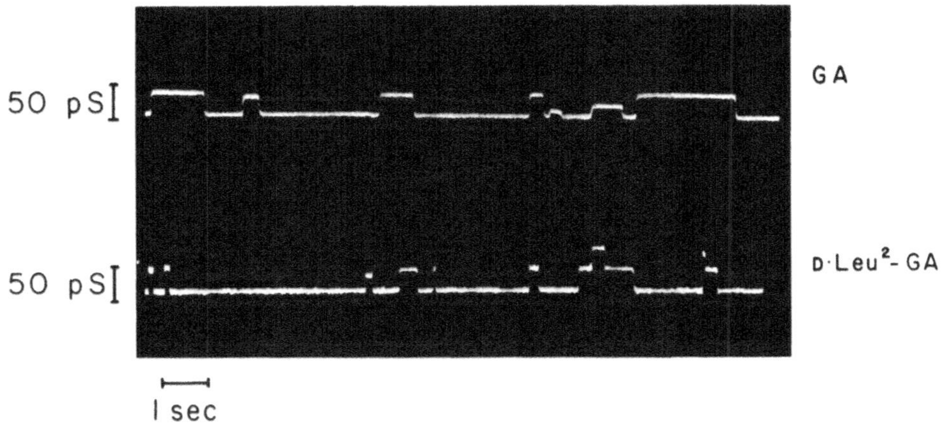

Figure 4. Traces of conductance steps, called single-channel conductances, of Gramicidin A (upper trace) and D · Leu^2 Gramicidin A (lower trace) in diphytanoyl L-α-lecithin–*n*-decane membranes 2 M KCl, 23°C, and 100 mV. While the step heights are similar (see Figure 5) the channel durations (lifetimes) are much shorter for the D · Leu^2 GA analog (see Figure 6). Values of the channel properties are given in Table 5. Reproduced with permission from Bradley *et al.* (1981a).

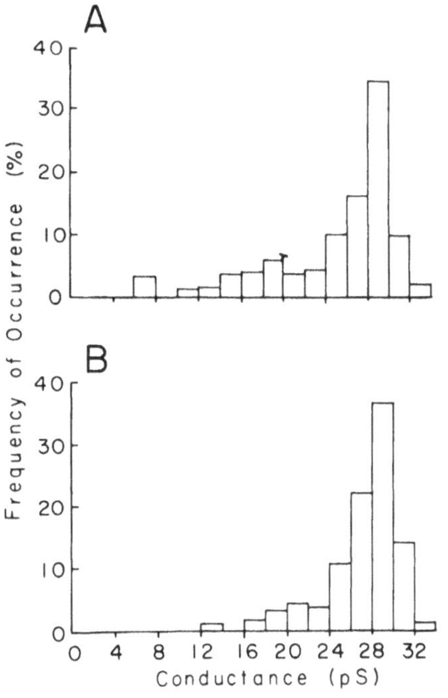

Figure 5. Histograms of the frequency of occurrence of single-channel conductances for (A) Gramicidin A and (B) D · Leu^2 Gramicidin A. Note that the most probable conductance is the same and that the distributions are similar, yet as shown in Figure 4 and Table 5 the channel lifetimes differ markedly. Reproduced with permission from Bradley *et al.* (1981a).

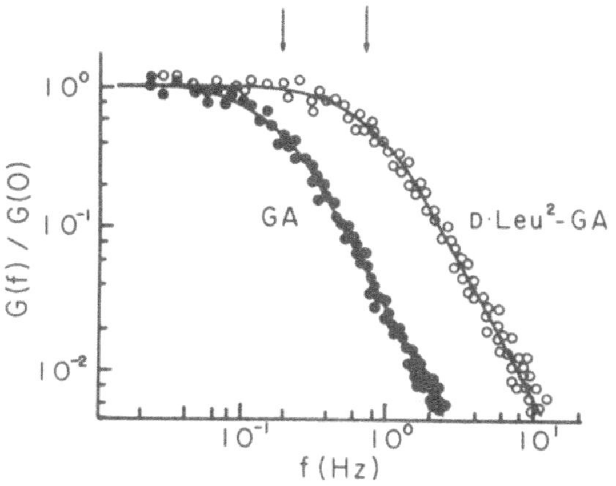

Figure 6. Normalized power spectra of Gramicidin A (GA) and D · Leu[2] Gramicidin A. The arrows mark the cut-off frequency, f_c, from which the channel mean lifetime, τ, is calculated, i.e., $\tau = 1/2\pi f_c$. The mean channel lifetime in this membrane (diphytanoyl L-α-lecithin–*n*-decane, 2 M KCl, 23°C, and 100 mV applied potential) are 800 msec for GA and 210 msec for D · Leu[2] GA. Reproduced with permission from Bradley *et al.* (1981a).

histograms of single-channel conductances (Figure 5) and the mean channel lifetimes become fingerprints characterizing a given channel and are the result of channel structure. As will be seen below, a comparison of these properties for a number of derivatives and analogues of Gramicidin as well as current–voltage data allow conclusions as to channel structure.

B. *Structural Implications of the Multiplicity of Single-Channel Conductances*

As is apparent in Figures 4 and 5, there is not one size of conductance step. There is a multiplicity of single-channel conductances. This was observed in the original characterizations of Hladky and Haydon (1970, 1972) on Gramicidin and in the studies of Läuger and his colleagues which included purified analogs of Gramicidin (Bamberg *et al.*, 1976; Kolb and Bamberg, 1977). The question is whether the dispersity of values derives from many different structures, such as those represented in Figure 3 and with the possible differences in numbers of residues per turn, whether it derives from analogs, their dimeric combinations, and other impurities, or whether it derives from more subtle differences in one or a few of the structures of Figure 3.

Figure 7 demonstrates that, in glyceryl monooleate–hexadecane membranes at 103 mV, 23°C with 1 M KCl, computer-sorted single-channel conductances of high pressure liquid chromatography purified Val[1] Gramicidin A and wholly solid phase peptide synthesized Val[1] Gramicidin A gave essentially identical results (Prasad *et al.*, 1982a). It can be concluded that the multiplicity of single-channel conductances is not due to the presence of analogs and impurities. Most significantly, Busath and Szabo (1981) have observed the same channel change its conducting state, e.g., change to a lower conductance, and then return to the original state. This occurred without any evidence of the channel turning on and off. Since interconversion between single- and double-stranded structures, between structures with different numbers of residues per turn, between left- and right-handed structures, or between parallel and antiparallel

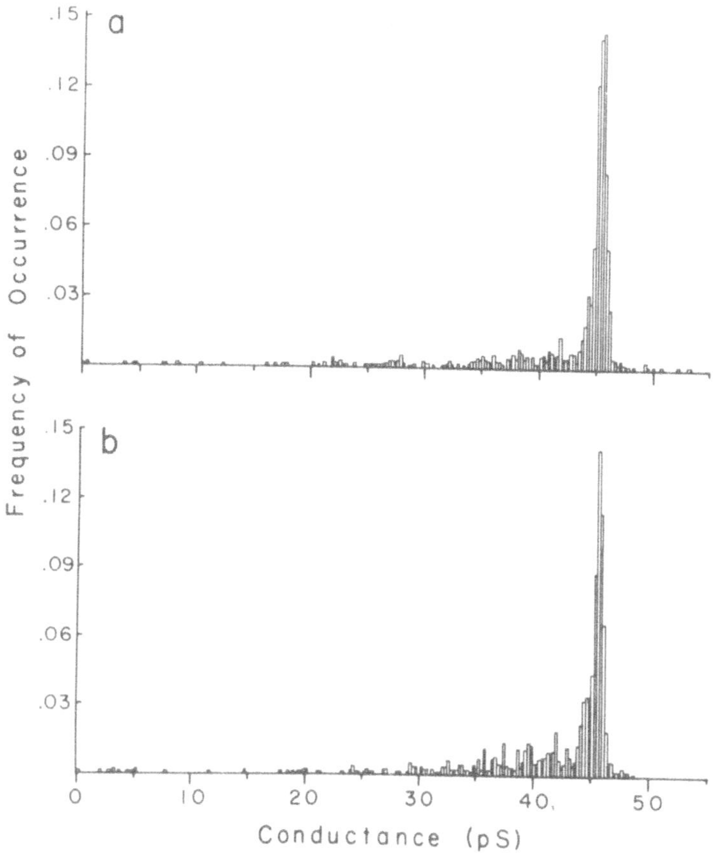

Figure 7. Histogram of the frequency of occurrence of single-channel conductances for (A) HPLC-purified Val[1] Gramicidin A and (B) a wholly synthetic Val[1] Gramicidin A. Recognizing that each histogram is the result of only about 1000 events, the results can be considered virtually identical. This means that the less probable lower conducting states are not the result of contaminating analogs and other impurities. Reproduced with permission from Prasad *et al.* (1982a).

alignment is quite improbable without the channel turning off, it can be concluded that *a single structure in Figure 3 is capable of being in different conducting states.* Furthermore, Busath and Szabo (1981) also report that the kinetics for turning on and off are similar and not a function of the magnitude of the single-channel current. This argues that *the different conductance states have the same on-off mechanism.* The issue as to what factors do vary the on-off rates, i.e., the mean channel lifetime, will be treated below.

C. Structural Implications of Current/Voltage Curves

Again, in the significant study of Busath and Szabo (1981), it was shown that the current/voltage (*i/V*) curves were symmetric for the dominant, most probable conductance states of the channel, and that less probable lower conducting states generally exhibited nonsymmetric *i/V* curves. This has important structural implications as demonstrated below.

1. Free Energy Profile for Na^+ Transport through the Malonyl Gramicidin Channel and the Relevant Current Equation

Using a set of synthetic 1-^{13}C-enriched GA molecules, it has been demonstrated that there are two symmetrically related binding sites in the channel (Urry *et al.*, 1982a). Furthermore, employing nuclear magnetic resonance relaxation (Urry *et al.*, 1980a,b) and dielectric relaxation (Henze *et al.*, 1982), methods estimates of all five of the experimentally required rate constants have been obtained. Eyring rate theory is used to introduce voltage dependence which removes the two-fold symmetry of the channel transport and yields net transport. In so doing, the five rate constants are converted to the requisite ten rate constants which, with the binding sites located, are all the information necessary for calculating sodium single-channel currents, i_{Na}. The single-channel currents were successfully calculated for three decades of ion activity and for 50, 100, 150, and 200 mV applied potentials. This allows calculations of the free energy profile for the malonyl Gramicidin channel as shown in Figure 8. The equation for the single channels current is

$$i_{Na} = \frac{ze}{2d}\{(d - a_1)\, C_x\, (k_1 - k_{-2})\, \chi(oo) + [2a_1 k_5 - (d - a_1)(k_{-1} + C_x k_{-3})]\, \chi(xo) + [(d - a_1)(k_2 + C_x k_4) - 2a_1 k_{-5}]\, \chi(ox) + (d - a_1)(k_3 - k_{-4})\, \chi(xx)\} \qquad (1)$$

The identification of the rate constants is

$$\begin{aligned} k_1 &= k^t_{on} X^{l}{}_1 \;;\; k_{-1} = k^t_{off} X^{-l}{}_2 \\ k_2 &= k^t_{off} X^{l}{}_2 \;;\; k_{-2} = k^t_{on} X^{-l}{}_1 \\ k_3 &= k^w_{off} X^{l}{}_2 \;;\; k_{-3} = k^w_{on} X^{-l}{}_1 \\ k_4 &= k^w_{on} X^{l}{}_1 \;;\; k_{-4} = k^w_{off} X^{-l}{}_2 \\ k_5 &= k_{cb} X^{a}{}_1 \;;\; k_{-5} = k_{cb} X^{-a}{}_1 \end{aligned} \qquad (2)$$

where $X \equiv \exp(zFV/2dRT)$, $l_1 = (d - b_1)$ and $l_2 = (b_1 - a_1)$ with the distances defined in Figure 8A. The quantities z, F, V, and R are the charge on the ion, the Faraday, the applied potential, and the gas constant, respectively. The quantities X_i are the probabilities of the occupancy states, e.g., the empty channel *oo*, the singly occupied channel states, *xo* and *ox*, and the doubly occupied channel *xx*. C_x is the activity of the ion in the solution. The values for the five experimental rate constants are: $k^t_{off} = 3 \times 10^5$/sec, $k^t_{on} = 3 \times 10^7$/sec, $k^w_{off} = 2 \times 10^7$/sec, $k^w_{on} = 2 \times 10^7$/sec and $k_{cb} = 3.2 \times 10^6$/sec. The superscripts t and w stand for the tight and weak sites and the subscript cb stands for the central barrier. By means of Eyring rate theory, the free energies of activation, $\Delta G^{\ddagger}_i$, are calculated, i.e.

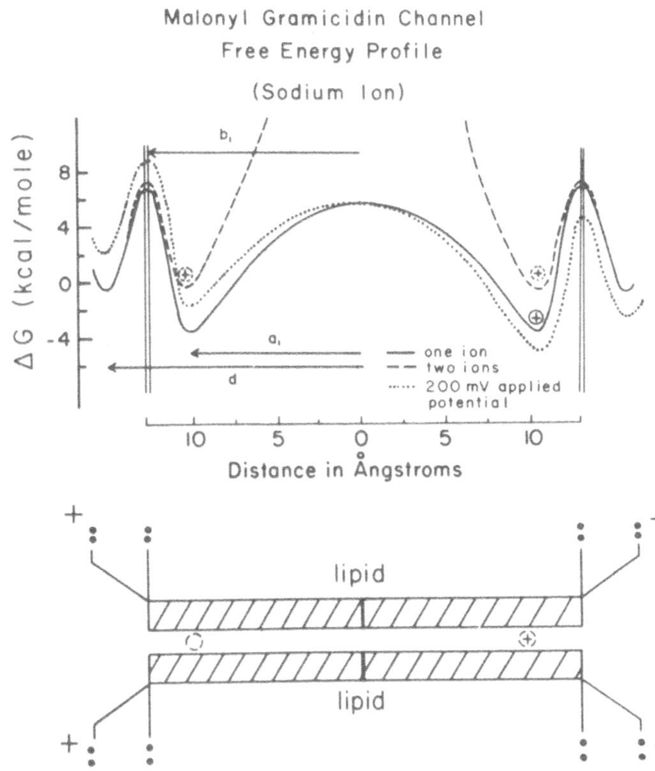

Figure 8. Free energy profile for sodium ion transport through the malonyl-bis-desformyl Gramicidin transmembrane channel. (A) The solid curve is the profile for single-ion occupancy with no applied potential. Note that the profile is symmetric in the absence of an applied field. The dashed curve is for double-ion occupancy. The dotted curve is the free energy profile for a 200 mV applied potential. The distances a_1, b_1, and d are the distance from the center of the membrane (midway point through the channel) to the binding site, the entry–exit barrier, and the aqueous solution, respectively. Reproduced with permission from Urry (1982). This free energy profile provides all the information to calculate the single-channel current for Na^+ transport through the malonyl GA channel over three decades of ion activity and for 50, 100, 150, and 200 mV applied potentials (Urry *et al.*, 1980b).

$$k' = \frac{kT}{\mathrm{h}} e^{-\Delta G_i^{\ddagger}/\mathrm{RT}} \tag{3}$$

such that the free energy profile is obtained from the rate constants and the location of binding sites and barriers. Accordingly, the associated free energies are $\Delta G^{l}_{\mathrm{off}} = 10$ kcal/mole, $\Delta G^{l}_{\mathrm{on}} = 7.25$ kcal/mole, $\Delta G^{w}_{\mathrm{off}} = 7.5$ kcal/mole, $\Delta G^{w}_{\mathrm{on}} = 7.5$ kcal/mole, and $\Delta G_{cb} = 8.6$ kcal/mole.

2. *Calculation of Current/Voltage Curves*

The above development of the free energy profile for Na^+ transport through the malonyl Gramicidin channel will be used as an example for calculating the single-

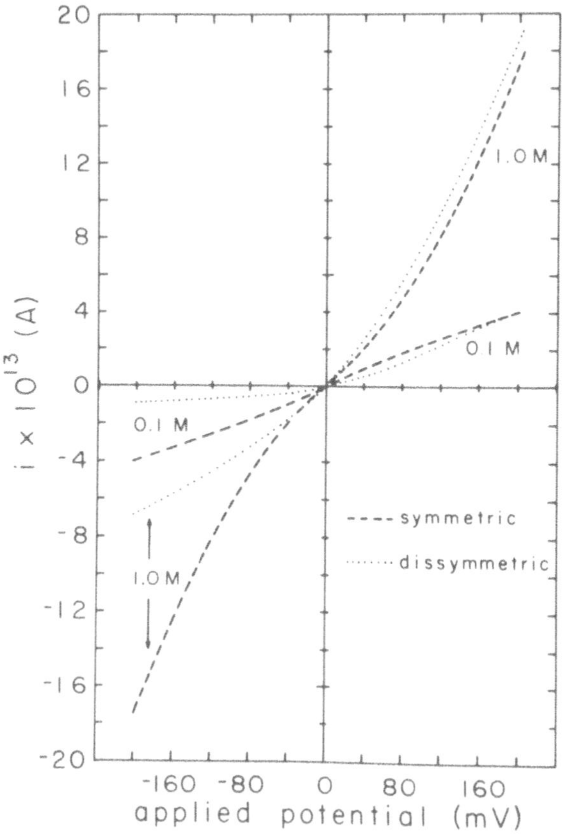

Figure 9. Calculated single-channel currents, using Eqs. (1), (2), and (3), as a function of applied potential. These are referred to as *i/V* curves. The dashed curves (for both 1 M and 0.1 M sodium ion activities) are for the symmetrical free energy profile of Figure 8 but with varying tilt due to the changing applied potential. The ±200 mV free energy profiles are given in Figure 10. The dotted curves are the result of adding 1 kcal/mole to the free energy of the right-hand barrier and binding site as shown in Figure 10A for +200 mV and Figure 10B for −200 mV. The effect of this small asymmetry in the free energy profile is to cause dramatic asymmetry in the *i/V* curves. One point to be made with this result relates to the observation of Busath and Szabo (1981) that the *i/V* curves of the dominant Gramicidin A channel were found to be symmetric. This requires a rigorously symmetric free energy profile which in turn requires an equally rigorously symmetric channel structure.

channel current/voltage (*i/V*) curves. An arbitrary free energy profile could have been used to make the point of interest but it seems most appropriate to use the free energy profile which is capable of calculating the actual channel transport. Accordingly, it should be appreciated that the conclusion to be made from the following analysis is independent of the development of the free energy profile.

In Figure 9, the current/voltage curves for 1 M salt and for 0.1 M salt, using Eq. (1) and the associated definitions, are plotted. These curves, shown as dashes, are

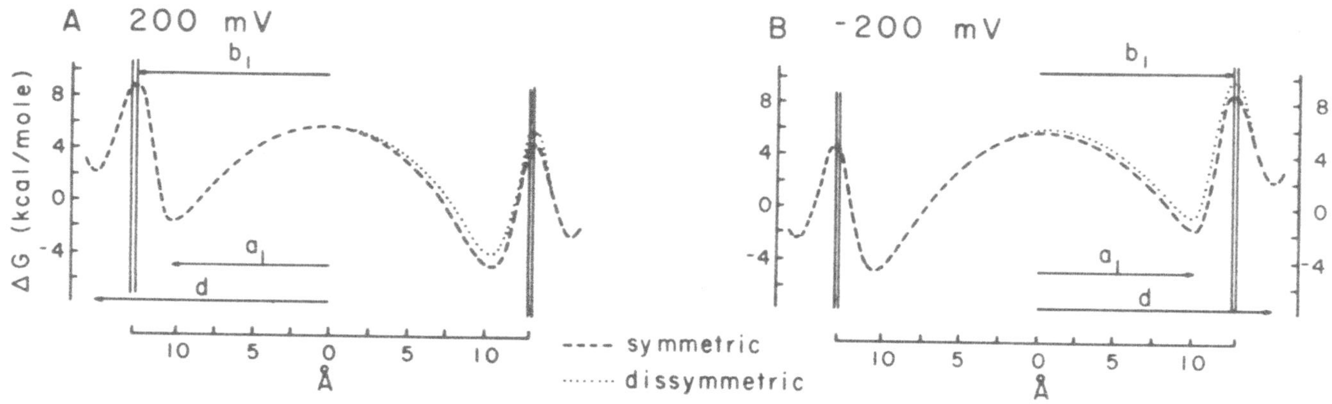

Figure 10. Free energy profiles with an applied +200 mV (A) and −200 mV (B). The dashed curves are based on the symmetric free energy profile in the absence of an applied field of Figure 8. The dotted curves are based on a small asymmetry of +1 kcal/mole added to the right side binding site and barrier. This very small change in free energy gives rise to the asymmetric *i/V* curves of Figure 9.

symmetrical with respect to ±voltage because the free energy profile is symmetrical in the absence of an applied field. The actual free energy profiles for +200 mV and −200 mV are shown in Figures 10A and B as the dashed curves. If the free energy of the right-hand barrier and binding site are increased by only 1 kcal/mole, as shown by the dotted free energy profiles in Figures 10A and B, the *i*/*V* curves change remarkably. This is seen in Figure 9 as the dotted curves for 1 M and 0.1 M salt. For positive potentials, the left-hand barrier is the rate-limiting barrier, and there is only a small change in current. But for negative potentials, the right-hand barrier becomes rate limiting and the small increase of 1 kcal/mole causes a dramatic decrease in the current. These are the sort of changes observed by Busath and Szabo (1981) for a single channel changing its conductance state.

The point to be made here is that symmetric single channel *i*/*V* curves require symmetric free energy profiles which require transmembrane channels with symmetrical structures. Thus, in Figure 3, head-to-tail single-stranded β-helices and parallel double-stranded β-helices can be eliminated as possible channel structures.

D. Structural Deductions from Derivatives and Analogs

Because the mean channel lifetime, τ, is lipid membrane dependent and particularly lipid layer thickness dependent and because the mean or most probable single-channel conductances are ion and concentration dependent, the conductance properties of derivatives and analogs of Gramicidin A have been included in six different tables. As seen in Tables 1 and 2, the analogs in which the change is in position 11 or 9 do little to the values of τ. Similarly, those derivatives listed in Tables 3 and 4, that are at the tail or ethanolamine end, also do very little to alter the values of γ and τ.

On the other hand, even the slightest change at the head or formyl end has dramatic effects (see Tables 2, 3, 4, 5, and 6, and Bamberg *et al.*, 1979). The simple change of formyl hydrogen for methyl (Table 2) causes a 40% drop in conductance but most strikingly a greater than 50-fold decrease in mean channel lifetime (Szabo and Urry, 1979). The traces comparing Gramicidin A and *N*-acetyl desformyl Gramicidin A are

Table 1. Conductance Properties of Natural Analogs of Gramicidin A in Dioleoyl Phosphatidyl Choline Membranes, 1 M KC1, 25°C[a]

	Bilayer activity	
	γ	τ
Gramicidin A HCO ▬▬▬▬▬▬ $NH(CH_2)_2OH$	50 pS	0.22
Gramicidin B HCO ▬▬▬▬▬▬ L · Phe[11] ▬▬ $NH(CH_2)_2OH$	35 pS	0.20
Gramicidin C HCO ▬▬▬▬▬▬ L · Tyr[11] ▬▬ $NH(CH_2)_2OH$	48 pS	0.21

[a] From Bamberg *et al.* (1976).

Table 2. Conductance Properties of Synthetic Analogs and the N-Acetyl Derivative in Glyceryl Monooleate–n-Hexadecane Membranes, 1 M KCl ~25°C

	Bilayer activity		Comments
	γ	τ	
Gramicidin A[a]			
HCO —— $NH(CH_2)_2OH$	45 pS	3 sec	
L · Phe^9 Gramicidin A[b]			
HCO —— L · Phe^9 —— $NH(CH_2)_2OH$	32 pS	2.4 sec	
L · Phe^{11} Gramicidin A[c]			
HCO —— L · Phe^{11} —— $NH(CH_2)_2OH$	33 pS	2.5 sec	
N-Acetyl desformyl Gramicidin A[a]			
CH_3CO —— $NH(CH_2)_2OH$	27 pS	.056 sec	Dramatic change in τ, some change in γ

[a] From Szabo and Urry (1979).
[b] From Prasad *et al.* (1983).
[c] From Prasad *et al.* (1982a,b).

Table 3. Conductance Properties of Pyromellityl Derivatives in Glyceryl Monooleate–n-Hexadecane Membranes, 1 M CsCl, 25°C

	Bilayer activity	
	γ	τ
Gramicidin A[a]		
HCO ——— $NH(CH_2)_2OH$	82 pS	2.0 sec
O-Pyromellityl Gramicidin A[a]		
HCO ——— $NH(CH_2)_2$ 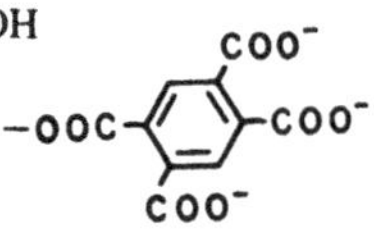	82 pS	2.2 sec
N-Pyromellityl desformyl Gramicidin A[b]		
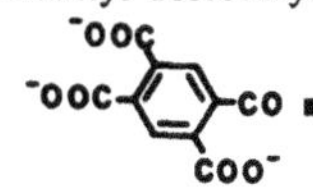 ——— $NH(CH_2)_2OH$	No observable channels	

[a] From Apell *et al.* (1977).
[b] From Bamberg *et al.* (1977).

given in Figure 11. This is an unlikely result for double-stranded structures, particularly since the *N*-acetyl derivative has no significant effect on the equilibrium constant for double-stranded structure formation in organic solvents (Veatch and Blout, 1974). The rapid on–off kinetics, i.e., monomer–dimer equilibrium for channel formation, in itself is difficult to perceive for the double-stranded state because the rate constants for double-stranded β-helix formation range from 10^{-2} to 10^{-7}/sec in the organic solvents ranging from methanol to dioxane (Veatch and Blout, 1974). It is difficult to consider the same kinetic process being as fast in the lipid layer of a membrane where the values run 10^1–10^2/sec. Since both channel lifetime and the frequency of channel formation are lipid layer thickness dependent (Hladky and Haydon, 1972), the structure of the transition state for the monomer dimer equilibrium must be in contact with both sides of the membrane simultaneously (Urry *et al.*, 1975) and the monomer–dimer interconversion must, therefore, occur while spanning the lipid layer of the membrane.

Table 4. Conductance Properties of Succinyl Derivatives in Glyceryl Monooleate–n-Decane Membranes, 1 M CsCl, 25°C

	Bilayer activity	
	γ	τ
Gramicidin A[a] HCO ——— $NH(CH_2)_2OH$	53 pS	~2 sec
O-Succinyl desformyl Gramicidin A[a] HCO ——— $NH(CH_2)_2OOC\text{-}(CH_2)_2COO\text{-}$	48 pS	1.3 sec
N-Succinyl desformyl Gramicidin A[a] $HOOC(CH_2)_2COO$ ——— $NH(CH_2)_2OH$ 10% $CH_3OOC(CH_2)_2COO$ ——— $NH(CH_2)_2OH$ 90%	19 pS	0.22 sec

[a] From Bradley *et al.* (1978).

Table 5. Conductance Properties of D · Leu² Analog in Diphytanoyl L · α-Lecithin–n-Decane Membranes, 2 M KCl, 23°C

	Bilayer activity	
	γ	τ
Gramicidin A[a]		
HCO ━━━━━━━━━━ $NH(CH_2)_2OH$	28 pS	0.8 sec
D · Leu² Gramicidin A[a]		
HCO ━━ D · Leu² ━━━━━━━━ $NH(CH_2)_2OH$	28 pS	0.21 sec

[a] From Bradley *et al.* (1981a).

The bilayer activity of other derivatives also combine to give the same result. The *O*-pyromellityl Gramicidin A derivative only forms channels when placed on both sides of the membrane (Apell *et al.*, 1977). This is because the multiply charged pyromellityl moiety cannot pass through the low dielectric lipid layer. The result requires that the channel structure be either the head-to-head single-stranded β-helices (structures 1 or 4 of Figure 3) or the antiparallel double-stranded β-helices (structures VIII or X of Figure 3). On the other hand the finding that *N*-pyromellityl desformyl Gramicidin A (see Table 3) does not form channels (Bamberg *et al.*, 1977) argues against the antiparallel double-stranded β-helix as a possible structure. And the potent activity of the covalent head-to-head dimer, malonylbis-desformyl Gramicidin (see Table 6), also indicates that the antiparallel structure is not the obligatory channel (Urry *et al.*, 1971; Bamberg and Janko, 1977). The malonyl dimer result coupled with the required two-fold symmetry of the dominant channel species (see Section II-C), in itself, eliminates the double-stranded structures from consideration as the dominant channel structure. In addition, the examination of some ten analogs, all at residues 1, led Morrow *et al.* (1979) to concur with the head-to-head single-stranded structure.

Table 6. Conductance Properties of Bis-Desformyl Derivatives in Phosphatidyl Choline–Cholesterol–n-Decane Membranes, 1 M CsCl, 25°C

	Bilayer activity	
	γ	τ
Gramicidin A[a]		
HCO ━━━━━━━━━━ $NH(CH_2)_2OH$	29 pS	1.7 sec
Malonyl-bis-desformyl Gramicidin A[a]		
CO ━━━━━━━━━━ $NH(CH_2)_2OH$		
|		
CH_2	14–19 pS	~200 sec
|		
CO ━━━━━━━━━━ $NH(CH_2)_2OH$		

[a] From Bamberg and Janko (1977).

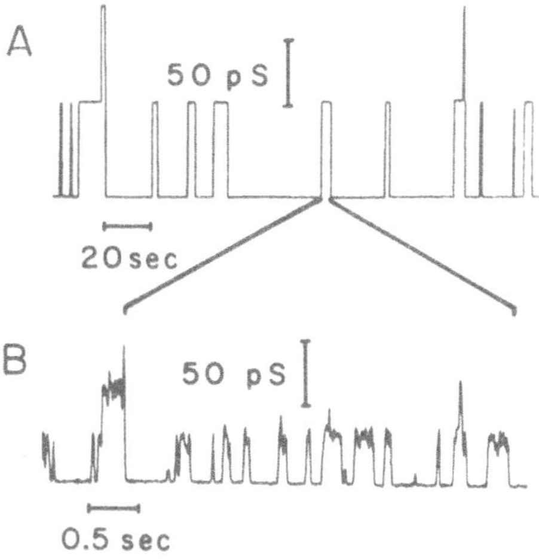

Figure 11. Traces of single-channel conductances for (A) Gramicidin A and (B) *N*-acetyl desformyl Gramicidin A in glyceryl monooleate–*n*-decane membranes at 1 M RbCl, 23°C, and 100 mV applied potential. Note the reduction in single-channel current height, but most dramatically the reduction in channel lifetime on replacement of the formyl hydrogen with a methyl moiety. This subtle change has no effect on the equilibrium and rate constants for forming double-stranded structures in solution (Veatch and Blout, 1974) but it has striking effects on the properties of the Gramicidin channel. Reproduced with permission from Szabo and Urry (1979).

Deciding among the single-stranded β-helices of Figure 3 is rather straightforward. The insensitivity of the channel conductance properties to tail or ethanolamine end modification, the high sensitivity to head or formyl end modifications, the effect of the D · Leu^2 analog which adds some steric crowding at the head-to-head junction, and all of the other derivatives allow the conclusion that the head-to-head single-stranded β-helix is the dominant transmembrane of the channel structure, i.e., structures I and/or IV of Figure 3. Of course, the head-to-head single-stranded structure also meets the criteria of the symmetric *i/V* curves, the relatively rapid on–off kinetic requirements, and the requirement that the transition state for the monomer–dimer equilibrium simultaneously sense both interfaces. Accordingly, the electrical transport studies on planar lipid bilayers alone allow the conclusion that structures I and/or IV of Figure 3 is the correct structure, and it is reasonable on grounds of conformational energetics and on ion transport energetics that the structure with 6.3 residues per turn shown in Figure 1 would be correct. What is not resolved by the preceding studies is whether the structure would be left- or right-handed. In the next section, the issue of structure will again be approached and the result will be arrived at by entirely independent methods.

III. SPECTROSCOPIC CHARACTERIZATION OF THE LIPID-INCORPORATED CHANNEL STATE

In this section, the molecular structure of Gramicidin incorporated into phospholipid structures is derived by spectroscopic methods. In order to relate the result to the transmembrane channel of the preceding electrical transport studies, exhaustive evidence must be developed which demonstrates that the structure in the incorporation studies is the same as in the transport studies. Once this is achieved, then the accessibility of chain termini and the identification of residues at the ion binding sites allow the conclusion as to which of the structures in Figure 3 is the transmembrane channel and also as to what the approximate number of residues per turn would be.

A. Criteria for the Channel State in Lysolecithin Structures

The first requirement for channel interaction with a lipid is that the channel span from one side to the other of the lipid component. Such spanning at appropriate lipid : Gramicidin molecular ratios, e.g., 10 : 1, should preferentially affect the mobility of the aliphatic lipid carbons. Using carbon-13 nuclear magnetic resonance, it was found on adequate heat incubation of Gramicidin with L-α-lysolecithin that mobilities of the lipid CH_2 and CH_3 moieties were markedly reduced (Urry *et al.*, 1979a). When the heat incubation was followed as a function of time by means of circular dichroism (CD), it was found that the CD spectrum underwent dramatic changes during the time course of heating and approached a unique CD spectrum which was stable on standing and on further heating. This is shown in Figure 12. The next step was to

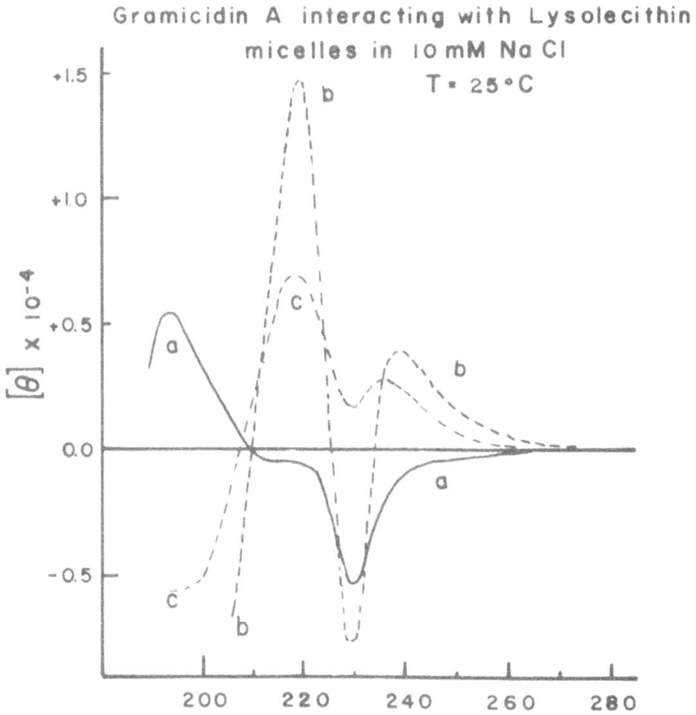

Figure 12. Effect of heating on the circular dichroism pattern of Gramicidin associated with lysolecithin at a lysolecithin : Gramicidin molar ratio of 10 : 1. Curve a, sample heated for 20 min at 45°C; curve b, sample heated for 8 hr at 70°C; curve c, sample heated for 15 hr at 70°C. Once achieved, curve c is stable on standing for weeks and is unchanged by further heating. This is the CD pattern of the channel state. Reproduced with permission from Urry *et al.* (1979a).

determine which of the states observed in Figure 12 demonstrated interaction with an ion that is conducted by the channel. Sodium-23 nuclear magnetic resonance longitudinal relaxation, T_1, studies demonstrated that only as the unique state, curve c of Figure 12, was approached did significant sodium ion interaction occur (Urry *et al.*, 1979a,b).

With an ion interaction observable, it becomes possible to characterize it in ways relevant to channel transport. By means of the temperature dependence of T_1 and of resonance line width, it is possible to determine an energy of activation for exchange with the conformation-dependent sodium ion binding site (Urry *et al.*, 1979a,b, 1980b). Interestingly, the values obtained for the energy of activation ranged from 6.8 kcal/mole to 7.4 kcal/mole and compared favorably with the energy of activation, 7.3 kcal/mole, obtained for sodium ion transport through the Gramicidin channel in dioleoyl lecithin membranes (Bamberg and Läuger, 1974). It is also possible to carry out competition studies with ions that are known from transport studies to have greater affinity for the channel than does Na^+ such as Ag^+ (McBride and Szabo, 1978) and Tl^+ (Neher, 1975). When these ions were added to the heat-incubated lysolecithin–Gramicidin system, they inhibited competitively the Na^+ interaction as demonstrated by changes in T_1 and line width (Urry *et al.*, 1979a,b). Furthermore, as Ca^{2+} and Ba^{2+} reduce the magnitude of the single-channel conductances of monovalent ions while themselves being impermeable (Bamberg and Läuger, 1977), the interaction of these divalent cations with the lysolecithin packaged state of Gramicidin would be of interest. Indeed, binding constants by monitoring 1-^{13}C-enriched residues of synthetic Gramicidin heat incubated with lysolecithin are in the same range, 1–10 M^{-1} (Urry *et al.*, 1982b,c), as those determined from the transport studies. Finally, as briefly discussed in Section II-C, sodium-23 nuclear magnetic resonance relaxation studies and dielectric relaxation studies have been able to determine five rate constants which could then be used to calculate the single-channel currents to within an accuracy of a factor of two or three, which is the uncertainty in the relaxation-derived values. Thus, the above results are an exhaustive demonstration that the state obtained on prolonged heat incubation of Gramicidin with lysolecithin is the channel state, and that the unique CD spectrum given as curve c in Figure 12 is that of the channel state.

B. Relationship between the Lysolecithin–Gramicidin Heat-Incorporated Channel State and the State of Gramicidin in Lipid Vesicles

It has been shown by means of circular dichroism that the channel state, i.e., the same CD pattern as curve c of Figure 12, is obtained for Gramicidin incorporated into lecithin liposomes (Masotti *et al.*, 1980). In fact, the same CD pattern has been independently identified as that of the channel state in dimyristoyl phosphatidyl choline (DMPC) and dilauryl phosphatidyl choline vesicles (Wallace *et al.*, 1981). This identification of the channel state was based on the comparison of activities of analogs and derivatives in transport studies with the CD pattern obtained on incorporation into vesicles. For example, desformyl Gramicidin and desformyl valyl Gramicidin, which have conductances relative to Gramicidin of 0.002 and 0.0001, respectively, do not give the CD spectrum of curve c in Figure 12 on association with DMPC, whereas *O*-acetyl and *N*-acetyl Gramicidins do (Wallace *et al.*, 1981).

Another relationship between the lysolecithin- and lecithin-packaged channel states of Gramicidin can be found in electron microscopy studies of the heat-incubated lysolecithin–Gramicidin system. Interestingly, the negatively-stained samples show vesicle-like structures (see Figure 13). In fact, careful characterization using negative-

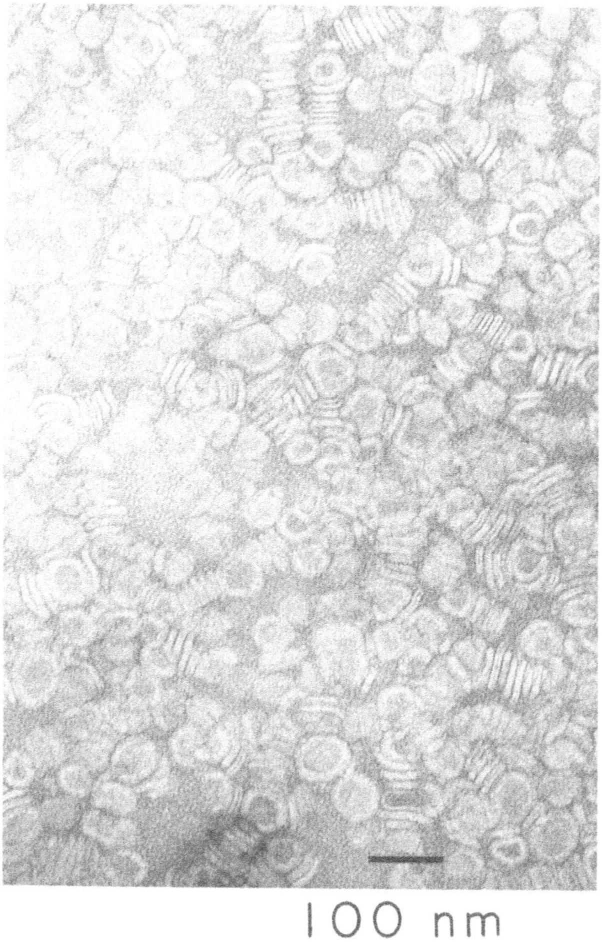

Figure 13. Electron micrograph of a sample of Gramicidin–L-α-lysolecithin (1 : 15) which had been heat incubated at 70°C for 20 hr and negatively stained with 1% uranyl acetate, pH 4.5. Vesicle-like structures are observed and many structures are simply-folded sheets of single-bilayer thickness (Pasquali-Ronchetti and Urry, unpublished). More extensive characterizations using freeze fracture and optical diffraction demonstrate that Gramicidin induces L-α-lysolecithin on heating to form bilayers (Pasquali-Ronchetti *et al.*, 1983).

staining and freeze-fracture techniques demonstrate that Gramicidin induces lysolecithin to form lipid bilayers (Pasquali-Ronchetti *et al.*, 1983). It is also apparent in Figure 13, however, that many of the vesicles are unsealed, i.e., that they are folded bilayer sheets. This lack of forming well-sealed vesicles has the advantage that no electrochemical potential will develop during ion titration. With monovalent cation selective channels in well-sealed vesicles, as the salt is added to the suspension during a titration, only the monovalent cation will enter the vesicle making it positively charged inside. This potential gradient acts to decrease binding within the channel.

C. Orientation of Gramicidin Chains in the Lipid Bilayer

The orientation of Gramicidin chains in DMPC vesicles has been shown by Weinstein *et al.* (1979) to be one in which the formyl end is buried deep within the lipid layer and in which the ethanolamine end only is exposed to the aqueous solution. The approach utilized ^{13}C and ^{19}F nuclear magnetic resonance of ^{13}C and ^{19}F labels at both the formyl and ethanolamine ends and looked for T_1 and chemical shift effects of paramagnetic probes in the aqueous solution and in the lipid. Thus, the double-stranded structures were excluded as they would have both ends at the membrane aqueous interface and neither end buried within the lipid. On the other hand, the Weinstein *et al.* (1979) result is exactly what is to be expected for the head-to-head dimerized β-helices, i.e., structures I and IV of Figure 3.

Realizing that Gramicidin incorporated into DMPC vesicles and Gramicidin incorporated into lysolecithin phospholipid structures both give the same unique CD pattern (Urry *et al.*, 1979a,b; Masotti *et al.*, 1980; Wallace *et al.*, 1981), it can be argued that the same orientation of chains would obtain in the lysolecithin case, but this can be demonstrated independently. The approach is to use several different synthetic isotopically substituted Gramicidin A molecules in which one carbonyl carbon of each has been enriched with ^{13}C (Urry *et al.*, 1983a), and to look for divalent ion-induced chemical shifts. The channel is impermeable to divalent ions such as Ca^{2+} and Ba^{2+} due to the effect of the low dielectric constant of the lipid. This means that only sites near the aqueous phase can be approached by these ions. It has been demonstrated that Ca^{2+} and Ba^{2+} can induce chemical shifts in the carbonyl carbon resonances of the Trp^9, Trp^{11}, Trp^{13}, Leu^{14}, and Trp^{15} residues (Urry *et al.*, 1982b,c). These residues, therefore, are all approachable and near the aqueous interface. On the other hand, Ba^{2+} does not induce a chemical shift in the carbonyl carbon resonance of the Val^1 residue. Figure 14 contains a comparison of the effect of 1 M $BaCl_2$ on $(1\text{-}^{13}C)$ L · Val^1 GA and on $(1\text{-}^{13}C)$ L · Trp^{15} GA. Thus, the same arguments apply as in the Weinstein *et al.* (1979) study, and of the structures in Figure 3 only I and IV are possible.

These observations of Gramicidin chain orientation in lipid bilayers provide one piece of information, which when considered in terms of the structures of Figure 3, allow the above conclusions. It would be more satisfying to add this information to additional data which would directly derive one of the structures in Figure 3. In the studies described below, the structural argument is carried directly to structure I of Figure 3 and even to the structure of Figure 1 which has approximately six residues per turn.

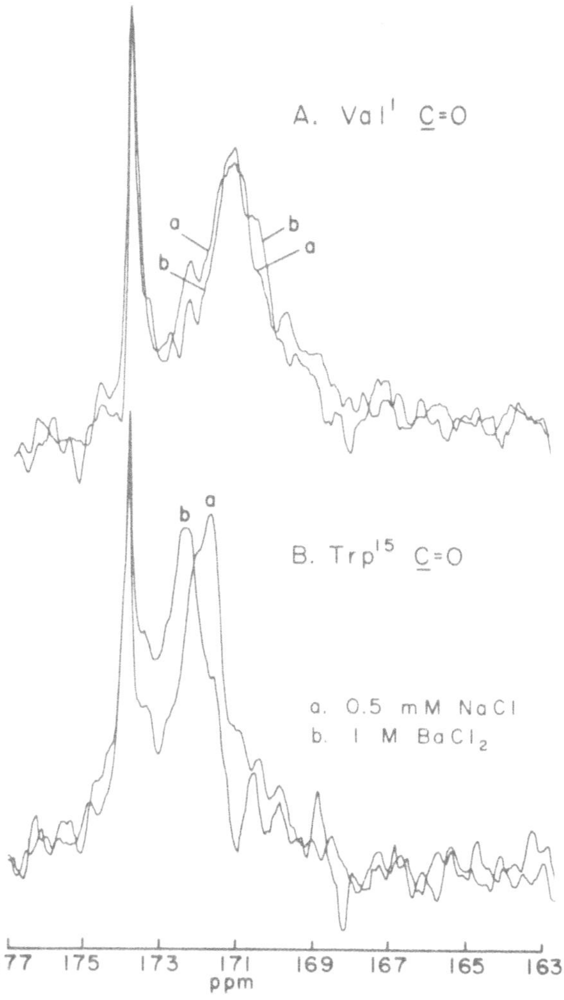

Figure 14. Carbon-13 nuclear magnetic resonance spectra of the carbonyl region of (A) (1-^{13}C) L · Val^1 Gramicidin A and (B) (1-^{13}C) L · Trp^{15} Gramicidin A, incorporated into phospholipid bilayers showing that Ba^{2+} ion induces a chemical shift in the Trp^{15} carbonyl carbon resonance but not in the Val^1 carbonyl carbon resonance. Since Ba^{2+} does not pass through the channel but does interact at the mouth of the channel, this indicates an accessible (tail) ethanolamine end and an inaccessible (head) formyl end.

D. Determining the Channel Conformation from Ion-Induced Carbonyl Carbon Chemical Shifts

As shown in Figure 15, ion interaction with the Gramicidin channel does not detectably alter the circular dichroism pattern (Urry *et al.*, 1982b; Henze *et al.*, 1982). This was also demonstrated for Cs^+ by Wallace *et al.* (1981) in vesicles and means that ion binding causes, at most, only small changes in the channel conformation. This result, and the fact that the structure is regular, as in a helical state, has the following

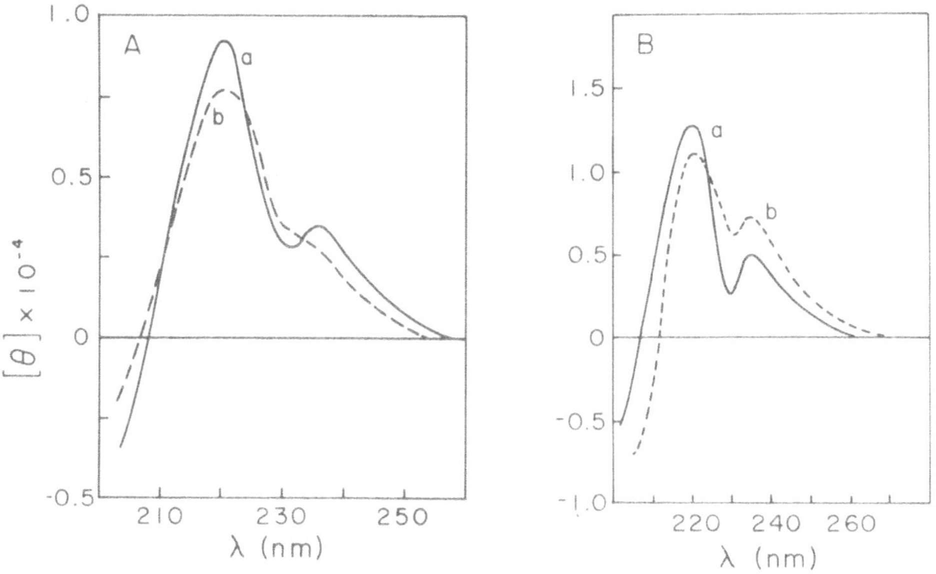

Figure 15. Circular dichroism spectra of Gramicidin incorporated into phospholipid (lysolecithin) bilayers. (A) Ba^{2+} binding causes only small changes in the CD pattern indicating only limited effect on conformation. Reproduced with permission from Urry *et al.* (1982b). (B) Tl^{+} binding also causes only small changes in the CD pattern, again indicating that ion binding results in only minor perturbation of the channel structure.

consequence. If a cation induces a chemical shift at one position in the channel, it should also do so at any symmetry-related position, i.e., any helically equivalent position, in the channel should it spend an equivalent fraction of time there. The backbone atoms of all L-residues would be helically equivalent as would the backbone atoms of all D-residues, but L-residues are not structurally equivalent to D-residues. This means that the ion-induced chemical shift of an L-residue carbonyl carbon resonance at one part of the channel and its absence in another part provide the location of a binding site in the channel.

1. *Location of Binding Sites in the Gramicidin A Channel*

Eight Gramicidin A syntheses have been carried out, in each of which the L-residue carbonyl carbon was isotopically enriched; these are (1-^{13}C)L · Val^{1} GA, (1-

Figure 16. Carbon-13 nuclear magnetic resonance of the carbonyl region of eight different synthetic Gramicidin A molecules incorporated into phospholipid bilayers. (A) (1-^{13}C) L · Val^{1} Gramicidin A, (1-^{13}C) L · Ala^{3} Gramicidin A, (1-^{13}C) L · Ala^{5} Gramicidin A, and (1-^{13}C) L · Val^{7} Gramicidin A. Reproduced with permission from Urry *et al.* (1983b). (B) (1-^{13}C) L · Trp^{9} Gramicidin A, (1-^{13}C) Trp^{11} Gramicidin A, (1-^{13}C) Trp^{13} Gramicidin A, and (1-^{13}C) Trp^{15} Gramicidin A. Reproduced with permission from Urry *et al.* (1982a). There are no detectable L-residue carbonyl carbon chemical shifts in the presence of 83 mM Tl^{+} for residues 1, 3, 5, and 7 but there are significant shifts observed for residues 9, 11, 13, and 15. These data are plotted for the double-stranded structures in Figure 17 and for single-stranded structures in Figure 18.

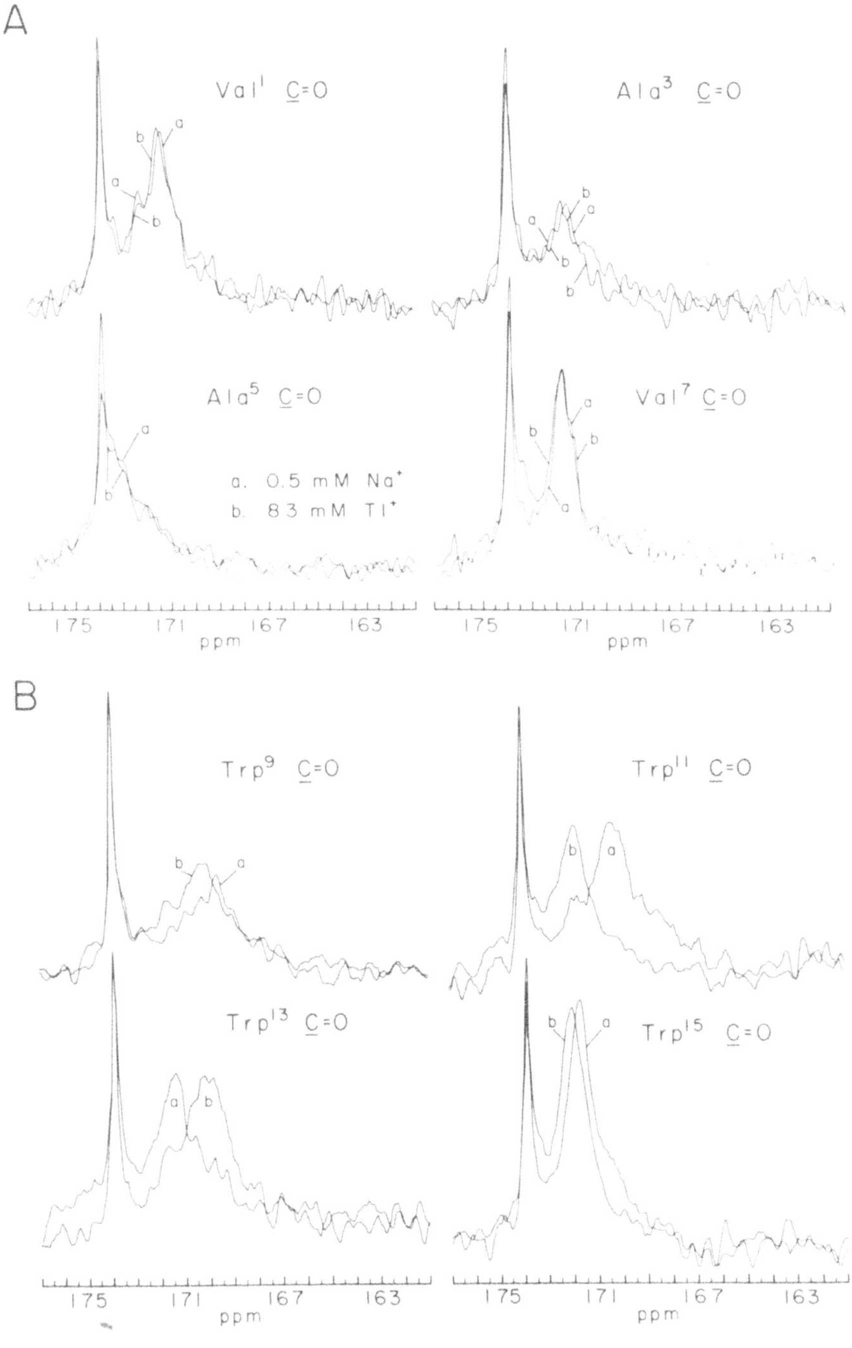
A
Val¹ C=O
Ala³ C=O
Ala⁵ C=O
Val⁷ C=O
a. 0.5 mM Na⁺
b. 83 mM Tl⁺
175
171
167
163
ppm
B
Trp⁹ C=O
Trp¹¹ C=O
Trp¹³ C=O
Trp¹⁵ C=O

^{13}C)L · Ala^{3}GA, (1-^{13}C)L · Ala^{5}GA, (1-^{13}C)L · Val^{7}GA, (1-^{13}C)L · Trp^{9}GA, (1-^{13}C)L · Trp^{11}GA, (1-^{13}C)L · Trp^{13}GA, and (1-^{13}C)L · Trp^{15}GA. The data for ion-induced chemical shifts of each of the L-residue carbonyl carbons of the Gramicidin A channel state in lysolecithin bilayers are in Figure 16. These data are plotted for the antiparallel double-stranded β-helix in Figure 17A and for the parallel double-stranded β-helix in Figure 17B. In Figure 17A, there is the contradiction that the carbonyl carbons of residues 9, 11, 13, and 15 show chemical shifts, whereas the helically-equivalent carbonyl carbons of residues 1, 3, 5, and 7, which are located in the same segment of the structure, do not exhibit ion-induced chemical shifts. These are all structurally equivalent carbonyls located in the same structural segment, and if several show chemical shifts then equally proximal equivalent carbonyl carbons should show essentially similar chemical shifts. Since this is not the case when assuming the antiparallel double-stranded β-helix, that structure is eliminated as a possible channel structure. In Figure 17B, the ion-induced chemical shifts for an assumed parallel double-stranded β-helix are plotted. In this case, the binding site is in the ethanolamine half with no binding site at the formyl end. This structure is excluded

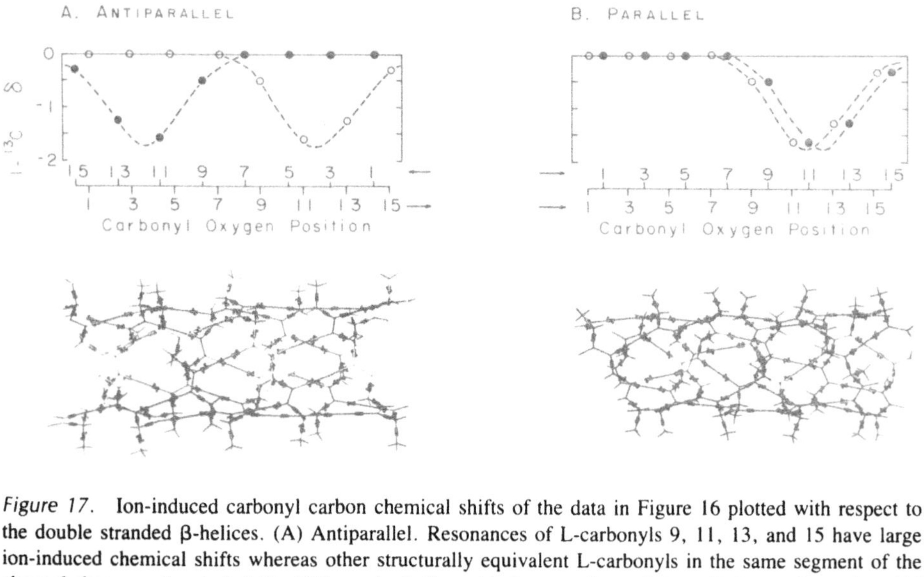

Figure 17. Ion-induced carbonyl carbon chemical shifts of the data in Figure 16 plotted with respect to the double stranded β-helices. (A) Antiparallel. Resonances of L-carbonyls 9, 11, 13, and 15 have large ion-induced chemical shifts whereas other structurally equivalent L-carbonyls in the same segment of the channel show no chemical shifts. This contradiction with the experimental data eliminates the antiparallel double-stranded structure from being the channel structure. (B) Parallel. The ion-induced chemical shifts show a structure which has a single broad ion-binding site in the ethanolamine half of the structure. Since the dominant channel has symmetric *i/V* curves such a structure with asymmetric binding is not possible. This structure is inconsistent with the finding of two binding sites (Urry *et al.*, 1980a,b), with the finding of a dielectric relaxation at 2.2 Hz (Henze *et al.*, 1982), and with the lack of exposure of the formyl end to the aqueous solution, all data obtained on the same system. Accordingly, this structure can be eliminated as a channel state with any more than a minor frequency of occurrence. Adapted with permission from Urry *et al.*, 1983b.

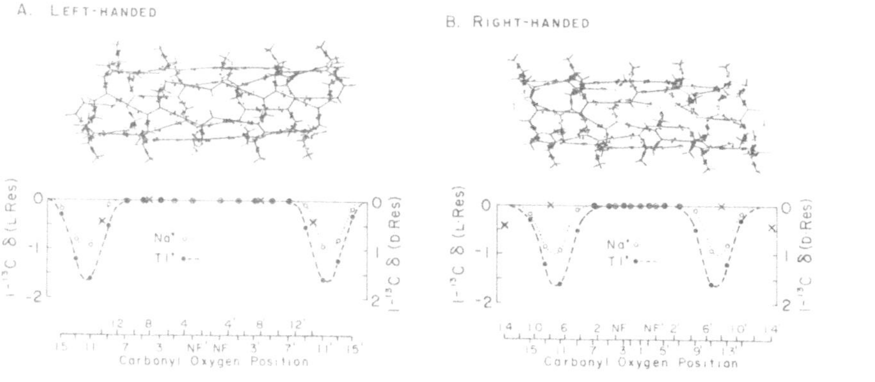

Figure 18. Ion-induced carbonyl carbon shifts of Figure 16 plotted as the dashed curve for the single-stranded, head-to-head dimerized $\beta^{6.3}_{3.3}$-helix of Gramicidin A. (A) Left-handed. Two binding sites are observed separated by about 23 Å. In this structure, the D-Leu[14] carbonyl is also in the binding site and the D · Val[8] carbonyl is well outside of the binding site. The Xs indicate the ion-induced chemical shifts for these D-residues. (B) Right-handed. Two binding sites are separated by about 16 Å, the D · Leu[14] carbonyl is outside of the binding site, and the D · Val[8] carbonyl is optimally placed with respect to the binding site. The ion-induced chemical shifts of these residues do not correspond to the L-carbonyl chemical shifts. It is concluded that the structure of the channel is the left-handed, single-stranded, head-to-head dimerized β-helix. Adapted with permission from Urry *et al.* (1982d).

because it would have an asymmetric free energy profile and would necessarily give an asymmetric *i/V* curve as discussed in Section II-C. Also, of course, as shown in Figure 14, the amino end is not exposed to Ba^{2+}, and as in the studies of Weinstein *et al.* (1979), the amino end is not exposed to the aqueous interface.

On the other hand, when the data of Figure 16 are plotted for a single-stranded β-helix as shown in Figure 18, the data show two binding sites separated by somewhat different distances depending on whether the structure is left-handed (structure I of Figure 3) or right-handed (structure IV of Figure 3). This structure is entirely consistent with the carbonyl carbon chemical shift data and with the orientation of chains discussed in Section III-C. The next question to resolve is the helix sense.

2. *Helix Sense of the Gramicidin A Channel*

The locations of two D-residue carbonyl oxygens, those of D · Val[8] and D · Leu[14], are shown as vertical dotted lines in Figure 18. If the structure is left-handed (Figure 18A), the D · Leu[14] carbonyl oxygen is within the binding site and the D · Val[8] is outside of the binding site, whereas if the structure is right-handed, just the inverse is the case with the D · Val[8] carbonyl oxygen being optimally located in the binding site. It should be noted that since the D-residue carbonyl carbons are not structurally equivalent to the L-residue carbonyl carbons, the ion-induced chemical shifts need not be the same but the presence or absence of an ion-induced chemical shift is what is sought. The Xs on the plots in Figure 18 indicate the ion-induced chemical shifts

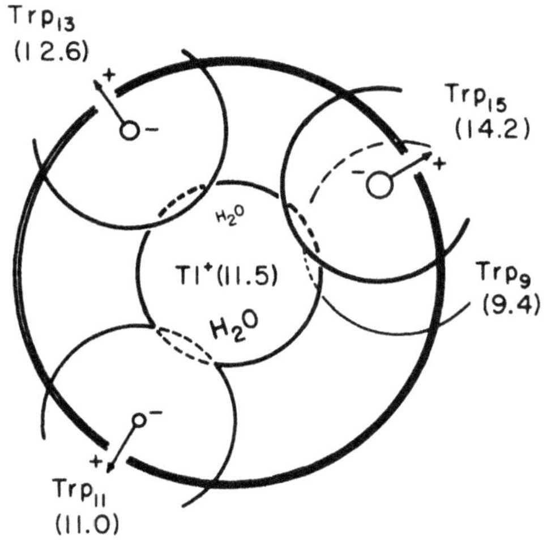

Figure 19. Channel axis perspective from solution of the interaction of the carbonyls of residues 9, 11, 13, and 15 with the thallium ion, showing the relative sizes of ion and carbonyl oxygens, their relative positions along the channel axis, and the small movement of the carbonyl oxygen into the channel to coordinate the bare ion. These relative positions and minor librations of carbonyls are met by the $\beta_{3.3}^{6.3}$-helix but are not met by other single-stranded β-helices with different numbers of residues per turn.

exhibited by the two D-residue carbonyl carbons, and they are plotted with respect to the ordinate on the right side of the figures. The absence of an ion-induced chemical shift for the D · Val[8] residue (Urry *et al.*, 1982d) brings the conclusion that the channel structure is a head-to-head dimerized, left-handed, single-stranded β-helix, i.e., structure I of Figure 3.

3. *Residues per Turn of the Gramicidin A Channel*

The final question to be resolved is the number of residues per turn. In this regard, it is important to realize that the magnitude of the observed ion-induced chemical shifts requires that the carbonyl oxygen be in direct contact with the cation. This is based on the magnitude of chemical shifts observed for enniatin B which is a cyclic analog of Gramicidin (Urry, 1976). The observation that the CD pattern is little affected by the presence of an ion in the channel (Wallace *et al.*, 1981; Henze *et al.*, 1982; Urry *et al.*, 1982c) requires that any structural modification be minor on ion binding. Another relevant point is that ions and water single file through the channel (Levitt *et al.*, 1978; Rosenberg and Finkelstein, 1978). Given the ionic radius and the van der Waals radius of the carbonyl oxygen as shown in Figure 19, these conditions are met by the $\beta_{3.3}^{6.3}$-helix. They are not met by the structures with approximately 4.4, 8, and 10 residues per turn. Thus, it is demonstrated that the structure of the Gramicidin transmembrane channel is as published in 1971 and as shown in Figure 1.

ACKNOWLEDGMENTS

This work was supported in part by NIH Grant GM-26898. The author wishes to thank present and past members of the Laboratory of Molecular Biophysics whose cumulative efforts of more than a decade have made this review possible, and in particular to thank K. U. Prasad for his extraordinary peptide syntheses contributions, T. L. Trapane for her contributions to the NMR studies, and C. M. Venkatachalam for his contributions to the calculation efforts.

REFERENCES

Apell, H.-J., Bamberg, E., Alpes, H., and Läuger, P., 1977, Formation of ion channels by a negatively charged analog of Gramicidin A, *J. Membr. Biol.* **31**:171–188.

Bamberg, E., and Benz, R., 1976, Voltage induced thickness changes of lipid bilayer membranes and the effect of an electric field on GA channel formation, *Biochim. Biophys. Acta* **426**:570–580.

Bamberg, E., and Janko, K., 1977, The action of a carbon suboxide dimerized Gramicidin A on lipid bilayer membranes, *Biochim. Biophys. Acta* **465**:486–499.

Bamberg, E., and Läuger, P., 1973, Channel formation kinetics of Gramicidin A in lipid bilayer membranes, *J. Membr. Biol.* **11**:177–194.

Bamberg, E., and Läuger, P., 1974, Temperature-dependent properties of Gramicidin A channels, *Biochim. Biophys. Acta* **367**:127–133.

Bamberg, E., and Läuger, P., 1977, Blocking of the Gramicidin channel by divalent cations, *J. Membr. Biol.* **35**:351–375.

Bamberg, E., Noda, K., Gross, E., and Läuger, P., 1976, Single-channel parameters of Gramicidin A, B, and C, *Biochim. Biophys. Acta* **419**:223–228.

Bamberg, E., Apell, H.-J., and Alpes, H., 1977a, Structure of the Gramicidin A channel: Discrimination between the $\pi_{L,D}$ and the β-helix by electrical measurements with lipid bilayer membranes, *Proc. Natl. Acad. Sci. USA* **74**(6):2402–2406.

Bamberg, E., Apell, H.-J., Bradley, R., Härter, B., Quelle, M.-J., and Urry, D. W., 1979, Formation of ionic channels in black lipid membranes by succinic derivatives of Gramicidin A, *J. Membr. Biol.* **50**:257–270.

Bradley, R. J., Urry, D. W., Okamoto, K., and Rapaka, R. S., 1978, Channel structures of Gramicidin: Characterization of succinyl derivatives, *Science* **200**:435–437.

Bradley, R. J., Prasad, K. U., and Urry, D. W., 1981a, Single channel properties of D-Leu^2-Gramicidin A: Side chain modulation of channel lifetime, *Biochim. Biophys. Acta* **649**:281–285.

Bradley, R. J., Urry, D. W., Parenti-Castelli, G., and Lenaz, G., 1981b, Effects of halothane on channel activity on N-acetyl Gramicidin, *Biochem. Biophys. Res. Commun.* **101**(3):963–969.

Busath, D., and Szabo, G., 1981, Gramicidin forms multi-state rectifying channels, *Nature* **294**:371–373.

DeFelice, L. J., 1977, Fluctuation analysis in neurobiology, *Int. Rev. Neurobiol.* **20**:169–208.

Fontana, A., 1972, Solid-phase synthesis of the pentadecapeptide valine-Gramicidin A, in: *Peptides, Proceedings of the 12th European Peptide Symposium* (J. Meienhofer, ed.), Ann Arbor Science Publishers, Ann Arbor, Michigan, pp. 229–234.

Goodall, M. C., 1970a, Structural effects in the action of antibiotics on the ion permeability of lipid bilayers. I. Tyrocidine B, *Biochim. Biophys. Acta* **203**:28–33.

Goodall, M. C., 1970b, Structural effects in the action of antibiotics on the ion permeability of lipid bilayers. III. Gramicidins "A" and "S," and lipid specificity, *Biochim. Biophys. Acta* **219**:471–478.

Henze, R., Neher, E., Trapane, T. L., and Urry, D. W., 1982, Dielectric relaxation studies of ionic processes in lysolecithin packaged Gramicidin channels, *J. Membr. Biol.* **64**(3):233–239.

Hladky, S. B., and Haydon, D. A., 1970, Discreteness of conductance change in bimolecular lipid membranes in the presence of certain antibiotics, *Biochim. Biophys. Acta* **225**:451–453.

Hladky, S. B., and Haydon, D. A., 1971, *Symposium of Molecular Mechanisms of Antibiotic Action on Protein Biosynthesis and Membranes* (E. Munoz, F. Garcia-Ferrandiz, and D. Vazquez, eds.), Elsevier, Amsterdam, pp. 738–753.

Hladky, S. B., and Haydon, D. A., 1972, Ion transfer across lipid membranes in the presence of Gramicidin A. I. Studies of the unit conductance channel, *Biochim. Biophys. Acta* **274**:294–312.

Kolb, H.-A., and Bamberg, E., 1977, Influence of membrane thickness and ion concentration on the properties of the GA channel, *Biochim. Biophys. Acta* **464**:127–141.

Kolb, H.-A., Bamberg, E., and Läuger, P., 1975, Correlation analysis of electrical noise in lipid bilayer membranes: Kinetics of Gramicidin A channels, *J. Membr. Biol.* **20**:133–154.

Levitt, D. G., Elias, S. R., and Hautman, J. M., 1978, Number of water molecules coupled to the transport of sodium, potassium and hydrogen ions via Gramicidin, nonactin, or valinomycin, *Biochim. Biophys. Acta* **512**:436–451.

Lotz, B., Colonna-Cesari, F., Heitz, F., and Spach, G., 1976, A family of double helices of alternating poly(y-benzyl-D-L-glutamate), a stereochemical model for Gramicidin A, *J. Mol. Biol.* **106**:915–942.

Masotti, L., Spisni, A., and Urry, D. W., 1980, Conformational studies on the Gramicidin A transmembrane channel in lipid micelles and liposomes, *Cell Biophys.* **2**(3):241–251.

McBride, D., and Szabo, G., 1978, Influence of double-layer and dipolar surface potentials on ionic conductance of Gramicidin channels, *Biophys. J.* **21**:A25.

Morrow, J. S., Veatch, W. R., and Stryer, L., 1979, Transmembrane channel activity of Gramicidin A analogs: Effects of modification and deletion of the amino-terminal residue, *J. Mol. Biol.* **132**:733–738.

Mueller, P., and Rudin, D. O., 1967, Development of $K^+ - Na^+$ discrimination in experimental biomolecular lipid membranes by macrocyclic antibiotics, *Biochem. Biophys. Res. Commun.* **26**:398–404.

Neher, E., 1975, Ionic specificity of the Gramicidin channel and the thallous ion, *Biochim. Biophys. Acta* **401**:540–544.

Noda, K., and Gross, E., 1972, Solid-phase synthesis of the pentadecapeptides Valine-Gramicidin B and C, in: *Chemistry and Biology of Peptides, Proceedings of the 3rd American Peptide Symposium* (J. Meienhofer, ed.), Ann Arbor Science Publishers, Ann Arbor, Michigan, pp. 241–250.

Pasquali-Ronchetti, I., Spisni, A., Casali, E., Masotti, L., and Urry, D. W., 1983, Gramicidin A induces lysolecithin to form bilayers, *Biosci. Rep.* **3**:127–133.

Prasad, K. U., Trapane, T. L., Busath, D., Szabo, G., and Urry, D. W., 1982a, Synthesis and characterization of 1-^{13}C-D · Leu^{12,14} Gramicidin A, *Int. J. Pep. Protein Res.* **19**(2):162–171.

Prasad, K. U., Trapane, T. L., Busath, D., Szabo, G., and Urry, D. W., 1982b, Solid phase (9-fluorenylmethyloxycarbonyl) synthesis and characterization of [(1-^{13}C) Phe^{11}] Gramicidin B, *J. Protein Chem.* **1**(3):191–202.

Prasad, K. U., Trapane, T. L., Busath, D., Szabo, G., and Urry, D. W., 1983, Synthesis and characterization of 1-^{13}C Phe^9 Gramicidin A: Effects of side chain variations, *Int. J. Pep. Protein Res.* **22**:341–347.

Ramachandran, G. N., and Chandrasekharan, R., 1972a, Studies on dipeptide conformation and on peptides with sequences of alternating L and D residues with special reference to antibiotic and ion transport peptides, in: *Progress in Peptide Research,* Vol. II (S. Lande, ed.), Gordon and Breach, Science Publishers, New York, pp. 195–215.

Ramachandran, G. N., and Chandrasekharan, R., 1972b, Conformation of peptide chains containing both L- and D-residues: Part I. Helical structures with alternating L- and D-residues with special reference to the LD-ribbon and the LD-helices, *Ind. J. Biochem. Biophys.* **9**:1–11.

Rosenberg, P. A., and Finkelstein, A., 1978, Interaction of ions and water in Gramicidin A channels, *J. Gen. Physiol.* **72**:327–340.

Sarges, R., and Witkop, B., 1964, Gramicidin A. IV. Primary sequence of valine and isoleucine Gramicidin A, *J. Am. Chem. Soc.* **86**:1862–1863.

Sarges, R., and Witkop, B., 1965a, Gramicidin A. V. The structure of valine- and isoleucine-gramicidin A, *J. Am. Chem. Soc.* **87**:2011–2020.

Sarges, R., and Witkop, B., 1965b, Gramicidin. VIII. The structure of valine- and isoleucine-Gramicidin C, *Biochemistry* **4**:2491–2494.

Szabo, G., and Urry, D. W., 1979, N-acetyl gramicidin: Single-channel properties and implications for channel structure, *Science* **203**: 55–57.

Tosteson, D. C., Andreoli, T. E., Tieffenbery, M., and Cook, P., 1968, The effect of macrocyclic compounds on cation transport in sheep red cells and thin and thick lipid membranes, *J. Gen. Physiol.* **51**:373–384.

Urry, D. W., 1971, The Gramicidin A transmembrane channel: A proposed $\pi_{(L,D)}$ helix, *Proc. Natl. Acad. Sci. USA* **68**:672–676.

Urry, D. W., 1972, A molecular theory of ion conducting channels: A field dependent transition between conducting and nonconducting conformations, *Proc. Natl. Acad. Sci. USA* **69**:1610–1614.

Urry, D. W., 1973, Polypeptide conformation and biological function of β-helices ($\pi_{L,D}$-helices) as permselective transmembrane channels, in: *Conformation of Biological Molecules and Polymers—The Je-*

rusalem Symposia on Quantum Chemistry and Biochemistry V, Israel Academy of Sciences, Jerusalem, pp. 723–736.

Urry, D. W., 1976, Nuclear magnetic resonance and the conformation of membrane active peptides, in: *Enzymes of Biological Membranes*, Vol. 1 (A. Martonosi, ed.), Plenum Publishing Corporation, New York, pp. 31–69.

Urry, D. W., 1982, On the molecular structure and ion transport mechanism of the Gramicidin transmembrane channel, in: *Membranes and Transport*, Vol. 2 (A. Martonosi, ed.), Plenum Publishing Corporation, New York, pp. 285–294.

Urry, D. W., Goodall, M. C., Glickson, J. D., and Mayers, D. F., 1971, The Gramicidin A transmembrane channel: Characteristics of head to head dimerized $\pi_{(L,D)}$ helices, *Proc. Natl. Acad. Sci. USA* **68:**1907–1911.

Urry, D. W., Long, M. M., Jacobs, M., and Harris, R. D., 1975, Conformation and molecular mechanisms of carriers and channels, *Ann. N.Y. Acad. Sci.* **264:**203–220.

Urry, D. W., Spisni, A., and Khaled, M. A., 1979a, Characterization of micellar-packaged Gramicidin A channels, *Biochem. Biophys. Res. Commun.* **88**(3):940–949.

Urry, D. W., Spisni, A., Khaled, M. A., Long, M. M., and Masotti, L., 1979b, Transmembrane channels and their characterization in phospholipid structures, *Int. J. Quantum Chem., Quantum Biol. Symp. No. 6*, pp. 289–303.

Urry, D. W., Venkatachalam, C. M., Spisni, A., Läuger, P., and Khaled, M. A., 1980a, Rate theory calculation of Gramicidin single channel currents using NMR-derived rate constants, *Proc. Natl. Acad. Sci. USA* **77:**2028–2032.

Urry, D. W., Venkatachalam, C. M., Spisni, A., Bradley, R. J., Trapane, T. L., and Prasad, K. U., 1980b, The malonyl Gramicidin channel: NMR-derived rate constants and comparison of calculated and experimental single channel currents, *J. Membr. Biol.* **55:**29–51.

Urry, D. W., Prasad, K. U., and Trapane, T. L., 1982a, Location of monovalent cation binding sites in the Gramicidin channel, *Proc. Natl. Acad. Sci. USA* **79:**390–394.

Urry, D. W., Trapane, T. L., and Prasad, K. U., 1982b, Molecular structure and ionic mechanisms of an ion selective transmembrane channel: Monovalent vs. divalent cation selectivity, *Int. J. Quantum Chem., Quantum Biol. Symp. No. 9*, 31–40.

Urry, D. W., Trapane, T. L., Walker, J. T., and Prasad, K. U., 1982c, On the relative membrane permeability of Na^+ and Ca^{2+}: A physical basis for the messenger role of Ca^{2+}, *J. Biol. Chem.* **257:** 6659–6661.

Urry, D. W., Walker, J. T., and Trapane, T. L., 1982d, Ion interactions in (1-^{13}C)D · Val^8 and D · Leu^{14} analogs of Gramicidin A, the helix sense of the channel and location of ion binding sites, *J. Membr. Biol.* **69:**225–231.

Urry, D. W., Trapane, T. L., Romanowski, S., Bradley, R. J., and Prasad, K. U., 1983a, Use of synthetic Gramicidins in the determination of channel structure and mechanism, *Int. J. Pep. Protein Res.* **21:**16–23.

Urry, D. W., Trapane, T. L., and Prasad, K. U., 1983b, Is the Gramicidin A transmembrane channel a single-stranded or double-stranded helix? A simple unequivocal determination. *Science* **221:** 1064–1067.

Veatch, W. R., and Blout, E. R., 1974, The aggregation of Gramicidin A in solution, *Biochemistry* **13:**5257–5264.

Veatch, W. R., Fossel, E. T., and Blout, E. R., 1974, The conformation of Gramicidin A, *Biochemistry* **13:**5249–5256.

Veatch, W. R., Mathies, R., Eisenberg, M., and Stryer, L., 1975, Simultaneous fluorescence and conductance studies of planar bilayer membranes containing a highly active and fluorescent analog of Gramicidin A, *J. Mol. Biol.* **99:**75–92.

Wallace, B. A., Veatch, W. R., and Blout, E. R., 1981, Conformation of Gramicidin A in phospholipid vesicles: Circular dichroism studies of effects of ion binding chemical modification and lipid structure, *Biochemistry* **20:**5754–5760.

Weinstein, S., Wallace, B. A., Blout, E. R., Morrow, J. S., and Veatch, W., 1979, Conformation of Gramicidin A channel in phospholipid vesicles: A ^{13}C and ^{19}F nuclear magnetic resonance study, *Proc. Natl. Acad. Sci. USA* **76**(9):7230–7234.

Zingsheim, H. P., and Neher, E., 1974, The equivalence of fluctuation analysis and chemical relaxation measurements: A kinetic study of ion pore formation in thin lipid membranes, *Biophys. Chem.* **2:**197–207.

Conventional ESR Spectroscopy of Membrane Proteins: Recent Applications

Philippe F. Devaux

I. INTRODUCTION

Conventional electron spin resonance (ESR) spectroscopy of nitroxide radicals is sensitive to motions corresponding to correlation times in the range 10^{-11}–10^{-7} sec. If the motion is anisotropic, with a fast component in the 10^{-9}–10^{-11} sec. range, an order parameter can be measured; the latter is indicative of the angular domain associated with the fast motion. These properties are well suited to the investigation of lipid viscosity. On the other hand, saturation transfer ESR (ST-ESR), which is an extension of the spin-labeling method, is sensitive to motions corresponding to the so-called slow motion regime: 10^{-3}–10^{-7} sec. The latter technique is adapted to the investigation of the motion of proteins and in particular of membrane proteins. In this book, D. T. Thomas reviews the applications of ST-ESR to membrane proteins, while the present chapter deals only with the applications of conventional ESR. Previous reviews have exposed the details of this technique which was introduced first in 1965 by Stone *et al.* The two volumes, edited by Berliner (1976, 1979), on the topic of "spin-labeling" contain the most useful references for the background necessary for an understanding of the present review. In the 1976 edition of *The Enzymes of Biological Membranes* edited by A. Martonosi, two brief chapters were devoted to the use of spin labels to study membrane bound enzymes, receptors, and transport systems (Gaffney and Chang Lin, 1976; Vignais and Devaux, 1976). The present edition will emphasize the newest applications of spin-labeling to the investigation of membrane-bound enzymes.

We feel that the most important applications in this field in recent years deals

Philippe F. Devaux • Institut de Biologie Physico-Chimique, 13, rue Pierre et Marie Curie, 75005 Paris, France.

with the subject of lipid–protein interactions. Influence of proteins on the mobility of lipids, as well as specificity of lipid–protein interactions, are two domains which have progressed largely because of applications of magnetic resonance. ESR and NMR have been used. The disadvantage of ESR as opposed to NMR is a possible probe perturbation; this is compensated by (1) the sensitivity and (2) the time scale of ESR. Thus, a large part of this review is devoted to the subject of lipid–protein interactions in membranes.

Additional applications of ESR to membranes are described more briefly at the end of the chapter. Spin-labeling has been extensively used for the determination of membrane fluidity, however, this does not constitute a new application and will be only briefly mentioned. Among the more recent applications is ESR time-resolved measurements of electric potential. The latter technique is described below together with some recent application of spin-labeling as a means to monitor enzyme conformation.

II. THE TIME SCALE OF PHOSPHOLIPID EXCHANGE AT THE BOUNDARY OF NONAGGREGATED INTRINSIC PROTEINS

Early descriptions of biological systems emphasized their spatial organization in such a way that a rather static picture emerged. So, for example, the organization of lipids and proteins in membranes may have appeared as a frozen structure. About ten years ago, ESR has served this description by emphasizing the difference between phospholipids at the boundary of intrinsic proteins and phospholipids in a bulk lipid phase. Indeed, in 1973, Jost and co-workers showed that a spin-labeled fatty acid, in partially delipidated membranes and enriched in cytochrome oxidase, gives two ESR components: a "strongly-immobilized component" and a "bilayer-type component." A tentative determination of the amount of "rigid" lipids led to the proposal that one layer of immobilized lipids surrounds each protein. Results obtained by other workers with membrane-containing cytochrome oxidases, Ca^{2+}ATPase, rhodopsin, lipophilin, and gramicidin A were all consistent. However, these results required a lipid-to-protein ratio below the physiological threshold. Thus, they were obtained under conditions where protein aggregation is a likely event. The following question then arises: do the two states exist in a real biological membrane containing enough lipids to disperse the proteins?

The relevance of Jost and collaborators' experiment was first questioned by Vignais and Devaux (1976), and later by Chapman *et al.* (1977). It is not trivial to answer this question since increasing the lipid-to-protein ratio results in an increase of the bilayer component. The latter, corresponding to narrow lines, renders very difficult the detection and quantification of any broad component associated with immobilized boundary lipids. In 1978, Marsh and Barrantes showed that a spin-labeled fatty acid in intact membranes from the electric organ of *Torpedo marmorata* gives an immobilized component, quite visible at low temperature (−4°C). However, this raised another controversy since two different laboratories showed later that a fatty acid spin label behaves differently from a spin-labeled phospholipid in this membrane. Apparently, free fatty acids have a particular affinity for the acetylcholine receptor (Rousselet *et al.*, 1979; McNamee *et al.*, 1982). In 1979, Watts and collaborators were able to show convincingly that a spin-labeled fatty acid or a phospholipid with the probe at the 14th position of a fatty acid chain, incorporated into rod outer segment disks, gives rise to a two-component ESR spectrum (Figure 1). The two components are best seen

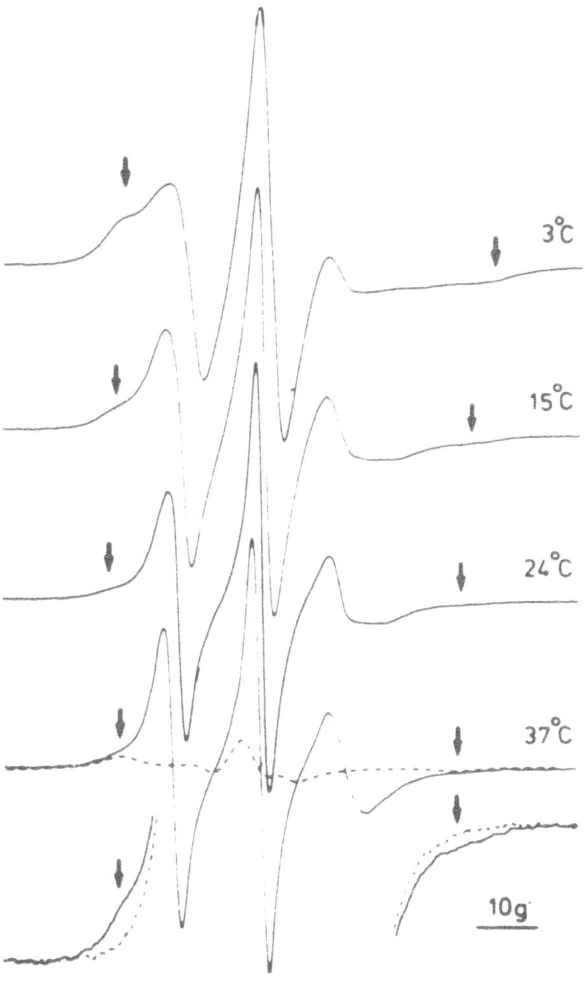

Figure 1. ESR of 14-doxyl stearic acid in rod outer segment membranes at 3, 15, 24, and 37°C. The outer wings of the 37°C spectrum are expanded five times to display the immobilized components, and the dotted lines in the expanded spectrum are from an extracted lipid spectrum which has the same spectral parameters and shape spectrum (dotted line) superimposed upon the 37°C membrane spectrum from an integrated intensity of 37% of the membrane spectra. Arrows indicate the outermost wings of the immobilized spectrum, which are absent in the extracted lipids. Similar spectra are obtained with spin-labeled phospholipids (from Watts *et al.*, 1979).

Table 1. Spin-Labeled Lipids Reporting on the Boundary Layer of Intrinsic Proteins

I

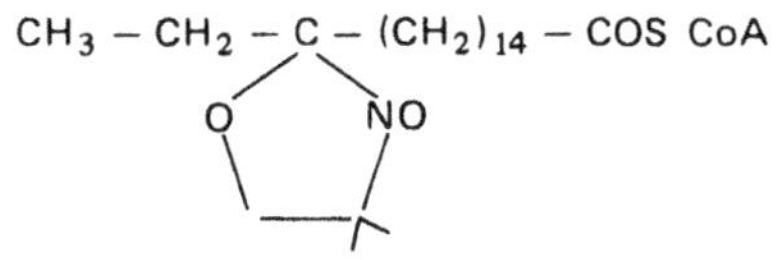

II

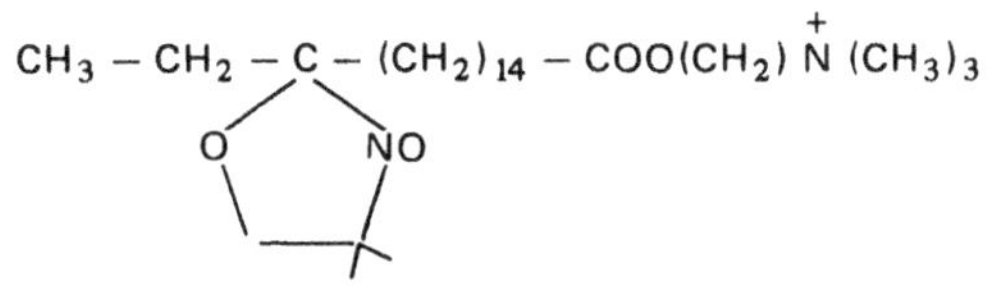

III

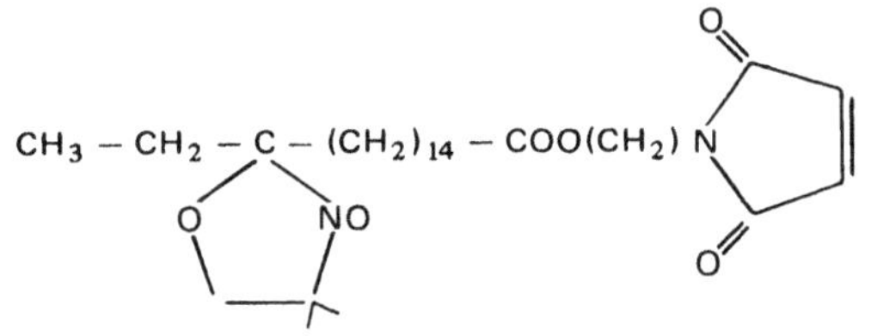

IV

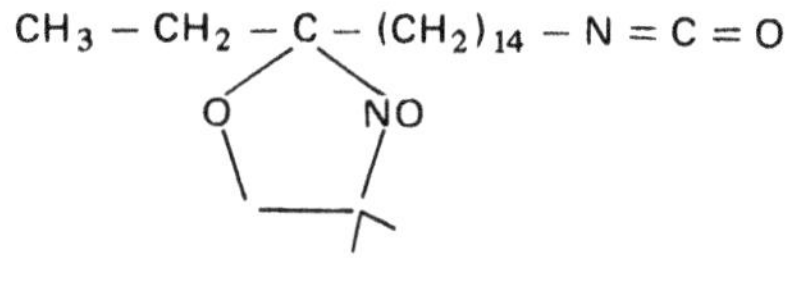

V

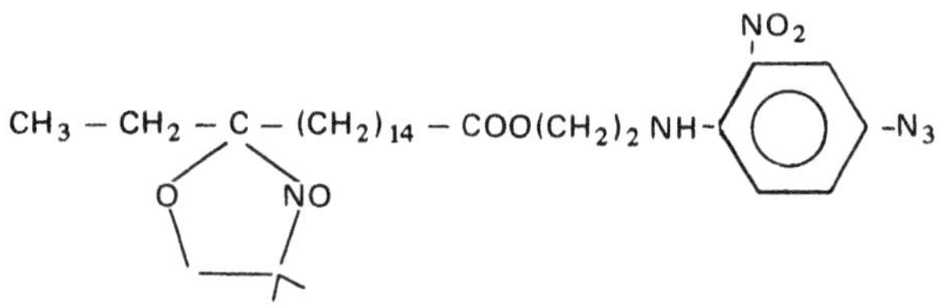

at low temperature (3°C) but seem to exist at higher temperatures providing one assumes a temperature-dependent lineshape for the immobilized component (hyperfine splitting of 64 G at 0°C and 54 G at 37°C). Thus, boundary lipids can be detected by ESR in a native biological membrane. However, it should be emphasized that these lipids experience a motion on a time scale where ESR spectra of nitroxides are motionally averaged, that is in the range $\tau_{c\,\mathrm{effective}} < 10^{-7}$ sec.

A simple way to enhance the spectral contribution of boundary lipids is to link a spin-labeled lipid to a hydrophobic protein. A very high ratio of lipid to protein is not an obstacle anymore to the lineshape determination of the broad component because the contribution of the spin labels in the bulk is reduced in principle to zero. Some of the spin labels which have been utilized for this purpose in our laboratory are shown in Table 1. They bind to the proteins in the polar region either reversibly or covalently. The nitroxide positioned near the ω-2 region of the fatty acid chain is buried in the hydrophobic core of the membrane, as can be verified by assaying the accessibility of the probe with ascorbate or with a paramagnetic ion. In 1975, Devaux *et al.* showed that a spin-labeled acyl CoA which binds with high affinity to the ADP carrier in mitochondria gives rise to a spectrum indicative of a high mobility, provided that the probe is located near the methyl terminal of the fatty acid chain. Similar results were obtained with a long chain spin-label acylcholine at the boundary of the acetylcholine receptor (Bienvenüe *et al.*, 1977). Later Davoust *et al.* (1979) investigated, at various temperatures, the motion of a chain at the boundary of rhodopsin in reconstituted systems containing either high or low ratio of lipid to protein. For physiological ratios of lipid to protein, the spectrum of spin label III bound to rhodopsin in egg lecithin vesicles resembles, at 37°C, the spectrum of a fatty acid in a fluid bilayer. Yet, at low temperature, the spectrum looks composite (see Figure 2). The overall splitting of the "immobilized component" is approximately 60 G at 5°C. In 1982, Davoust and Devaux showed that it is possible to account for the spectrum at any temperature by assuming an exchange between two states (protein associated and lipid bilayer). The model implies that the covalently-attached long chain spin label explores more than just one layer of lipid around the protein (see Figure 3). The boundary layer, or first layer, is associated with a high order and a strong immobilization of the probe on the time scale of conventional ESR ($\tau_c \gtrsim 10^{-7}$ sec.). In the bulk lipid phase, the order is low and the effective correlation time is approximately 10^{-9} sec. The two states are spec-

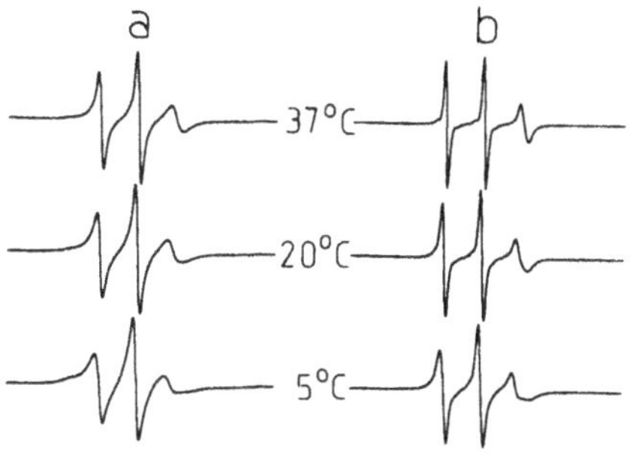

Figure 2. ESR spectra of spin-label III (a) attached to rhodopsin in rhodopsin-egg lecithin vesicles (lipid to protein ratio 500/1); (b) spectra of spin-label III in lecithin liposomes (from Davoust *et al.*, 1979).

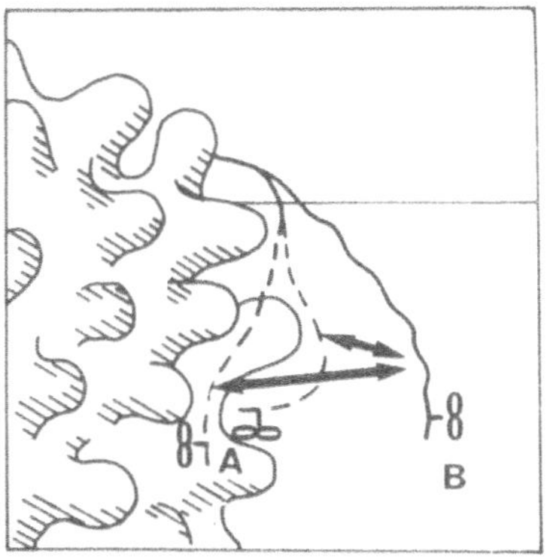

Figure 3. Schematic representation of the binding of spin-label III at the boundary of rhodopsin. In this model, the probe exchanges rapidly between the protein surface and the bulk lipid phase. At each collision with the protein rough surface, the label takes a new orientation (from Davoust and Devaux, 1982).

troscopically well differentiated at low temperature (5°C) in the slow exchange regime. As the temperature increases, exchange effects tend to average the boundary layer spectrum with that of the fluid lipid spectrum. This phenomenon gives rise at 20°C to an "apparent" boundary layer state with an effective correlation time of 10^{-8} sec. This latter value is equal to the inverse of the exchange frequency at 20°C. At 35°C, both components collapse, indicating that the exchange frequency is greater than the anisotropy of the hyperfine tensor of a nitroxide. In summary, the reversible disappearance of the low field shoulder in the ESR spectrum (Figure 2) is an indication that the time scale of the on- and off-movement of spin label III at the surface of the protein matches exactly the time scale of the spectral modification seen when using conventional ESR with nitroxides. Sweeping through the temperature (0–40°C) corresponds to the exploration of the sensitivity domain of the method. By applying this model, the rate of exchange is found to be comparable to the hopping frequency of freely diffusible phospholipids. In the latter calculation, the accuracy is less because of the large contribution of the bilayer component at all temperatures. Nevertheless, the broad component has a temperature-dependent splitting which can be accounted for by increasing the phospholipid exchange rate between bulk lipids and boundary lipids from $\sim 10^7$ sec^{-1} at 4°C to $\sim 10^8$ sec^{-1} at 45°C.

In conclusion, it is better to speak of "motionally-restricted components" associated with boundary lipids in nonaggregated membranes since the "boundary spectrum" is not a powder type spectrum with native membranes nor with reconstituted systems containing a high ratio of lipid to protein.

The "motionally restricted component" is not to be confused with the "strongly-immobilized component" which is seen when proteins are aggregated, for example, by delipidation. In such a case, the splitting is about 65 G and is barely influenced by temperature. In a recent note, Watts *et al.* (1981) have emphasized the two types of immobilized components which can be detected with ESR in rhodopsin-containing membranes. In all instances, "complete immobilization" means only that there is no motion within a characteristic time of 10^{-6}–10^{-7} sec or faster. As we shall see, ST-ESR as well as NMR prove that lipid motion does exist within these aggregates.

III. LIPIDS TRAPPED BETWEEN PROTEIN AGGREGATES OR PROTEIN OLIGOMERS

Proteins spontaneously aggregate in a few naturally occurring membranes. Bacteriorhodopsin in the purple patches of *Halobacterium halobium* forms two-dimensional crystals, containing a very low ratio of lipid to protein. A spin-labeled fatty acid in such membranes shows essentially a broad component (Chignell and Chignell, 1975). To obtain the same lineshape with most membranes requires a partial delipidation. As shown originally by Jost *et al.* (1973), with cytochrome oxidase, the amount of strongly immobilized component is inversely proportional to the amount of lipids left. We think that the immobilized component with a large splitting does not reflect boundary lipids exchanging freely but rather trapped lipids within clusters of aggregated proteins.

In some systems, it is possible to induce protein aggregation by lowering the temperature. Usually, a phospholipid transition from liquid crystal to gel is accompanied by the exclusion of the proteins from the gel. Thereby, the probability of protein aggregation in the lipid-poor region is high. The segregation of lipids is also accompanied by the exclusion of freely diffusible spin-labeled lipids from the protein environment. Thus, spin labels must be covalently bound to the proteins in order to report on their hydrophobic environment below the phospholipid transition temperature. Using the long chain spin label IV covalently attached to purified rhodopsin, Davoust *et al.* (1980) have investigated rhodopsin boundary in various lipids undergoing a phase transition. These authors were able to show that below the phospholipid phase transition, the residual motion of a fraction of the lipid chains at the boundary of the protein is larger (lower order parameter) than in the pure lipid gel phase. This probably indicates that proteins segregate with a substantial shell of lipids which are prevented from undergoing the phase transition by the interaction with the protein surface.

In conclusion, there are different degress of protein aggregation. The lipids squeezed between these proteins can experience various types of residual motion in the range of correlation time $\tau_c \gtrsim 10^{-7}$ sec. The technique best adapted to investigate these motions is not conventional ESR but rather ST-ESR. ST-ESR of rhodopsin labeled with spin label IV shows that the chain at the boundary of rhodopsin in dimyristoyl phosphatidylcholine undergoes a motion on the time scale of 10^{-6} sec at 4°C, while the protein rotational correlation time is of the order of 10^{-3} sec (Davoust and Devaux, unpublished; Devaux, unpublished).

Lipid chains can also be trapped within small protein oligomers. Such oligomers can be detected by binding covalently a spin-labeled fatty acid to intrinsic proteins (see Figure 4). This technique was employed to show the oligomericity of Ca^{2+} ATPase in sarcoplasmic reticulum (Andersen *et al.*, 1981; Thomas *et al.*, 1982), Na^+K^+ ATPase in torpedo membranes (Zachowski and Devaux, unpublished; Devaux, unpublished), and band 3 in erythrocytes (Bittman *et al.*, 1984).

Figure 5A shows the spectrum of spin label III bound to solubilized Ca^{2+} ATPase. The protein after solubilization is a mixture of monomers and oligomers. A direct proof that the spectrum reflects this heterogeneity can be obtained by the actual physical

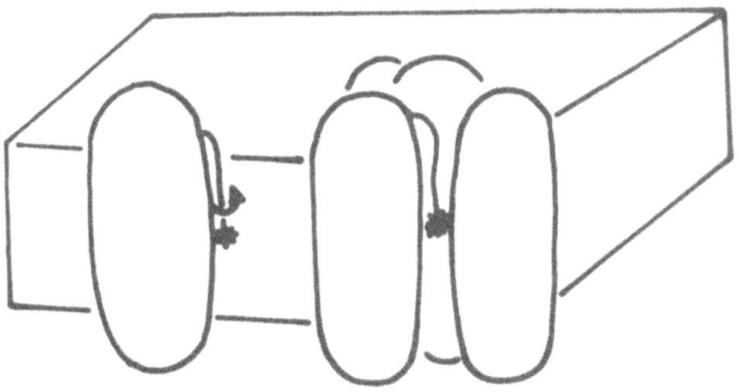

Figure 4. When a spin-labeled fatty acid is covalently attached to the boundary of an oligomeric protein, the motion of the probe is restricted $\tau_c \gtrsim 10^{-7}$ sec (right). If the protein dissociates into monomers, the probe experiences a much faster motion ($\tau_c \sim 10^{-9}$ sec) (left). Thus, conventional ESR enables one to differentiate oligomeric proteins from monomeric proteins.

separation of the molecules associated with each component. This is achieved by ultracentrifugation. According to the sedimentation velocity, the supernatant, whose spectrum is B, contains only monomers, while the resuspended pellet (spectrum C) is associated with large oligomers (Andersen *et al.*, 1981). The quantitative determination of the amount of strongly immobilized component can be used to monitor the state of oligomerization of specific proteins in membranes. For example, in erythrocyte ghosts, it can be shown that alkaline treatment or Ca^{2+} induces aggregation of band 3 protein, while increasing temperature results in its partial dissociation. This is in agreement with the previous conclusions obtained by Nigg and Cherry from the investigation of the rotational correlation time of band 3 by transient dichroïsm (Nigg and Cherry, 1979). Spin label III was also used to show that cytoskeletal proteins (spectrin–ankyrin–band 4.1) have no influence on the state of oligomerization of band 3 (Bittman *et al.*, 1984). The oligomeric states of Ca^{2+}ATPase, $Na^{+}K^{+}$ATPase, or band 3 seem to be physiological states, the function of which is not known. On the

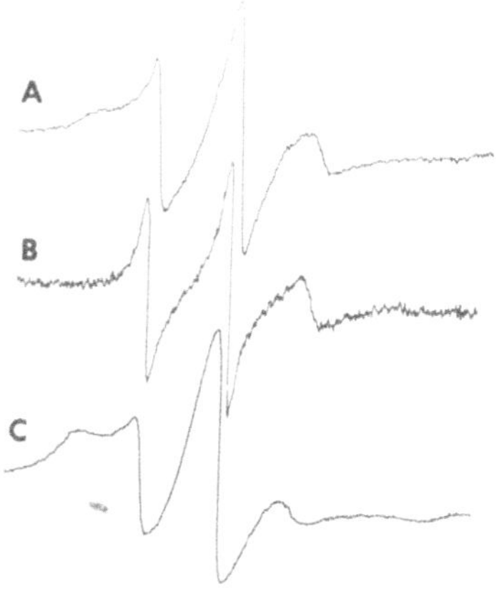

Figure 5. Solubilized Ca^{2+}ATPase labeled with spin-label III. (A) Solubilized ATPase before centrifugation. (B) Supernatant after 40 min centrifugation at 130,000*g*. (C) Pellet resulting from the centrifugation, resuspended in the elution buffer. Analytical centrifugation shows that only (B) sample corresponds to a monomer (from Andersen *et al.*, 1981).

other hand, the aggregation of cytochrome oxidase which is obtained in most reconstituted systems *even when high ratios of lipid to protein are used* appears to be an artifact of reconstitution. Using spin label III (Swanson *et al.*, 1980) showed that the spectrum of spin-labeled cytochrome oxidase reconstituted in soybean lecithin was mainly a broad spectrum or mainly narrow lines depending on the type of detergent and stage of purification of the protein. In parallel ST-ESR experiments they showed, using a short chain maleimide spin label, that the rotational correlation time of the protein indicated in the former case no submillisecond rotation, while in the latter case $\tau_c \sim 10^{-4}$ sec. Finally, bleached rhodopsin forms stable oligomers if exposed to prolonged illumination. The formation of these oligomers (of nonphysiological function) can again be monitored using spin label III (Davoust and Devaux, unpublished; Devaux, unpublished).

At this point, it is interesting to compare the results obtained by ESR on lipid–protein interactions and those obtained by NMR. Since the averaging time in NMR is of the order of 10^{-4}–10^{-5} sec, one should not expect to see a two-component NMR spectrum corresponding to bulk and bound lipids. Even if proteins are aggregated, for example when low molar ratios of lipids are used, the motion of lipids remains fast on the NMR time scale. Indeed, all investigators have found, for example with ^{2}H-NMR, that proteins only slightly broaden the lines without significantly affecting the quadrupole splitting which for ^{2}H-NMR is proportional to the molecular order parameter (for a review see Seelig *et al.*, 1982; Oldfield *et al.*, 1982). On the other hand, T_2 are sensitive to fast motions, and relaxation measurements demonstrate that the lipids are indeed in the proximity of the proteins. Very likely, T_2 is affected by changes in the rate of transisomerization of the chains. Quantitative information on molecular motions from the relaxation measurements are very difficult. Thus, to summarize, one can consider that ESR is more informative than NMR on lipid motions at the proximity of proteins. Considering in addition the larger sensitivity of ESR, it is not surprising that the problem of lipid–protein specificity has been approached more successfully with ESR than NMR, in spite of the probe perturbation.

IV. SPECIFICITY OF LIPID–PROTEIN INTERACTIONS AS INVESTIGATED WITH SPIN LABELS

A. First Approach: Estimation of the Relative Percentage of Immobilized Component

Although reconstitution experiements have not proven clearly the necessity of particular lipid compositions for the function of membrane-bound enzymes, it is tempting to assume that particular proteins require particular lipids. In addition, if proteins attract specific lipids, one would expect lipid differentiation between various membranes or within a membrane. Because we have emphasized the rapid exchange of phospholipids at the boundary of a protein, does not mean that proteins cannot experience preferential interaction for a certain class of lipids. If protein affinity for lipids is affected by head groups, chain lengths, and chain unsaturation, the proteins may be preferentially surrounded by one class of lipids even though the lipids are exchanging rapidly. Thus, there is a need for systematic measurements of the local lipid composition around specific membrane proteins under physiological conditions.

Several spectroscopic techniques can be used for this purpose. To recognize "boundary lipids" by a spectroscopic technique requires that the time scale of the technique be fast enough, so that averaging between bulk and boundary does not take place. Such is the case for ESR and fluorescence; usually it is not the case for NMR. ESR permits different approaches. One approach is developed in this section, while another is presented in Section IV-B. We postpone for Section V spin-label studies related to extrinsic proteins.

Table 2. Spin-Labeled Phospholipids[a]

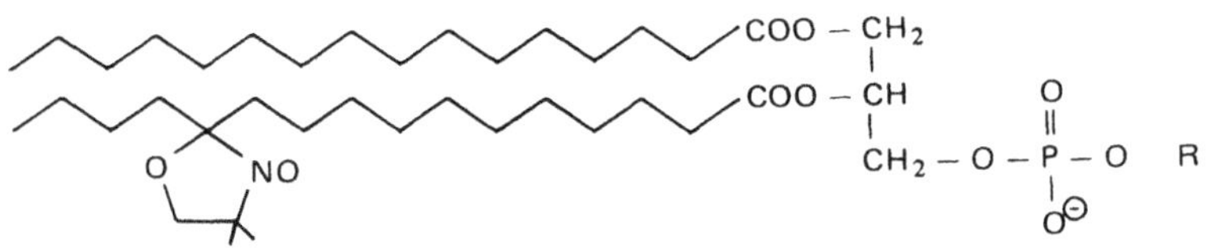

VI	R = H	PA
VII	R = $(CH_2)_2 - N^+H_3$	PE
VIII	R = $(CH_2)_2\ N^+(CH_3)_3$	PC
IX	R = $CH_2 - CH(COO^-) - N^+H_3$	PS
X	R = $CH_2 - CHOH - CH_2OH$	PG

XI

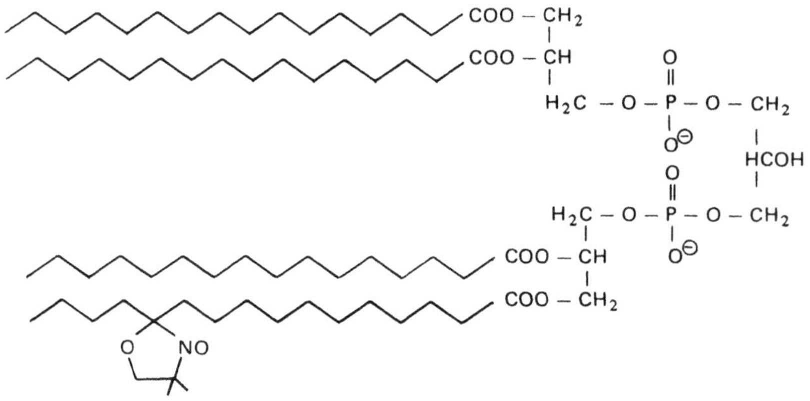

[a] From Knowles *et al.* (1979). Note that the probe is at the 12th position of the fatty acid, whereas in Table 1, the probe is at the 16th position.

The technique discussed here implies the comparison of the ESR spectra recorded after the incorporation of different spin-labeled lipids into membrane fragments or reconstituted systems. Generally, for spin-labeled phospholipids, the probe is carried by an acyl chain, and different head groups are tested (Table 2). As pointed out in the previous sections, a spin-labeled lipid interacting with an intrinsic protein gives rise to a broad ESR component corresponding either to a motionally restricted component or a strongly immobilized component.

Ohnishi and his collaborators were the first to show that different spin-labeled phospholipids give different spectra in erythrocyte membranes. Spin-labeled phosphatidylcholine was found more immobilized than spin-labeled phosphatidylethanolamine or phosphatidylserine (Tanaka and Ohnishi, 1976). In this experiment, the spin labels were introduced in erythrocytes by fusion. Later, the same group studied *E. coli* membranes in which they biosynthetically incorporated spin-labeled phospholipids (Takeuchi *et al.*, 1978, 1981). Differences were observed among the various phospholipids; addition of ions suppressed most of the differences. Unfortunately, in such membranes, it is impossible to assess which protein is involved.

In 1980, Cable and Powell showed that spin-labeled cardiolipin has a strong interaction with cytochrome oxidase in reconstituted systems, while Bosterling *et al.* (1981) showed that cytochrome p-450 in reconstituted vesicles interacts preferentially with negatively charged phosphatidic acid. More quantitative investigations have been carried out by Marsh and collaborators in other membranes: disk membranes (Watts *et al.*, 1979), *Torpedo marmorata* membrane fragments (Marsh *et al.*, 1981), yeast cytochrome oxidase in reconstituted vesicles (Knowles *et al.*, 1981), and Na^+K^+ATPase in reconstituted vesicles (Marsh *et al.*, 1982a,b). Figure 6 shows a series of spectra corresponding to different head group spin labels introduced into cytochrome oxidase–dimyristoyl phosphatidylcholine complexes (Knowles *et al.*, 1981). The contribution of the immobilized component in spectra obtained with cardiolipin and phosphatidic acid is significantly larger than that of spectra obtained with other phospholipids. By intersubtraction of spectra, the shape of a "pure immobilized" component and a "pure mobile" component can be generated. This allows a quantitative estimation of the fraction of "protein-bound" label.

Let n_b, n_f, and n_t be, respectively, the number of bound, of free, and the total number of phospholipids *per protein;* while n_b^*, n_f^*, and n_t^* are the corresponding numbers of spin-labeled lipids. If one assumes an equal partition of spin-labeled lipids between "bound" and "free" states

$$n_f/n_b = n_f^*/n_b^*$$

then from simple conservation law, one can write the following equation:

$$n_f^*/n_b^* = n_t/n_b - 1 \qquad (1)$$

The quantity n_f^*/n_b^* can be determined experimentally, while n_t is known, thus allowing n_b to be determined from the slope of n_f^*/n_b^* vs. n_t. Figure 7 illustrates an application from Knowles *et al.* (1979) to cytochrome oxidase–dimyristoyl phosphatidylcholine

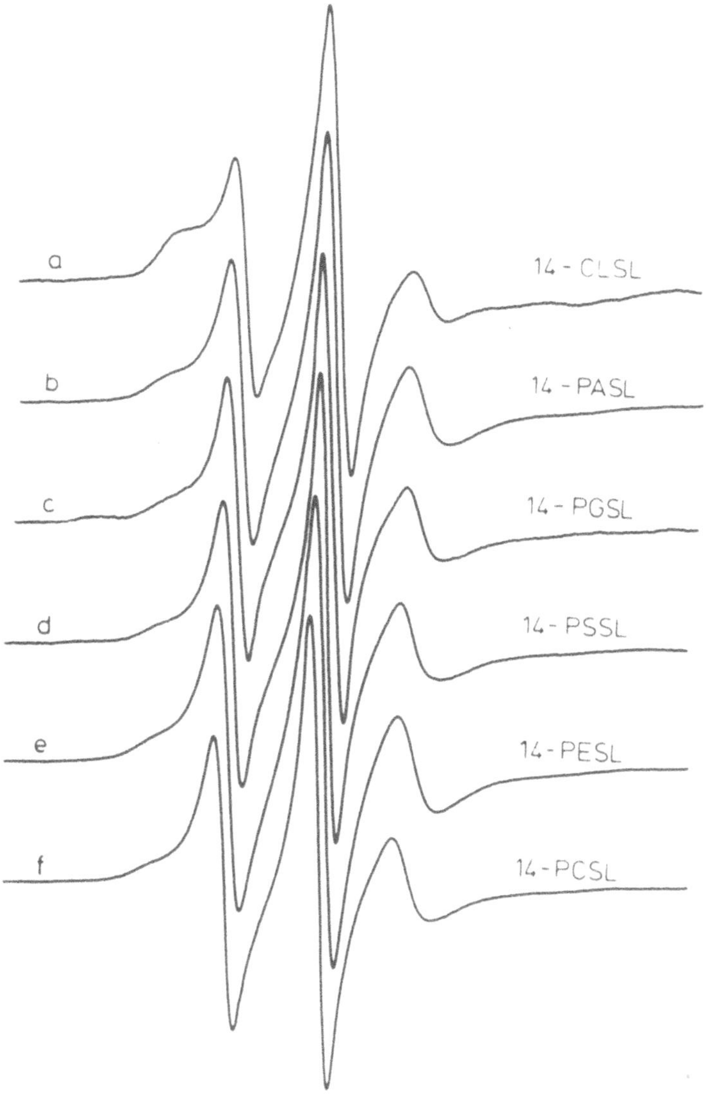

Figure 6. ESR spectra of different 14-doxyl phospholipid head group spin-labels (see Table 2) added exogenously to separate aliquots taken from the same cytochrome oxidase–DMPC complex; lipid to protein ratio, 95/1; temperature, 32°C (from Knowles *et al.*, 1981).

complexes. Griffith *et al.* (1982) and Brotherus *et al.* (1981) have developed a more elaborate model which takes into account multiple binding sites. The analysis, although somewhat more sophisticated, leads essentially to the same formula, Eq. (1), after simplification. As a matter of fact, too elaborate models of lipid–protein equilibrium are difficult to verify with spin labels for the following reasons: (1) the determination of the fraction of immobilized component is a difficult task, often partially arbitrary and always associated with a large uncertainty, and (2) the probability of heterogeneity of the samples increases as the ratio of lipid to protein decreases. Thus, as the immobilized component becomes easy to detect, its meaning becomes less obvious. For example, in Figure 7, n_f^*/n_b^* deviates from linearity in the only region where it can be accurately measured.

The results can be expressed in terms of the relative affinities of the phospholipids for the proteins, the affinity of phosphatidylcholine being taken as 1. Knowles *et al.* (1981) found, in the case of cytochrome oxidase, $K_r^{CL} = 5.5$ for cardiolipin, $K_r^{PA} = 2$ for phosphatidic acid, while $K_r^{PG} = K_r^{PE} = 1$ (respectively for phosphatidylglycerol and phosphatidylethanolamine). The number of "sites" calculated per protein is relatively high, >90 for cardiolipin, 70 for phosphatidic acid, and 55 for phosphatidyl-

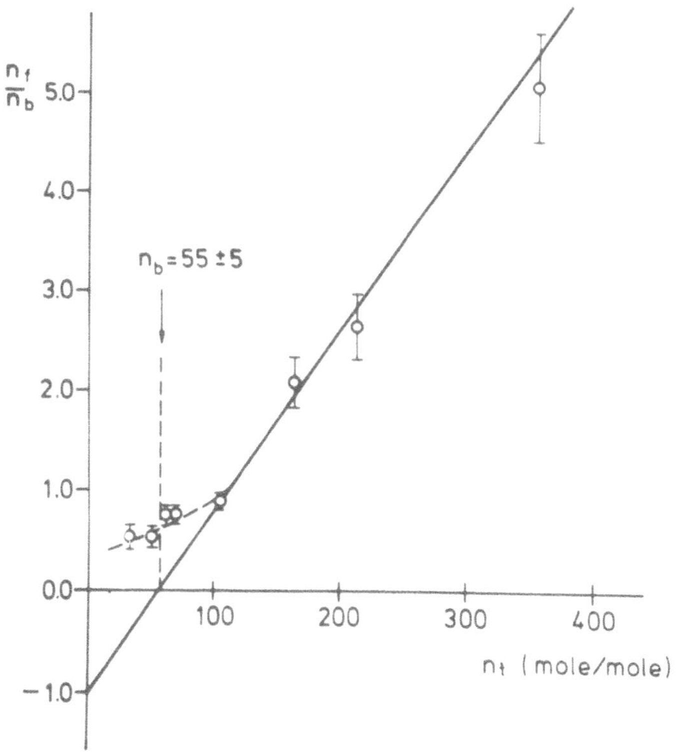

Figure 7. Lipid–protein titration of yeast cytochrome oxidase–DMPC reconstituted complexes, obtained from ESR difference spectra of 12-doxyl phosphatidylcholine. n_t is the total lipid–protein ratio in the complex. n_f^*/n_b^* is the ratio of spin-labeled free over spin-labeled bound. n_f^*/n_b^* is determined from the fraction of total integrated intensity which is removed from spectra of the complex on subtracting an arbitrary immobilized component. The intercept of the straight line obtained by joining points for $n_t \geq 200$ and the abcissa determines the number of protein-bound sites per protein (see Eq. (1) in the text; from Knowles *et al.*, 1979).

choline per cytochrome oxidase molecule (200,000 daltons) (Knowles *et al.*, 1981). The effective occupancy number of nonspecific lipid sites next to Na^+K^+ATPase is 60, 15 for myelin proteolipid, and 24 for rhodopsin, according to Marsh *et al.* (1982b). Whether the "sites" correspond to the entire protein surface or are restricted to protein invaginations is not known. They nevertheless must correspond to large fractions of the protein.

In conclusion, some intrinsic proteins bear a preferential interaction to a given class of phospholipids.

B. Second Approach: Spin–Spin Interaction between Nitroxide Radicals

Detection of the motionally restricted component in natural membranes or membranes with a high lipid-to-protein ratio can be a difficult task, particularly at high temperature. Thus, it is interesting to mention other techniques allowing one to detect lipid specificity with spin labels. The two methods developed in the present section take advantage of spin–spin interaction between nitroxides.

In the first method, a comparative estimation of the line broadening is carried out with spin-labeled phospholipids incorporated at relatively high concentration in membrane. A crude estimation of the extent of spin–spin interaction can be obtained by a plot of the linewidth as a function of spin-labeled phospholipid molar ratio. If all lipids diffuse freely, the slope which is roughly proportional to the spin exchange frequency is determined by the rate of lateral diffusion. However, if a fraction of spin-labeled lipids is trapped, even temporarily, by the proteins, the apparent diffusion constant of this lipid will be diminished. Hence, comparison of the relative slopes indicates preferential interactions.

For this type of study, the following type of spin labels are very convenient:

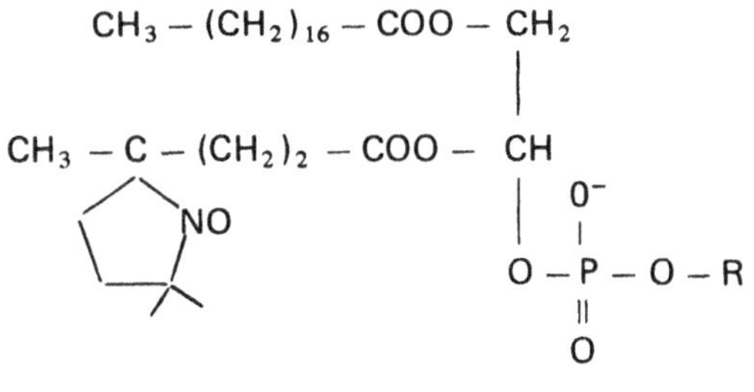

The advantage of one short chain is that the molecule is relatively soluble in water. It can be incorporated readily in the membranes by direct addition to the aqueous medium containing the membrane suspension. Such derivatives of phosphatidylcholine, phosphatidylethanolamine, and phosphatidylserine have been added successively to membranes rich in Na^+K^+ATPase from the electric organ of *Torpedo marmorata*. A preferential interaction of the phosphatidylserine molecule for the Na^+K^+ATPase was found (Zachowski and Devaux, 1983). This technique confirms the finding of

Brotherus *et al.* (1981) and Marsh *et al.* (1982b) concerning the lipid environment of Na^+K^+ATPase.

An alternative method for the investigation of lipid specificity has been proposed recently (Seigneuret *et al.*, 1981; Davoust *et al.*, 1983a,b). The method takes advantage of a double labeling. The protein is labeled with a [^{15}N]nitroxide, while a specific phopholipid is labeled with the ordinary [^{14}N]nitroxide (see Figure 8). The amount of spin–spin interaction between the two species can be determined accurately and reflects the collision frequency between the proteins and the lipids. Figure 9 illustrates the theoretical effect of increasing the spin exchange frequency between ^{14}N–^{15}N lines. The amplitude of the low field line of ^{15}N ($m = +1/2$) in the first derivative spectrum can be used to monitor the amount of interaction. The use of different isotopes for the labeling of the proteins and the lipids enables one to separate on the one hand spin–spin interaction within phospholipids and on the other hand spin–spin interaction between lipids and proteins.

Experiments with rhodopsin reconstituted in a lipid mixture containing phosphaditylcholine, phosphatidylethanolamine, and phosphatidylserine have been performed. They show comparable interaction for the three types of phospholipids at 40°C, again in confirmation of the work of Watts *et al.* (1979). However, at low temperature, in the presence of Ca^{2+}, phosphatidylserine seems to be partially excluded from the protein environment, very likely because of the formation of domains of phosphatidylserine which are induced by Ca^{2+}.

The technique of double labeling should be extended to other systems. In principle, it permits one to detect interactions between a specific protein and a specific lipid in a complex membrane. For example, one can label the ADP carrier in mitochondria and detect its particular lipid environment without having to purify the protein and reconstitute it in artificial vesicles. In addition, this technique permits one to differentiate between specific interactions due to trapped lipids or motionally restricted lipids.

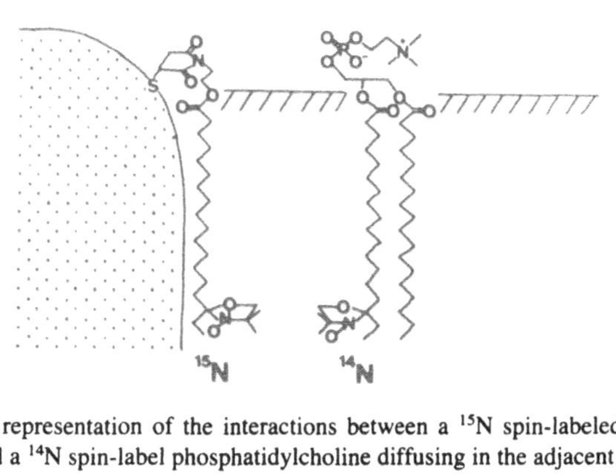

Figure 8. Schematic representation of the interactions between a ^{15}N spin-labeled fatty acid covalently bound to rhodopsin and a ^{14}N spin-label phosphatidylcholine diffusing in the adjacent bilayer (from Davoust *et al.*, 1983b).

Figure 9. Calculated spectra corresponding to a mixture of ^{15}N and ^{14}N spin-labels with increasing spin exchange frequency. The exchange rate is indicated on the right-hand side. The decrease in peak intensity of the low field ^{15}N line can be used to monitor the rate of collision between the two labels (from Davoust *et al.*, 1983a).

V. INTERACTIONS BETWEEN EXTRINSIC PROTEINS AND LIPIDS

The interaction between extrinsic proteins and phospholipids has been investigated by ESR. Various methods have been employed, some of which combine spin labels and NMR. ESR and NMR have been utilized to determine how deep the proteins penetrate into the bilayer. They are also utilized to study the specificity of phospholipid head groups involved in the binding. Finally they are utilized to study lipid phase diagrams and the extent of lipid segregation induced by extrinsic proteins.

A. Protein Penetration

Van and Griffith (1975), using spin-labeled fatty acids, showed that cytochrome *c* interacts only superficially with the lipid bilayer. They used 5-, 12-, and 16-doxylstearic acids incorporated into mixtures of cardiolipin–lecithin (1 : 4 by weight). The binding of cytochrome *c* is *not* accompanied by the appearance of an immobilized component even at low lipid-to-protein ratio. The result was confirmed by Brown and Wüthrich in 1977. These authors investigated the perturbation on the ^{1}H-NMR relaxation of the lipids, caused by the binding of spin-labeled cytochrome *c* to the membranes. They concluded that the spin label appears to penetrate only slightly into the membrane.

Similarly, Boggs and Moscarello (1978) investigated the penetration of myelin-basic protein in lipid bilayers. An ESR study, using a series of fatty acid spin labels, showed that the order parameter of the lipids was increased by the protein near the polar head group of the phospholipids; little or no effect could be detected on the order of motion in the interior or the bilayer. Thus, basic protein only penetrates part way into the bilayer. However, a small difference was observed between phosphatidylglycerol and phosphatidic acid, suggesting a deeper penetration in the former lipid.

It should be pointed out that partial penetration of proteins into the bilayer must be accompanied by destabilization of the bilayer which might trigger a phase change. De Kruijff and Cullis (1980) have suggested that lipids binding to cytochrome *c* form local micellar structure.

B. Protein Induction of Lateral Phospholipid Separation

Extrinsic proteins are generally charged proteins. They bind via electrostatic interaction to the head group of charged lipids. Thus, they are capable of clustering lipids into domains in a way comparable to divalent ions. This effect can be demonstrated, using ESR spectroscopy, by analysis of the spin–spin interaction which results from the clustering of specific spin-labeled phospholipids. Ohnishi and Ito (1973) have shown, by this technique, Ca^{2+}-induced segregation of phosphatidylserine or phosphatidic acid. More recently, Hartmann and Galla (1978) used mixtures of spin-labeled phosphatidylcholine and unlabeled phosphatidic acid. The addition of polylysine, considered as a model of extrinsic protein, resulted in an increased spin–spin broadening between phosphatidylcholine molecules.

Using the TEMPO-partitioning method of McConnell and co-workers (1972), Galla and Trudell (1980) have carried out experiments showing that the phase transition curve of phosphatidic acid is modified by the addition of charged polypeptides. In these experiments, they used polymixin, a decapeptide antibiotic, possessing five positive charges. The modification of the whole phase diagram implies large domains of bound polymixin.

C. Protein Induction of Transverse Phospholipid Separation

Charged phospholipids, such as phosphatidylserine, phosphatidic acid, or cardiolipin, are required for the binding of extrinsic proteins. Conversely, the presence of extrinsic proteins on one side of a membrane can stabilize charged phospholipids and therefore induce an anisotropic distribution of phospholipids. This idea was suggested some years ago by Heast *et al.* (1978). These authors used phospholipase to test the sidedness of the endogenous phospholipids.

Very recently, we have been able to show that spin-labeled phospholipids incorporated into erythrocyte ghosts orient themselves on the inner or outer side of the membrane depending on their head group (Seigneuret and Devaux, 1984). For these experiments we used the spin labels XII described in Section IV-B. Because of their solubility in water, they incorporate readily as monomers in the outer face of erythrocyte. Their localization in the membranes can be assayed by ascorbate reduction because the probe is not deeply buried within the hydrophobic core; only the spin labels on the outer face are reduced by ascorbate. By this technique, we have shown that in intact red blood cells, phosphatidylserine derivative is stabilized in the inner face of the cell membrane while phosphatidylcholine remains on the outer leaflet. Extrinsic proteins may be involved in the stabilization mechanism. We have found that the formation of a disymmetric distribution of spin-labeled phospholipids is an ATP-dependent process both in erythrocytes and resealed ghosts.

VI. OTHER APPLICATIONS OF CONVENTIONAL ESR SPECTROSCOPY TO THE INVESTIGATION OF MEMBRANE-BOUND ENZYMES

As pointed out above, the time scale of conventional ESR spectroscopy with nitroxide radicals is best suited to the study of lipid motions. Proteins can be studied indirectly if correlations between lipid viscosity and enzyme activity can be established. There is a vast literature on the use of spin labels to study membrane fluidity in conjunction with normal or abnormal protein activity. The reader is referred to the recent review of Butterfield (1983) for the medical applications of spin labeling, which are mostly based on spin-label determinations of cell membrane viscosity. In the following, we expose briefly other applications of conventional ESR which correlate either directly or indirectly membrane-to-protein activity.

A. Measurement of Surface Potentials and Intermembrane Potentials

Spin-label lipids are not only useful to measure membrane fluidity. In 1972, Kornberg and co-workers showed that spin labels can be used to determine membrane potential. Since then, several laboratories have designed a variety of nitroxide derivatives whose partition coefficient between lipids and water is a function of the transmembrane potential or of the surface potential. Table 3 shows the most commonly used molecules. All molecules are charged. Molecules XIII–XV are impermeable to phospholipid vesicles, while molecules XVI–XVIII can traverse a lipid barrier. Several groups have independently shown that the relative distribution of these amphiphilic molecules between the hydrophobic phase and the aqueous phase can be used to determine membrane surface potential with the former molecules (Castle and Hubbell, 1976; Gaffney and Mich, 1976; Mehlhorn and Packer, 1979; Tokutomi *et al.*, 1981), while transmembrane potential can be determined with the latter (Cafiso and Hubbell, 1978).

The general approach is to determine the partitioning of these charged molecules from the difference in lineshape of bound and free molecules. The ratio of label bound–label free is a function of the amphiphilic character of each molecule *and* of the local electric potential. For example, for the nonpermeant molecules, the surface potential and the partition coefficient K between membrane and water are related by

$$\psi_s = -(RT/zF) \log K/K_0$$

where K_0 is the partition coefficient in the absence of surface potential. The other symbols have their usual meaning. The Gouy–Chapman theory can be used to relate ψ_s to the surface charge density σ. The following general relation is obtained:

$$(\text{label bound–label free})^{\pm 1} = A + B\,\sigma^2$$

For negatively charged lipids, the positive exponent applies to positive spin probes and vice versa. In this formula, A and B are constants which depend upon salt concentration and the ratio of the volume of the membrane phase to that of the aqueous phase.

Determination of the transmembrane potential with probes XVI–XVIII is somewhat more elaborate (Cafiso and Hubbell, 1978). The principle of the experiment is also to determine the partition coefficient.

In both types of determination (surface potential or transmembrane potential), one takes advantage of the difference in lineshape of the membrane-bound and free-spin label. Since the spin label in water gives narrow lines, it is easy to quantize (see Figure 10) and rapid variations in surface potential or transmembrane potential can be monitored by ESR. Influence of the ionic strength and the pH can be tested. Paramagnetic probes are particularly useful when potential variations can be triggered by light. Fluorescent probes which are used as alternatives to spin labels for the determination of surface changes cannot be employed in such cases. On the other hand,

Table 3. Spin-Labeled Lipids Reporting on Membrane Surface Potential or Intermembrane Potential

XIII

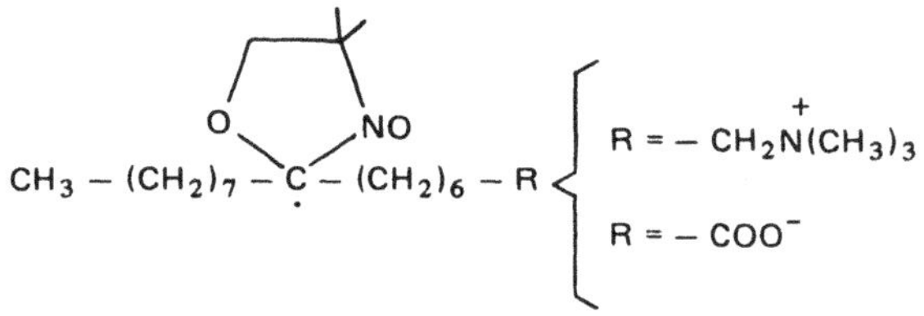

XIV

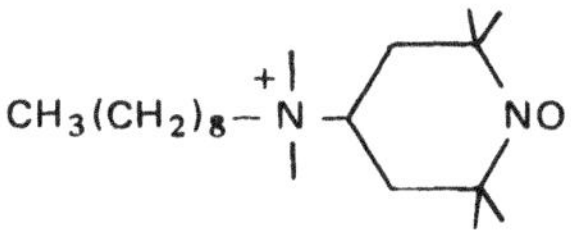

XV

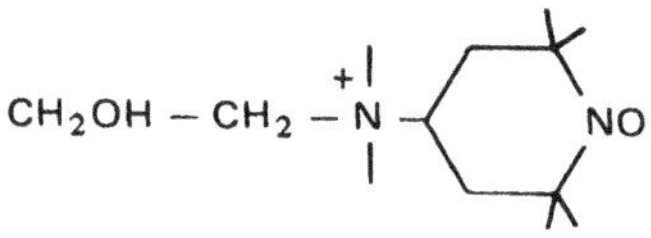

XVI

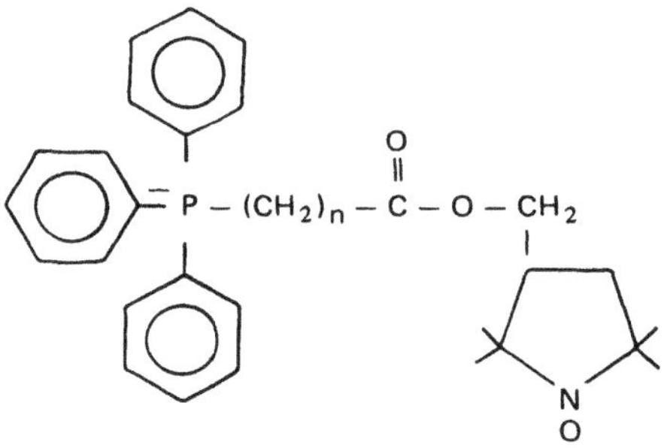

XVII

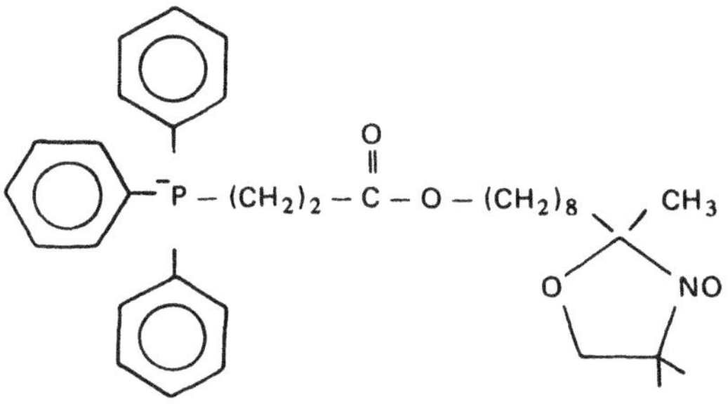

XVIII

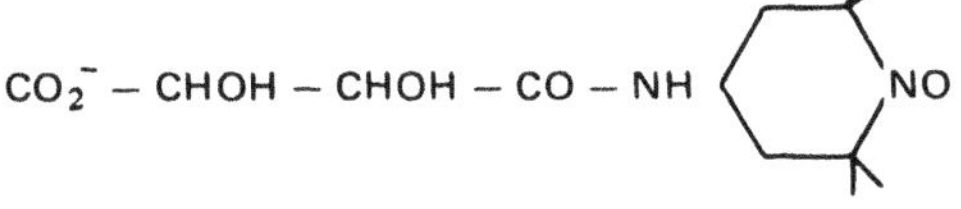

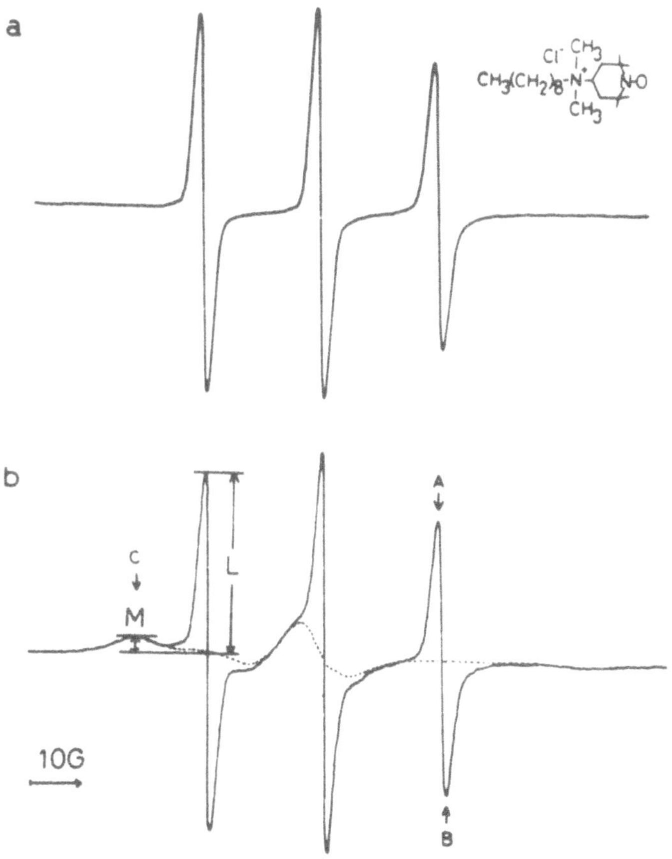

Figure 10. (a) ESR spectra of a charged amphilphilic molecule (see inset) in medium (pH 6.5) containing 100 mM KCl and 20 mM phosphate at 2°C. (b) ESR spectra of the same spin-label in aqueous suspension of light-adapted purple membranes at 2°C. The dotted line spectrum was obtained by subtracting spectrum (a) from spectrum (b). The ratio L/M can be used to monitor the partition coefficient of the probe. On flash illumination, a transient hyperpolarization of the surface potential results in temporary change of L/M (from Tokutomi *et al.*, 1981).

time-resolved ESR allows, with standard equipment, a resolution of about 10 ms (Tokutomi *et al.*, 1981).

Applications are found in the literature for the investigation of biochemical mechanisms involved in thylakoid membranes (Quintanilha and Mehlhorn, 1978), mitochrondrial inner membranes (Quintanilha and Packer, 1977), purple membranes (Carmeli *et al.*, 1980; Tokutomi *et al.*, 1981), and disk membranes (Cafiso and Hubbell, 1982).

Several investigators have attempted to differentiate viscosity effects from charge effects in membranes by following with appropriate spin labels the modifications which take place in the course of an enzyme activity. A very elegant technique to monitor simultaneously the fluidity and the surface potential of thylakoids was proposed by Quintanilha and Mehlhorn (1978). These authors used a ^{14}N label to follow the pH changes while a ^{15}N label allowed them to monitor the lipid viscosity in the same membranes.

B. Conformation of Membrane-Bound Enzymes

Tightly-bound spin probes can report on the overall rotational correlation time of a membrane protein (see D. T. Thomas, this book) while loosely-bound probes can report on the local environment at the binding site. Because a loosely-bound probe can give rise to a wide variety of spectral lineshapes (by conventional ESR), it becomes possible to differentiate various sites of binding or protein conformations. Thus, in principle, spin labels and conventional ESR can be used to monitor a change of conformation, provided that this change of conformation affects the region of the protein where the label is attached. Finally, if one takes advantage of the spin–spin interaction which takes place when probes are 10–15 Å apart it is possible to determine proximity relationships from conventional ESR.

There is an appreciable literature on spectra obtained with spin-labeled SH groups of sarcoplasmic reticulum membranes and the Ca^{2+}ATPase (Yu *et al.*, 1977; Hidalgo and Thomas, 1977; Champeil *et al.*, 1978; Champeil *et al.*, 1980; Andersen *et al.*, 1980; Nakamura and Martonosi, 1981; earlier references are found in Gaffney and Chang Lin, 1976). These studies have involved both the maleimide and iodo acetamide classes of spin labels whose syntheses are described in Gaffney's review article (1976).

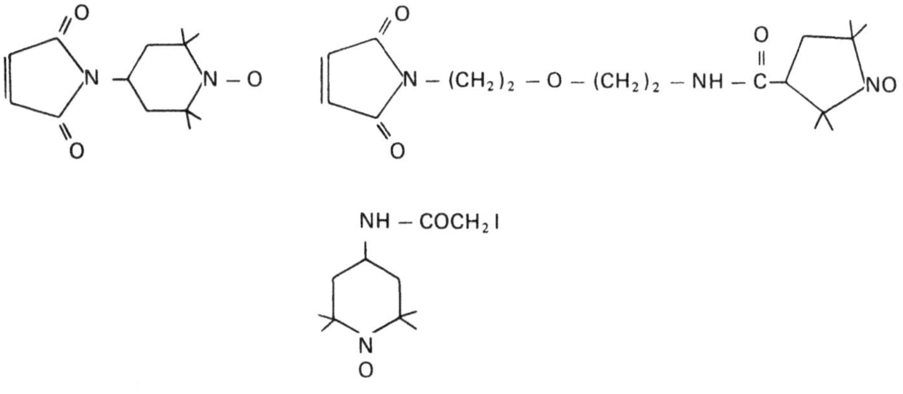

The reactivity of these labels under various conditions of protein activity and conformation have been tested. For example, Champeil *et al.* (1980) have monitored the noninhibitory binding of the iodoacetamide spin label by measuring the steady reduction in free label spectrum intensity during the labeling reaction. The reaction rate was highly sensitive to pH and temperature. Ca^{2+} chelation in the presence of magnesium accelerated the reaction slightly, and nucleotides accelerated it several fold in the presence of Ca^{2+}. The different kinetics show the difference in exposure of the

SH groups accompanying protein turnover, thus, indirectly, conformation changes are revealed. A more direct proof of these conformation changes is obtained by analyzing the spectra of the covalently-bound labels. Most investigators, in what might be an oversimplifying view, decompose the ESR spectrum in a "weakly-bound" and a "strongly-bound" component (see Figure 11). An appropriate peak height ratio is used to follow the interconversion between the two supposedly existing states. This procedure, although crude, was successful to prove conformational changes of the Ca^{2+}ATPase induced by Ca^{2+} (Champeil *et al.*, 1978), by temperature (Andersen *et al.*, 1980), or by lipid replacement (Nakamura and Martonosi, 1981). The same approach has been used by other authors on different membrane-bound proteins (see for example Ginsburg *et al.*, 1981, for their study of the anion transport protein of human erythrocytes).

More specific probes can be synthetized. Azzi and collaborators have labeled the proteolipid subunit of chloroplast ATP complex (Sigrist-Nelson and Azzi, 1979) or the mitochondrial cytochrome oxidase proton pump (Casey *et al.*, 1981) with a spin label derivative of a specific inhibitor. Both proteins bind covalently and specifically a spin label analog of *N;N'*-diclyclohexylcarbodiimide (DCCD). When solubilized in detergent, the proteins yield a highly mobile type of spectrum, whereas the membrane-

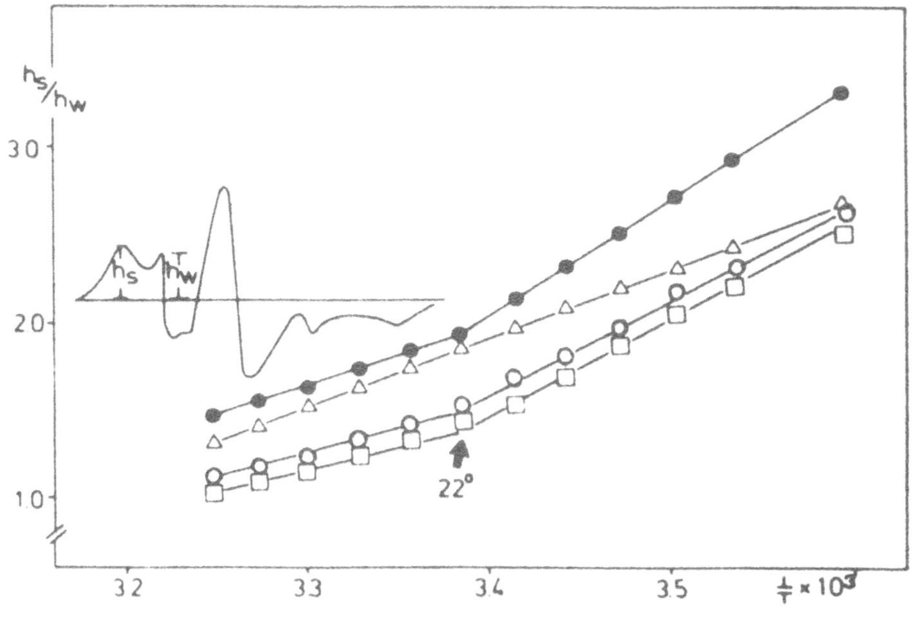

Figure 11. Arrhenius plot of h_s/h_w for membranous and solubilized ATPases labeled with a short chain maleimide spin-label (XIX). The definition of h_s and h_w is indicated in the inset. (●——●) Membranous ATPase, 0.1 mM $CaCl_2$; (△——△) heat-denaturated solubilized ATPase, 0.1 mM $CaCl_2$, 2 mg$C_{12}E_8$/mg ATPase; (○——○) active solubilized ATPase, 0.1 mM $CaCl_2$, 2 mg $C_{12}E_8$/mg ATPase; (□——□) inactivated solubilized ATPase, 1 mM EGTA, 2 mg $C_{12}E_8$/mg ATPase. This figure shows that a conformational change takes place around 22°C which is not due to lipids (from Andersen *et al.*, 1980).

bound proteins gave strongly-immobilized spectra. When high concentrations of the spin label analog of DCCD were employed for labeling of the chloroplast membranes, spin–spin interaction between labeled proteins could be demonstrated. Authors concluded that the DCCD-binding proteolipid of chloroplasts exists in a polymeric form in which some of the monomers are located at a maximal distance of 15–20 Å from each other. Spin–spin interaction between bound probe molecules was also found in oxidase, if solubilized in sodium cholate. Interaction was attributed to the labeling of two different subunits: subunit III and IV of cytochrome oxidase.

More recently, Park *et al.* (1982) have synthetized NAD analogs with a deuterated nitroxide radical at the C-8 or N-6 position of the adenine ring. NAD is a coenzyme for glyceraldehyde-3-phosphate dehydrogenase (GAPDH). GAPDH is a tetrameric enzyme; it can be in a soluble form or it can bind to band 3 protein of red blood cell. Dipole–dipole interactions were observed as the NAD analog was stoichiometrically bound to the soluble enzyme. When the protein-active site was oxidized, the dipole–dipole interaction was obliterated indicating that this site is involved in positioning the coenzyme. Finally, when GAPDH was attached to red blood cell membranes only three instead of four moles of NAD were bound per mole of tetrameric enzyme. This work took advantage of the enhanced sensitivity produced by deuterium substitution (Beth *et al.*, 1981).

ACKNOWLEDGMENTS

The author is thankful to J. Davoust, M. Seigneuret, and A. Zachowski for providing results prior to publication. This work was supported by grants from the "Centrē National de la Recherche Scientifique" (ERA 690), the "Délégation Générale à la Recherche Scientifique et Technique," and the "Université Paris VII."

REFERENCES

Andersen, J. P., Le Maire, M., and Møller, J. V., 1980, Properties of detergent-solubilized and membranes (Ca^{2+} + Mg^{2+})-activated ATPase from sarcoplasmic reticulum as studied by sulfhydryl reactivity and ESR spectroscopy, *Biochim. Biophys. Acta* **603**:84–100.

Andersen, J. P., Fellmann, P., Møller, J. V., and Devaux, P. F., 1981, Immobilization of a spin label fatty acid chain covalently attached to Ca^{2+} ATPase from sarcoplasmic reticulum suggests an oligomeric structure, *Biochemistry* **20**:4928–4936.

Berliner, L. J. (ed.), 1976, *Spin Labeling: Theory and Applications*, Vol. I, Academic Press, New York.

Berliner, L. J. (ed.), 1979, *Spin Labeling: Theory and Applications*, Vol. II, Academic Press, New York.

Beth, A. H., Balasubramanian, K., Walder, R. T., Venkataramu, S. D., Robinson, B. H., Dalton, L. R., Pearson, D. E., and Park, J. H., 1981, Structural and motional changes in glyceraldehyde 3-phosphate dehydrogenase upon binding to the band 3 protein of the erythrocyte membrane examined with ^{15}N-^{2}H maleimide spin labels and EPR, *Proc. Natl. Acad. Sci. USA* **78**:4955–4959.

Bienvenüe, A., Rousselet, A., Kato, G., and Devaux, P. F., 1977, Fluidity of the lipids next to the acetylcholine receptor protein of torpedo membrane fragments. Use of amphiphilic reversible spin labels, *Biochemistry* **16**:841–848.

Bittmann, R., Sakaki, T., Tsuji, A., Devaux, P. F., and Ohnishi, S. I., 1984, Spin label studies of the oligomeric structure of band 3 protein in erythrocyte membranes and in reconstituted systems, *Biochem. Biophys. Acta* **769**:85–95.

Boggs, J. M., and Moscarello, M. A., 1978, Effect of basic protein from human central nervous system myelin on lipid bilayer structure, *J. Membr. Biol.* **39**:75–96.

Bosterling, B., Trudell, J. R., and Galla, H. J., 1981, Phospholipid interaction with cytochrome P 450 in reconstituted vesicles: Preference for negatively charged phosphatidic acid, *Biochim. Biophys. Acta* **643**:547–556.

Brotherus, J. R., Griffith, O. H., Brotherus, M. O., Jost, P. C., and Silvius, J. R., 1981, Lipid–protein multiple binding equilibria in membranes, *Biochemistry* **20**:5261–5267.

Brown, L. R., and Wüthrich, K., 1977, NMR and ESR studies of the interactions of cytochrome *c* with mixed cardiolipin-phosphatidylcholine vesicles, *Biochim. Biophys. Acta* **468**:389–410.

Butterfield, D. A., 1983, Spin labeling in disease, in: *Biological Magnetic Resonance*, Vol. 4 (L. J. Berliner, ed.), Plenum Press, New York pp. 1–78.

Cable, M. B., and Powell, G. L., 1980, Spin labeled cardiolipid: Preferential segregation in the boundary layer of cytochrome *c* oxidase, *Biochemistry* **19**:5679–5686.

Cafiso, D. S., and Hubbell, W. L., 1978, Estimation of transmembrane potentials from phase equilibria of hydrophobic paramagnetic ions, *Biochemistry* **17**:187–195.

Cafiso, D. S., and Hubbell, W. L., 1982, Transmembrane electrical currents of spin labeled hydrophobic ions, *Biophys. J.* **39**:263.

Carmeli, C., Quintanilha, A. T., and Packer, L., 1980, Surface charge changes in purple membranes and the photoreaction cycle of bacteriorhodopsin, *Proc. Natl. Acad. Sci. USA* **77**:4707–4711.

Casey, R. P., Broger, C., and Azzi, A., 1981, Structural studies on the cytochrome *c* oxidase proton pumps using a spin label probe, *Biochim. Biophys. Acta* **638**:86–93.

Castle, J. D., and Hubbell, W. D., 1976, Estimation of membrane surface potential and charge density from the phase equilibrium of a paramagnetic amphiphile, *Biochemistry* **15**:4818–4831.

Champeil, P., Büschlen-Boucly, S., Bastide, F., and Garybobo, C. M., 1978, Sarcoplasmic reticulum ATPase. Spin labeling detection of ligand induced changes in the relative reactivities of certain sulfhydryl groups, *J. Biol. Chem.* **253**:1179–1186.

Champeil, P., Rigaud, J. L., and Garybobo, C. M., 1980, Ca translocation mechanism in sarcoplasmic Reticulum vesicles deduced from localisation studies of protein bound spin labels, *Proc. Natl. Acad. Sci. USA* **77**:2405–2409.

Chapman, D., Cornell, B. A., Eliasz, A. W., and Perry, A., 1977, Interactions of helical polypeptide segments which span the hydrocarbon region of lipid bilayers. Studies of the gramicidin A lipid–water system, *J. Mol. Biol.* **113**:517–538.

Chignell, C. F., and Chignell, D. A., 1975, A spin label study of purple membrane from Halobacterium halobium, *Biochem. Biophys. Res. Commun.* **62**:136–143.

Davoust, J., and Devaux, P. F., 1982, Simulation of electron spin resonance spectra of spin labeled fatty acids covalently attached to the boundary of an intrinsic membrane protein. A chemical exchange model, *J. Magn. Res.* **48**:475–494.

Davoust, J., Schoot, B. M., and Devaux, P. F., 1979, Physical modifications of rhodopsin boundary lipids in lecithin–rhodopsin complexes: A spin label study, *Proc. Natl. Acad. Sci. USA* **76**:2755–2759.

Davoust, J., Bienvenüe, A., Fellmann, P., and Devaux, P. F., 1980, Boundary lipids and protein mobility in rhodopsin–phosphatidylcholine vesicles. Effect of lipid phase transitions, *Biochim. Biophys. Acta* **596**:28–42.

Davoust, J., Seigneuret, M. Hervé, P., and Devaux, P. F., 1983a, Collisions between nitrogen 14 and nitrogen 15 spin labels. 1. Lipid-lipid interactions in model membranes, *Biochemistry*, **22**:3137–3145.

Davoust, J., Seigneuret, M., Hervé, P., and Devaux, P. F., 1983b, Collisions between nitrogen 14 and nitrogen 15 spin labels. 2. Investigations on the specificity of the lipid environment of rhodopsin, *Biochemistry*, **22**:3146–3151.

De Kruijff, B., and Cullis, P. R., 1980, Cytochrome *c* specifically induces non-bilayer structures in cardiolipin-containing model membranes, *Biochim. Biophys. Acta* **602**:477–490.

Devaux, P. F., Bienvenüe, A., Lauquin, G., Brisson, A. D., Vignais, P. M., and Vignais, P. V., 1975, Interaction between spin-labeled acyl-coenzyme A and the mitochondrial adenosine diphosphate carrier, *Biochemistry* **14**:1271–1280.

Gaffney, B. J., 1976, The chemistry of spin labels, in: *Spin Labeling, Theory and Applications* (L. J. Berliner, ed.), Academic Press, New York, pp. 183–238.

Gaffney, B. J., and Chang Lin, D., 1976, Spin-label measurements of membrane-bound enzymes, in: *The Enzymes of Biological Membranes*, Chap. 3 (A. Martonosi, ed.), Plenum Press, pp. 71–90.

Gaffney, B. J., and Mich, R. J., 1976, A new measurement of surface charge in model and biological lipid membranes, *J. Am. Chem. Soc.* **98**:3044–3045.

Galla, H. J., and Trudell, J. R., 1980, Pressure induced changes in the molecular organization of a lipid–peptide complex. Polymyxin binding to PA membranes, *Biochim. Biophys. Acta* **602**:522–530.

Ginsburg, H., O'Connor, S., and Grisham, C., 1981, Evidence for electron paramagnetic resonance for function related conformation changes in the anion transport protein of human erythrocytes, *Eur. J. Biochem.* **114**:533–538.

Griffith, O. H., Brotherus, J., and Jost, P. C., 1982, Equilibrium constants and number of binding sites for lipid–protein interaction in membranes, in: *Lipid-Protein Interactions,* Vol. 2, Chap. 6 (P. C. Jost and O. H. Griffith, eds.), John Wiley and Sons, New York, pp. 225–237.

Hartmann, W., and Galla, H. J., 1978, Binding of polylysine to charged bilayer membrane molecular organization of a lipid–peptide complex, *Biochim. Biophys. Acta* **509**:474–490.

Heast, C. W., Plasa, G., Kamp, D., and Deuticke, B., 1978, Spectrin as a stabilizer of the phospholipid asymmetry in the human erythrocyte membrane, *Biochim. Biophys. Acta* **509**:21–32.

Hidalgo, C., and Thomas, D. T., 1977, Heterogeneity of SH groups in sarcoplasmic reticulum, *Biochem. Biophys. Res. Commun.* **78**:1175–1182.

Jost, P. C. Griffith, O. H., Capaldi, R. A., and Van der Kooi, G., 1973, Evidence of boundary lipid in membranes, *Proc. Natl. Acad. Sci. USA* **70**:480–484.

Knowles, P. F., Watts, A., and Marsh, D., 1979, Spin-label studies of lipid immobilization in dimyristoylphosphatidylcholine-substituted cytochrome oxidase, *Biochemistry* **18**:4480–4487.

Knowles, P. F., Watts, A., and Marsh, D., 1981, Spin-label studies of head-group specificity in the interaction of phospholipids with yeast cytochrome oxidase, *Biochemistry* **20**:5888–5894.

Kornberg, R. G., McNamee, M. G., and McConnell, H. M., 1972, Measurement of transmembrane potentials in phospholipid vesicles, *Proc. Natl. Acad. Sci. USA* **69**:1508–1513.

Marsh, D., and Barrantes, F. J., 1978, Immobilized lipid in acetylcholine receptor-rich membranes from T. marmorata, *Proc. Natl. Acad. Sci. USA* **75**:4329–4333.

Marsh, D., Watts, A., and Barrantes, F. J., 1981, Phospholipid chain immobilization and steroid rotational immobilization in acetylcholine receptor-rich membranes from Torpedo marmorata, *Biochim. Biophys. Acta* **645**:97–101.

Marsh, D., Watts, A., Pates, R. D., Uhl, R., Knowles, P. F., and Esmann, M., 1982a, ESR spin label studies of lipid-protein interactions in membranes. *Biophys. J.* **37**:265–274.

Marsh, D., Horvath, L., Pates, R. D., Brophy, P. J., and Esmann, M., 1982b, Spin label studies of lipid–protein interactions in membranes by conventional and saturation transfer ESR, Xth International Conference on Magnetic Resonance in Biological Systems, Abst. S39, Stanford, Connecticut.

McConnell, H. M., Wright, K. L., and McFarland, B. G., 1972, The fraction of the lipid in a biological membrane that is in a fluid state: A spin label assay, *Biochem. Biophys. Res. Commun.* **47**:272–281.

McNamee, M. G., Ellena, J. F., and Dalziel, A. W., 1982, Lipid–protein interactions in membranes containing the acetylcholine receptor, *Biophys. J.* **37**:103–104.

Mehlhorn, R. J., and Packer, L., 1979, Membrane surface potential measurements with amphiphilic spin labels, *Meth. Enzymol.* **56**:515–526.

Nakamura, H., and Martonosi, A. N., 1981, Effect of phospholipid substitution on the mobility of spin labels bound to the ATPase of sarcoplasmic reticulum, *J. Biochem. (Tokyo)* **89**:21–28.

Nigg, E. A., and Cherry, R. J., 1979, Influence of temperature and cholesterol on the rotational diffusion of band 3 in the human erythrocyte membrane, *Biochemistry* **18**:3457–3465.

Ohnishi, S. I., and Ito, T., 1973, Clustering of lecithin molecules in phosphatidylserine membranes induced by calcium ion binding to phosphatidylserine, *Biochem. Biophys. Res. Commun.* **51**:132–138.

Oldfield, E., Janes, N., Kinsey, R., Kintanar, A., Lee, R. W. K., Rothgeb, M., Schram, S., Skarjune, R., Smith, R., and Tsai, M. D., 1981, Protein crystals, membrane proteins and membrane lipids, *Biochem. Soc. Symp.* **46**:155–181.

Park, J. H., Beth, A. H., Venkataramu, S. D., Balasubramanian, K., Wilder, R., Park, C. R., Glogger, K., and Trommer, W. E., 1982, Correlation of motional and catalytic properties of soluble and membrane bound enzymes using ^{15}N and/or ^{2}H spin labels, Xth International Conference on Magnetic Resonance in Biological Systems, Abst. P56, Stanford, California.

Quintanilha, A. T., and Mehlhorn, R. J., 1978, pH gradients across thylakoïd membranes measured with a spin labeled amine, *FEBS Lett.* **91**:104–108.

Quintanilha, A. T., and Packer, L., 1977, Surface potential changes on energization of the mitochondrial inner membrane, *FEBS Lett.* **78**:161–165.

Rousselet, A., Devaux, P. F., and Wirtz, K. W., 1979, Free fatty acids and esters can be immobilized by receptor rich membranes from Torpedo marmorata but not phospholipid acyl chains, *Biochem. Biophys. Res. Commun.* **90**:871–877.

Seelig, J., Seelig, A., and Tamm, L., 1982, Nuclear magnetic resonance and lipid–protein interactions, in: *Lipid–Protein Interactions*, Vol. 2, Chap. 3 (P. C. Jost and O. H. Griffith, eds.), John Wiley and Sons, New York, pp. 127–148.

Seigneuret, M., Davoust, J., Hervé, P., and Devaux, P. F., 1981, Lipid–lipid and lipid–protein collision rates in membranes. Determination by evaluation of spin–spin interactions between ^{15}N and ^{14}N spin labels. *Biochimie* **63**:867–870.

Seigneuret, M., and Devaux, P. F., 1984, ATP dependent asymmetric distribution of spin labeled phospholipids in the erythrocyte membrane. Relation to shape changes, *Proc. Natl. Acad. Sci. U.S.A.*, in press.

Sigrist-Nelson, K., and Azzi, A., 1979, Proteolipid subunit of chloroplast ATP complex. Mobility, accessibility and interaction studied by a spin labeled technique, *J. Biol. Chem.* **254**:4470–4474.

Stone, T. J., Buckman, T., Nordio, P. L., and McConnell, H. M., 1965, Spin labeled biomolecules, *Proc. Natl. Acad. Sci. USA* **54**:1010–1017.

Swanson, M. S., Thomas, D. D., and Quintanilha, A. T., 1980, Protein rotational mobility and lipid fluidity of purified and reconstituted cytochrome *c* oxidase, *J. Biol. Chem.* **255**:7494–7502.

Takeuchi, Y., Ohnishi, S. I., Ishinaga, M., and Kito, M., 1978, Spin-labeling of Escherichia coli membrane by enzymatic synthesis of phosphatidylglycerol and divalent cation-induced interaction of phosphatidylglycerol with membrane proteins, *Biochim. Biophys. Acta* **506**:54–63.

Takeuchi, Y., Ohnishi, S. I., Ishinaga, M., and Kito, M., 1981, Dynamic states of phospholipids in *Escherichia coli* B membranes. Electron spin resonance studies with biosynthetically generated phospholipid spin labels, *Biochim. Biophys. Acta* **646**:119–125.

Tanaka, K. I., and Ohnishi, S. I., 1976, Heterogeneity in the fluidity of intact erythrocyte membrane and its homogenization upon hemolysis, *Biochim. Biophys. Acta* **426**:218–231.

Thomas, D. D., Bigelow, D. J., Squier, T. C., and Hidalgo, C., 1982, Rotational dynamics of protein and boundary lipid in sarcoplasmic reticulum membranes, *Biophys. J.* **37**:217–225.

Tokutomi, S., Iwasa, T., Yoshizawa, T., and Ohnishi, S. I., 1981, Flash induced fast change in surface potential and fluidity of purple membrane studied by spin label method, *Photochem. Photobiol.* **33**:467–474.

Van, S. P., and Griffith, O. H., 1975, Bilayer structure in phospholipid-cytochrome *c* model membranes, *J. Membr. Biol.* **20**:155–170.

Vignais, P. M., and Devaux, P. F., 1976, The use of spin labels to study membrane bound enzymes, receptors and transport systems, in: *The Enzymes of Biological Membranes*, Vol. 1 (A. Martonosi, ed.), Plenum Press, New York, pp. 91–115.

Watts, A., Volotovski, I. D., and Marsh, D., 1979, Rhodopsin–lipid associations in bovine rod outer segment membranes. Identification of immobilized lipid by spin-labels, *Biochemistry* **18**:5006–5012.

Watts, A., Davoust, J., Marsh, D., and Devaux, P. F., 1981, Distinct states of lipid mobility in bovine rod outer segment membranes. Resolution of spin label results, *Biochim. Biophys. Acta* **643**:673–676.

Yu, B. P., Mason, E. J., Down, J., and Wharton, D., 1977, Spin label study of sulfhydryl groups of sarcoplasmic reticulum, *J. Biol. Chem.* **252**:5262–5266.

Zachowski, A., and Devaux, P. F., 1983, Non uniform distribution of phospholipids in (Na^+ + K^+) ATPase-rich membranes from *Torpedo marmorata* electric organ evidenced by spin-spin interactions between spin-labeled phospholipids, *FEBS Lett.* **163**:245–249.

8

Saturation Transfer EPR Studies of Microsecond Rotational Motions in Biological Membranes

David D. Thomas

I. INTRODUCTION

In order to elucidate the molecular details of dynamic processes in biological membranes, it is necessary to make direct measurements of molecular motions and to correlate these motions with biological function. It has been largely in response to this need that nitroxide spin-labels and electron paramagnetic resonance (EPR) techniques have been developed. It was precisely in the study of membranes, however, that a crucial shortcoming of EPR became evident. The conventional EPR experiment on a nitroxide radical is sensitive to rotational motion only if it occurs in the nsec time range or faster. This time range is appropriate for the rapid rotational motions of fluid lipid hydrocarbon chains and protein side chains, and for the rapid tumbling of small proteins in aqueous solution. However, the relatively high effective viscosity of membranes (usually at least 100 times that of water) results in many motions in the μsec range or slower, particularly those of integral membrane proteins and gel-phase lipids. Partly as a result of this limitation of conventional EPR to the detection of nsec motions, an unbalanced description of membrane dynamics emerged, weighted heavily toward the lipids, despite the primary importance of proteins in determining membrane function.

The need for sensitivity to the slow rotational motions of membrane proteins was

David D. Thomas • Department of Biochemistry, University of Minnesota Medical School, Minneapolis, Minnesota 55455.

a primary motivation for the development of saturation transfer EPR (ST-EPR), a method with maximum sensitivity to μsec rotational motions of nitroxide spin-labels. In the summer of 1972, Hyde and Dalton (1972) published the seminal paper in the saturation transfer literature, in which they described an EPR experiment (dispersion, detected under conditions of moderate saturation, with the 100 kHz phase-sensitive detector set 90° out-of-phase with respect to the applied 100 kHz field modulation, later referred to as U_1') that showed marked line-shape sensitivity to the μsec rotational motions of nitroxide radicals in a supercooled organic glass. H. McConnell, who had led the development of conventional EPR for membrane studies, became interested in using ST-EPR to study slow motions in membranes. Accordingly, he encouraged me, his graduate student, to choose ST-EPR as my thesis topic. I worked with Hyde on the experimental methodology, seeking procedures useful in biological studies (Hyde and Thomas, 1973), and with McConnell on the theory, seeking verification that the changes observed in experimental ST-EPR spectra of model systems were those predicted for rotational diffusion (Thomas and McConnell, 1974). The methodological developments were summarized in a paper in 1976 (Thomas *et al.*, 1976), and applications of ST-EPR to membranes began to appear in print in 1977 (discussed below).

The primary purpose of the present review is to discuss the published applications of ST-EPR to membranes (as of December 1982), focusing on studies of membrane-bound enzymes. Because the technology of ST-EPR is new and developing, the first section is a condensed review of saturation transfer methodology, focusing on recent developments. In the sections that follow, the applications to membrane enzymes are discussed, followed by applications to other membranes.

ST-EPR has been reviewed comprehensively by Hyde and Thomas (1980), and in other papers mentioned in that article. Applications of ST-EPR to biological membranes have been briefly reviewed (Thomas, 1982). Applications of conventional EPR to biological membranes have been reviewed in the first edition of this volume (Gaffney and Lin, 1976; Vignais and Devaux, 1976). EPR studies of spin-labels have been reviewed comprehensively in two volumes edited by Berliner (1976, 1979), and were recently reviewed by Marsh (1981).

II. ST-EPR METHODOLOGY

A. General Principles of ST-EPR

ST-EPR experiments are performed on the same types of nitroxide spin-labels used in conventional EPR experiments (Figure 1). For both conventional and saturation transfer EPR, the sensitivity of the experiment to rotational motion arises from the high degree of orientational resolution in the nitroxide EPR spectrum. Due to anisotropic magnetic interactions, each position in the spectrum of a nitroxide radical corresponds to a narrow range of orientations of the spin-label relative to the applied D.C. magnetic field. Thus, scanning the spectrum is equivalent to scanning probe orientation, and rotational motion transfers spins from one part of the spectrum to another. If the minimum time required for this transfer (roughly equal to 0.1 τ, where

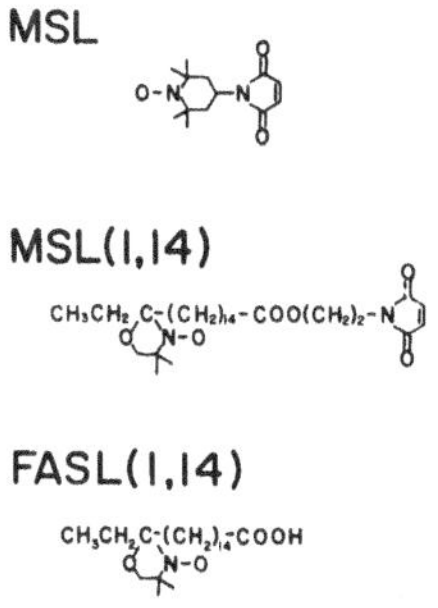

Figure 1. Typical nitroxide spin-labels used in ST-EPR studies. The probe most often used to monitor protein rotational motion is the maleimide derivative of TEMPO, usually designated MSL (N-[1-oxyl-2,2,6,6-tetramethyl-4-piperidinyl]-maleimide). The five-membered pyrrolidinyl derivation (designated 5MSL in this article) is also used. At or below pH 7, MSL reacts selectively with protein SH groups, and selectivity can be enhanced by blocking fast-reacting groups with diamagnetic maleimide derivatives (Hidalgo and Thomas, 1977) or by treating with ferricyanide (Graceffa and Seidel, 1980; Swanson *et al.*, 1980). Selective reaction with a particular protein is not sufficient; the probe must be strongly immobilized, so that its spectra depend on large-scale protein motions, not on local probe motions. Lipid hydrocarbon chain motions are monitored by using fatty acid spin-labels like the one pictured. FASL (1,14) is a derivative of stearic acid with the nitroxide attached at position 16, so it is often designated 16-SASL or 16-NS. The chain length and position of the nitroxide can be varied, and either the free fatty acid or its phospholipid derivative can be used. Maleimide derivatives, such as MSL (1,14), have been synthesized by Devaux and co-workers, for the purpose of probing the mobility of lipid chains adjacent to protein (boundary lipid). Except for samples containing gel-phase lipids, these lipid probes usually undergo nsec motions in lipids, making spectral analysis more appropriate with conventional EPR than with ST-EPR.

τ is the rotational correlation time) is comparable to (or less than) the inverse of the frequency resolution in the spectrum (rougly equal to T_2, the transverse relaxation time), the spectrum will be narrowed by the rotational motion. For a typical nitroxide radical, T_2 is about 20 nsec, so motional narrowing occurs only when the rotational correlation time is less than 1 μsec. The term "conventional" EPR usually refers to an experiment that probes the linear response of a spin system, i.e., an experiment in which the microwave field intensity is low enough that saturation is negligible. Under these conditions, the spectrum of a nitroxide is determined almost exclusively by motional narrowing, and is thus sensitive only to rotational correlation times less than 1 μsec.

However, if the microwave radiation is intense enough to cause saturation, so that the response of the spin system is no longer linear, a rotation-dependent transfer of spins that occurs within a time comparable to the longitudinal relaxation time (T_1) will cause a transfer of saturation from one part of the spectrum to another. A rotation-dependent transfer of spins occurs within about 0.1 τ (just as in conventional EPR), and T_1 is about 10 μsec for a slowly rotating nitroxide. Therefore, saturation transfer occurs whenever the rotational correlation time becomes less than 1 msec, optimum sensitivity to τ occurs in the range of 10–100 μsec, and saturation transfer becomes maximal when τ is less than 1 μsec. The development of ST-EPR technology has been a process of searching for an EPR experiment that is optimally sensitive to saturation, hence sensitive to saturation transfer, and hence sensitive to μsec rotational motions.

B. Methodology Used in Most Published Applications

The early development of the instrumentation, experimental methodology, and theory of saturation transfer EPR was summarized in a paper by Thomas *et al.* (1976), and virtually all applications of ST-EPR have been based closely on the techniques

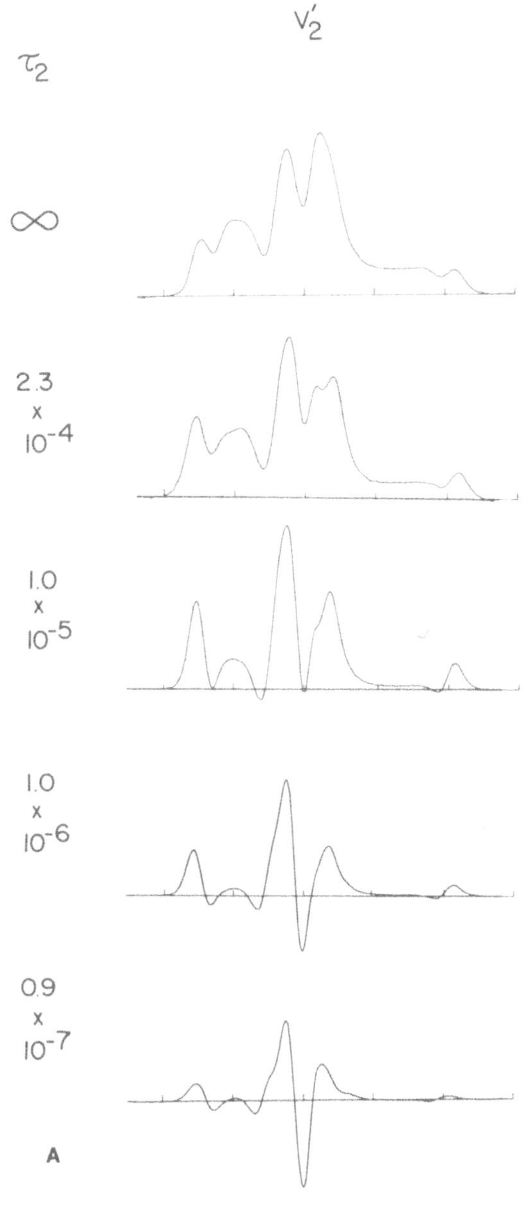

Figure 2. ST-EPR (V_2') reference spectra (A) and parameter plot (B) obtained from MSL-labeled hemoglobin in aqueous glycerol solutions. These data, or equivalent data, have been used to analyze most V_2' spectra to date. The rotational correlation time (τ_2) was determined from the measured viscosity, temperature, and the known molecular volume (Thomas *et al.*, 1976). The microwave power was set at 63 mW, resulting in an amplitude (H_1) of 0.25 gauss for an aqueous sample at 20°. Spectra are plotted so that their intensities are approximately equal. Adapted from Thomas *et al.* (1976) and Thomas (1978).

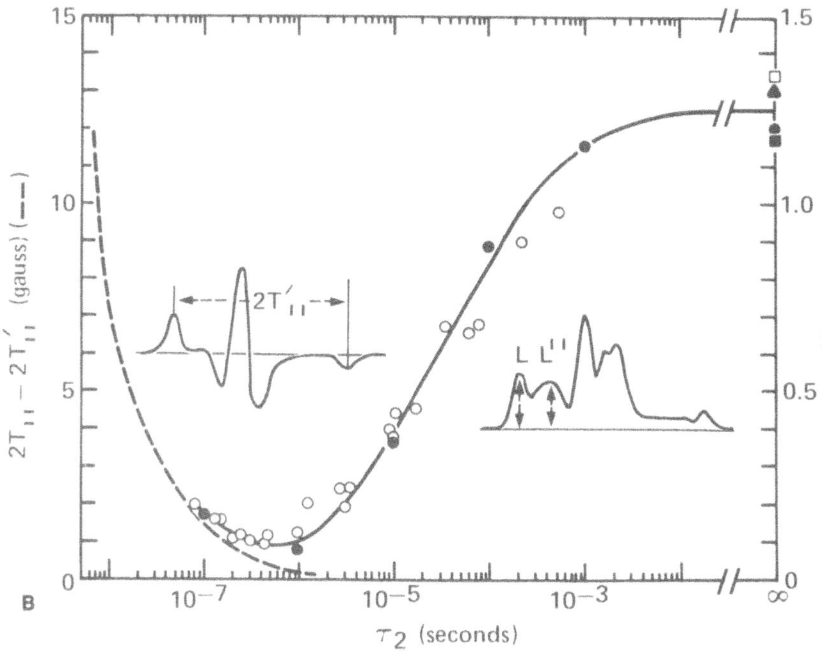

Figure 2. *(Continued)*

described in that paper. Those techniques are sometimes designated "passage" ST-EPR, because they are based on the use of high-frequency (50 kHz or 100 kHz) field modulation to pass the saturated spin system rapidly through resonance. In the absence of saturation (in conventional EPR), the spin system follows the modulation faithfully, but in the presence of saturation, the spin system lags behind the field modulation (if the modulation is rapid relative to the relaxation rate $1/T_1$), and a signal is observed 90° out of phase with respect to the modulation. By observing this out-of-phase signal, the investigator observes selectively a signal that depends on saturation, and hence on saturation transfer. In general, saturation transfer will decrease the saturation at the resonance position, and hence will decrease the amplitude of the out-of-phase signal. In addition, because rotation causes more saturation transfer in some regions of the spectrum than others, rotational motion changes the shape of the spectrum (Thomas *et al.*, 1976). In order to assess the utility of different instrumental approaches for detecting saturation transfer, aqueous glycerol solutions of spin-labeled hemoglobin (for which the rotational correlation times were known) were used to generate reference spectra. Two modes of detection and display were found to be particularly sensitive, in both amplitude and shape, to rotational motion. One was U_1' (dispersion, first harmonic, out-of-phase), using 100 kHz field modulation and detection, the method introduced in the first ST-EPR paper by Hyde and Dalton (1972). The U_1' display, however, was not easily applied to biological systems, because its low signal-to-noise prevented the study of spin-label concentrations below about 1 mM. The second display discussed in detail by Thomas *et al.* (1976) was V_2' (absorption, second harmonic, out-of-phase), using 50 kHz field modulation and 100 kHz detection, the method introduced in the first biological application of ST-EPR by Hyde and Thomas (1973). Although the V_2' spectrum was found to be more complex in shape and more difficult to analyze theoretically, it offered a much higher signal-to-noise ratio than U_1', permitting experiments on relatively dilute (as low as 10 μM in nitroxide) aqueous samples. To permit a quantitative analysis of V_2' spectra, Thomas *et al.* (1976) plotted selected spectral peak ratios against rotational correlation time for the hemoglobin model system. These plots were verified by computer simulation of the spectra, using the diffusion-coupled Bloch equations of Thomas and McConnell (1974). These reference spectra and plots (Figure 2) have been used to analyze most of the published ST-EPR spectra of spin-labeled membranes. The remainder of this section will briefly outline some of the recently proposed advances in ST-EPR technology that will probably be used in future applications.

C. Recent Developments in ST-EPR Methodology

1. The V_2' Experiment

Most methodological developments have focused on refining the V_2' method, attempting to improve the precision and accuracy of measurements and to avoid ambiguities in interpretation of data. The V_2' experiment is performed under conditions (saturation, high modulation amplitude, out-of-phase detection) that make it more sensitive than V_1 to slow rotational motion, but also increase the sensitivity to other experimental variables, such as microwave power, T_1, modulation amplitude, modulation phase, and the size, shape, and dielectric constant of the sample. Changes in these variables can cause spectral changes similar to those caused by motional changes, so these experimental conditions must be carefully controlled.

One of the most important parameters to be controlled and monitored is the amplitude of the microwave field (H_1) at the sample, since an increase in H_1 can cause spectral effects similar to those of a shorter correlation time (Thomas *et al.*, 1976). It is crucial, therefore, to perform experiments at the same value of H_1 if their spectra are to be compared. The reference spectra of Thomas *et al.* (1976) were recorded at a microwave power of 63 mW, chosen so that H_1 was 0.25 gauss for an aqueous sample at room temperature, as determined from saturation studies on a sample with known saturation properties. Subsequently, some workers have used the same power setting (63 mW) without calibrating the instrument to determine H_1 accurately. While the V_2' spectra are still sensitive to μsec rotational motions, correlation times determined with the reference spectra of Thomas *et al.* (1976) are not likely to be accurate unless the same H_1 values are used. A more subtle problem arises from the variation of dielectric constant (and hence H_1) caused by the variation in temperature and glycerol concentration in the hemoglobin solutions used to obtain the reference spectra, or caused by the variation in lipid and protein content in membrane samples. Kusumi *et al.* (1980) and Fajer and Marsh (1982) developed procedures for monitoring H_1 and adjusting the microwave power to compensate for changes in dielectric constant, and published reference spectra obtained at constant H_1. Fajer and Marsh (1982) also pointed out that the use of very small samples (5 mm long or less) reduces problems due to field inhomogeneities within the EPR cavity. Considering the high sensitivity of V_2' spectra to instrumental parameters, it is advisable in each laboratory to prepare spin-labeled hemoglobin as a standard and to produce reference spectra (Figures 1 and 3).

A number of investigators have explored other ways of parameterizing V_2' spectra, besides those suggested by Thomas *et al.* (1976). One of the most striking features of the V_2' spectrum is that a correlation time of 10 μsec or greater results in a spectrum that is entirely above the baseline, while a correlation time of 100 nsec results in a spectrum that is nearly symmetrical about the baseline (having equal areas above and below the baseline). Evans (1981) used a digital computer to quantitate this observation, showing that the integral of the V_2' spectrum, appropriately normalized, can be used to estimate the effective rotational correlation time (illustrated in Figure 3). Although integration might obscure some features of the line shape that are sensitive to the details of the motion, Evans (1981) showed that it suppresses spectral components corresponding to weakly immobilized spin-labels ($\tau < 10^{-8}$ sec), thus permitting the selective study of strongly immobilized labels ($\tau > 10^{-7}$ sec).

Although most methods of estimating correlation times from V_2' spectra have

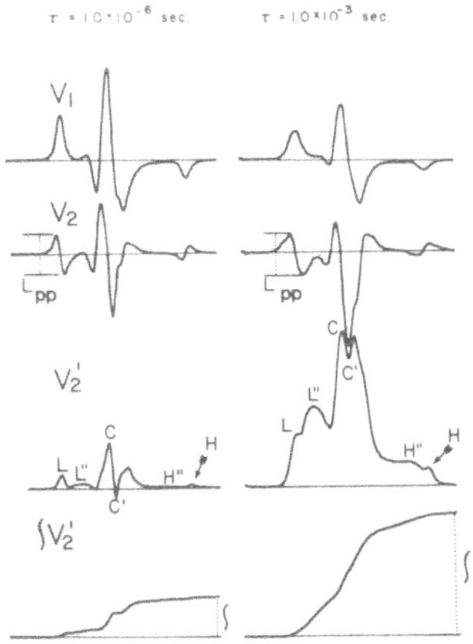

Figure 3. Conventional (V_1') and ST-EPR (V_2') reference spectra obtained from MSL-labeled hemoglobin in aqueous glycerol solutions. The spectra on the left are from a sample in 50% glycerol at −12°C, corresponding to a τ value of 10^{-6} sec (Thomas *et al.*, 1976). The spectra on the right are from a sample in 90% glycerol at −32°C, corresponding to a τ value of 10^{-3} sec (Thomas *et al.*, 1976). Power was adjusted so that H_1 was 0.25 G, using the method of Fajer and Marsh (1982). V_2' (out-of-phase) spectra are normalized to correspond to the same number of spins (by dividing by the double integral of V_1) so that the effects of motion on both shape and absolute amplitude are illustrated. V_2 (in-phase) spectra are shown at a gain ten times less than V_2'. The most commonly used parameter is $L''(V_2')/L(V_2')$, which measures the change in shape of the spectrum (Thomas *et al.*, 1976). However, *both* $L''(V_2')$ and $L(V_2')$ increase with τ. It is clear that greater sensitivity to τ is achieved if the parameter $L''(V_2')/L(V_2)$ is used, since the V_2 spectrum is relatively insensitive to motion (and has higher S/N). Plots of this parameter vs. τ are currently used in the author's laboratory to determine effective τ values with greater precision than previously possible (Squier and Thomas, to be published). Maximum precision can be obtained by measuring the normalized integral of V_2' (bottom), as proposed by Evans (1981), although information about possible anisotropic motion is lost.

been based on changes in the shape of the spectrum, the overall intensity of the spectrum is affected as well. All points in the V_2' spectrum decrease in intensity as rotational motion increases, while V_2 (in-phase) changes very little with motion (Figure 3). Thus, the ratio of two peaks in V_2' is less sensitive to motion than the ratio of a peak in V_2' to a peak in V_2. This principle is illustrated in Figure 3, using spectra from spin-labeled hemoglobin.

In addition to spectral integration, several other digital manipulations of V_2' spectra have been proposed. Hemminga and De Jager (1981) showed that a "magnitude" ST-EPR spectrum, equal to the root-mean-square of the digitized V_2 and V_2' spectra, is sensitive in shape to τ and has the advantage of being insensitive to errors in setting the reference phase on the phase-sensitive detector. Another group (Watanabe *et al.*, 1980, 1982; Sasaki *et al.*, 1980) has used a transient recorder and a Fourier transform to detect V_2 and V_2' digitally, without the need for the standard analog phase-sensitive detector.

Several methods have been used recently to obtain increased resolution in ST-EPR studies, including the use of ^{15}N- and ^{2}H-substituted spin-labels (Beth *et al.*, 1981b; Johnson *et al.*, 1982b) and a higher microwave frequency (Johnson and Hyde, 1981).

2. *Other ST-EPR Methods*

As mentioned above, the U_1' (out-of-phase dispersion) experiment has not been used much in biological experiments, because of the high noise levels observed when using a conventional resonant cavity. Recent studies using alternative cavity structures have reduced these noise levels significantly (Mailer *et al.*, 1980; Froncisz and Hyde, 1982). The most promising of these devices is the "loop-gap resonator" (Froncisz and Hyde, 1982), which yields U_1' spectra having better signal-to-noise ratios than can be achieved with V_2' spectra. Considering the simpler data analysis possible with U_1', these devices could make U_1' the ST-EPR experiment of choice in the near future. As discussed below, steady-state ST-EPR experiments (V_2' and U_1') suffer from some ambiguity in interpretation, especially when motions are anisotropic. Time-resolved measurements, such as saturation recovery or pulsed electron double resonance, should remove much of this ambiguity, as in the case of time-resolved fluorescence.

3. *Theoretical Analysis*

Although the dependence of ST-EPR line shapes on rotational motion can be qualitatively understood by making physical arguments as discussed above and in other papers (Thomas *et al.*, 1976; Thomas, 1978; Hyde and Thomas, 1980), the quantitative analysis of spectra requires spectral simulation by computer. Recent work has focused on extending the theory beyond the model of isotropic Brownian motion (Thomas and McConnell, 1974; Thomas *et al.*, 1976) to consider the types of anisotropic motions more likely to occur in biological systems, particularly in membranes. Simulations are even more important for anisotropic motion than for isotropic, since it is much more difficult to find experimental model systems for generating reliable reference spectra (Gaffney, 1979). In general, as in the case of conventional EPR, spectra depend not only on the rates (diffusion coefficients) and amplitudes of motions, but also on the orientation of the nitroxide relative to the axes of diffusion. If, for example, the principal nitroxide axis is approximately parallel to the axis of diffusion, the effect on the spectrum will be much less (the effective τ value, using isotropic motion reference spectra, will be greater) than if it is nearly perpendicular. Robinson and Dalton (1980, 1981) have simulated ST-EPR spectra (mainly U_1') corresponding to the unrestricted motion of rigid ellipsoids of revolution, tumbling in solution. Lindahl and Thomas (1982) have considered the effects of restricted-amplitude motions on V_2' spectra, reporting that restriction of the angular amplitude of motion, without a change in the rate, can cause a substantial increase in the effective τ (obtained from isotropic reference spectra). As in the case of conventional EPR, extremely anisotropic motions can sometimes be recognized directly from the ST-EPR line shape, that is, there is no isotropic motion that could give rise to the same shape. This is equivalent to the observation that different spectral parameters (L″/L, C′/C, H″/H) yield different effective correlation times when analyzed with parameter plots obtained with isotropic motion reference spectra. However, because of the lack of motional narrowing effects in the μsec time range, line-shape changes are usually more subtle and less uniquely interpretable for ST-EPR than for conventional EPR, so the interpretation of spectra is more ambiguous. In most cases, it remains necessary to compare the observed spectrum with reference spectra corresponding to isotropic motion (or some other model) and to refer to the resulting τ value as an *effective* correlation time. Caution must be used when comparing these effective τ values from different systems or in the same system under different conditions, since changes in τ can be due to changes in the rate of motion, angular amplitude of motion, or probe orientation. Spectra obtained from oriented membranes can help reduce this ambiguity by defining the orientation of the spin-label relative to the membrane (Sasaki *et al.*, 1982). Time-resolved ST-EPR, e.g., pulsed saturation-recovery experiments, should also be important in reducing the ambiguity of data interpretation.

III. MEMBRANE-BOUND ENZYMES

A. Sarcoplasmic Reticulum Calcium Transport ATPase

The first membrane enzyme to be studied by ST-EPR was the Ca-ATPase of sarcoplasmic reticulum (SR). This 100,000-dalton integral membrane protein pumps Ca^{2+} across the membrane (into the SR) to relax the muscle, in a reaction that is coupled to ATP hydrolysis and requires Mg^{2+}. As in the case of most other membrane systems, previous physical studies (including EPR) had focused almost exclusively on the lipids, resulting in proposals that enzyme activity depends on a fluid lipid environment (Warren *et al.*, 1974; Hidalgo *et al.*, 1976). A series of ST-EPR studies have been carried out to investigate the rotational dynamics of both proteins and lipids, and to determine whether these motions are correlated with enzymatic activity in this system (Thomas and Hidalgo, 1978; Hidalgo *et al.*, 1978; Thomas *et al.*, 1982; Squier, 1982), and independent studies have been carried out by others (Kirino *et al.*, 1978; Kaizu *et al.*, 1980; Pringle and Hidalgo, 1982). As in most other ST-EPR studies, the probe chosen to study protein motion was a maleimide spin-label (MSL). Hidalgo and Thomas (1977) developed a procedure for labeling the Ca-ATPase in SR specifically with this probe, with only a slight effect on enzymatic activity (Thomas *et al.*, 1982). The next step was to determine whether the spin-label was rigidly bound to the protein, enough so that the V_2' spectrum could be used to characterize the protein's overall rotational motion. This question often arises in ST-EPR studies, and Figure 4 shows how it was addressed (Thomas and Hidalgo, 1978). The conventional (V_1) spectrum of SR vesicles (top left) is characteristic of "strongly immobilized" probes, implying that there are no significant nsec rotational motions. The V_2' spectrum (Figure 4A, right) has a shape characteristic of μsec rotational motion. The value of L″/L is 0.75, corresponding to an effective rotational correlation time of 60 μsec (using Figure 2). This was the first measurement of the rotational motion of this protein, and it was later confirmed using

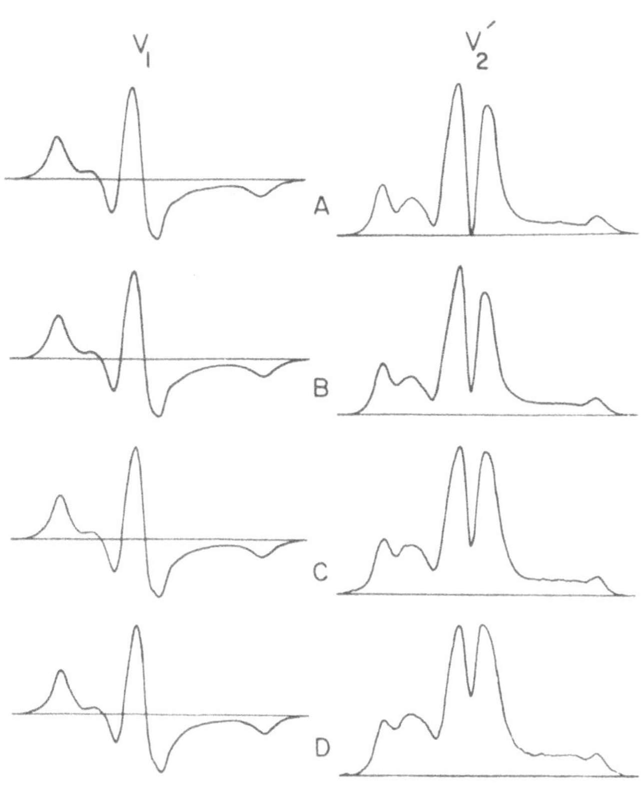

Figure 4. Conventional (V_1) and ST-EPR (V_2') spectra of MSL attached to the Ca-ATPase of SR at 4°C. (A) A suspension of membrane vesicles (L″/L = 0.75, τ = 60 μsec). (B) A membrane pellet (L″/L = 0.75). (C) Vesicles treated with 80 mM glutaraldehyde for 10 min (L″/L = 0.91). Gels showed that the proteins were partially crosslinked, with about half of the Ca-ATPase remaining monomeric. (D) SR vesicles treated with 80 mM glutaraldehyde for 12 hr (L″/L = 1.1). Gels showed that the proteins were completely crosslinked. The V_1 spectra were unchanged by crosslinking, implying that the inhibited motions are in the μsec time range. From Thomas and Hidalgo (1978).

optical spectroscopy. The spectra are unaffected by pellet formation (Figure 4B), implying that the observed motion is not that of the membrane vesicles themselves. Protein–protein crosslinking with glutaraldehyde does inhibit the motion (Figure 4C and D). The most likely interpretation of these observations is that the effective correlation time of 60 μsec characterizes the overall rotational motion of the protein within the membrane at 4°. However, since glutaraldehyde can cause intramolecular as well as intermolecular crosslinking, the glutaraldehyde-sensitive motions could also be slow internal protein motions.

Kirino *et al.* (1978) labeled SR by a similar procedure and published a V_2' spectrum taken at 2° (Figure 5) that was similar to that obtained by Thomas and Hidalgo (1978) at 4° (Figure 4A). However, other data reported by Kirino *et al.* (1978) indicated somewhat slower motion than reported by Thomas and Hidalgo (1978), and it is not clear whether this was due to differences in instrumentation, sample preparation, or data interpretation. One clear difference was the appearance of a weakly immobilized component in the spectra of Kirino *et al.* (1978) that was not present in the spectra of Thomas and Hidalgo (1978). This component tends to cause errors in the direction of longer effective correlation times.

Having established a method for detecting the μsec rotational motions of the Ca-ATPase, the investigators in these two groups proceeded to vary conditions and to correlate the protein motions with lipid motions and enzymatic activity. The motion of the protein probe (MSL) was shown to increase with temperature, having a temperature dependence qualitatively similar to that of Ca-ATPase enzymatic activity (Thomas and Hidalgo, 1978; Kirino *et al.*, 1978; Kaizu *et al.*, 1980). However, the presence of weakly immobilized probes in some preparations (Kirino *et al.*, 1978; Kaizu *et al.*, 1980) appeared to cause artifactual increases in the effective τ as the temperature increased above 20°.

Hidalgo *et al.* (1978) studied the effect of gel-phase lipid (dipalmitoyl phosphatidyl choline at 4°) on the enzyme's activity and rotational mobility. When the endogenous SR lipids were replaced by DPPC (yielding a preparation designated DPL-ATPase), conventional EPR spectra of a stearic acid spin-label (Figure 5, left) showed that lipid hydrocarbon chain mobility was greatly decreased, to the point where motion was

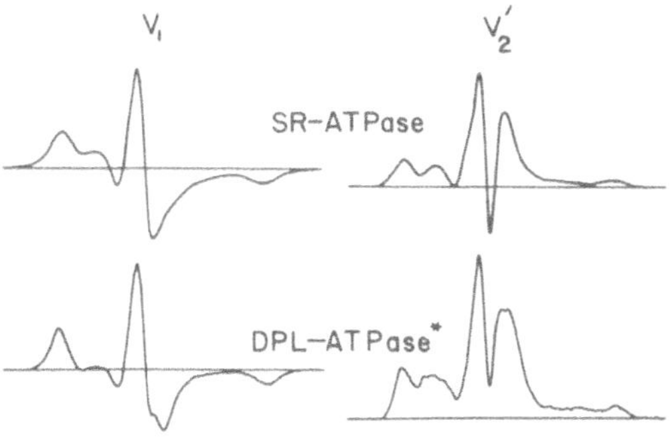

Figure 5. Conventional (V_1) and ST-EPR (V_2') spectra of a stearic acid spin-label (having the doxyl group attached to C5) in membranes containing the purified Ca-ATPase and either endogenous SR lipids (top) or DPPC (bottom), at 4°C. Spectra are 100 G wide. Note that V_1 is affected by the lipid change, implying that nsec motions are decreased. However, since the V_1 spectrum of DPL-ATPase has the characteristic shape of "strongly immobilized" spin-labels, the motion in this case is too slow to be detected by V_1 ($\tau > 10^{-7}$ sec), so V_2' must be used to determine the effective τ (10^{-4} sec, using Figure 1). From Hidalgo *et al.* (1978).

detectable only by saturation transfer EPR (Figure 5, right). The effective rotational correlation time (using Figure 2) is 100 μsec. These were the first published ST-EPR spectra of lipid spin-labels, and they demonstrate the utility of this technique for studying strongly restricted lipid motions, particularly in the gel phase.

Figure 6 shows that the rigid lipid environment in DPL-ATPase strongly inhibits protein rotational mobility. L″/L increases from 0.75 (in SR lipid, top left) to 1.13 (in DPPC, center left), corresponding to at least a ten-fold increase in the effective rotational correlation time (using Figure 2). It is not clear whether this decrease in protein mobility is due directly to the decrease in microviscosity or to protein aggregation that might accompany lipid gel formation. Note that the protein motion (Figure 6) is considerably slower than the lipid motion (Figure 5) in each preparation. The decreased protein mobility in DPL-ATPase is accompanied by a strong inhibition of enzymatic activity. The addition of the detergent Triton X-100 to DPL-ATPase fluidizes the lipid environment (Hidalgo *et al.*, 1978), reversing the inhibition of both protein mobility (Figure 6, bottom left) and enzymatic activity. Similarly, lipid fluidity, enzymatic activity, and protein mobility are all restored in DPL-ATPase above the lipid phase transition temperature (Hidalgo *et al.*, 1978). Finally, enzymatic activity and protein mobility are both inhibited by the addition of high (10 mM) Ca^{2+} to SR (Figure 6, top), or to Triton-treated DPL-ATPase Figure 6, bottom), even though high Ca^{2+} has no detectable effect on lipid chain mobility (Hidalgo *et al.*, 1978). In summary, enzymatic activity shows a remarkable correlation with protein rotational mobility, and the requirements of the enzyme for a fluid lipid environment and for submillimolar Ca^{2+} may arise from the requirement for protein mobility.

In recent ST-EPR studies on SR, reduction of the lipid–protein ratio, achieved by extraction with deoxycholate, has been used as a means of reducing the enzyme's rotational mobility (Thomas *et al.*, 1982; Squier, 1982). In the first study (Thomas *et al.*, 1982), it was found that reducing the lipid–protein ratio by 55% results in a substantial reduction of the protein mobility (monitored by ST-EPR of MSL attached

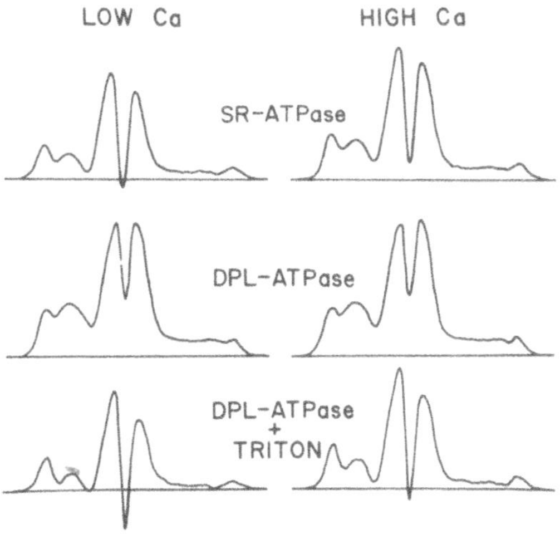

Figure 6. ST-EPR (V_2') spectra of a maleimide spin-label (MSL) attached to the Ca-ATPase in endogenous lipids (top), DPPC (center), and in DPPC plus Triton X-100 (1.5 mg/mg protein, bottom), at 4°C. Left, 0.1 mM Ca^{++}; Right, 10 mM Ca^{++}. From Hidalgo *et al.* (1978).

to the Ca-ATPase), an increase in the fraction of restricted lipid chains (monitored by conventional EPR on a freely diffusing fatty acid spin-label; see Figure 1), and a decrease in the mobility of the lipid chains adjacent to the protein (monitored by conventional EPR of MSL (1,14); see Figure 1). The reduced protein mobility most likely arises from increased protein–protein interactions in the crowded membrane, which in turn may play a role in the increased restriction of lipid chain mobility. In another study (Squier, 1982), the effect of this decreased protein mobility on enzymatic activity was studied over a wide range of lipid–protein ratios. ST-EPR spectra showed that reducing the lipid–protein ratio below a critical level of about 40 lipids per protein reduces protein rotational mobility, and that the fractional decrease in protein mobility ($1/\tau$) correlates well with the fractional decrease in Ca-ATPase activity. Since it had been previously shown that delipidation inhibits phosphoenzyme decomposition (Warren *et al.*, 1974), the step most likely to be associated with transport, this result provided further evidence that protein mobility (or at least protein disaggregation) is required for a crucial step in the Ca^{2+} transport process.

B. Mitochondrial Electron Transport Chain

Although the difficult study of protein motions in mitochondrial membranes is in its early stages, questions about protein motion and function seem especially important in the electron transport chain, the function of which involves the dynamic interaction of several integral enzyme complexes. While most probe studies of membrane proteins have been done on systems having one predominant protein, facilitating selective labeling of the protein in native membranes, the complex composition of mitochondrial membranes has required the purification of the enzymes before spin-labeling with MSL.

Swanson *et al.* (1980) carried out an ST-SPR study of purified cytochrome *c* oxidase (complex IV, ferrocytochrome *c* : oxygen oxidoreductase, EC 1.9.3.1), the terminal component of the electron transport chain, from beef heart mitochondria. These workers purified the enzyme using detergent (either deoxycholate or Triton), spin-labeled the protein with MSL (using ferricyanide and cysteine to eliminate signal from weakly immobilized spin-labels, as suggested by Graceffa and Seidel, 1980), and recombined the purified enzyme with lipid to increase the lipid–protein ratio. Spin-labeling did not affect enzyme activity, and the lipid addition reconstituted the full enzyme function (including respiratory control). V_2' spectra of MSL were used to assess protein mobility, V_1 spectra of fatty acid spin-labels were used to monitor lipid chain mobility, and V_1 spectra of MSL (1,14) were used to monitor the mobility of lipid chains adjacent to the protein. The V_2' spectrum of the monomeric enzyme in detergent solution and of glutaraldehyde-crosslinked membranes both indicated that the spin-label was rigidly bound, reporting the overall motion of the protein. Membranes with a low lipid–protein ratio showed low protein mobility and low lipid mobility, as in the case of the Ca-ATPase (discussed above). However, in contrast with the results of the studies on Ca-ATPase motion, it was found that a high fluid lipid–protein ratio was not always sufficient to produce μsec protein rotational mobility. One method of preparation resulted in high protein mobility ($L''/L = 0.65$, effective

τ = 40 μsec, in agreement with an independent study by Ariano and Azzi, 1980), while another resulted in much lower mobility (L″/L > 1, implying little or no submillisecond motion; Figure 7). Thus, cytochrome oxidase appears to have a greater tendency to aggregate than does the Ca-ATPase. Since the two cytochrome oxidase preparations were both found to have high enzyme activity, the activity of this enzyme does not appear to require submillisecond rotational mobility, in contrast with the Ca-ATPase (discussed above). The nsec mobility of MSL (1,14) did correlate with μsec protein mobility, suggesting that the protein aggregation that decreases protein mobility also decreases boundary lipid mobility.

Cytochrome *c* reductase (complex III of the electron transport chain in beef heart mitochondria, ubiquinol : cytochrome *c* oxidoreductase, EC 1.10.2.2) has been studied by ST-EPR (Quintanilha *et al.*, 1982). Spin-labeling with MSL resulted in a V_1 spectrum that indicated only strongly immobilized probes, so treatment with ferricyanide was unnecessary. The V_2' spectra showed relatively high protein mobility in solution (τ = 400 nsec) and in membranes containing a high lipid–protein ratio (τ = 70 μsec). Since this preparation was treated essentially as the cytochrome oxidase preparation that showed much slower protein motion, this result suggests that cytochrome reductase

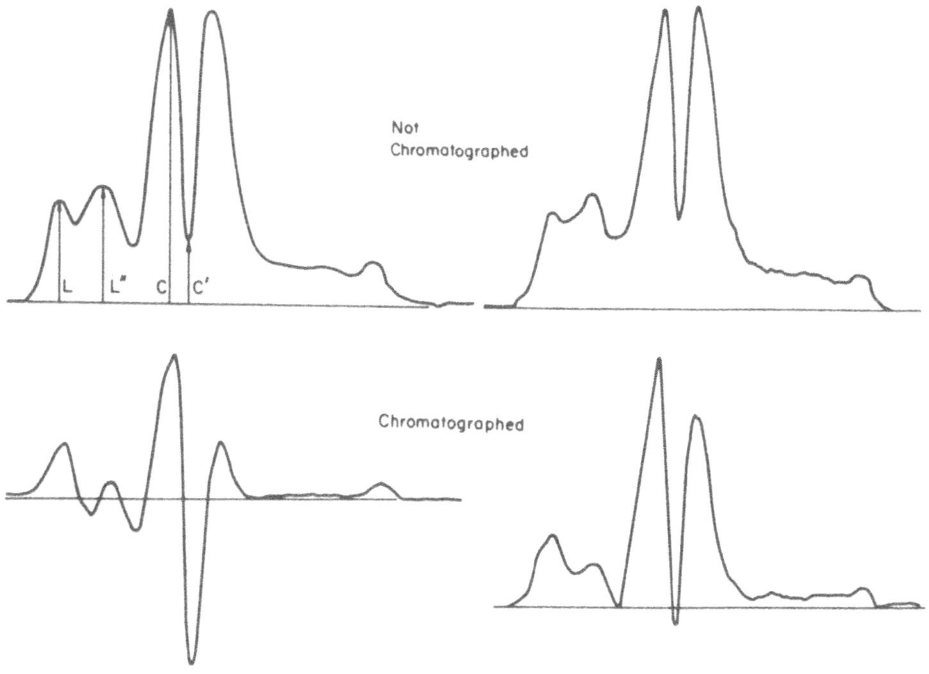

Figure 7. ST-EPR (V_2') spectra of purified (left) and reconstituted (right) cytochrome *c* oxidase, at 4°C. Top and bottom rows show the results of two different preparation procedures, starting from the same MSL-labeled enzyme preparation. From Swanson *et al.* (1980).

has a lower tendency to self-aggregate. The boundary lipid probe MSL (1,14) was also mobile, consistent with unaggregated protein.

An ST-EPR study on both cytochrome *c* reductase (complex III) and NADH-ubiquinone oxidoreductase (complex I) has been performed by Poore *et al.* (1982). They recombined both enzymes in the same membrane with varying lipid content, and then measured protein mobility (with ST-EPR), lipid chain mobility (with conventional EPR), and enzymatic activity. They found μsec protein mobility at high levels of fluid lipid, and much slower motion when the lipid–protein ratio was low or when the lipid was in the gel state. Enzyme kinetics showed an interesting correlation with molecular dynamics, providing insight into models for electron transport function.

C. Cytochrome P-450

Cytochrome P-450 (EC 1.14.14.1) from rabbit liver microsomes has been studied by ST-EPR (Schwarz *et al.*, 1982a). As in the case of mitochondrial membranes, this is an interesting system in which to study the interaction of different membrane-bound enzymes. In these microsomes, the two major enzymes are P-450 and its reductase, and alternative structural models have been proposed in which the enzymes diffuse freely or are in a specific and rigid cluster. These workers used 5MSL to label either intact microsomes or a water soluble purified enzyme designated P-450 LM2. The labeling reaction in microsomes was 70% specific for P-450, and NADPH-dependent reduction of the enzyme was still possible after labeling. Conventional (V_1) EPR spectra indicated that the probe was strongly immobilized at 20°, ST-EPR (V_2') spectra of the microsomes (Figure 8) were unaffected by pelleting, and glutaraldehyde produced a V_2' spectrum in which L″/L = 1.2 ($\tau > 1$ msec). Thus the V_2' spectra appeared to reflect protein motion within the membrane. The V_2' spectrum of the soluble enzyme indicated an effective τ of about 220 nsec (C′/C = −0.7), much longer than predicted for a monomer, and consistent with previous studies showing the soluble enzyme to be a hexamer. The effective τ in microsomes was 480 μsec (L″/L = 1.0), much slower than has been observed for most other integral membrane proteins in fluid lipids. The

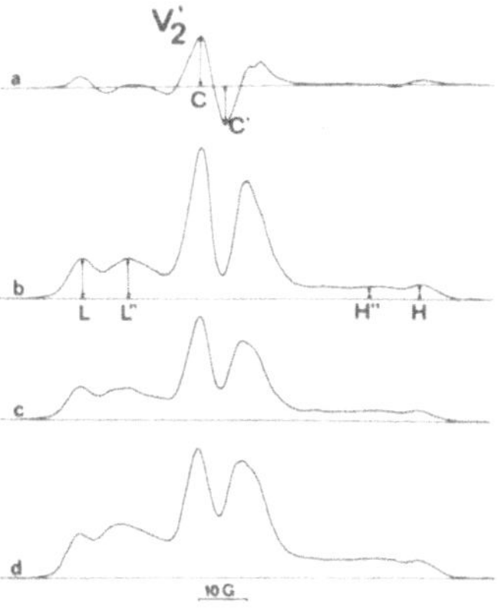

Figure 8. ST-EPR (V_2') spectra of 5MSL-labeled cytochrome P-450 at 20°C. (a) The water-soluble P-450 LM2 in buffer solution. (b) Suspension of microsomes. (c) Pellet of microsomes, obtained by centrifugation at 100,000*g* for 1 hr. (d) Glutaraldehyde-treated microsomes. From Schwarz *et al.* (1982a).

authors concluded that the enzyme is probably in an aggregated state in the membrane. Although this conclusion seems plausible, it was based on assuming an effective microviscosity of ten poise and using the Stokes–Einstein equation (which applies to an isotropic fluid) to estimate the predicted τ for a monomer (21 μsec). Besides the large uncertainty in predicting the rotational diffusion coefficient of a membrane protein from theory, there is the added uncertainty in the relative orientation of the probe's principle axis and the axis of diffusion, which determines the relationship between the diffusion coefficient and the effective correlation time. In short, it is difficult to attach much quantitative significance to the absolute value of an effective correlation time determined from a V_2' spectrum of a membrane protein. The conclusion that proteins are aggregated would be much more convincing if spectra could be obtained for comparison under conditions where the proteins are more mobile at the same microviscosity, as was done for cytochrome oxidase (Swanson *et al.*, 1980) and for acetyl choline receptor (discussed below).

Schwarz *et al.* (1982b) have demonstrated that $K_3Fe(CN)_6$ may be used to selectively eliminate spectral components due to weakly immobilized spin-labels in P-450, for purposes of simplifying the interpretation of V_2' spectra. This is a reversible line-broadening effect, and should not be confused with the chemical destruction of the nitroxide group in the presence of ferricyanide and nearby SH groups (Graceffa and Seidel, 1980).

D. GAPDH

The only membrane-bound enzyme studied by ST-EPR that is a peripheral membrane protein has been glyceraldehyde-3-phosphate dehydrogenase (GAPDHase, D-glyceraldehyde-3-phosphate : NAD^+ oxidoreductase, EC 1.2.1.12). Beth *et al.* (1981a,b) labeled the enzyme with $[^{15}N,^{2}H]$-MSL at a level of one spin-label per tetramer, resulting in a 28% decrease in enzymatic activity. Thus, it appears that the activity of the labeled subunit is abolished by labeling, but the remaining three subunits have normal activity. ST-EPR (V_2') spectra were obtained at 2°C from the enzyme free in aqueous solution (Figure 9, top) and bound to red blood cell membranes (Figure 9, center), where it is known to bind to the cytoplasmic segment of band 3, an integral membrane protein. Although these spectra have different features from those of ^{14}N spin-labels, they can be analyzed by a similar method, i.e., by comparing them with reference spectra obtained from GAPDHase in glycerol (Figure 9, bottom), or from computer simulations (Figure 9, dashed curves). The result is that the rotational mobility of the probe is substantially slowed by this binding; the effective τ increases from 160 nsec to about 20 μsec. Nevertheless, this motion is much faster than that of band 3 (measured by optical methods), indicating motion of the probe with respect to the membrane. It is not clear whether this motion is due to motion within GAPDHase, motion of GAPDHase with respect to band 3, or motion of a flexible segment of band 3. Spectra were found to be indistinguishable from reference spectra corresponding to isotropic motion.

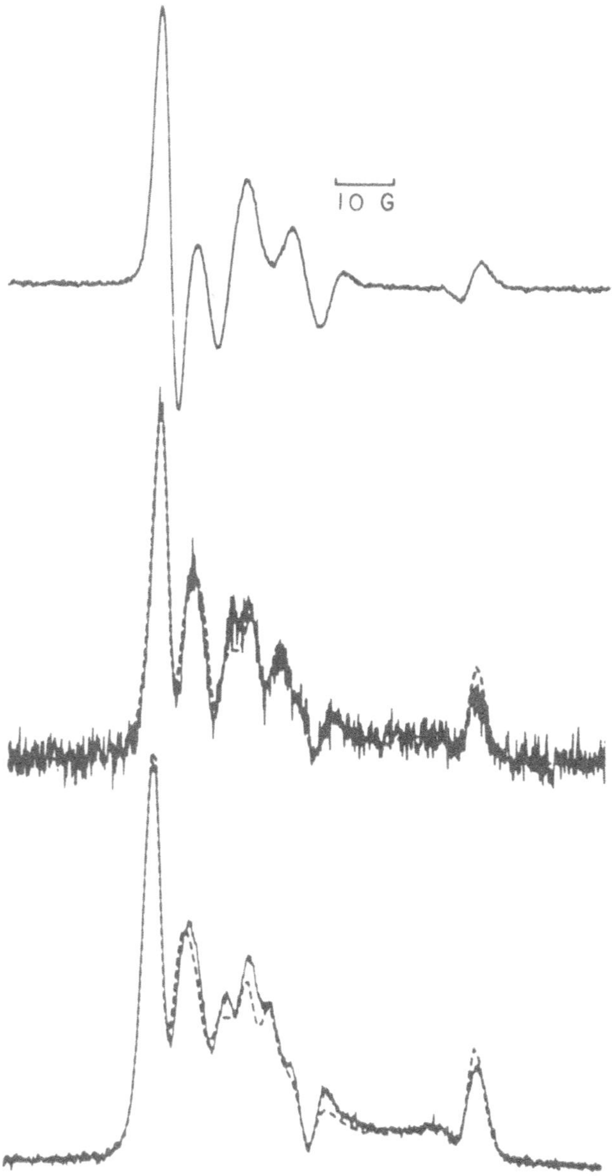

Figure 9. ST-EPR (V_2') spectra of [$^{14}N,^{2}H$]-MSL-labeled GAPDHase at 2° in solution (top), bound to the cytoplasmic segment of band 3 in red blood cell membranes (center), and in 78% glycerol (bottom). Dashed curves show the results of computer simulations for isotropic rotational diffusion (τ = 20 μsec). From Beth *et al* (1981a).

IV. OTHER MEMBRANE PROTEINS

A. Rhodopsin

Rhodopsin, the photoreceptor protein of retinal rod disk membranes, has been studied by ST-EPR in the laboratories of Devaux and Ohnishi. This work has been reviewed recently by Devaux (1982). Disk membranes, like SR membranes, are easily isolated and contain one predominant protein, making them an excellent system in which to study the rotational motion of spin-labeled membrane proteins, either in native or recombinant membranes. Furthermore, recent models of visual transduction involve light-induced protein–protein interactions within these membranes. Workers in Devaux's laboratory (Baroin *et al.*, 1977) labeled rhodopsin with 5MSL in native disk membranes. AT 20°C, considerable rotational motion was observed, corresponding to an effective correlation time of 20 μsec (Figure 10, left). Most of the submillisecond motion was stopped when the proteins were crosslinked with glutaraldehyde, indicating that the observed motion was probably that of the protein as a whole (Figure 10, right). These results were in agreement with previous time-resolved optical measurements, and were thus important in establishing the utility of ST-EPR in studying membrane protein motions. The EPR experiments went beyond the optical experiments, because they could be performed in both the presence and the absence of light. A brief exposure to light had no effect on the motion of rhodopsin, indicating that light-induced aggregation of rhodopsin (which would be expected to decrease rotational mobility) is not involved in visual transduction. Ohnishi's group (Kusumi *et al.*, 1978) obtained similar results with ST-EPR, except that they used MSL and observed much less weakly-immobilized signal. They observed some residual motion after glutaraldehyde crosslinking, suggesting some segmental motion within rhodopsin. They also observed a slight decrease in mobility after prolonged (and presumably irrelevant physiologically) exposure to light, as later confirmed by Devaux's group (Baroin *et al.*, 1979).

Devaux and co-workers (Baroin *et al.*, 1979; Favre *et al.*, 1979; Davoust *et al.*, 1980) have carried out studies to correlate the rotational mobility of the protein with that of lipid chains. In disk membranes of normal composition, in which the protein is quite mobile (presumably, therefore, not aggregated), they found that the boundary-lipid probe [MSL(1,14)] was nearly as mobile as a probe in the bulk of the lipid (and much more mobile than the protein probe). However, when the lipid content was reduced by phospholipase treatment, the protein (5MSL) and boundary-lipid [MSL(1,14)] probes were both immobilized, probably due to protein–protein interactions. These workers concluded that the concept of a strongly immobilized boundary layer of lipid coating an intrinsic membrane protein is not valid for rhodopsin, and that the evidence for strongly immobilized lipid in the presence of integral membrane proteins may often be an indication that lipid has been trapped by protein aggregation. As in the case of subsequent studies on cytochrome oxidase and Ca-ATPase, this finding does not contradict models in which slight restriction of lipid chain motions must occur at protein surfaces.

Several ST-EPR studies have been done on recombinant membranes containing

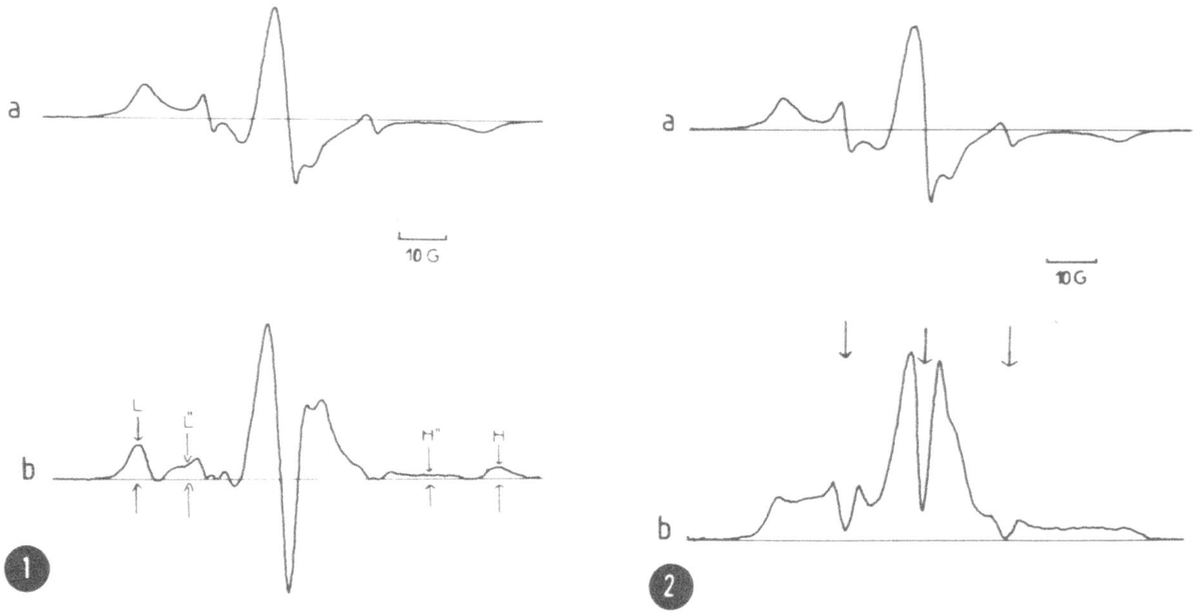

Figure 10. Conventional [V_1(a)] and ST-EPR [V_2(b)] EPR spectra of 5MSL attached to rhodopsin in retinal rod outer segment disk membranes before (1) and after (2) crosslinking with glutaraldehyde. From Baroin *et al.* (1977).

purified rhodopsin and exogenous lipid (Davoust *et al.*, 1980; Kusumi *et al.*, 1980, Kusumi and Hyde, 1982). Protein mobility was found to increase with temperature, but the transition was much more gradual than that of the lipid chain mobility, and significant submillisecond protein mobility was still observed below the transition temperature. The immediate cause of decreased protein mobility at low temperature may be protein aggregation, not increased lipid microviscosity around the protein (Devaux, 1982). As is the case of native disk membranes, protein mobility decreased when the lipid–protein ratio was decreased, presumably due to protein–protein interactions (Kusumi *et al.*, 1980; Kusumi and Hyde, 1982). It was suggested that phospholipid chain length is important in determining protein aggregation (Davoust *et al.*, 1980; Kusumi and Hyde, 1982).

B. Acetyl Choline Receptor

The rotational motion of acetylcholine receptors, from Torpedo marmorata membranes, has been studied by ST-EPR using either 5MSL (Rousselet and Devaux, 1977; Rousselet *et al.*, 1981) or a spin-labeled α-toxin (Rousselet *et al.*, 1982). In native membrane fragments, the labeled receptor was found to have little or no μsec rotational mobility (Rousselet and Devaux, 1977), suggesting protein aggregation, and thus immobilization, was not affected much by temperature variation or fusion of exogenous lipids (Rousselet *et al.*, 1981). These results suggest the presence of persistent receptor clusters, as verified by electron microscopy. Detectable μsec rotational motion ($L''/L = 0.65$, $\tau = 40$ μsec) was observed upon alkaline extraction, which removes a 43,000-dalton peripheral protein, and this increased mobility was reversed when this protein was added back (Rousselet *et al.*, 1981). The mobility of receptors in the alkaline-extracted membranes, in contrast with the native membranes, was sensitive to temperature and lipid changes. Thus, protein mobility in native acetylcholine receptor membranes appears to be controlled primarily by protein–protein interactions, not by lipid–protein interactions.

C. Red Blood Cell Membranes

ST-EPR has been used to study the motions of both peripheral and integral proteins of the red blood cell membrane, as well as the lipids. Fung *et al.* (1979) have used MSL to study both intact red cell membranes and the isolated spectrin–actin complex, which forms part of the peripheral cytoskeleton on the inside surface of red cells. Although weakly immobilized spin-labels (nsec motions detected by conventional EPR) made spectral analysis difficult at neutral pH, the probes were strongly immobilized at pH 4.5 (where the proteins precipitate) making it possible to use ST-EPR to study μsec rotational motions. Lemaigre-Dubreil *et al.* (1980) purified spectrin, labeled it with MSL, and found no weakly immobilized labels even at pH 7 (where spectrin is not highly aggregated). They observed rapid μsec motions that were only slightly restricted upon binding to red cell membranes, suggesting that spectrin is flexible and that its flexibility is relatively independent of membrane attachment.

Workers in Ohnishi's laboratory have isolated band 3, an integral membrane protein, from red blood cells, recombined it with DMPC, labeled it with MSL, and studied its rotational motion with ST-EPR. They found that the effective correlation time (from V_2' spectra) was about 100 μsec at 30° (above the lipid phase transition temperature) and increased to about 300 μsec at 10° (below the transition temperature). The response of the protein mobility to the phase transition was significant but less pronounced than that of the lipid, as also observed for the Ca-ATPase (Hidalgo *et al.*, 1978) and for rhodopsin (Davoust *et al.*, 1980; Kusumi *et al.*, 1980). The addition of peripheral ("cytoskeletal") membrane proteins resulted in decreased protein mobility, consistent with protein–protein crosslinking. These workers prepared partially oriented membranes, permitting them to estimate the orientation of the probe relative to the membrane, and hence to estimate the rotational diffusion coefficient (3.9×10^4 s^{-1} at 37°).

Fung (1981) studied both protein motions (using MSL) and lipid motions using a fatty acid spin-label [FASL(10,3) or 5NS] and both conventional EPR (V_1) and saturation transfer EPR (both V_2' and U_1'), in red blood cell membranes (ghosts). U_1' spectra appeared more sensitive than V_2' to lipid motions (probably in the range around 10^{-7} sec), while V_2' spectra appeared more sensitive than U_1' to the slower protein motions. Other peripheral membrane proteins that have been studied by ST-EPR are GAPDHase (Beth *et al.*, 1981a, discussed above), hemoglobin (Cassoly, 1982), and ankyrin (Cassoly, 1982).

V. LIPID PROBES

In several of the above-mentioned saturation transfer studies on membrane proteins, spin-labels have also been used to probe lipid chain motions. In most cases, it has been found that lipid motions are sufficiently rapid (in the nsec range) that conventional EPR suffices to detect them. However, ST-EPR was needed when lipids were in the gel-phase (Hidalgo *et al.*, 1978) and when the membranes were so highly delipidated that the remaining lipid probes were immobilized by protein (Favre *et al.*, 1979). It is interesting to note that, in both of these cases, the ST-EPR spectra indicated that the lipid probes, although immobile on the nsec time scale, were more mobile than the protein probes.

ST-EPR has been used recently to study lipid model membranes in the gel phase, i.e., at temperatures below the gel-to-liquid crystal phase transition. These studies indicate the presence of substantial submillisecond rotational mobility of hydrocarbon chains and steroids in gel-phase lipid. Spectra of a phosphatidylcholine spin-label (labeled at the C5 position, designated 5-PCSL) in DPPC dispersions showed a marked increase in rotational motion as the temperature increased above 25°C, a temperature that is well below the main phase transition and that probably corresponds to the calorimetric "pretransition" (Marsh, 1980; Marsh and Watts, 1980; Watts and Marsh, 1981). Due to the well-known orientation of the nitroxide group relative to the membrane, the pronounced spectral changes in the center of the spectrum were assigned to long-axis rotation of the lipid probe (Figure 11). Phosphatidyl ethanolamine, which does not display a pretransition, showed much smaller V_2' effects than PC below the

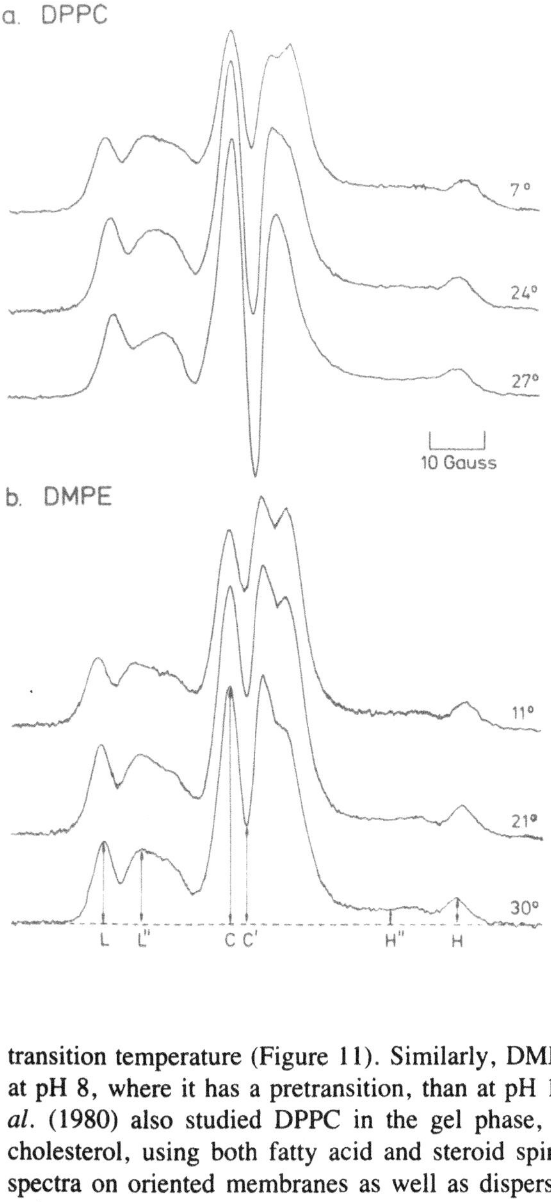

Figure 11. ST-EPR (V_2') spectra of phosphatidylcholine spin-label (doxyl group on C5 of β-chain) in dispersions of DPPC or DMPE, at the indicated temperatures. From Marsh (1980).

transition temperature (Figure 11). Similarly, DMPG showed much larger V_2' effects at pH 8, where it has a pretransition, than at pH 1.5, where it does not. Delmelle *et al.* (1980) also studied DPPC in the gel phase, including studies on the effect of cholesterol, using both fatty acid and steroid spin-labels, and recording V_1 and V_2' spectra on oriented membranes as well as dispersions. Koole *et al.* (1981) reported V_2' spectra of a steroid spin-label in oriented membranes containing DMPC and cholesterol. Johnson *et al.* (1982a) showed that V_2' spectra obtained at a higher frequency (35 GHz) should have increased sensitivity to distinguish among different anisotropic motional models. In most of these V_2' studies on lipids in the gel phase, these systems have been discussed as useful model systems for understanding the effects of anisotropic motions on ST-EPR spectra. The orientation of the nitroxide principal axis relative to the likely diffusion axes are well known, thus greatly reducing the ambiguity of spectral interpretation. However, the interpretation of spectra in terms of slow (μsec), anisotropic rotations is complicated by the fact that conventional (V_1) EPR spectra show some evidence of rapid (nsec) anisotropic motions in almost all of these systems, even in the gel phase (Delmelle *et al.*, 1980). Until the effects of such motions on ST-EPR spectra are understood, it remains true that the unambiguous interpretation of V_2' spectra in terms of anisotropic motion requires that V_1 spectra show no motional narrowing effects.

Although the quantitative interpretation of lipid V_2' spectra is difficult, it is still true that V_2' spectra are often more sensitive to changes than are V_1 spectra, and this increased sensitivity can be exploited in qualitative studies where precise physical interpretation is not crucial (Wilkerson *et al.*, 1978; Swift *et al.*, 1980; Fung, 1981).

VI. SUMMARY

A little more than 10 years after the origination of saturation transfer EPR, this technique has become established as one of the most important tools in the study of molecular dynamics in membranes, particularly in the study of μsec protein motions. Although the majority of ST-EPR studies have been done on the development of methodology using simple model systems, an increasing number of applications are focusing on the relationship between molecular dynamics and enzyme function. The required equipment is widely available in commerical spectrometers, and many investigators have learned to take the extra care required to obtain reliable data. The major limitation of the technique, as in any molecular probe method, is the labeling problem—specificity, rigidity, and preservation of function. More controls are required than in conventional EPR, the theory is more complex, and the temptation to overinterpret data is not always resisted. Future refinements in theory and instrumentation (especially time-resolved EPR) along with increasingly detailed information about the protein structures that are undergoing the detected motions, will continue to reduce the ambiguity of data interpretation and make ST-EPR a more powerful technique in future studies of membrane molecular dynamics.

ACKNOWLEDGMENT

I thank T. C. Squier for advice and assistance in preparing this manuscript.

REFERENCES

Ariano, B. H., and Azzi, A., 1980, Rotational motion of cytochrome c oxidase in phospholipid vesicles, *Biochem. Biophys. Res. Commun.* **93**:478–485.

Baroin, A., Thomas, D. D., Osborne, R., and Devaux, P. F., 1977, Saturation transfer electron paramagnetic resonance on membrane-bound proteins. I-Rotational diffusion of rhodopsin in the visual receptor membrane, *Biochem. Biophys. Res. Commun.* **78**:442–449.

Baroin, A., Bienvenue, A., and Devaux, P. F., 1979, Spin-label studies of protein–protein interactions in retinal rod outer segment membranes. Saturation transfer electron paramagnetic resonance spectroscopy, *Biochemistry* **18**:1151–1155.

Berliner, L. J. (ed.), 1976, *Spin Labeling*, Academic Press, New York.

Berliner, L. J. (ed.), 1979, *Spin Labeling II*, Academic Press, New York.

Beth, A. H., Balasubramanian, K., Wilder, R. T., Venkataramu, S. D., Robinson, B. H., Dalton, L. R., Pearson, D. E., and Park, J. H., 1981a, Structural and motional changes in glyceraldehyde-3-phosphate dehydrogenase upon binding to the band-3 protein of the erythrocyte membrane examined with [15N, 2H]maleimide spin-label and electron paramagnetic resonance, *Proc. Natl. Acad. Sci. USA* **78:**4955–4959.

Beth, A. H., Venkataramu, S. D., Balasubramanian, K., Dalton, L. R., Robinson, B. H., Pearson, D. E., Park, C. R., and Park, J. H., 1981b, ^{15}N- and ^{2}H-substituted maleimide spin-labels: Improved sensitivity and resolution for biological EPR studies, *Proc. Natl. Acad. Sci. USA* **78:**967–971.

Cassoly, R., 1982, Interaction of hemoglobin with the red blood cell membrane. A saturation transfer electron paramagnetic resonance study, *Biochim. Biophys. Acta* **689:**203–209.

Davoust, J., Bienvenue, A., Fellmann, P., and Devaux, P. F., 1980, Boundary lipids and protein mobility in rhodopsin-phosphatidylcholine vesicles. Effect of lipid phase transitions, *Biochim. Biophys. Acta* **596:**28–42.

Delmelle, M., Butler, K. W., and Smith, I. C. P., 1980, Saturation transfer electron spin resonance spectroscopy as a probe of anisotropic motion in model membrane systems, *Biochemistry* **19:**698–704.

Devaux, P. F., 1982, Lipid-Protein interactions: Saturation transfer electron paramagnetic resonance of spin-labeled rhodopsin, *Meth. Enzymol.* **81:**703–709.

Evans, C. A., 1981, Use of the integral of saturation transfer electron paramagnetic resonance spectra to determine molecular rotational correlation times. Slowly tumbling spin-labels in the presence of rapidly tumbling spin-labels, *J. Magn. Reson.* **44:**109–116.

Fajer, P., and Marsh, D., 1982, Microwave and modulation field inhomogeneities and the effect of cavity Q in saturation transfer ESR spectra. Dependence on sample size, *J. Magn. Reson.* **49:**212–224.

Favre, E., Baroin, A., Bienvenue, A., and Devaux, P. F., 1979, Spin-label studies of lipid-protein interactions in retinal rod outer segment membranes. Fluidity of the boundary layer, *Biochemistry* **18:**1156–1162.

Froncisz, W., and Hyde, J. S., 1982, The loop-gap resonator: A new microwave ESR sample structure, *J. Magn. Reson.* **47:**515–521.

Fung, L. W. M., 1981, Spin-label studies of the lipid and protein components of erythrocyte membranes. A comparison of electron paramagnetic resonance and saturation transfer electron paramagnetic resonance methods, *Biophys. J.* **3:**253–262.

Fung, L. W. M., Soo-Hoo, M. J., and Meena, W. A., 1979, Molecular dynamics of spectrin-actin at low pH, saturation transfer EPR studies, *FEBS Lett.* **105:**379–383.

Gaffney, B. J., 1979, Spin label-thiourea adducts. A Model for saturation transfer EPR studies of slow anisotropic rotation, *J. Phys. Chem.* **83:**3345–3349.

Gaffney, B. J., and Lin, D. C., 1976, Spin-label measurements of membrane-bound enzymes, in: *The Enzymes of Biological Membranes*, Vol. 1 (A. Martonosi, ed.), Plenum Press, New York, pp. 71–90.

Graceffa, P., and Seidel, J. C., 1980, A reaction involving protein sulfhydryl groups, a bound spin-label, and $K_3Fe(CN)_6$ as a probe of SH proximity in myosin, *Biochemistry* **19:**33–39.

Hemminga, M. A., and De Jager, P. A., 1981, Magnitude saturation transfer electron paramagnetic resonance spectroscopy: A new ST-EPR technique insensitive to the null-phase setting, *J. Magn. Reson.* **43:**324–327.

Hidalgo, C., and Thomas, D. D., 1977, Heterogeneity of SH groups in sarcoplasmic reticulum, *Biochem. Biophys. Res. Commun.* **78:**1175–1182.

Hidalgo, C., Ikemoto, N., and Gergely, J., 1976, *J. Biol. Chem.* **251:**4224–4232.

Hidalgo, C., Thomas, D. D., and Ikemoto, N., 1978, Effect of the lipid environment on protein motion and enzymatic activity of sarcoplasmic reticulum calcium ATPase, *J. Biol. Chem.* **253:**6879–6887.

Hyde, J. S., and Dalton, L. R., 1972, Very slowly tumbling spin-labels: Adiabatic rapid passage, *Chem. Phys. Lett.* **16:**568–572.

Hyde, J. S., and Thomas, D. D., 1973, New EPR methods for the study of very slow motion: Application to spin-labeled hemoglobin, *Ann. N.Y. Acad. Sci.* **222:**680–692.

Hyde, J. S., and Thomas, D. D., 1980, Saturation transfer spectroscopy, *Annu. Rev. Phys. Chem.* **31:**293–317.

Johnson, M. E., and Hyde, J. S., 1981, 35-GHZ (Q-band) saturation transfer electron paramagnetic resonance studies of rotational diffusion, *Biochemistry* **20:**2875–2880.

Johnson, M. E., Lee, L., and Fung, L. W. M., 1982a, Models for slow anisotropic rotational diffusion in saturation transfer electron paramagnetic resonance at 9 and 35 GHZ, *Biochemistry* **21**:4459–4467.

Johnson, M. E., Thiyagarajan, P., Bates, B., and Currie, B. L., 1982b, A comparison of resolution-enhancement methods in saturation-transfer EPR. Nitrogen-15 isotopically substituted spin-labels and 35 GHZ high-frequency operation, *Biophys. J.* **37**:553–557.

Kaizu, T., Kirino, Y., and Shimizu, H., 1980, A saturation transfer electron spin resonance study on the break in the Arrhenius plot for the rotational motion of calcium ion-dependent adenosine triphosphatase molecules in purified and lipid-replaced preparations of rabbit skeletal muscle sarcoplasmic reticulum, *J. Biochem.* **88**:1837–1843.

Kirino, Y., Ohkuma, T., and Shimizu, H., 1978, Saturation transfer electron spin resonance study on the rotational diffusion of calcium- and magnesium-dependent adenosine triphosphatase in sarcoplasmic reticulum membranes, *J. Biochem.* **84**:111–115.

Koole, P., Dijkema, C., Casteleijn, G., and Hemminga, M. A., 1981, Spin-label saturation transfer ESR study of very slow anisotropic motion in oriented multibilayers of lecithin and cholesterol in the gel phase, *Chem. Phys. Lett.* **79**:360–365.

Kusumi, A., and Hyde, J. S., 1982, Spin-label saturation transfer electron spin resonance detection of transient association of rhodopsin in reconstituted membranes, *Biochemistry* **21**:5978–5983.

Kusumi, A., Ohnishi, S., Ito, T., and Yoshizawa, T., 1978, Rotational motion of rhodopsin in the visual receptor membrane as studied by saturation transfer spectroscopy, *Biochim. Biophys. Acta* **507**:539–543.

Kusumi, A., Sakaki, T., Yoshizawa, T., and Ohnishi, S., 1980, Protein-lipid interaction in rhodopsin recombinant membranes as studied by protein rotational mobility and lipid alkyl chain flexibility measurements, *J. Biochem.* **88**:1103–1111.

Lemaigre-Dubreil, Y., Henry, Y., and Cassoly, R., 1980, Rotational dynamics of spectrin in solutions and ankyrin bound in human erythrocyte membrane, *FEBS Lett.* **113**:231–234.

Lindahl, K., and Thomas, D. D., 1982, Effect of limited rotational motion on simulated conventional and saturation transfer EPR spectra of nitroxide spin labels, *Biophys. J.* **37**:71a.

Mailer, C., Thomann, H., Robinson, B. H., and Dalton, L. R., 1980, Crossed TM110 bimodal cavity for measurement of dispersion electron paramagnetic resonance and saturation transfer electron paramagnetic resonance signals for biological materials, *Rev. Sci. Instrum.* **51**:714–721.

Marsh, D., 1980, Molecular motion in phospholipid bilayers in the gel phase: Long axis rotation, *Biochemistry* **19**:1631–1637.

Marsh, D., 1981, Electron spin resonance: Spin-labels, in: *Membrane Spectroscopy* (E. Grell, ed.), Springer-Verlag, Berlin, pp. 1–92.

Marsh, D., and Watts, A., 1980, Molecular motion in phospholipid bilayers in the gel phase: Spin-label saturation transfer ESR studies, *Biochem. Biophys. Res. Commun.* **94**:130–137.

Poore, V., Fitzsimons, J. T. R., and Ragan, C. I., 1982, The effects of lipid fluidity on the rotational diffusion of complex I and complex III in reconstituted NADH-cytochrome c oxidoreductase, *Biochim. Biophys. Acta* **693**:113–124.

Pringle, M. J., and Hidalgo, C., 1982, Lipid-protein interactions in SR are not perturbed by ionophore A23187, *Biophys. J.* **37**:633–636.

Quintanilha, A. T., Thomas, D. D., and Swanson, M., 1982, Protein-lipid interactions within purified and reconstituted cytochrome *c* reductase and oxidase, *Biophys. J.* **37**:68–69.

Robinson, B. H., and Dalton, L. R., 1980, Anisotropic rotational diffusion studied by passage saturation transfer electron paramagnetic resonance, *J. Chem. Phys.* **72**:1312–1324.

Robinson, B. H., and Dalton, L. R., 1981, Approximate methods for the fast computation of EPR and ST-EPR spectra. V. Application of the perturbation approach to the problem of anisotropic motion, *Chem. Phys.* **54**:253–259.

Rousselet, A., and Devaux, P. F., 1977, Saturation transfer electron paramagnetic resonance on membrane bound proteins. II. Absence of rotational diffusion of the cholinergic receptor protein in *Torpedo marmorata* membrane fragments, *Biochem. Biophys. Res. Commun.* **78**:448–454.

Rousselet, A., Cartaud, J., and Devaux, P. F., 1981, Effects of temperature, lipid modification, and pH on the mobility of the major proteins of the receptor-rich membranes from *Torpedo marmorata*, *Biochim. Biophys. Acta* **648**:169–185.

Rousselet, A., Cartaud, J., Devaux, P. F., and Changeux, J., 1982, The rotational diffusion of the acetylcholine receptor in *Torpedo marmorata* membrane fragments studied with a spin-labeled α-toxin: Importance of the 43,000 protein(s), *EMBO J.* **1**:439–445.

Sasaki, T., Kanaoka, Y., and Watanabe, T., 1980, A digitized EPR system and its application to saturation transfer electron paramagnetic resonance spectroscopy, *J. Magn. Reson.* **38**:385–390.

Sasaki, T., Tsuji, A., Chang, C. H., and Ohnishi, S., 1982, Rotational mobility of an erythrocyte membrane integral protein band 3 in dimyristoylphosphatidylcholine reconstituted vesicles and effect of binding of cytoskeletal peripheral proteins, *Biochemistry* **21**:2366–2372.

Schwarz, D., Pirrwitz, J., and Ruckpaul, K., 1982a, Rotational diffusion of cytochrome P-450 in the microsomal membrane—evidence for a clusterlike organization from saturation transfer electron paramagnetic resonance spectroscopy, *Arch. Biochem. Biophys.* **216**:322–328.

Schwarz, D., Pirrwitz, J., Rein, H., and Ruckpaul, K., 1982b, Selective removal of spectral components in complex ST-EPR spectra of spin-labeled cytochrome P-450, *J. Magn. Reson.* **47**:375–382.

Squier, T., 1982, Effects of partial delipidation on the Ca-ATPase of sarcoplasmic reticulum, *Biophys. J.* **37**:139a.

Swanson, M. S., Quintanilha, A. T., and Thomas, D. D., 1980, Protein rotational mobility and lipid fluidity of purified and reconstituted cytochrome c oxidase, *J. Biol. Chem.* **255**:7494–7502.

Swift, L. L., Atkinson, J. B., and Perkins, R. C., Jr., 1980, Electron paramagnetic resonance and saturation transfer electron paramagnetic resonance studies on erythrocytes from goats with and without heritable myotonia, *J. Membr. Biol.* **52**:165–172.

Thomas, D. D., 1978, Large-scale rotational motions of proteins detected by electron paramagnetic resonance and fluorescence, *Biophys. J.* **24**:439–462.

Thomas, D. D., 1982, Saturation transfer EPR studies of rotational dynamics in membranes, in: *Membranes and Transport,* Vol. 1 (A. Martonosi, ed.), Plenum Press, New York, pp. 135–139.

Thomas, D. D., and Hidalgo, C., 1978, Rotational motion of the sarcoplasmic reticulum Ca^{2+}-*ATPase*, *Proc. Natl. Acad. Sci. USA* **75**:5488–5492.

Thomas, D. D., and McConnell, H. M., 1974, Calculation of paramagnetic resonance spectra sensitive to very slow rotational motion, *Chem. Phys. Lett.* **25**:470–475.

Thomas, D. D., Dalton, L. R., and Hyde, J. S., 1976, Rotational diffusion studied by passage saturation transfer electron paramagnetic resonance, *J. Chem. Phys.* **65**:3006–3024.

Thomas, D. D., Bigelow, D. J., Squier, T. C., and Hidalgo, C., 1982, Rotational dynamics of protein and boundary lipid in sarcoplasmic reticulum membrane, *Biophys. J.* **37**:217–225.

Vignais, P. M., and Devaux, P. F., 1976, The use of spin-labels to study membrane-bound enzymes, receptors, and transport systems, in: *The Enzymes of Biological Membranes,* Vol. 1 (A. Martonosi, ed.), Plenum Press, New York, pp. 91–115.

Warren, G. B., Toon, P. A., Birdsall, N. J. M., Lee, A. G., and Metcalfe, J. C., 1974, Reversible lipid titrations of the activity of pure ATPase-lipid complexes, *Biochemistry* **13**:5501–5507.

Watanabe, T., Sasaki, T., and Sawatari, K., 1980, A new detection system of saturation transfer electron paramagnetic resonance spectroscopy by a Fourier transformation technique, *Appl. Spectrosc.* **34**:456–460.

Watanbe, T., Sasaki, T., and Fujiwara, S., 1982, Phase dependence of saturation transfer *EPR* signals and estimated rotational correlation times, *Appl. Spectrosc.* **36**:174–178.

Watts, A., and Marsh, D., 1981, Saturation transfer ESR studies of molecular motion in phosphatidylglycerol bilayers in the gel phase, effects of pretransitions and pH titration, *Biochim. Biophys. Acta* **642**:231–241.

Wilkerson, L. S., Perkins, R. C., Jr., Roelofs, R., Swift, L., Dalton, L. R., and Park, J. H., 1978, Erythrocyte membrane abnormalities in Duchenne muscular dystrophy monitored by saturation transfer electron paramagnetic resonance spectrosocpy, *Proc. Natl. Acad. Sci. USA* **75**:838–841.

9

Dye Probes of Cell, Organelle, and Vesicle Membrane Potentials

Alan S. Waggoner

I. INTRODUCTION

The membrane potential is an important property of almost all cells and organelles. Three methods are commonly used to determine membrane potentials. Microelectrodes are used if the cells are large enough. The distribution of radioactive permeant ions between the medium and the cells can yield an estimate of membrane potential. Measurements of the fluorescence or light absorption of certain dye molecules associated with cells, organelles, or vesicles will often provide this information. This brief review covers what is known about the mechanisms of potential-sensitive dyes, their advantages and disadvantages, and examples of how they have been used in the past three years. Reviews of the earlier literature are available (Cohen and Hoffman, 1982; Freedman and Laris, 1981; Waggoner, 1979).

II. TYPES OF POTENTIAL-SENSITIVE DYES

Examples of potential-sensitive dyes are shown in Figure 1. They may be divided into two classes, "slow dyes" and "fast dyes," based on the speed of their response. Slow dyes respond in times of seconds. The cyanine dyes, diSC$_3$(5), diOC$_5$(3), and diIC$_3$(5); the oxonal dyes, diBAC$_4$(5) and OXVI; and the tricyclic chromophores, rhodamine 123, ethidium, and safranin O are examples of slow dyes. The fast dyes respond in times of microseconds to milliseconds. Examples are merocyanine 540, WW375, WW781, RH160, and diASP5.

Alan S. Waggoner • Center for Fluorescence Research in Biomedical Sciences and Department of Biological Sciences, Carnegie-Mellon University, Pittsburgh, Pennsylvania 15213.

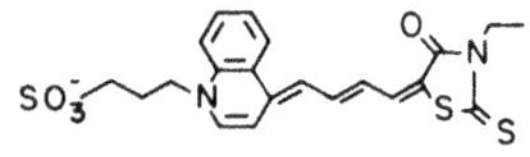

WW 375

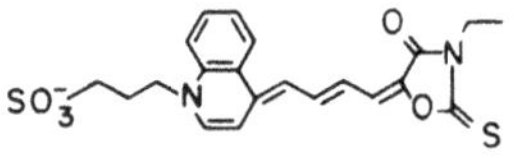

NK 2367

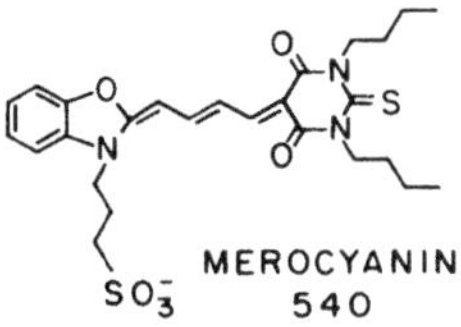

MEROCYANINE 540

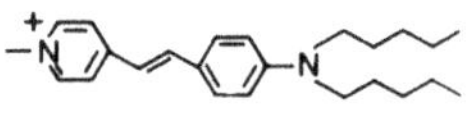

di-C_5-ASP

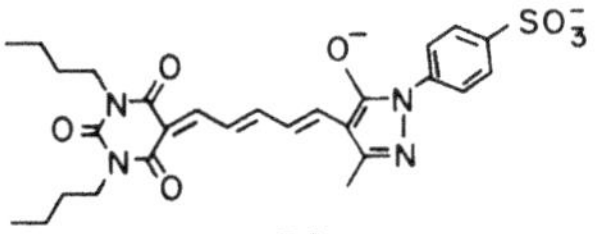

WW 781

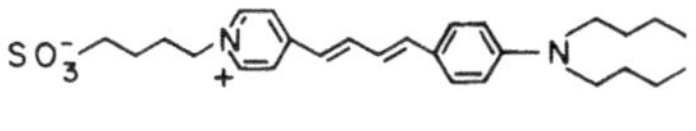

RH 160

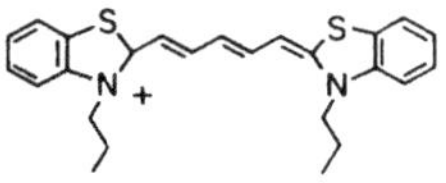

di S-C_3-(5)

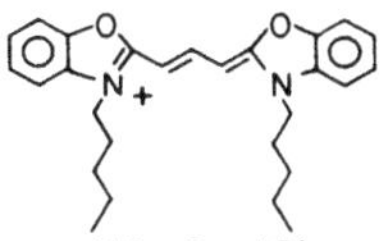

di O-C_5-(3)

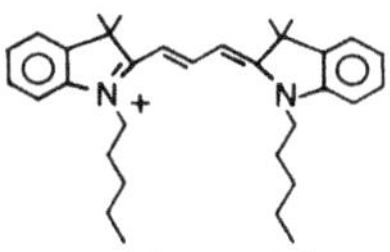

di I-C_5-(3)

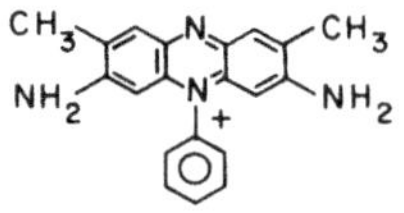

SAFRANIN O

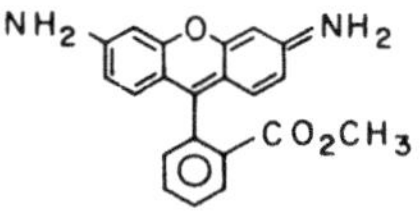

RHODAMINE 123

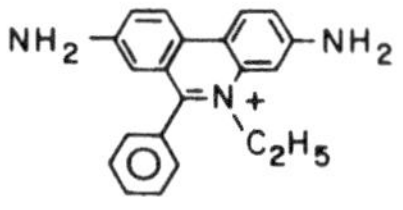

ETHIDIUM

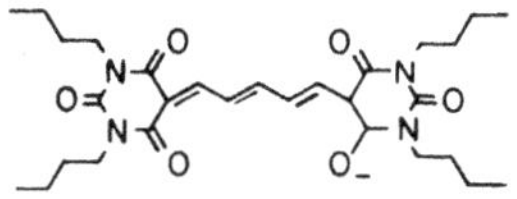

di BA-C_4-(5)

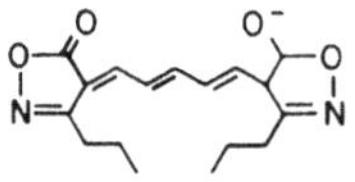

OX-VI

III. SLOW DYES

A. Mechanism of Slow Dyes

The slow dyes are membrane permeant anions and cations. Therefore, their distribution between the inside and outside of a cell, organelle, or vesicle is membrane potential dependent. The positively charged cyanine dyes will be accumulated when cells reach a more negative internal potential (become hyperpolarized) and are driven out of cells that are depolarized. The negatively charged permeant oxonol dyes move in the opposite direction from the cyanine dyes. The fluorescence and absorption changes that occur upon dye redistribution are not straightforward, however, and depend on the concentration of dye used in the experiment. Since this has been a source of confusion in the literature, it is worth considering how the amount of cyanine or oxonal dye accumulated by a cell affects the absorption and fluorescence properties of both the individual cells and suspensions of cells.

First consider experiments carried out at very low dye concentrations, say 10^{-8}–10^{-7} M. I define this situation as a "nonquenching" condition. Red shifts in the absorption spectra of cyanine or oxonal dyes that have been accumulated by cells after a potential change indicate that most of the cell-associated dye is bound to hydrophobic sites. Therefore, increases in dye fluorescence, excited state lifetime, and polarization that are observed when dye is accumulated, under nonquenching conditions, are undoubtedly attributable to association with hydrophobic binding sites (Smith *et al.*, 1980; Smith *et al.*, 1981; Guillet and Kimmich, 1981; Sims *et al.*, 1974).

At much higher dye concentrations, typically micromolar, the fluorescence of accumulated cyanine, safranin O, and oxonol dyes becomes quenched. Quenching occurs when nonfluorescent aggregates are formed or when collisional quenching or energy transfer between dye molecules shortens the excited state lifetime (Sims *et al.*, 1974; Guillet and Kimmich, 1981; Freedman and Hoffman, 1979). The dye concentration that must be reached with a cell to produce quenching depends on the dye structure. $DiSC_3(5)$ forms nonfluorescent aggregates readily even at relatively low concentrations, but $diOC_5(3)$, $diIC_2(5)$, $diBAC_4(3)$, safranin O, and oxonol VI must become considerably more concentrated for quenching to occur. I define this situation to be a "quenching" condition.

From the above information, one can generally relate the fluorescence of a cell suspension (when appropriate excitation and emission wavelengths are used) to the direction of the cell membrane potential change. The relationship can be reasoned as follows.

When cells are hyperpolarized in the presence of cyanine dye (or depolarized in the presence of oxonol dye), the dye molecules move from the medium to the inside of the cells. If low dye concentrations are used (nonquenching conditions), the sus-

←

Figure 1. Representative selection of potential-sensitive probes. Membrane impermeant fast dyes include WW375, NK2367, merocyanine 540 (merocyanines), WW781 (oxonol), and diC_5-ASP, RH160 (styryls). Membrane permeant slow dyes include $diSC_3(5)$, $diOC_5(3)$, $diIC_5(3)$ (cationic cyanines), Safranin O, Rhodamine 123, Ethidium (tricyclic cations), and $diBAC_4(5)$, OX-VI (anionic oxonols).

pension fluorescence will increase as the dye molecules bind to sites where their fluorescence is somewhat greater than their fluorescence in the medium. The suspension fluorescence increase may be small if the quantum yield increase upon binding is small. The fluorescence of the individual cells, which can be observed in a flow cytometer or under a microscope, will increase to a much greater degree, however, because the fluorescence is proportional to the number of dye molecules associated with the cell rather than to the quantum yield change that occurs upon binding.

At high enough dye concentrations (quenching conditions), the suspension fluorescence will diminish upon hyperpolarization because fluorescent dye molecules from the medium become quenched after accumulation within the cells. The fluorescence change of single cells is difficult to predict under quenching conditions so this situation should be avoided.

The opposite responses to a membrane potential change under quenching and nonquenching conditions are illustrated in Figure 2. Neutrophil depolarization produced by the chemotactic factor f-Met-Leu-Phe leads to an increase in suspension fluorescence with diSC$_3$(5), but to a decrease with diOC$_5$(3) (Seligmann and Gallin, 1983). Other examples can be found (Juretić and Kotnik-Karvza, D., 1981; Sims *et al.*, 1974).

For studies involving a fluorometer and cell suspensions, it is usually convenient to use quenching conditions (micromolar) and the probe diSC$_3$(5) because large signals are obtained. For example, fluorescence decreases of 80% are observed for hyper-

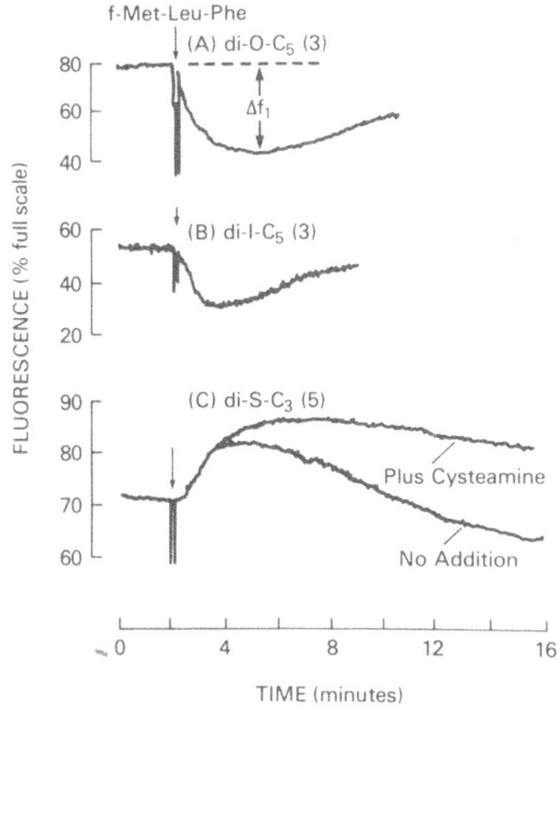

Figure 2. (A) Time course of diOC$_5$(3) fluorescence changes. Points of addition of cells and stimuli [gramicidin D (1 ug/ml) and valinomycin (5×10^{-7}M)] are shown. The spike in the tracing at each point of addition is an artifact. (B) Time course using diIC$_5$(3) as the fluorescent probe. (C) Using diSC$_3$(5) as the fluorescent probe.

polarized red blood cells (Sims *et al.*, 1974). However, because most potential-sensitive dyes are toxic at micromolar concentrations, there are advantages to using $diOC_n(3)$, $diIC_n(3)$, $diIC_n(5)$ or oxonols at nonquenching concentrations (10^{-8}–10^{-7} M). Single-cell fluorescence can be determined by flow or microscopic techniques or suspension fluorescence can be followed with a spectrofluorometer.

B. Examples of the Use of Slow Dyes

Table 1 lists most of the published studies with slow dyes that have appeared since 1980. Two hundred earlier publications can be found in Table 1 of Freedman and Laris (1981). Some recent examples are discussed below because they illustrate advantages and disadvantages of membrane potential probes.

Blood cells are frequently studied with membrane potential probes. Seligmann and Gallin (1983), for example, followed fluorescence responses of $diOC_5(3)$, $diIC_5(3)$, and $diSC_3(5)$ in suspensions of neutrophils that were stimulated with the chemotactic peptide f-Met-Leu-Phe (Figure 2). The fluorescence responses of these dyes as well as measurements of the redistribution of membrane permeant cation TPMP (in the presence of TPB^-) were taken as evidence of a peptide-stimulated depolarization of the neutrophils. As discussed earlier, the depolarization produced an increase in $diSC_3(5)$ fluorescence but decreased the fluorescence of the other two probes (Figure 2). There are sometimes surprises when potential-sensitive dyes are used. For example, in the neutrophil study the $diSC_3(5)$ fluorescence signal was biphasic but the other probes gave monophasic changes. The membrane potential-independent decrease in the fluorescence of $diSC_3(5)$ was due to oxidation of the dye by products of the respiratory burst. Experiments with superoxide dismutase indicate that H_2O_2 is the culprit, as originally suggested by Whitin *et al.* (1981). $DiOC_5(3)$ and $diIC_5(3)$ are not affected by peroxide.

The neutrophil study of Seligmann and Gallin (1983) also demonstrates the usefulness of single-cell fluorescence measurements to detect membrane potential changes. By studying cells with a flow cytometer, it was possible to reduce the $diOC_5(3)$ concentration to 5×10^{-9} M, thereby eliminating toxic effects seen with this and other cyanine and oxonol dyes used at micromolar concentrations (Conover and Schneider, 1981, and references therein). The cyanine dye concentration may not have to always be lowered to this extreme. Campisi and Waggoner (unpublished) found that cultured cells grew four generations with only a 20% increase in cell cycle time at 10^{-7} M $diIC_3(5)$. Growth curves were identical to controls at 10^{-8} M dye. It is not known whether compensations in metabolic pathways occurred to produce the apparently normal growth kinetics.

The second advantage of single cell analysis is the opportunity to detect populations of neutrophils that have different fluorescence responses to f-Met-Leu-Phe stimulation (Seligmann and Gallin, 1983). Apparently, the history or age of the neutrophils affects the magnitude and direction of the membrane potential changes that follow stimulation.

Often, it is desirable to know the exact membrane potential of a cell. Dye fluorescence usually varies in a complex way with membrane potential. Assuming, for

Table 1. Membrane Permeant Slow Dyes

Erythrocytes	WW781	Freedman and Novak (1983)
	$diSC_3(5)$	Pape (1982)
		Rink and Hladky (1982)
Lymphocytes	Rhod 123	Darzynkiewicz *et al.* (1981,1982)
	$diOC_6(3)$	Shapiro *et al.* (1879)
Neutrophils	$diOC_6(3)$	Korchak *et al.* (1982)
	$diOC_5(3)$	Seligmann and Gallin (1980, 1983)
	$diIC_5(3)$	Seligmann *et al.* (1980)
	$diSC_3(5)$	Sklar *et al.* (1981)
	$diSC_3(5)$	Whitin *et al.* (1981)
	$diSC_3(5)$	Tatham *et al.* (1980)
	$diSC_3(5)$	H. J. Cohen *et al.* (1981)
	$diSC_3(5)$	Simchowitz *et al.* (1982)
Mast cells	$diSC_3(5)$	Sugiyama and Utsumi (1979)
Platelets	$diSC_3(5)$	Horne *et al.* (1981)
	$diSC_3(5)$	Whitin *et al.* (1982)
	$diSC_3(5)$	Sepersky and Simons (1981)
	$diSC_3(5)$	Larsen and Simons (1981)
Ascites cells	$diSC_3(5)$	Okimasu *et al.* (1979)
	$diSC_3(5)$	Hacking and Eddy (1981)
Hepatocytes	Safranin	Åkerman and Järvisalo (1980)
Cornea	$diSC_3(5)$	Graves and Sachs (1982)
Cultured cells		
Glioma hybrid	$diSC_3(5)$	Kondo *et al.* (1981)
Neuroblastoma	$diSC_3(5)$	Milligan and Strange (1982)
Cultured cells, adherent		
Fibroblasts	$diOC_7(3)$	R. L. Cohen *et al.* (1981)
Fibroblasts	Rhod 123	Goldstein and Korczack (1981)
Transformed mink	Rhod 123	Johnson *et al.* (1982)
Fibroma	Rhod 123	Johnson *et al.* (1981)
Fibroma	Rhod 123	Johnson *et al.* (1980)
Carcinoma	Rhod 123	Lampidis *et al.* (1982)
	Rhod 123	Bernal *et al.* (1982a,b)
Promyelocytes	Rhod 123	James and Bohman (1981)
Promyelocytes	Rhod 123	Collins and Foster (1983)
Muscle and carcinoma	Rhod 123	Summerhayes *et al.* (1982)
Yeast	Tricyclic cations	Peña *et al.* (1979)
	Ethidium	Peña *et al.* (1980)
	$diSC_3(5)$	Kováč and Varečka (1981)

(Continued)

Table 1. (Continued)

Yeast (*continued*)	$diSC_3(5)$	Kováč and Poliachová (1981)
	$diSC_3(5)$	van den Broek *et al*. (1982)
Plant cells		
Haemanthus	$diOC_5(3)$	Wolniak *et al*. (1983)
Bacteria		
E. coli	OXVI	Armitage and Evans (1981)
S. typhimurium	$diSC_3(5)$	Laszlo and Taylor (1981)
B. subtilis	$diSC_3(5)$	Steensma (1981)
R. sphaeroides	OXVI	Armitage and Evans (1981)
B. subtilis	$diSC_3(5)$	Zaritsky *et al*. (1981)
Mycoplasma	$diSC_3(5)$	Schiefer and Schummer (1982)
E. coli	$diIC_1(5)$	Letellier and Shechter (1979)
E. coli	$dIIC_1(5)$	Labedan and Letellier (1981)
E. coli	$diOC_5(3)$	Bayer and Bayer (1981)
Mitochondria	$diSC_3(5)$	Okimasu *et al*. (1979)
	Safranin	Åkerman and Järvisalo (1981)
	$TriS\text{-}C_4(5)$	Terada *et al*. (1981)
Vesicles from		
Corn root	OX-VI	Bennett and Spanswick (1983)
Retinal rod	$diSC_3(5)$	Bennett *et al*. (1980)
Microsomes	$diOC_5(3)$	Meissner and Allen (1981)
Synaptosomes	$diOC_5(3)$	Baba and Cooper (1979)
Renal	$diSC_3(5)$	Wright *et al*. (1981)
Nerve endings	$diSC_3(5)$	Nordmann *et al*. (1982)
Mitochondria	OX-VI	Kiehl and Hanstein (1981)
Mitochondria	OX-VI	
Mitochondria	$diSC_3(5)$	Smith *et al*. (1981)
Mitochondria	Cyanines	Smith *et al*. (1980)
Mitochondria	Merocyanines	
Mitochondria	Oxonols	
Sarcolemma	$diSC_3(5)$	Bartschat *et al*. (1980)
Sarcoplasmic reticulum	$diIC_1(5)$	Oetliker (1981)
	M540	Haeyaert *et al*. (1980)
	$diOC_5(3)$	Meissner and Young (1980)
	$diSC_3(5)$	Beeler (1980)
	Cyanine	Beeler *et al*. (1981)
	Merocyanine	
	Oxonol	
	Tricyclic	
	$diOC_5(3)$	Meissner (1981)
	$diOC_5(3)$	Young *et al*. (1981)
	$diSC_3(5)$	Yamamoto and Kasai (1980)
Chromaffin cells	OX-V	Scherman and Henry (1980)
	$diSC_3(5)$	Friedman *et al*. (1980)
	$diSC_3(5)$	Lelkes *et al*. (1982)
	$diSC_3(5)$	Salama *et al*. (1980)

(Continued)

Table 1. (Continued)

Mechanisms and properties		
Interference	$diSC_3(5)$ M540	Lavie and Sonenberg (1980)
Mechanisms	$diOC_3(5)$ $diSC_3(5)$ $diOC_5(3)$	Guillet and Kimmich (1981) Juretić and Kotnik-Karuza (1981)
Mitochondria	$diSC_3(5)$ Oxonols Safranin O Cyanine dyes Rhodamine dyes Merocyanine dyes	Howard and Wilson (1979) Smith *et al.* (1981) Conover and Schneider (1981)
	Rhodamine 123 Cyanines Merocyanines	Morita *et al.* (1982) Kinnally and Tedeschi (1978)
Calibration	$diSC_3(5)$ $diSC_3(5)$	Smith (1982) Johnstone *et al.* (1982)
Lipid interaction	Oxonols Oxonols Merocyanines	Bashford *et al.* (1979) Smith *et al.* (1980)
Bilayers	$diSC_n(5)$	Krasne (1980a,b)

the time being, that mitochondria or other intracellular organelles do not contribute to distribution of dye between the cell and the medium, it is possible to calibrate the fluorescence with membrane potential if the cell membrane potential can be set to a known value. This usually requires accurate knowledge of the intracellular concentration of potassium and other ions involved in membrane transport, their permeability coefficients, and the contribution of electrogenic pumps to the membrane potential. These parameters are worked out reasonably well for red blood cells (Freedman and Hoffman, 1979; Guillet and Kimmich, 1981; Cohen and Hoffman, 1982; Freedman and Laris, 1981; Rink and Hladky, 1982). Neutrophil ion transport is less characterized so that Simchowitz *et al.* (1982) had to measure sodium and potassium permeability coefficients in order to calibrate $diSC_3(5)$ fluorescence. After using ouabain to inhibit the Na^+-K^+ pump, they used the K^+ ionophore, valinomycin, to set the membrane potential at values calculated from the Goldman, Hodgkin, Katz equation. $DiSC_3(5)$ fluorescence was measured at the set membrane potentials. The membrane potential of the neutrophils was estimated in this way to be 59 mV in a physiological medium and 50 mV when the electrogenic Na^+-K^+ ATPase was inhibited. The possible interference of mitochondria was neglected but neutrophils have few of this organelle.

Calibration of dye fluorescence signals has been addressed in other places, e.g., Freedman and Laris (1981). Particularly interesting in this regard are the back-to-back pro vs. con articles by Smith (1982) and Johnstone *et al.* (1982).

A potentially significant source of fluorescence from cells studied with permeant cationic dyes is the mitochondria. Fluorescence micrographs of fibroblasts stained with the so-called mitochondria-specific stain rhodamine 123 make this point abundantly clear (Johnson *et al.*, 1980). Cyanine dyes used at low enough concentrations also accumulate in the mitochondria as shown in Figure 3. The amount of cationic dye accumulated by the mitochondria depends on the mitochondrial membrane potential and the plasma membrane potential. The latter is important because it regulates the cytoplasmic dye concentration which in turn affects distribution of dye between the cytoplasm and the matrix. It is possible that other factors like surface charge and mitochondrial structure may to some degree affect mitochondrial dye content.

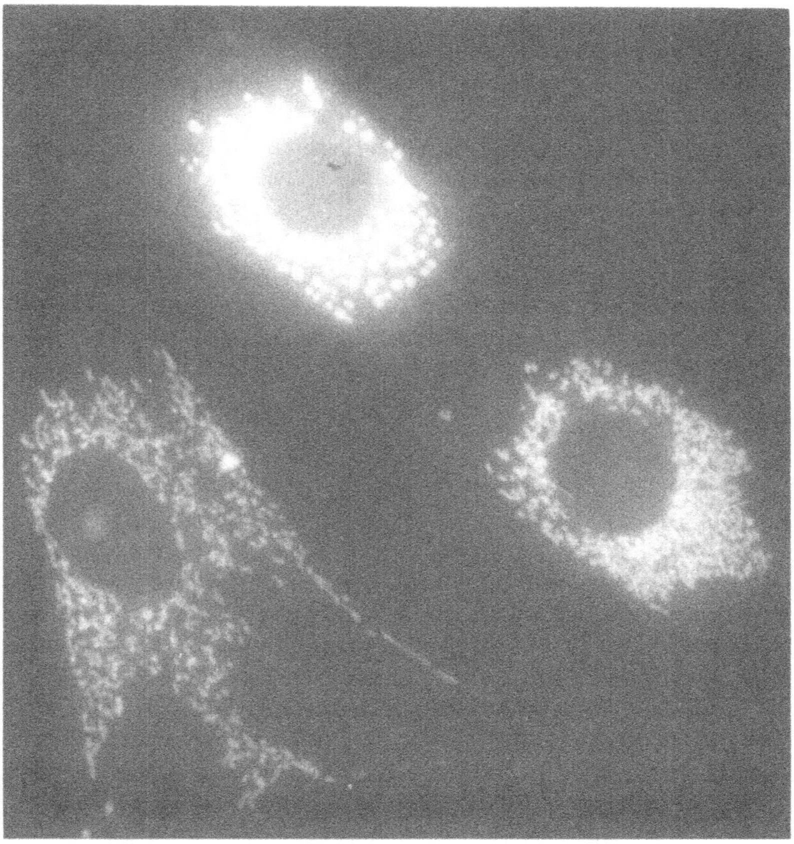

Figure 3. Adherent mouse A-31 cells stained with $diIC_3(3)$. Note wide variation in the fluorescence intensity and shape of the mitochondria. Little cytoplasmic fluorescence can be seen.

If the goal of an experiment is to determine plasma membrane potential changes, one way to minimize fluorescence contributions from mitochondria is to add appropriate inhibitors of energy-driven H^+ transport across the mitochondrial membrane. This approach was taken by Philo and Eddy (1978a,b) in a study of ascites tumor cells with $diOC_3(5)$. Taking the opposite approach, others have assumed that mitochondrial fluorescence in certain cells is large enough so that it can be followed independently of cytoplasmic fluorescence. Åkerman and Järvisalo (1980), following the later view, have shown that safranin fluorescence in suspensions of hepatocytes responds to mitochondrial drugs in a predictable fashion. While these authors suggest that cyanine dyes may not be as sensitive to mitochondrial fluorescence, we find that at low concentrations all the permeant cationic dyes including cyanines produce significant fluorescence from the mitochondria of fibroblasts.

In short, it is clear that with permeant cationic dyes, the cell fluorescence will depend not only upon the plasma membrane potential but also upon the number of mitochondria in the cell and their state of energization. The results of many of the published studies with whole cells using cyanine dyes are probably correct because the cells have very few mitochondria or because the mitochondrial fluorescence remains constant during the experiments. Other work where the mitochondria were not considered may be questionable. Bacteria, of course, present no problem in this respect. The dye concentration is also an important factor to consider because at moderate dye concentrations, fluorescence quenching may occur in the mitochondria but not in the cytoplasm. It is also likely that at high dye concentrations the mitochondria will be inhibited and uncoupled (Morita *et al.*, 1982; Conover and Schneider, 1981; Howard and Wilson, 1979; Kinnally and Tedeschi, 1978). Thus, the relative contribution of mitochondria to total cell fluorescence may be complicated. With quantitative microscopic techniques, it may be possible to measure the distribution of cationic dyes between the medium, cytoplasm, and mitochondria, and therefore detect simultaneously changes in plasma membrane potential and mitochondrial energization.

IV. FAST DYES

A. General Properties

"Fast dyes" respond in times of microseconds to membrane potential changes. Of the 20 or so most sensitive fast dyes, each faithfully reflects the shape and magnitude of the action potential as judged by squid giant axon experiments in which the nerve impulse can simultaneously be followed by electrical means (Ross *et al.*, 1977). Thus, fast dyes are optimal for studying activity of excitable cells.

Several fast dyes are shown in Table 2. Some of these probes produce large fluorescence changes while for others, transmission (or absorption) measurements give better signals. The type of excitable tissue to be studied usually guides probe selection since it is difficult to do transmission measurements through thick samples.

Fast dyes are membrane impermeant. One or more charged functional groups on the dye structures prevent their movement to the cell interior. This property affects

Table 2. Membrane Impermeant Fast Dyes

Nerve		
Salivary gland	NK2367	Senseman *et al.* (1983)
Exocrine gland	NK2367	Senseman and Salzberg (1980)
Olfactory bulb	NK2367	Orbach and Cohen (1983)
Invertebrate ganglia	Merocyaninesand oxonols	Grinvald *et al.* (1981a)
Squid axon	Merocyanines, oxonols and styryls	Gupta *et al.* (1981)
Cultured neurons	Merocyanines and oxonols	Grinvald *et al.* (1981b) Grinvald and Farber (1981)
Cultured neurons	Styryls	Grinvald *et al.* (1982a, 1983)
Hippocampal slice	Merocyanine	Grinvald *et al.* (1982b)
Muscle		
Skeletal muscle	Cyanine	Baylor *et al.* (1981)
Skeletal muscle	Merocyanine	Heiny and Vergara (1982)
T-system	Merocyanine	Nakajima and Gilai (1980a)
Single fibers	Merocyanine	Nakajima and Gilai (1980b)
Heart	Merocyanine	Fujii *et al.* (1980, 1981a)
Heart	Merocyanine	Kamino *et al.* (1981)
Heart	Oxonol	Dillon and Morad (1981)
Heart	Merocyanine	Hill and Courtney (1982)
Mechanisms and properties		
Lipid bilayer	Styryl	Loew and Simpson (1981)
General	Styryl	Loew (1982)

both the maximum signal size and the mechanisms of the fast dyes. For example, only the dye molecules associated with the plasma membrane surface are able to respond to membrane potential changes. Therefore, fast dyes give fractional absorbance and fluorescence changes that are usually much smaller than the slow dyes yield. Fractional changes between 10^{-3} and 10% per 100 mV are easily recorded from fast dyes on excitable cells. Slow dye signals from different cells can be as large as 90% per 100 mV.

Signals from fast probes can be explained by several mechanisms, each of which depends on the substantial change in electric field strength ($>10^5$ V/cm) within the membrane during the action potential. Merocyanine 540, for example, has a dipolar chromophore that reorients in the changing electric field (Ross *et al.*, 1977; Dragsten and Webb, 1978; Waggoner, 1979). One orientation favors formation of dimers which are relatively nonfluorescent. Permeant cyanine and oxonol dyes are driven electrophoretically on or off lipid bilayer membranes during potential changes (Waggoner *et al.*, 1977). The same mechanism may account for the signals of sensitive impermeant oxonols like WW 781. Loew and Simpson (1981) have suggested that electrochromic shifts may at least partially account for the signals of some of the styryl dyes which they designed on the basis of molecular orbital calculations. In all mechanistic studies, great care is necessary to separate all the possible dye charge movements, membrane charge movements, and electrostrictive effects that can occur during large potential changes.

B. Examples of the Use of Fast Dyes

Orbach and Cohen (1983) have demonstrated the feasibility of mapping neuronal organization at the surface of the visual cortex using membrane potential probes. The magnified image of a stained intact salamander olfactory bulb was formed on a 124-element photodiode array. Relatively large absorption and fluorescence signals were found in response to olfactory nerve stimulation (see Figure 4). Tentative assignments of cellular origins of some signals could be made from results of paired volley experiments. Pharmacological effects produced by some dyes were minimized by reducing the dye concentration to an optimal level. Using similar instrumentation, Grinvald *et al.* (1982b) were able to visualize the spread of electrical activity in rat hippocampal slices.

Dillon and Morad (1981) have developed a rapid laser scanning system to map the spread of excitation in amphibian and mammalian hearts stained with WW 781. Isochronic maps of conduction were constructed by timing the upstroke of the optical action potential. At this stage of instrument development, 128 sites could be scanned every 4 msec. The accuracy of the method was verified by recording simultaneously from 16 electrodes placed in different areas of the heart. Studies of the propagation of action potentials in ischemic cat heart suggest that this scanning system is capable of monitoring normal and abnormal electrical activity in heart.

Grinvald *et al.* (1981b) have shown that diode arrays and laser microbeams can be used to study electrical activity from multiple sites of single cultured neurons. A 5-mW HeNe laser served as a light source for simultaneous optical recording of calcium action potentials from growth cones and somata of neuroblastoma cells (Grinvald and Farber, 1981). Senseman *et al.* (1983) have used linear and square diode arrays and a merocyanine dye to monitor the interaction between synaptic and electronic excitation of cells from the salivary gland of the snail *Helisoma trivolvis*. Laterally opposed acini exhibited highly synchronous electrical activity suggesting a correspondingly high degree of electrical coupling.

The radical propagation of muscular action potentials along the tubular system has been examined with potential-sensitive dyes by Nakajima and Gilai (1980a,b). After subtraction of the surface potential component from the total optical signal, the authors were able to estimate a radial conduction velocity for the action potential along the T-system of 6.4 cm/sec at 24.5°C.

Efforts to use optical methods to detect the presence of sarcoplasmic reticulum potential changes in intact muscle fibers have been more frustrating (Oetliker, 1981). As demonstrated by Beeler *et al.* (1981), calcium concentration changes alone can produce absorption and fluorescence changes in SR-membrane suspensions containing potential sensitive dyes. Furthermore, none of the experiments with isolated SR (see Table 1) definitively prove that membrane potentials occur with these membranes.

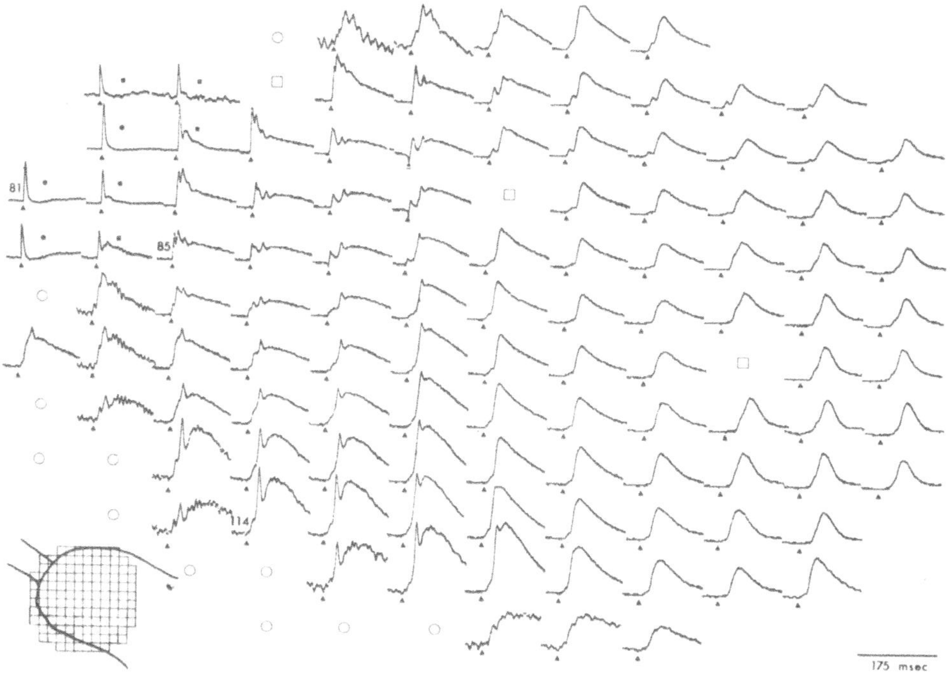

Figure 4. Absorption signals simultaneously measured by 111 diodes from an *in vitro* salamander olfactory bulb. An increase in intensity reaching the photodiodes is drawn upward. In this figure, the traces are arranged so that their relative positions in the figure correspond to the relative positions of the areas of the preparation imaged onto the detectors. The inset on the lower left illustrates the location of the image of the bulb on the array elements. The olfactory nerve, imaged on the upper left portion of the array, was stimulated via suction electrode with a 0.2-msec pulse of 2 mA at the time indicated by the triangles on each trace. Three sharp peaks and slower signals corresponding to electrical activity can be seen. Three of the amplifiers were not functioning; these array elements are indicatd by the open squares. Eleven of the elements, indicated by open circles, did not have signals; these detectors received light that did not pass through the preparation. All of the traces were recorded at the same amplifier and computer gains except for eight traces in the upper left hand corner. The traces with filled squares were reduced in size by a factor of two; the traces with filled circles were reduced by a factor of four. The resting light intensity reaching detector 87 was about 4 × 10^9 photons/msec. The bulb had been stained in 0.5 mg/ml of dye WW385 for 10 min and then rinsed. The image of the bulb was projected on the detector array with a Wild, 7x, 0.2 numerical aperture (na), objective; the condenser was set for Kohler illumination. Each detector received light from a 200-μm^2 area of the bulb. Filter 1 had a peak transmission at 750 nm (Orbach and Cohen, 1983).

Perhaps if an electrochromic potential probe is found that is insensitive to surface potential changes, it will be possible to unambiguously determine whether the SR undergoes transient potential changes during excitation–contraction coupling.

While numerous improvements have been made in development of dyes and instrumentation for monitoring electrical activity of excitable systems (Grinvald *et al.*, 1982a, 1983; Gupta *et al.*, 1981; Loew and Simpson, 1981), much work is needed. Further increases in dye and instrument sensitivity would allow more challenging and important experiments to be carried out more simply. Larger diode arrays and faster computers would permit mapping of activity in larger areas of brain, but would not help solve the problem of the limited depth that can be observed in thick neural tissues. Although photodynamic and pharmacologic effects produced by the dyes are less troublesome than they were several years ago, the problems have not been eliminated. Probes that are more photostable are needed.

REFERENCES

Åkerman, K. E. O., and Järvisalo, J. O., 1980, Effects of ionophores and metabolic inhibitors on the mitochondrial membrane potential within isolated hepatocytes as measured with the safranine method, *Biochem. J.* **192:**183–190.

Armitage, J. P., and Evans, M. C. W., 1981, Comparison of the carotenoid bandshift and oxanol dyes to measure membrane potential changes during chemotactic stimulation of *Rhodopseudomonas sphaeroides* and *Escherichia coli, FEBS Lett.* **126**(1):98–102.

Baba, A., and Cooper, J. R., 1979, The action of black widow spider venom on cholinergic mechanisms in synaptosomes, *J. Neurochem.* **34**(6):1369–1379.

Bartschat, D. K., Cyr, D. L., and Lindenmayer, G. E., 1980, Depolarization-induced calcium uptake vesicles in a highly enriched sarcolemma preparation from canine ventricle, *Biol. Abst.* **71**(7):4564.

Bashford, C. L., Chance, B., Smith, J. C., and Yoshida, T., 1979, The behavior of oxonol dyes in phospholipid dispersions, *Biophys. J.* **25:**63–85.

Bayer, M. E., and Bayer, M. H., 1981, Fast responses of bacterial membranes to virus adsorption; a fluorescence study, *Proc. Natl. Acad. Sci. USA* **78:**5618–5622.

Baylor, S. M., Chandler, W. K., and Marshall, M. W., 1981, Studies in skeletal muscle using optical probes of membrane potential, in: *The Regulation of Muscle Contraction: Excitation–Contraction Coupling* (A. Grinnel, ed.), Academic Press, New York, pp. 97–127.

Beeler, T. J., 1980, Ca^{2+} uptake and membrane potential in sarcoplasmic reticulum vesicles, *J. Biol. Chem.* **255**(19):9156–9161.

Beeler, T. J., Barmen, R. H., and Martonosi, A. N., 1981, The mechanism of voltage-sensitive dye responses on sarcoplasmic reticulum, *J. Membr. Biol.* **62:**113–137.

Bennett, A. B., and Spanswick, R. M., 1983, Optical measurements of ΔpH and $\Delta\psi$ in corn root membrane vesicles: Kinetic analysis of Cl^- effects on a proton-translocating ATPase, *J. Membr. Biol.* **71:**95–107.

Bennett, N., Michel-Villaz, M., and Dupont, Y., 1980, Cyanine dye measurement of light-induced transient membrane potential associated with metarhodopsin II intermediate in rod-outer-segment membranes, *Eur. J. Biochem.* **111:**105–110.

Bernal, S. D., Shapiro, H. M., and Chen, I. B., 1982a, Monitoring the effect of anti-cancer drugs on L1210 cells by a mitochondrial probe, rhodamine-123, *Int. J. Cancer* **30:**219–224.

Bernal, S. D., Lampidis, J., Summerhayes, C., and Chen, I. B., 1982b, Rhodamine-123 selectively reduces clonal growth of carcinoma cells *in vitro, Science* **218:**1117–1118.

Cohen, H. J., Newburger, P. E., Chovaniec, M. E., Whitin, J. C., and Simons, E. R., 1981, Opsonized zymosan-stimulated granulocytes—activation and activity of the superoxide-generating system and membrane potential changes, *Blood* **58**(5):975.

Cohen, L. B., and Hoffman, J. F., 1982, Optical measurements of membrane potential, in: *Techniques in Cellular Physiology,* Vol. 118, Amsterdam Elsevier/North Holland, pp. 1–13.

Cohen, R. L., Muirhead, K. A., Gill, J. E., Waggoner, A. S., and Horan, P. K., 1981, A cyanine dye distinguishes between cycling and non-cycling fibroblasts, *Nature* **290**:593–595.

Collins, J. M., and Foster, K. A., 1983, Differentiation of promyelocytic (HL-60) cells into mature granulocytes: Mitochondrial-specific rhodamine 123 fluorescence, *J. Cell Biol.* **96**:94–99.

Conover, T. E., and Schneider, R. F., 1981, Interaction of certain cationic dyes with the respiratory chain of rat liver mitochondria, *J. Biol. Chem.* **256**(1):402–408.

Darzynkiewicz, Z., Staiano-Coico, L., and Melamed, M. R., 1981, Increased mitochondrial uptake of rhodamine 123 during lymphocyte stimulation, *Proc. Natl. Acad. Sci. USA* **78**(4):2383–2387.

Darzynkiewicz, Z., Tranganos, F., Staiano-Coico, L., Kapuscinski, J., and Melamed, M. R., 1982, Interactions of rhodamine 123 with living cells studied by flow cytometry, *Cancer Res.* **42**:799–806.

Dillon, S., and Morad, M., 1981, A new laser scanning system for measuring action potential propagation in the heart, *Science* **214**:453–456.

Dragsten, P. R., and Webb, W. W., 1978, Mechanism of the membrane potential sensitivity of the fluorescent membrane probe merocyanine 540, *Biochemistry* **17**:5228–5240.

Freedman, J. C., and Hoffman, J. F., 1979, The relation between dicarbocyanine dye fluorescence and the membrane potential of human red blood cells set at varying Donnan equilibria, *J. Gen. Physiol.* **74**:187–212.

Freedman, J. C., and Laris, P. C., 1981, Electrophysiology of cells and organelles: Studies with optical potentiometric indicators, *Int. Rev. Cytol. Suppl.* **12**:177–246.

Freedman, J. C., and Novak, T. S., 1983, Membrane potentials associated with Ca-induced K conductance in human red blood cells: Studies with a fluorescent oxonol dye, WW781, *J. Membr. Biol.* **72**:59–74.

Friedman, J. E., Lelkes, P. I., Rosenheck, K., and Oplatka, A., 1980, The possible implication of membrane-associated actin in stimulus-secretion coupling in adrenal chromaffin cells, *Biochem. Biophys. Res. Commun.* **96**(4):1717–1723.

Fujii, S., Hirota, A., and Kamino, K., 1980, Optical signals from early embryonic chick heart stained with potential sensitive dyes: Evidence for electrical activity, *J. Physiol.* **304**:503–518.

Fujii, S., Hirota, A., and Kamino, K., 1981a, Optical recording of development of electrical activity in embryonic chick heart during early phases of cardiogenesis, *J. Physiol.* **311**:147–160.

Fujii, S., Hirota, A., and Kamino, K., 1981b, Optical indications of pace-maker potential and rhythm generation in early embryonic chick heart, *J. Physiol.* **312**:253–263.

Goldstein, S., and Korczack, L. B., 1981, Status of mitochondria in living human fibroblasts during growth and senescence *in vitro:* Use of the laser dye rhodamine 123, *J. Cell Biol.* **91**:392–398.

Graves, C., and Sachs, G., 1982, Quantitation of corneal endothelial potentials using a carbocyanine dye, *Biochim. Biophys. Acta* **685**:27–31.

Grinvald, A., and Farber, I. C., 1981, Optical recording of calcium action potentials from growth cones of cultured neurons with a laser microbeam, *Science* **212**:1164–1167.

Grinvald, A., Cohen, L. B., Lesher, S., and Boyle, M. B., 1981a, Simultaneous optical monitoring of activity of many neurons in invertebrate ganglia using a 124-element photodiode array, *J. Neurophysiol.* **45**:829–840.

Grinvald, A., Ross, W. N., and Farber, I., 1981b, Simultaneous optical measurements of electrical activity from multiple sites on processes of cultured neurons, *Proc. Natl. Acad. Sci. USA* **78**(5):3245–3249.

Grinvald, A., Hildesheim, R., Farber, I. C., and Anglister, L., 1982a, Improved fluorescent probes for the measurement of rapid changes in membrane potential, *Biophys. J.*, **39**:301–308.

Grinvald, A., Manker, A., and Segal, M., 1982b, Visualization of the spread of electrical activity in rat hippocampal slices by voltage-sensitive optical probes, *J. Physiol.* **333**:269–291.

Grinvald, A., Fine, A., Farber, I. C., and Hildesheim, R., 1983, Fluorescence monitoring of electrical responses from small neurons and their processes, *Biophys. J.* **42**:195–198.

Guillet, E. G., and Kimmich, G. A., 1981, DiO-C_3-(5) and DiS-C_3-(5): Interactions with RBC, ghosts and phospholipid vesicles, *J. Membr. Biol.* **59**:1–11.

Gupta, R. K., Salzberg, B. M., Grinvald, A., Cohen, L. B., Kamino, K., Lesher, S., Boyle, M. B., Waggoner, A. S., and Wang, C. H., 1981, Improvements in optical methods for measuring rapid changes in membrane potential, *J. Membr. Biol.* **58**:1–15.

Hacking, C., and Eddy, A. A., 1981, The accumulation of amino acids by mouse ascites-tumour cells, *Biochem. J.* **194:**415–426.

Haeyaert, P., Verdonck, F., and Wuytack, F., 1980, Fluorescence changes of the potential-sensitive merocyanine 540 during Ca transport in sarcoplasmic reticulum, *Arch. Int. Pharmacodyn. Ther.* **244**(2):333.

Heiny, J. A., and Vergara, J., 1982, Optical signals from surface and T system membranes in skeletal muscle fibers, *J. Gen. Physiol.* **80:**203–230.

Hill, B. C., and Courtney, K. R., 1982, Voltage-sensitive dyes, *Biophys. J.* **40:**255–257.

Horne, W. C., and Simons, E. R., 1978, Probes of transmembrane potentials in platelets: Changes in cyanine dye fluorescence in response to aggregation stimuli, *Blood* **51**(4):741–749.

Horne, W. C., Norman, N. E., Schwartz, D. B., and Simons, E. R., 1981, Changes in cytoplasmic pH and in membrane potential in thrombin-stimulated human platelets, *Eur. J. Biochem.* **120:**295–302.

Howard, P. H., Jr., and Wilson, S. B., 1979, Effects of the cyanine dye 3,3′-dipropylthiocarbocyanine on mitochondrial energy conservation, *Biochem. J.* **180:**669–672.

James, T. W., and Bohman, R., 1981, Proliferation of mitochondria during the cell cycle of the human cell line (HL-60), *J. Cell Biol.* **89:**256–260.

Johnson, L. V., Walsh, M. L., and Chen, L. B., 1980, Localization of mitochondria in living cells with rhodamine 123, *Proc. Natl. Acad. Sci. USA* **77**(2):990–994.

Johnson, L. V., Walsh, M. L., Bockus, B. J., and Chen, L. B., 1981, Monitoring of relative mitochondrial membrane potential in living cells by fluorescence microscopy, *J. Cell Biol.* **88:**526–535.

Johnson, L. V., Summerhayes, I. C., and Chen, L. B., 1982, Decreased uptake and retention of rhodamine 123 by mitochondria in feline sercoma virus-transformed mink cells, *Cell* **28:**7–14.

Johnstone, R. M., Laris, P. C., and Eddy, A. A., 1982, The use of fluorescent dyes to measure membrane potentials: A critique, *J. Cellular Physiol.* **112:**298–301.

Juretić, D., and Kotnik-Karuza, D., 1981, The fluorescent probe for the membrane potential of red blood cells, *Period. Biol.* **83**(1):156–158.

Kamino, K., Hirota, A., and Fujii, S., 1981, Localization of pace-making activity in early embryonic heart monitored using voltage-sensitive dye, *Nature* **290:**595–597.

Kiehl, R., and Hanstein, W. G., 1981, Oxonol response and energy transduction in an ATP-P_i exchange complex, in: *Vectorial Reactions in Electron and Ion Transport in Mitochondria and Bacteria* (F. Palmieri, E. Quagliariello, N. Siliprandi, and E. C. Slater, eds.), Elsevier/North-Holland Biomedical Press, Amsterdam, pp. 217–222.

Kinnally, K. W., and Tedeschi, H., 1978, Metabolic effects of some electrofluorimetric dyes, *Biochim. Biophys. Acta* **503:**380–388.

Kondo, K., Shimizu, T., and Hayaishi, O., 1981, Effects of prostaglandin D_2 on membrane potential in neuroblastoma X glioma hybrid cells as determined with a cyanine dye, *Biochem. Biophys. Res. Commun.* **98**(3):648–655.

Korchak, H. M., Rich, A. M., Wilkenfeld, C., Rutherford, L. E., and Weissmann, G., 1982, A carbocyanine dye, $diOC_6(3)$, acts as a mitochondrial probe in human neutrophils, *Biochem. Biophys. Res. Commun.* **108**(4):1495–1501.

Kováč, L., and Poliachová, V., 1981, Membrane potential monitoring cyanine dyes uncouple respiration and induce respiration-deficient mutants in intact yeast cells, *Biochem. Int.* **2**(5):503–507.

Kováč, L., and Varečka, L., 1981, Membrane potentials in respiring and respiration-deficient yeasts monitored by a fluorescent dye, *Biochim. Biophys. Acta* **637:**209–216.

Krasne, S., 1980a, Interactions of voltage-sensing dyes with membranes. I. Steady-state permeability behaviors induced by cyanine dyes, *Biophys. J.* **30:**415–440.

Krasne, S., 1980b, Interactions of voltage-sensing dyes with membranes. II. Spectrophotometric and electric correlates of cyanine-dye adsorption to membranes, *Biophys. J.* **30:**441–462.

Labedan, B., and Letellier, L., 1981, Membrane potential changes during the first steps of coliphage infection, *Proc. Natl. Acad. Sci. USA* **78:**215–219.

Lampidis, T. J., Bernal, S. D., Summerhayes, I. C., and Chen, L. B., 1982, Rhodamine-123 is selectively toxic and preferentially retained in carcinoma cells *in vitro*, *Ann. N.Y. Acad. Sci.* **397:**299–302.

Larsen, N. E., and Simons, E. R., 1981, Preparation and application of a photoreactive thrombin analogue: Binding to human platelets, *Biochemistry* **20:**4141–4147.

Laszlo, D. J., and Taylor, B. L., 1981, Aerotaxis in *Salmonella typhimurium:* Role of electron transport, *J. Bacteriol.* **145**(2):990–1001.

Lavie, E., and Sonenberg, M., 1980, Spectroscopic evidence for interactions of merocyanine 540 with valinomycin in the presence of potassium, *FEBS Lett.* **111**(2):281–284.

Lelkes, P. I., Naquira, D., Friedman, J. E., Rosenheck, K., and Schneider, A. S., 1982, Plasma membrane vesicles from bovine adrenal chromaffin cells: Characterization and fusion with chromaffin granules, *Adv. Biosci.* **36**:143–150.

Letellier, L., and Shechter, E., 1979, Cyanine dye as monitor of membrane potentials in *Escherichia coli* cells and membrane vesicles, *J. Biochem.* **102**:441–447.

Loew, L. M., 1982, Design and characterization of electrochromic membrane probes. *J. Biochem. Biophys. Meth.* **6**:243–260.

Loew, L. M., and Simpson, L. L., 1981, Charge shift probes of membrane potential. A probable electrochromic mechanism for ASP probes on a hemispherical lipid bilayer, *Biophys. J.* **34**:353–365.

Meissner, G., 1981, Calcium transport and monovalent cation and proton fluxes in sarcoplasmic reticulum vesicles, *J. Biol. Chem.* **256**(2):636–643.

Meissner, G., and Allen, R., 1981, Evidence for two types of rat liver microsomes with differing permeability to glucose and other small molecules, *J. Biol. Chem.* **256**(12):6413–6422.

Meissner, G., and Young, R. C., 1980, Proton permeability of sarcoplasmic reticulum vesicles, *J. Biol. Chem.* **255**(14):6814–6819.

Milligan, G., and Strange, P. G., 1982, The use of biochemical methods for estimating membrane potential, *Prog. Brain Res.* **55**:321–329.

Morita, T., Mori, M., Ikeda, F., and Tatibana, M., 1982, Transport of carbamyl phosphate synthetase I and ornithine transcarbamylase into mitochondria, *J. Biol. Chem.* **257**(18):10547–10550.

Nakajima, S., and Gilai, A., 1980a, Action potentials of isolated single muscle fibers recorded by potential-sensitive dyes, *J. Gen. Physiol.* **76**:729–750.

Nakajima, S., and Gilai, A., 1980b, Radial propagation of muscle action potential along the tubular system examined by potential-sensitive dyes, *J. Gen. Physiol.* **76**:751–762.

Nordmann, J. J., Desmazes, J. P., and Georgescauld, D., 1982, The relationship between the membrane potential of neurosecretory nerve endings, as measured by a voltage-sensitive dye, and the release of neurohypophysial hormones, *Neuroscience* **7**(3):731–737.

Oetliker, H., 1981, Does indodicarbocyanine fluorescence reflect membrane potential of the sarcoplasmic reticulum in skeletal muscle?, *Adv. Physiol. Sci.* **5**:345–361.

Okimasu, E., Akiyama, J., Shiraishi, N., and Utsumi, K., 1979, The mechanism of inhibition on the endogenous respiration of Erlich ascites tumor cells by the cyanine dye disS-C_3-(5), *Physiol. Chem. Phys.* **11**:425–433.

Orbach, H. S., and Cohen, L. B., 1984, Optical monitoring of activity from many areas of the *in vitro* and *in vivo* salamander olfactory bulb: A new method for studying functional organization in the vertebrate CNS, *J. Neurosci.*, in press.

Pape, L., 1982, Effect of extracellular Ca^{2+}, K^+, and OH^- on erythrocyte membrane potential as monitored by the fluorescent probe 3-3′-dipropylthiodicarbocyanine, *Biochim. Biophys. Acta* **686**:225–232.

Peña, A., Mora, M. A., and Carrasco, N., 1979, Uptake and effects of several cationic dyes on yeast, *J. Membr. Biol.* **47**:261–284.

Peña, A., Clemente, S. M., Borbolla, M., Carrasco, N., and Uribe, S., 1980, Multiple interactions of ethidium bromide with yeast cells, *Arch. Biochem. Biophys.* **201**(2):420–428.

Philo, R. D., and Eddy, A. A., 1978a, The membrane potential of mouse ascites-tumour cells studied with the fluorescent probe 3,3′-dipropyloxadicarbocyanine, *Biochem. J.* **174**:801–810.

Philo, R. D., and Eddy, A. A., 1978b, Equilibrium and steady-state models of the coupling between the amino acid gradient and the sodium electrochemical gradient in mouse ascites-tumour cells, *Biochem. J.* **174**:811–817.

Rink, T. J., and Hladky, S. B., 1982, Measurement of red cell membrane potential with fluorescent dyes, in: *Red Cell Membranes—A Methodological Approach* (J. C. Ellory and J. D. Young, eds.), Academic Press, New York, pp. 321–334.

Ross, W. N., Salzberg, B. M., Cohen, L. B., Grinvald, A., Davila, H. V., Waggoner, A. S., and Wang, C. H., 1977, Changes in absorption, fluorescence, dichroism, and birefringence in stained giant axons: Optical measurement of membrane potential, *J. Membr. Biol.* **33**:141–183.

Salama, G., Johnson, R. G., and Scarpa, A., 1980, Spectrophotometric measurements of transmembrane potential and pH gradients in chromaffin granules, *J. Gen. Physiol.* **75:**109–140.

Scherman, D., and Henry, J. P., 1980, Oxonol-V as a probe of chromaffin granule membrane potentials, *Biochim. Biophys. Acta* **599:**150–166.

Schiefer, H.-G., and Schummer, U., 1982, The electrochemical potential across mycoplasmal membranes, *Rev. Infect. Dis.* **4:**565.

Seligmann, B., and Gallin, J. I., 1980, Secretagogue modulation of the response of human neutrophils to chemoattractants: Studies with a membrane potential sensitive cyanine dye, *Mol. Immun.* **17:**191–200.

Seligmann, B., and Gallin, J. I., 1983, Comparison of indirect probes of membrane potential utilized in studies of human neutrophils, *J. Cell. Physiol.* **115:**105–115.

Seligmann, B. E., Gallin, E. K., Martin, D. L., Shain, W., and Gallin, J. I., 1980, Interaction of chemotactic factors with human polymorphonuclear leukocytes: Studies using a membrane potential-sensitive cyanine dye, *J. Membr. Biol.* **52:**257–272.

Senseman, D. M., and Salzberg, B. M., 1980, Electrical activity in an exocrine gland: Optical recording with a potentiometric dye, *Science* **208:**1269–1271.

Senseman, D. M., Shimizu, H., Horwitz, I. S., and Salzberg, B. M., 1984, Multiple site optical recording of membrane potential from a salivary gland: Interaction of synaptic and electronic excitation, *J. Gen. Physiol.*, in press.

Sepersky, S. M. G., and Simons, E. R., 1981, Effect of metabolic inhibitors of thrombin-induced membrane potential changes in washed human platelets, *Thromb. Res.* **24:**299–306.

Shapiro, H. M., Natale, P. J., and Kamentsky, L. A., 1979, Estimation of membrane potentials of individual lymphocytes by flow cytometry, *Proc. Natl. Acad. Sci. USA* **76:**(11):5728–5730.

Simchowitz, L., Spilberg, I., and DeWeer, P., 1982, Sodium and potassium fluxes and membrane potential of human neutrophils, *J. Gen. Physiol.* **79:**453–479.

Sims, P. J., Waggoner, A. S., Wang, C., and Hoffman, J. F., 1974, Studies on the mechanism by which cyanine dyes measure membrane potential in red blood cells and phosphatidyl choline vesicles, *Biochemistry* **23:**3315–3330.

Sklar, L. A., Jesaitis, A. J., Painter, R. G., and Cochrane, C. G., 1981, The kinetics of neutrophil activation, *J. Biol. Chem.* **256**(19):9909–9914.

Smith, J. C., Frank, S. J., Bashford, C. L., Chance, B., and Rudkin, B., 1980, Kinetics of the association of potential-sensitive dyes with model and energy-transducing membranes: Implications for fast probe response times, *J. Membr. Biol.* **54:**127–139.

Smith, J. C., Hallidy, L., and Topp, M. R., 1981, The behavior of the fluorescence lifetime and polarization of oxonol potential-sensitive extrinsic probes in solution and in beef heart submitochondrial particles, *J. Membr. Biol.* **60:**173–185.

Smith, T. C., 1982, The use of fluorescent dyes to measure membrane potentials: A response, *J. Cell. Physiol.* **112:**302–305.

Steensma, H. Y., 1981, Effect of defective phages on the cell membrane of *Bacillus subtilis* and partial characterization of a phage protein involved in killing, *J. Gen. Virol.* **56:**275–286.

Sugiyama, K., and Utsumi, K., 1979, Changes in membrane potential on histamine release from mast cells: Measurement with a fluorescent dye, *Cell Struct. Funct.* **4:**257–260.

Summerhayes, I. C., Lampidis, T. J., Bernal, S. D., Nadakavukaren, J. J., Nadakavukaren, K. K., Shepherd, E. L., and Chen, L. B., 1982, Unusual retention of rhodamine 123 by mitochondria in muscle and carcinoma cells, *Proc. Natl. Acad. Sci. USA* **79:**5292–5296.

Tatham, P. E. R., Delves, P. J., Shen, L., and Roitt, I. M., 1980, Chemotactic factor-induced membrane potential changes in rabbit neutrophils monitored by the fluorescent dye 3,3′-dipropylthiadicarbocyanine iodide, *Biochim. Biophys. Acta* **602:**285–298.

Terada, H., Nagamune, H., Osaki, Y., and Yoshikawa, K., 1981, Specific requirement for inorganic phosphate for induction of bilayer membrane conductance by the cationic uncoupler carbocyanine dye, *Biochim. Biophys. Acta* **646:**488–490.

Vacata, V., Kotyk, A., and Sigler, K., 1980, Membrane potential in yeast cells measured by direct and indirect methods, *Biochim. Biophys. Acta* **643:**265–268.

Van den Broek, P. J. A., Christianse, K., and Van Steveninck, J., 1982, The energetics of *d*-fucose transport in *Saccharomyces fragilis*. The influence of the protonmotive force on sugar accumulation, *Biochim. Biophys. Acta* **692:**231–237.

Waggoner, A. S., 1979, Dye indicators of membrane potential, *Annu. Rev. Biophys. Bioeng.* **8:**47–68.
Waggoner, A. S., 1982, Optical probes of the potential difference across membranes, in: *Membranes and Transport,* Vol. 1 (A. N. Martonosi, ed.), Plenum Publishing, New York, pp. 195–201.
Waggoner, A. S., Wang, C. H., and Tolles, R. L., 1977, Mechanism of potential-dependent light absorption changes of lipid bilayers in the presence of cyanine and oxonol dyes, *J. Membr. Biol.,* **33:**109–140.
Whitin, J. C., Clark, R. A., Simons, E. R., and Cohen, H. J., 1981, Effects of the myeloperoxidase system on fluorescent probes of granulocyte membrane potential, *J. Biol. Chem.,* **256:**8904–8906.
Whitin, J. C., Gordon, R. K., Corwin, L. M., and Simons, E. R., 1982, The effect of vitamin E deficiency on some platelet membrane properties, *J. Lipid Res.* **23:**276–282.
Wolniak, S. M., Hepler, P. K., and Jackson, W. T., 1983, Ionic changes in the mitotic apparatus at the metaphase/anaphase transition, *J. Cell Biol.* **96:**598–605.
Wright, S. H., Krasne, S., Kippen, I., and Wright, E. M., 1981, Na^+-dependent transport of tricarboxylic acid cycle intermediates by renal brush border membranes. Effects on fluorescence of a potential-sensitive cyanine dye, *Biochim. Biophys. Acta* **640:**767–778.
Yamamoto, N., and Kasai, M., 1980, Donnan potential in sarcoplasmic reticulum vesicles measured by using a fluorescent cyanine dye, *J. Biochem.* **88:**1425–1435.
Young, R. C., Allen, R., and Meissner, G., 1981, Permeability of reconstituted sarcoplasmic reticulum vesicles; Reconstitution of the K^+, Na^+ channel, *Biochim. Biophys. Acta* **640:**409–418.
Zaritsky, A., Kihara, M., and Macnab, R. M., 1981, Measurement of membrane potential in *Bacillus subtilis:* A comparison of lipophilic cations, rubidium ion, and a cyanine dye as probes, *J. Membr. Biol.* **63:**215–231.

10

Selective Covalent Modification of Membrane Components

Hans Sigrist and Peter Zahler

I. INTRODUCTION

Essential to an understanding of membrane-related biological phenomena at the molecular level is not only identification of the components involved but rather knowledge of their respective structural and functional characteristics. Such knowledge is attained by disintegrating biological membranes into particulate entities whose chemical properties can then be more specifically investigated. Successful reconstitution of isolated components in model systems leads to an improved understanding of transmembrane processes. Membrane architecture and the mechanism of transbilayer events are, however, most conclusively explored within a functionally active membrane. The latter approach therefore requires selective identification of defined membrane components *in situ*. As a consequence, the search for means to locate and specifically characterize defined membrane constituents has intentionally been pursued. This objective is most efficiently achieved by the use of chemical modification. In spite of inherent limitations, the approach offers obvious advantages in that (1) the extent and topology of label binding can be controlled chemically, and (2) label allocation is feasible at both the membrane and molecular level.

Within this review, emphasis has been placed on the discussion of diverging lines of research which effect selectivity in membrane modification. The large volume of literature involved has rendered it impossible to completely cite all individual papers. Selected studies which document original and promising investigations have therefore been chosen to provide a key to the appropriate literature.

Hans Sigrist and Peter Zahler • Institute of Biochemistry, University of Berne, CH-3012 Berne, Switzerland.

A. Aim and Purpose of Selective Covalent Modification

The incentive to selectively modify biological membranes is given by the information gained through (1) characterization of membrane constituents with respect to sided orientation and membrane topography, (2) exploration of lipid–protein interrelationships, (3) introduction of "reporter" groups such as fluorophores, spin labels, haptens, or heavy metal derivatives, and (4) identification of amino acid residues involved in membrane-related catalysis and binding events. In all cases, the universal objective is the attainment of the most decisive information possible, "communicated" by a selectively introduced label. Obviously, definition of the selectivity concept inherently depends upon the system from which knowledge is to be acquired. For example, investigations concerning the sidedness of membrane components require impermeant reagents as well as uniformly oriented membranes. Alternatively, specific labeling of either lipid or protein is mandatory in simultaneously analyzing mutual interrelationships of both membrane components. Finally, enzyme activities or signal transitions may be affected by introduced inhibitors or site-directed labels, respectively. In these studies, valid structural and mechanistic conclusions can only be drawn from a modifying reagent whose location has been accurately determined within the lipid or protein structure. *Selective* chemical modification of membrane components is therefore required. Achievement of reagent binding specificity is determined by the individual membrane system utilized. Reaction selectivity can, to a certain extent, be directed by deliberate selection of the modifying reagent and appropriate reaction conditions.

B. Biological Membranes as Reactants in Chemical Reactions

Membranes either of biological origin or artificially constituted are inherent multiphasic systems. They consist of an apolar domain, two aqueous phases, and membrane–water interphases present on each side of the system (Figure 1). Stability of the membrane is effected by complementary hydrophobic and hydrogen-bonding forces occurring among the membrane-constituting components and the aqueous phase. Electrostatic interactions additionally contribute to the membrane stability.

The *aqueous phase,* an essential part of every membrane system, is relatively homogeneous due to the polarity imposed by the presence of greater than 50 M H_2O. Generally, membrane bilayer systems tolerate fairly well-limited alterations in the ionic composition of the aqueous phase (osmolarity, pH, changes of ionic species). Subtle variations in solvent polarity or effects on the water structure, however, may disorganize hydrophobic binding forces within membrane protein complexes and weaken membrane stability.

The physicochemistry of *membrane–water interfaces,* where surface labeling occurs, has been minimally characterized. Due to the structural complexity of biological membrane systems, chemical properties of the polar–apolar interfacial domain are difficult to predict. Glycolipids and glycoproteins, which are major components of many membrane systems, extend their sugar moieties into these regions. Alternatively,

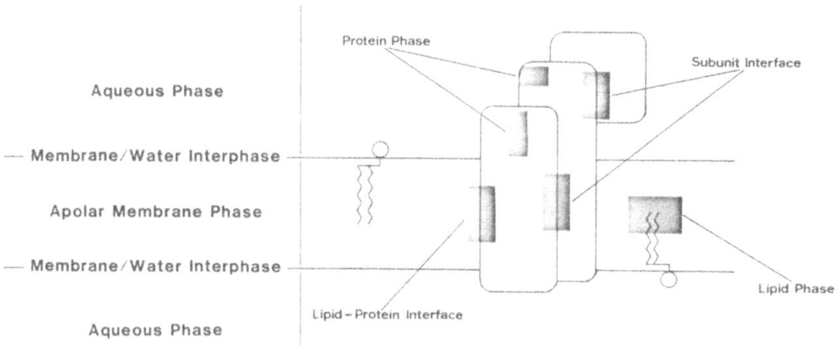

Figure 1. Hydrophobic membrane domains are indicated by hatched areas.

membrane-associated extrinsic proteins mediate between the polar aqueous phase and the hydrophobic membrane domain.

The *apolar membrane phase* is markedly heterogeneous. It is comprised of at least four distinguishable hydrophobic domains: the apolar lipid phase, the lipid–protein interphase, apolar regions within membrane proteins, and the domain in between subunits of membrane protein complexes (Figure 1). In chemical terms, the hydrophobic domain of a biological membrane cannot be considered uniform. The hydrophobicity profile of the lipid phase is discontinuous and may depend upon the extent of H_2O penetration, lipid composition and segregation, and the presence of sterols. Polarity gradients are locally revealed at the lipid–protein interphases. Individually characterized hydrophobic domains may further exist within membrane proteins (hydrophobic pockets) and at subunit interfaces.

Intrinsic parameters of the membrane–reagent system are of assistance in providing the intended binding specificity. Selective chemical modification of membrane components can be achieved by considering the existing membrane characteristics and versatility combining them with feasible chemical reactions.

C. Selectivity-Promoting Factors in Membrane-Labeling Studies

Chemical modification studies have generated extensive knowledge concerning the reaction of membrane components with a respectable number of reagents. Nevertheless, chemical labeling of biomembranes remains a challenge for the investigator, the methodology being in most cases unique to each individual membrane system. The intrinsic characteristics of each system are dictated by the reactive groups available for modification and by the reagent's compatibility. In Table 1, specificity-inducing factors are summarized.

Selective modification is significantly promoted by the *asymmetry* of biological membranes. Asymmetric distribution of membrane proteins and lipids along the axis perpendicular to the membrane is essential to every biological membrane. Whereas

Table 1. Procurement of Selective Membrane Modification

Promoter	Specificity-inducing factor	Examples and applications	References
Membrane	Sided orientation of membrane constituents	Carbohydrate topography	Gahmberg (1981)
		Lipid asymmetry	Op den Kamp (1979)
		Substrate accessibility	
Functional group	Accessibility	Enzyme-catalyzed surface labeling	Steck and Dawson (1974)
	Inherent ligand selectivity	Photolabeled substrate analogs	Boulay *et al*. (1983), Raftery *et al*. (1980)
	Enhancement of specific group modification	Differential labeling	Phillips (1977)
	Environmentally influenced pK of reactive groups	Hydrophobically located nucleophiles	Sigrist *et al*. (1984)
	Nucleophilicity of functional groups	pH-dependent reactivity	Glazer (1976), Drobnica *et al*. (1977)
Labeling reagent	Reagent polarity	Surface labeling/hydrophobic photolabeling	Hubbard and Cohn (1976), Bayley (1982), Gitler and Bercovici (1980)
	Structural compatibility	Inhibitors of erythrocyte anion exchange	Barzilay *et al*. (1979)
	Reaction mechanism	Hydrophobic labeling	Sigrist and Zahler (1982a)

the polypeptide chains of membrane proteins may extend on either side of the lipid bilayer, absolute asymmetry is ascribed to membrane glycoconjugates and in part to membrane lipids (Nicolson and Singer, 1974; Op den Kamp, 1979). Cell membrane carbohydrate is attached to protein and to *N*-acyl sphingosines, thus forming glycoproteins and glycolipids, respectively. The plasma membrane of the human erythrocyte has been especially well studied in this respect. It has been shown that carbohydrates of this membrane are exclusively associated with the external surface of the membrane (Steck and Dawson, 1974; Gahmberg and Andersson, 1977). In a similar manner, also utilizing the erythrocyte plasma membrane, lipid asymmetry has been investigated. Erythrocyte sphingomyelin (80–85%) and 65–70% of phosphatidylcholine are located on the outer surface of the membrane, while the reactive groups of phosphatidylserine (100%) and phosphatidylethanolamine (80%) are accessible almost exclusively from the cytoplasmic (inner) membrane surface (Verkleij *et al.*, 1973). With the topographical distribution known, selective modification of membrane constituents is feasible. Surface markers or haptens can be placed on a defined side of the biological membrane, providing selectivity with respect to membrane sidedness.

Selectivity of the modification reaction is strictly dependent upon a favorable binding energy between the site to be modified and the specific modifier molecule. The chemical properties of the reactive groups and their respective availability in membrane systems may influence the extent of label binding. Label *accessibility* to defined membrane reactants can be induced by removal of shielding entities, e.g., sugar residues, or by elimination of charge repulsion occurring between the membrane surface and the modifying reagent (Gahmberg and Hakomori, 1973).

Selectivity in label binding is likewise promoted in systems where high *affinity* ligands interact with a specific site. The structures of the normal ligands, substrates, and inhibitors of membrane proteins (enzymes) serve as guide to the design of site-specific reagents. The specificity of labeling depends upon the fact that complex formation between the reagent and the protein greatly increases the concentration of the reagent at the binding site as compared to the concentration in solution. Consequently, at low concentrations of reagent, specific reactions at the binding site are observed. For any membrane protein, the design of a covalent affinity label will represent an individual problem in organic chemistry. For example, covalent ligand (substrate) analog binding has been attained by nucleophile modification (Passow *et al.*, 1982) or Schiff-base reduction by borohydride (Bayley *et al.*, 1981). Alternatively, chemically reactive substrate analogs may include epoxides (Quaroni and Semenza, 1976). Through the application of photosensitive ligand analogs, selective covalent modification may be achieved (Chowdhry and Westheimer, 1979). Photoaffinity labels have been utilized for the isolation and characterization of various membrane receptors (Das and Fox, 1979; Raftery *et al.*, 1980; Yip *et al.*, 1982) and carrier proteins (Boulay *et al.*, 1983).

Enhancement of specific group modification is further achieved by *differential labeling*. In this procedure, selectivity is gained through the use of a nonreactive, tightly bound ligand (substrate, coenzyme, competitive inhibitor) which masks its own binding site against modification. Reactions can thereby occur only at other reactive sites. Removal of the ligand, followed by retreatment of the membrane system with the same reactive agent in a radioactive form, thus labels only the site originally protected by the ligand (Phillips, 1977).

Reactivity of a functional group is imposed by the *microenvironment* which may drastically influence its chemical reactivity. Aberrations in the ionization constants (pK_a) of a titratable functional group are effected by factors including local charge distribution (formation of ion pairs), hydrogen bonding, and solvation. Amino acid side chains located within hydrophobic membrane domains are therefore the most obvious candidates which may possess unusually reactive groups (Sigrist *et al.*, 1984).

A further distinctive element in selective labeling studies is the *nucleophilicity* of the reactive functional groups. Within a single class of nucleophiles there is very good correlation between nucleophilic reactivity and basicity (or pK) of the reacting nucleophiles. The pH-dependent reactivity is therewith implied (Glazer, 1976). An extended documentation on general nucleophile reactivity is compiled in a review by Bunnett (1963). In the specific case of nucleophile reactivity with isothiocyanates, it has been shown that in accordance with theoretical predictions, isothiocyanates react with OH^-, HS^-, RS^-, and RO^- ions. Amines react in the nonprotonated form $R\text{-}NH_2$. The reactivity of the investigated nucleophiles increases in the order of $OH^- < R^{-1}NH_2 < RO^- < HS^- \ll RS^-$ (Drobnica *et al.*, 1977).

Functional groups of membrane systems are generally nonvariable parameters in modification studies. The choice of the appropriate reagent is a variable parameter which presents, however, a challenging task for the investigator to creatively select a label from an abundant variety of modifiers differing in polarity, size, structure, and reaction mechanism.

The *polarity of a reagent* determines the residence time for which the label is in contact with distinct phases of a membrane system (Figure 1). Its partition coefficient approximately indicates the topology of preferential label interaction. For both membrane surface labeling (Hubbard and Cohn, 1976) and selective modification of intramembrane domains, it is advisable to thoroughly ascertain the degree of membrane permeability and the reagent's preference for the hydrophobic environment, respectively (Brunner, 1981; Bayley, 1982).

Exploration of a defined site of a membrane component requires sagacious selection of *compatible reagents* which differ in their chemical characteristics. The analysis of structure–activity relationship relies on several system parameters which may be deduced from preceding functional or binding studies. For example, the erythrocyte anion transport system, which is responsible for physiological Cl^-/HCO_3^- exchange, is inhibited by covalent stoichiometric binding of 4,4′-diisothiocyano-2,2′-stilbene disulfonic acid (DIDS). The chemical properties of this site, crucially involved in the transmembrane process, have been investigated by a variety of benzene sulfonic acid and stilbene disulfonic acid analogs (Barzilay *et al.*, 1979). It was concluded from these studies that the microenvironment of the substrate recognition site bears a positive multipolar character and possesses functionally essential groups with electron donor capacity embedded in a hydrophobic area. It is presumed that the optimally positioned sulfonyl groups of DIDS sterically fit with the anion (substrate) binding sites (Knauf, 1979). Thiocarbamylation of a lysine ε-amino group occurs subsequent to the charge neutralization process (Ramjeesingh *et al.*, 1981).

Selective chemical modification may critically depend upon the *reaction mechanism* by which a covalent bond between the reagent and the modified group is formed. Means and Feeney (1971) described the mechanistic implications for several modification reagents, a considerable number of which are routinely applied in membrane studies. Mechanistic characteristics in conjunction with the reagent's polarity may procure chemical differentiation between site-directed hydrophobic labeling and surface modification respectively. To illustrate this point, the phenylthiocarbamylation reaction and trinitrophenylation of amino groups are depicted in Figure 2. Both membrane modification reactions may proceed at neutral pH. The apolar phenylisothiocyanate

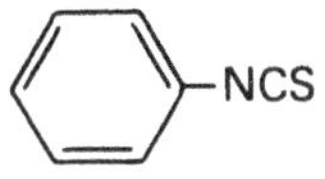

1

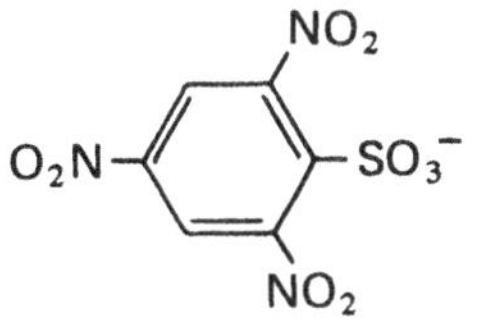

2

cyanate (1) interacts with the nucleophilic R-NH_2 by a nucleophilic addition reaction (Drobnica *et al.*, 1977). Thiocarbamylation can proceed in an apolar environment since the reaction, the reagent, and the reaction products are non-ionogenic and noncharged

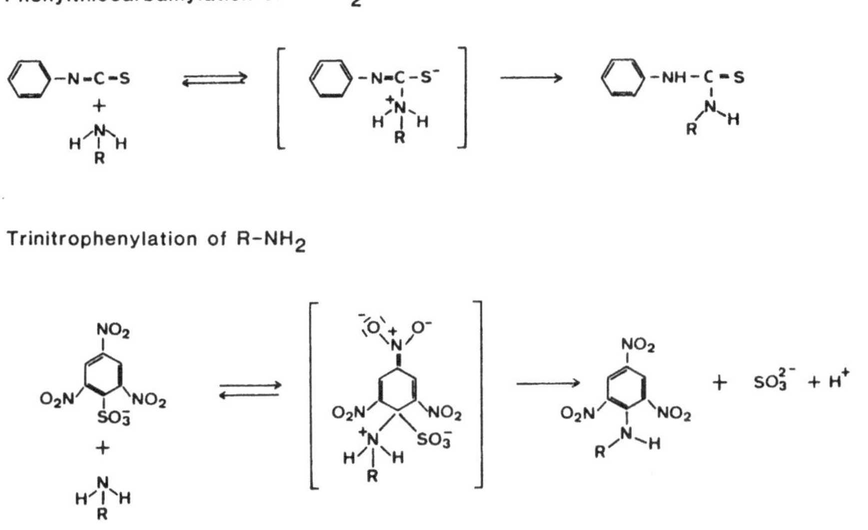

Figure 2. The overall electroneutral phenylthiocarbamylation reaction can proceed in an apolar environment. Trinitrophenylation of an amino group by 2,4,6-trinitrobenzene sulfonic acid is confined to polar (aqueous) domains.

(Sigrist *et al.*, 1984). 2,4,6-Trinitrobenzene sulfonic acid (TNBS) (2) also reacts with nucleophilic groups by a bimolecular mechanism similar to the displacement reaction at aliphatic carbons. As described for the thiocarbamylation reaction, TNBS requires the unprotonated nucleophile (R-NH_2) as reactant (Freedman and Radda, 1968). The presence of the sulfonyl group renders the label exceedingly polar. High polarity equally applies to the reaction intermediate and the reaction products sulfite (SO_3^{2-}) and H^+ or bisulfite (HSO_3^-), respectively. By comparing the two nucleophile modification reactions, it is evident that the former is applicable for modifications occurring in nonpolar domains (Sigrist and Zahler, 1982a), whereas excellent surface labeling is attained with 2,4,6-trinitrobenzene sulfonic acid (Gordesky, 1975).

II. COVALENT MODIFICATION OF LIPID COMPONENTS

The enzymatic degradation of choline-containing phospholipids by phospholipases is a unique procedure which provides selectivity for membrane lipid components. In contrast, nucleophile-modifying chemical reagents do not discriminate between membrane protein and membrane lipid, both of which react according to their chemical properties. Galactose oxidase is equally indifferent although selective for galactolipids and glycoproteins containing terminal galactose or *N*-acetyl galactosamine residues. In either case, however, separation of labeled membrane components is biochemically feasible. Label association with either membrane lipids and/or proteins can be assessed. Selected examples for polar and apolar modification of membrane lipids are summarized in Table 2.

A. Lipid Polar Head Group Modification

Chemical and enzyme-catalyzed modifications of those parts of membrane lipids which are exposed to the aqueous phase have successfully been applied in studies delineating lipid asymmetry in biological membranes (Op den Kamp, 1979; Van Deenen, 1981).

1. Chemical Modifications

In 1972, Bretscher suggested an inherent asymmetry in the lipid composition of the two halves of the erythrocyte lipid bilayer. Phosphatidylethanolamine was the main

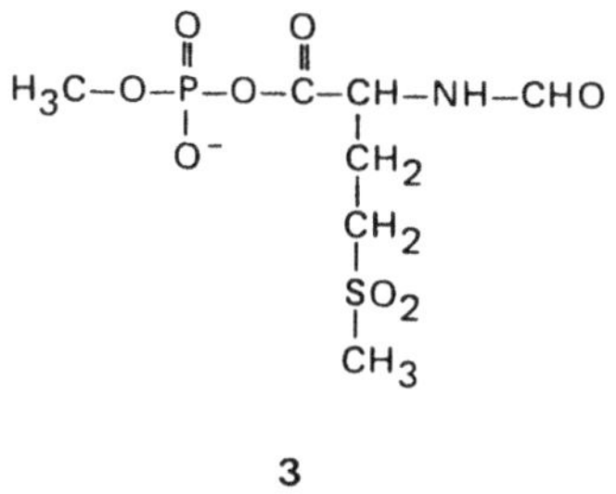

3

lipid component modified by the relatively impermeant reagent formyl-methionyl(sulfone)methyl phosphate (3). To some extent, protein modification did occur. Using 2,4,6-trinitrobenzene sulfonic acid (2) as a surface marker in erythrocyte membranes, partial labeling of phosphatidylethanolamine and to a minor extent phosphatidylserine was attained (Gordesky *et al.*, 1975). Trinitrobenzene sulfonic acid treatment of membranes introduces a bulky electronegative group both in membrane lipids and proteins. It can be envisioned that this may cause spatial restriction at the membrane surface and prevent the reaction from going to completion (Marinetti and Love, 1976; Bishop *et al.*, 1977). Monolayer experiments indicated that the increase in surface pressure of the lipids may be responsible for the incomplete reaction (Bishop *et al.*, 1979). In the same study, it has been found that the presence of negatively charged phospholipids such as cardiolipin or phosphatidylglycerol decreases the extent of phosphatidylethanolamine trinitrophenylation. Charge repulsion between the anionic reagent and the negatively charged membrane surface is most probably responsible for the observed effect. The same argument can be applied for the inefficient modification of phosphatidylserine.

Membrane lipid and lipid–protein association has been investigated with crosslinking probes which differ in hydrophobicity and charge. With the reagents 4,4′-difluoro-3,3′-dinitrophenyl-sulfone (4) and *N*-(3-fluoro-4,6-dinitrophenyl)-cysteamine (5), intermolecular crosslinks among the aminophospholipids have been observed. The

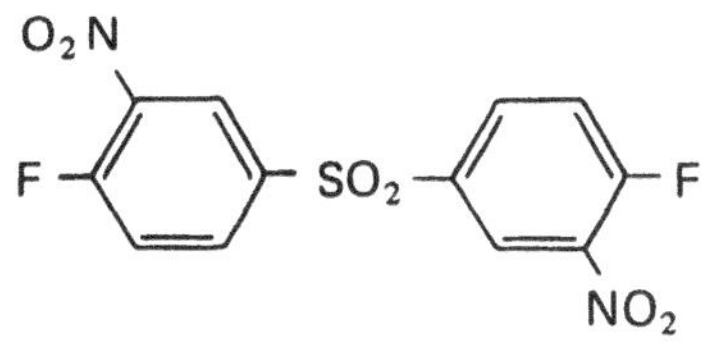

4

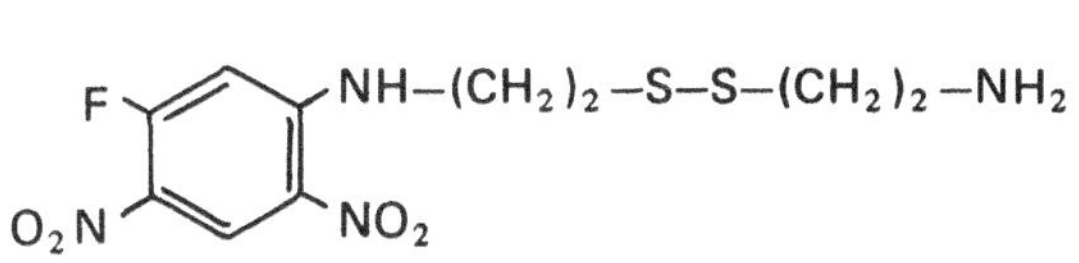

5

former reagent is electrically neutral and thus able to penetrate bilayer membranes. The reaction with amino groups of phospholipids consists of two consecutive nucleophilic substitutions (SN_2-type). The reaction of *N*-(3-fluoro-4,6-dinitrophenyl)-cysteamine with amino phospholipids is likely to be confined to the membrane surface. The monofunctional reagent is positively charged and therefore not expected to penetrate the membrane. Crosslinking occurs at the level of a pair of reduced sulfhydryl groups to a disulfide bond by air oxidation (Marfey and Tsai, 1975). Crosslinking of phosphatidylethanolamine in erythrocyte membranes has further been attained with methylacetimidate (6) and the polar analogue isethionyl acetimidate (7) (Shaw and Marinetti, 1980). In this reaction, the *N*-alkyl imidate is formed upon imidate binding to phosphatidylethanolamine. The *N*-alkyl imidate can then react with another appropriately positioned amino group to give a disubstituted amidine (Hand and Jenks,

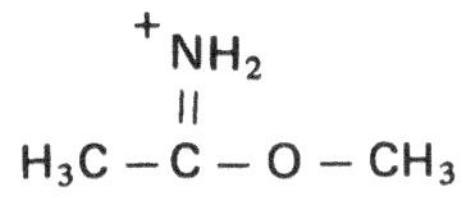

6

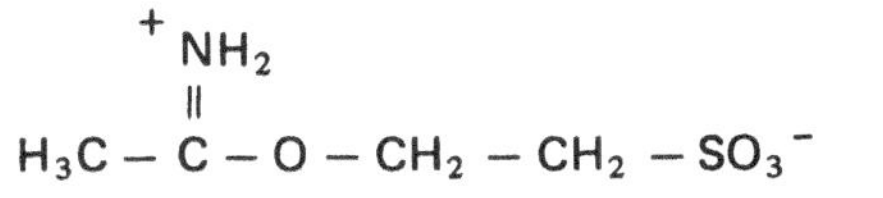

7

Table 2. Selected Examples for the Modification of Membrane Lipids

Reagent	Reactive site in membrane	Structure of modified site(s)	References
A. Polar head group modification			
N-formyl-methionyl-(sulfone)-methylphosphate: FMMP (3)	Aminophospholipids (nucleophilic groups of proteins)	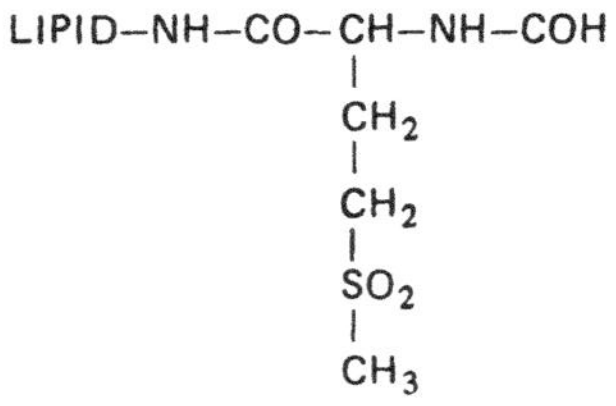	Bretscher (1972)
2,4,6-Trinitrobenzene sulfonic acid: TNBS (2)	Phosphatidylethanolamine > phosphatidylserine (nucleophilic groups of proteins)	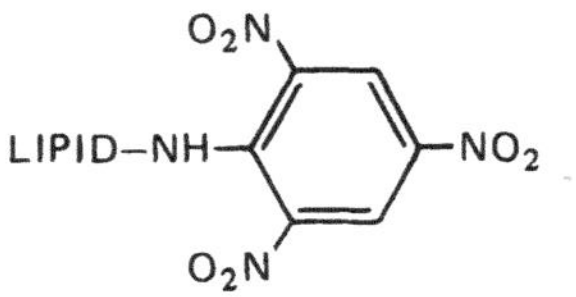	Marinetti and Love (1976); Bishop *et al.* (1977)
4,4′-Difluoro-3,3′- dinitrodiphenyl-sulfone: FFDS (4)	Phosphatidylethanolamine, phosphatidylserine	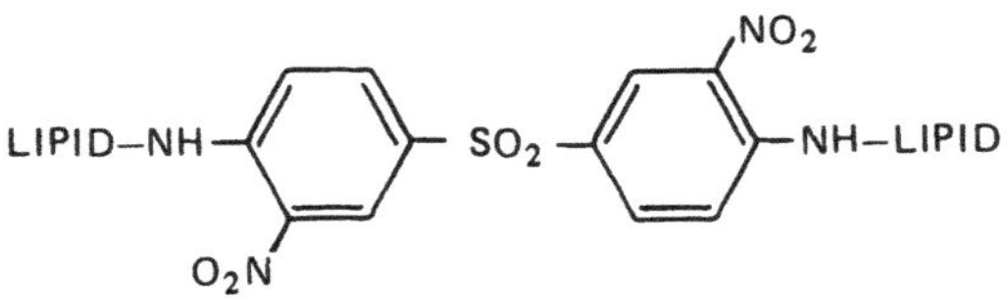	Marfey and Tsai (1975)
Phospholipases A_2, C, D	Glycerophospholipids	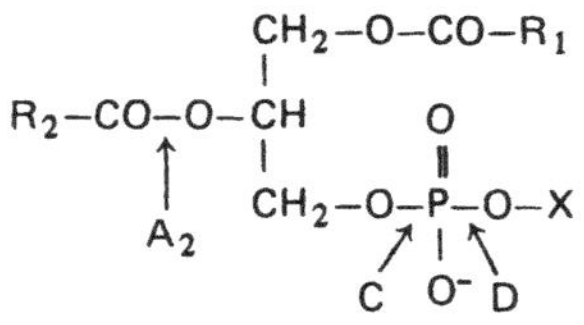	Zwaal and Roelofsen (1976)

Galactose oxidase/NaBH₄	Galactolipids (glycoproteins)	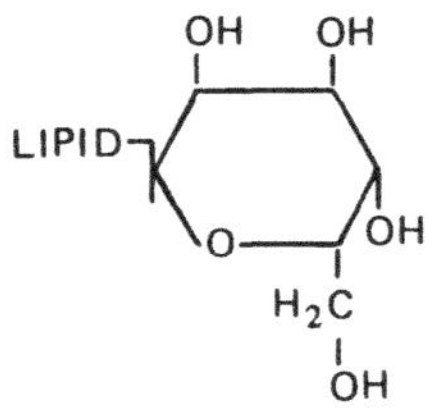	Gahmberg and Hakomori (1973), Steck and Dawson (1974)
B. Lipid labeling within the apolar membrane phase			
Phenylazide (8)	Phosphatidylcholine (membrane lipids, proteins)	Fatty acids ≫ diacylglycerides	Abu-Salah and Findlay (1977), Bayley and Knowles (1978a)
Phenyldiazirine (9)	Dioleylphosphatidylcholine (proteins) C═C, C—H bonds	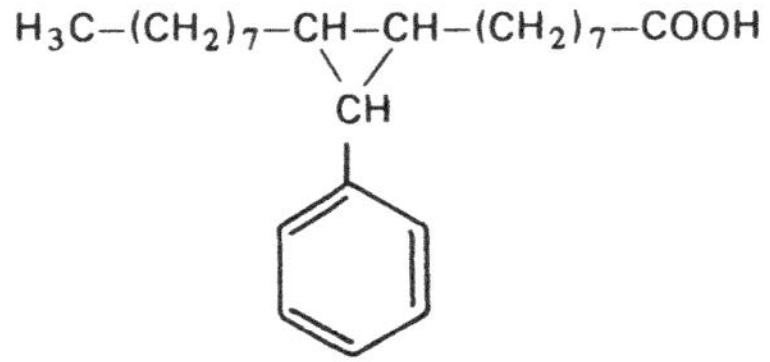	Bayley and Knowles (1978b)
1-Palmitoyl-2-(16-azidopalmitoyl)glycero-3-phosphocholine (12)	1,2-Dipalmitoylglycero-3-phosphocholine C-H bonds towards the ω-end	Palmitic acid	Stoffel *et al.* (1982)
1-Palmitoyl-2-ω-(diazirinophenoxy)-undecanoyl-glycero-3-phosphocholine (13)	Phosphytidylcholine C-H bonds	Palmitic acid	Gupta *et al.* (1979)

1962). Thus, an intermolecular crosslink can be produced with what is normally a monofunctional reagent. Nonenzymatic modification of phosphatidylglycerol has been attained by successive treatment of PM_2 viral membranes with diazotized sulfanilic acid and $NaBH_4$ (Schäfer *et al.*, 1974).

2. Enzyme-Catalyzed Modification

In order to obtain information about the localization of the phospholipids in membranes, the degradation by lipolytic enzymes has been studied in a number of membrane systems. The accessibility of membrane phospholipids *in situ* to exogenous phospholipases is governed by surface-pressure-dependent enzyme catalysis, the substrate specificity of the enzyme, the sided orientation of the phospholipid, the compactness of the lipid bilayer, and by the degree of shielding by membrane proteins or other membrane constituents (Zwaal and Roelofsen, 1976). Enzymatic modification of choline-containing phospholipids (phosphatidylcholine, sphingomyelin; Zwaal *et al.*, 1973), phosphatidylserine, and phosphatidylethanolamine is occurring from the aqueous phase. The products of the enzyme-catalyzed surface reaction are lysophospholipids and fatty acids (phospholipase A), diglycerides (phospholipase C), ceramides (sphingomyelinase), and phosphatidic acid (phospholipase D), respectively. Membrane lipid rearrangement may be caused by the lipolytic products mentioned (Van Deenen, 1981).

With respect to membrane topography, sided orientation has been ascertained for membrane-bound phospholipases (Van den Bosch, 1980). The enzymatic active site of the sheep erythrocyte phospholipase A_2 is allocated on the outer leaflet of the cell membrane (Frei and Zahler, 1979). Site-specific transfer of this phospholipase to other membranes is achieved by polyethylene glycol fusion (Zahler and Wolf, 1982). Both soluble and membrane-bound phospholipases bear the inherent specificity for sided enzymatic action on membrane lipids.

The galactose-oxidase approach is unique for enzyme-mediated modification of glycolipids and glycoproteins. The enzyme is used to oxidize the carbon 6 hydroxyl of terminal D-galactose and *N*-acetyl-D-galactosamine residues found in cell surface glycoconjugates (Hubbard and Cohn, 1976; Gahmberg, 1981). The resultant aldehydes are labeled by tritiated borohydride. This procedure provides selectivity for membrane components topologically allocated opposite to the cytoplasm (Gahmberg and Hakomori, 1973; Steck and Dawson, 1974).

B. Lipid Labeling within the Apolar Membrane Phase

Covalent chemical modification of the apolar lipid phase of bilayer membranes was initiated through the use of photolabeled hydrophobic reagents (Klip and Gitler, 1974; Chakrabarti and Khorana, 1975). Since then, a large number of apolar nitrene- or carbene-generating reagents have been applied in biological and artificial membrane systems. Utilizing these reagents, membrane lipid modification has been investigated, although the final purpose of most of these studies is the indiscriminate random modification of those parts of membrane proteins which are in close association with the apolar lipid phase (Bayley, 1982). The photoreagents used have been photolyzed in synthetic lipid bilayer systems to define preferential binding with respect to topography and nature of the covalent interaction. Selectivity of the labels (Table 2, section B) for the apolar membrane phase is implied in their hydrophobicity which results in preferential partitioning of the probe into the bilayer lipid phase. Azide- or diazirine-bearing fatty acids, either singly (Hu and Wisnieski, 1979) or as integrated parts of phospholipids bury the reactive group in the apolar lipid phase due to the amphipathic character of the molecule. The latter approach, when performed in liposomal systems, entails inter- and intramolecular crosslinking of membrane lipids.

Modification of membrane lipid fatty acids *in situ* is achieved by introducing apolar reagents into the bilayer. The reagents are expected to provide chemical and reactivity properties which procure covalent binding to C—H and C═C double bonds present almost exclusively in lipid acyl chains. Photogenerated nitrenes and carbenes, produced by illumination of the parent compounds within the membrane apolar domain, will insert into C—H bonds or undergo addition reactions with C═C double bonds (Nielsen and Buchardt, 1982; Iddon *et al.*, 1979; Kirmse, 1971).

1. Hydrophobic Photolabels

Phenylnitrene generated photochemically from phenylazide (8) within artificial phospholipid vesicles preferably labels the fatty acid chains of the lipids. A small amount of the reagent is recovered in the head group region of phospholipid molecules (Abu-Salah and Findlay, 1977) hence the yield of arylnitrene binding to the aliphatic groups is controversial (Bayley and Knowles, 1978a). Reasonable yields of labeled fatty acid chains have been recovered upon photolysis of phenyldiazirine (9) and adamantane-2,2′-diazirine (10) in vesicles prepared with either saturated or unsaturated phospholipids (Bayley and Knowles, 1978b). Carbene insertion into both C—H and C═C bonds was found to occur. In a side reaction, diazirines are known to rearrange to diazo compounds which are expected to react with nucleophiles. The diazo rear-

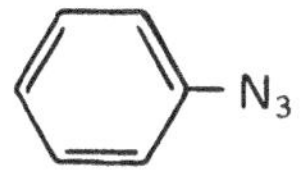

8

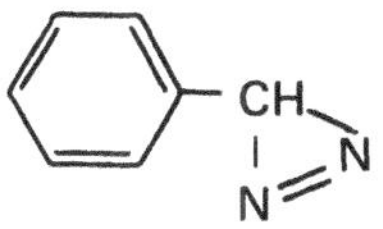

9

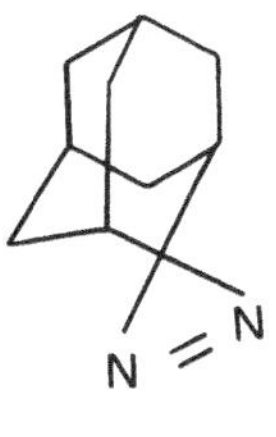

10

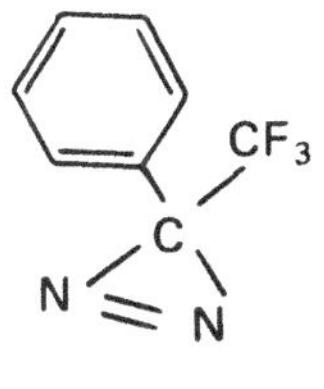

11

rangement products of trifluoromethyl phenyldiazirines, however, are exceptionally inert. High yields of C—H insertion have been achieved using 3-trifluoromethyl-3-phenyldiazirine (11) and photosensitive phospholipid analogues thereof (Brunner and Richards, 1980).

2. *Photolabeled Phospholipids*

The development of photoreactive phospholipids capable of identifying membrane components in specific regions of the bilayer provides a novel approach to the study of membrane organization. Lipid molecules tagged with a photosensitive group which on light irradiation generate a highly reactive nitrene or carbene, will crosslink to their nearest neighbor by addition to double bonds or insertion into C—H bonds. Photocrosslinking with alkylazido- or diazirino-phenoxy-labeled fatty acids incorporated into phospholipids (12, 13) yielded intermolecular crosslinks between acyl chains with no crosslinking to the glycerol backbone or choline head groups (Gupta *et al.*, 1979; Stoffel *et al.*, 1982). The products obtained by lipid crosslinking were investigated by gas chromatography–mass spectroscopy. Both studies provide evidence that pho-

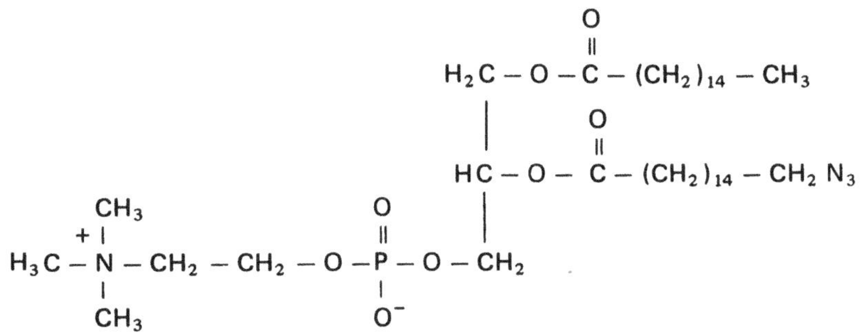

12

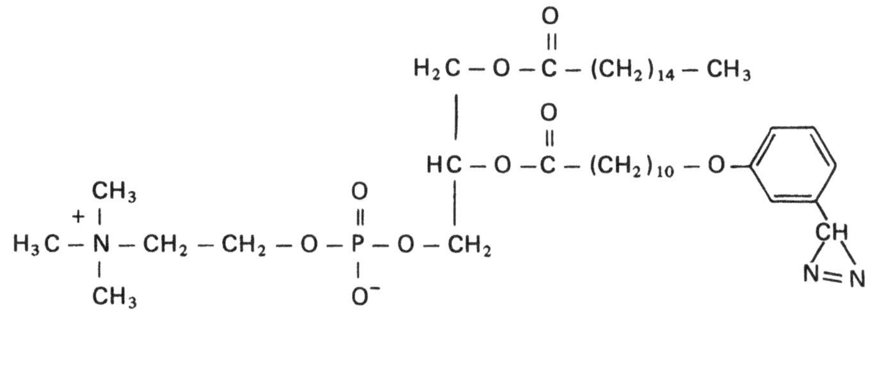

13

tolabeled phospholipids react selectively within the apolar lipid phase. The photolyzed group does not protrude to the membrane–water interface, a statement of crucial importance for the study of lipid–protein interaction.

III. SELECTIVE COVALENT MODIFICATION OF PROTEIN COMPONENTS

Membrane proteins take part in crucial transmembrane processes. They transmit cell signals through the hydrophobic membrane barrier, transport solutes and electrons, and act as specialized membrane receptor molecules. The urge to better understand the mechanisms effecting defined membrane phenomena have made membrane proteins the most intensively investigated membrane components. Modification of one selected protein or a group of proteins is optimally aspired to. The more specific a reagent can be placed into and allocated within the membrane (protein) structure the more conclusive is the information gained by the introduced agent. Binding selectivity is therefore intended in investigations concerning topography and organization, membrane protein sidedness, lipid association, structure, and mechanistic aspects of transmembrane processes.

The chemical modification procedures applied in membrane research projects are based, in many instances, on well-characterized reactions previously utilized for soluble proteins and enzymes (Means and Feeney, 1971; Sigman and Mooser, 1975; Vallee and Riordan, 1969; Glazer, 1976). Application of mono- and bifunctional reagents has been reviewed by Wold (1972), Carraway (1975), Hubbard and Cohn (1976), Peters and Richards (1977), Ji (1979), Das and Fox (1979), Chowdhry and Westheimer (1979), Freedman (1979), Brunner (1981), Sigrist and Zahler (1982a), and Bayley (1982). Some of the chemical modification reactions which imply enhanced specificity in either topographical, structural, or functional studies of membrane proteins are summarized in Table 3.

A. Protein Modification Attained by Polar Reagent–Membrane Interaction

1. Chemical Modification of Amino Groups

Sulfonated arylisothiocyanates are generally designated as being membrane-impermeant surface labels. Modification of an unprotonated amino group does occur by reacting 4,4′-diisothiocyano-2,2′-stilbene disulfonic acid (14) with the erythrocyte membrane. The anionic groups on the molecule promote electrostatic interactions with positively-charged functions of the anion exchange protein, band 3. Anion transport is concomitantly inhibited (Passow *et al.*, 1982). Enzymatic cleavage of modified band 3 and partial sequence analysis of the labeled peptide indicated that the site of covalent interaction is a lysine residue located in the 17,000-dalton transmembrane fragment of band 3 (Ramjeesingh *et al.*, 1980; Ramjeesingh *et al.*, 1981). Under suitable conditions, the two isothiocyanate groups of the anion transport inibitor reacted co-

Table 3. *Selected Examples for the Chemical Modification of Membrane Proteins*

Reagent	Reactive site in membrane	Structure of modified site(s)	References
A. Labeling of proteinaceous domains exposed to the aqueous phase			
4,4′-Diisothiocyano-2,2′-stilbene disulfonic acid: DIDS (14)	Lysine-$^{\epsilon}NH_2$ (band 3); nucleophilic groups	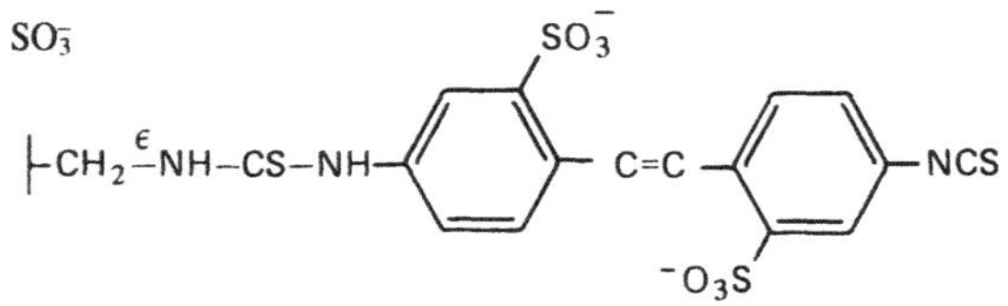	Cabantchik and Rothstein (1974), Ship *et al.* (1977), Ramjeesingh *et al.* (1981), Kampmann *et al.* (1982)
Dimethyl-3,3′-dithiobispropionimidate: DTPB	Lysine-$^{\epsilon}NH_2$, nucleophilic groups	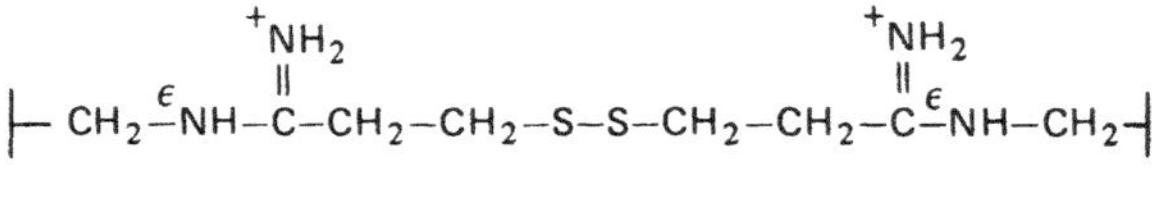	Peters and Richards (1977), Dellweg and Sumper (1978)
5,5′-Dithiobis(2-nitrobenzoic acid): DTNB (17)	Cysteine-SH (thiolate)	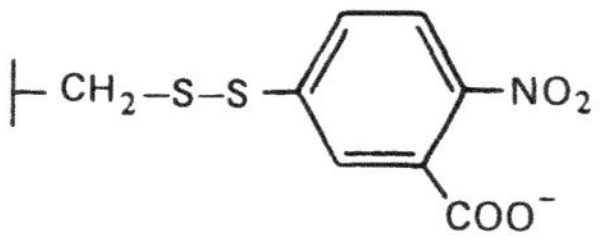	Habeeb (1972), Andersen and Møller (1977)
Diazotized sulfanilic acid: DSA (19)	Tyrosine-OH (phenolate), nucleophilic groups	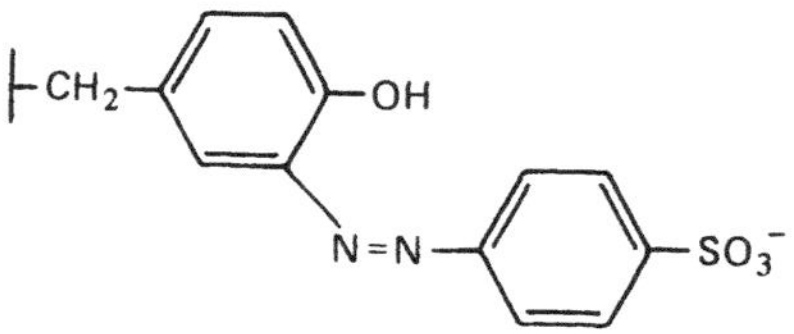	Ludwig *et al.* (1979), Edwards *et al.* (1979), Lemke *et al.* (1982)
N-(4-Azido-2-nitrophenyl)-2-aminoethylsulfonate: NAP-taurine (20)	Membrane C—H bonds (S—H, N—H, O—H)	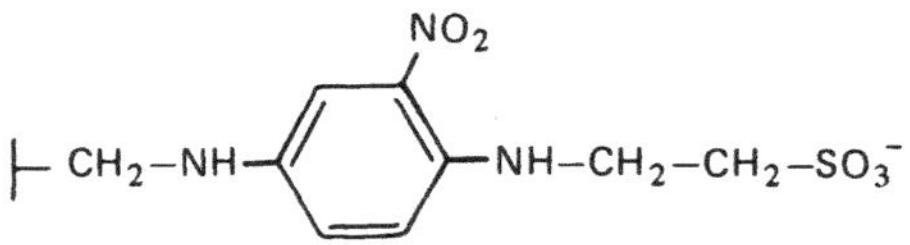	Knauf and Rothstein (1980)

2-Azido-adenosine-diphosphate: N_3-ATP	ADP binding site (ATPases)	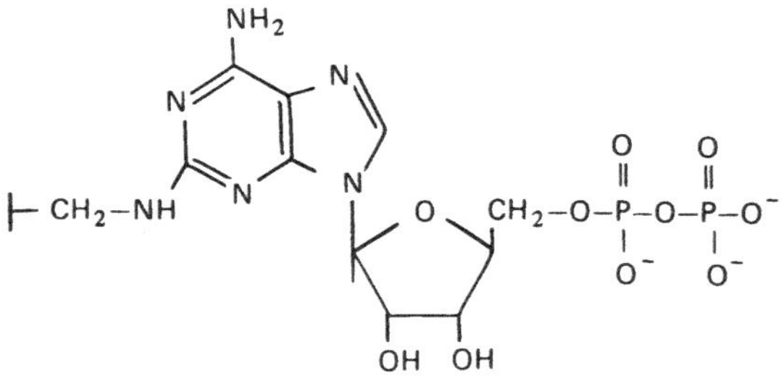	Hoyer *et al.* *(1980)*, *Czarnecki et al.* (1982)
Sodium periodate/$NaBH_4$	Glycoproteins, terminal sialic acid	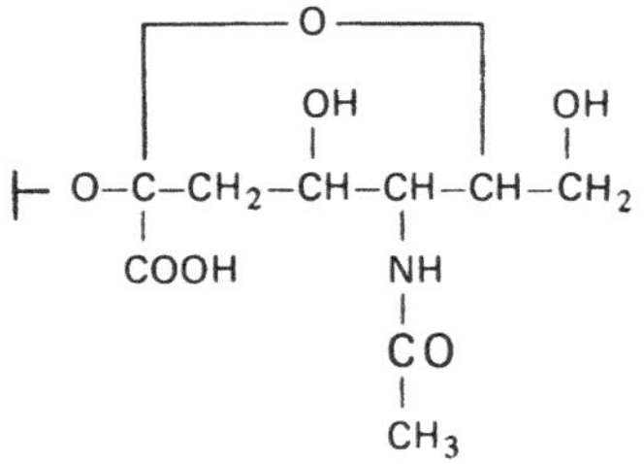	Gahmberg and Andersson (1977)
Lactoperoxidase/I^-	Tyrosine	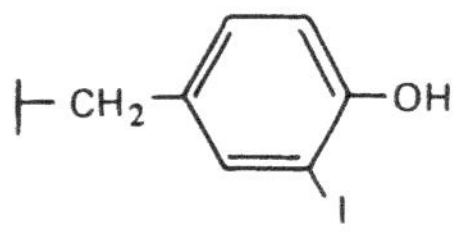	Hubbard and Cohn (1976), Katre and Stroud (1981)
B. Labeling of hydrophobic protein domains			
5-Iodonaphthyl-1-azide: INA (21)	Protein/lipid C—H bonds (S—H, N—H, O—H)	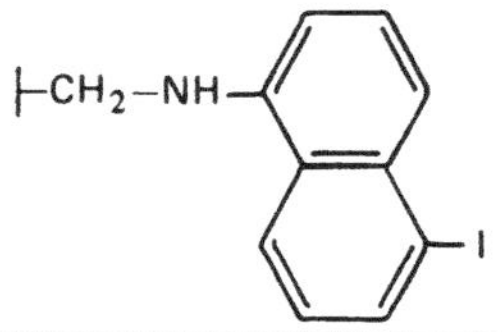	Bercovici and Gitler (1978), Gitler and Bercovici (1980) and references cited therein, Jørgensen *et al.* (1982)

(Continued)

Table 3. (Continued)

Reagent	Reactive site in membrane	Structure of modified site(s)	References
3-Trifluoromethyl-3-(*m*-iodophenyl)-diazirine: TID (22)	Protein/lipid C—H bonds (S—H, N—H, O—H)	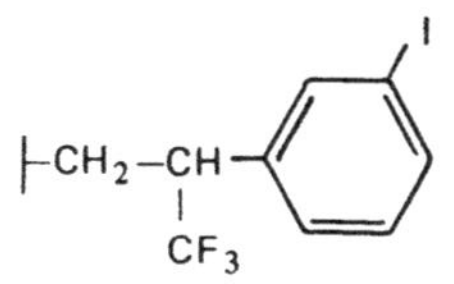	Brunner and Richards (1980), Brunner and Semenza (1981), Spiess *et al*. (1982), Frielle *et al*. (1982)
1-Palmitoyl-2-ω-(*m*-3 H-diazirinophenoxy)-undecanoylglycero-3-phosphorylcholine (13)	Glutamic acid-γCOOH (reconstituted glycophorin)	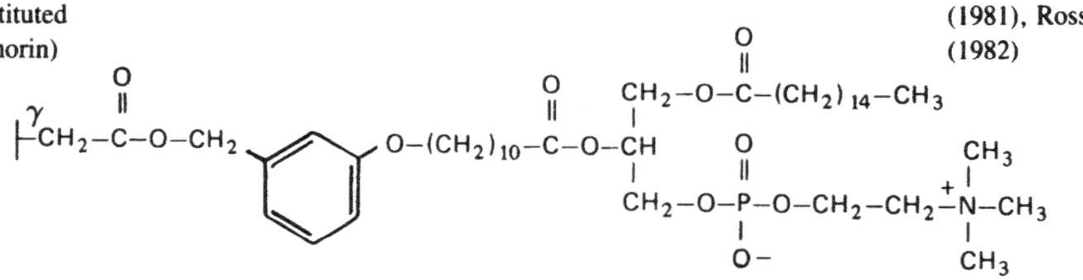	Radhakrishnan *et al*. (1981), Ross *et al*. (1982)
12-(4-Azido-2-nitrophenoxy)-stearoyl-glucosamine (23)	Protein/lipid C—H bonds (S—H, N—H, O—H)	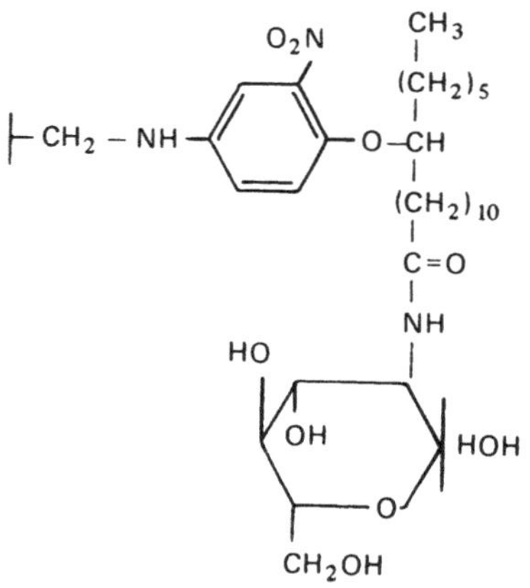	Hu and Wisnieski (1979), Wisnieski and Bramhall (1981)

4-*N*,*N*-Dimethylamino-azobenzene-4′-isothiocyanate: DABITC (24)	Nucleophilic groups, lysine-$^{\epsilon}NH_2$ (bacteriorhodopsin)	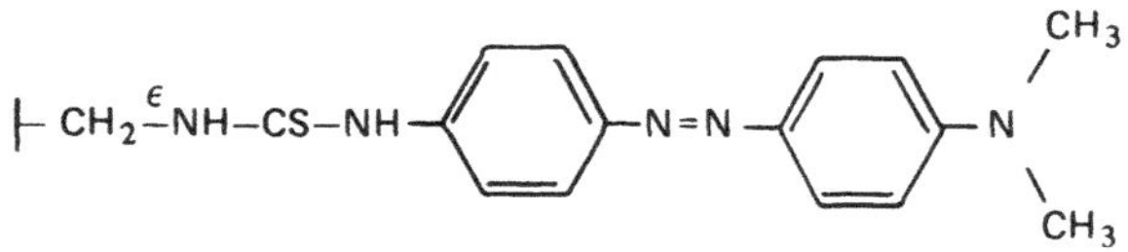	Stauffer *et al*. (1982), Sigrist *et al*. (1983)
7-Chloro-4-nitrobenz-2-oxa-1,3-diazole: NBD-Cl (26)	Tyrosine-OH (ATP′-ase), lysine-$^{\epsilon}NH_2$ (bacteriorhodopsin), cysteine-SH	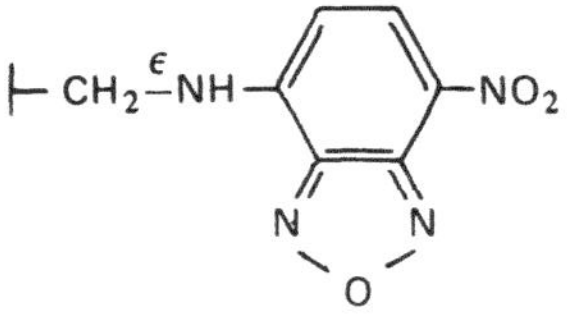	Ferguson *et al*. (1975), *Allegrini et al*. (1983)
N,*N*′-Dicyclohexylcarbodiimide: DCCD (29)	Glutamic acid-$^{\gamma}COOH$, aspartic acid-$^{\beta}COOH$	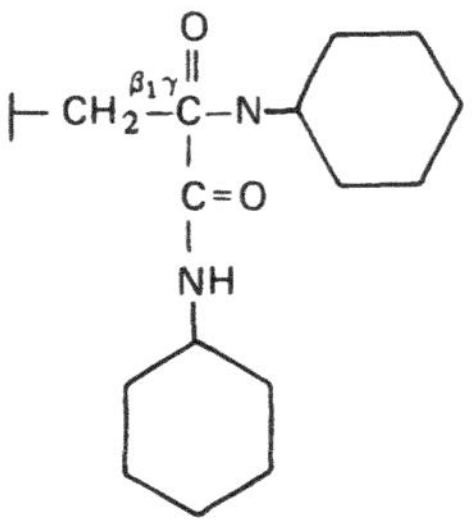	Pick and Racker (1979), Sigrist-Nelson and Azzi (1980), Hoppe and Sebald (1981), Ross *et al*. (1982)

valently with two lysine residues which reside on the chymotryptic 55,000-dalton (N-terminal) and 38,000-dalton (C-terminal) segment of band 3, respectively (Kampmann *et al.*, 1982).

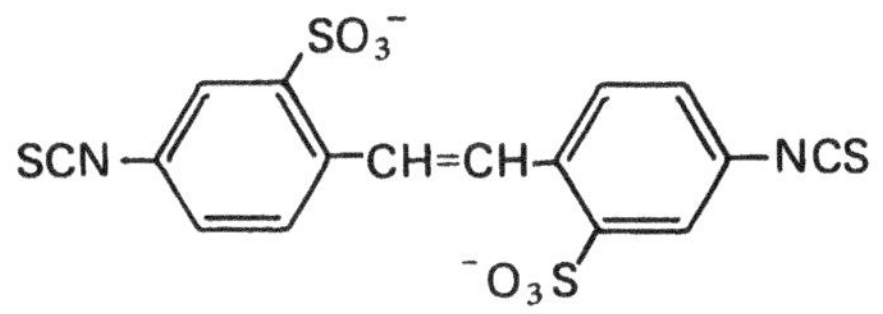

14

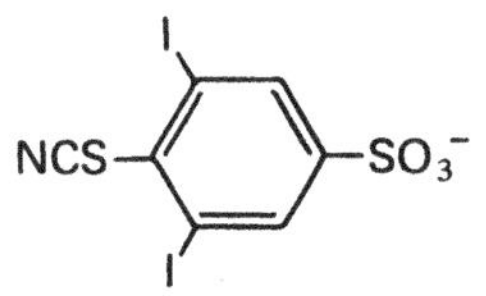

15

Johnstone and Crumpton (1979) described the synthesis and application of diiodosulfophenyl-isothiocyanate (15). This reagent which can be prepared with a high specific radioactivity [^{125}I] does not appear to penetrate cell surface membranes. The conditions used for covalent binding to lymphocyte cell surfaces did not significantly reduce cell viability. In erythrocyte membranes, the reagent has been found to bind within the presumed anion-binding domain of the anion transport protein (Mawby and Findlay, 1982).

Selective thiocarbamylation of a membrane protein has been reported by Mitchinson and collaborators in 1982. Following the modification of the sarcoplasmic Ca^{2+}-ATPase by fluorescein isothiocyanate, a single-labeled peptide was purified from the inhibited ATPase. It has been demonstrated that the ε-amino group of a lysine was modified.

Modification of protein amino groups can be achieved by imidoesters. It is assumed that the reaction takes place between the unprotonated amine and the protonated imidate (Hand and Jenks, 1962; Peters and Richards, 1977). Imidoesters are readily soluble in water and react rapidly with a half-life of several minutes to approximately half an hour depending on the pH. The main field of imidoester application is in crosslinking studies to delineate protein–protein interrelationships in membranes and membrane–protein complexes. Most imidates, including bisimidates, penetrate through membranes (Miyakawa *et al.*, 1978; Wang and Richards, 1975). The hydrophilic cleavable crosslinker diisethionyl-3,3′-dithiobis-propionimidate (16) described by Staros *et al.* (1981), however, is impermeant to biological membranes. Numerous cleavable and noncleavable homo- and heterobifunctional imidoesters are now available, varying in their linking range and the nature of the heterofunction (Ji, 1977, 1979; Peters and Richards, 1977; Freedman, 1979; Vanin and Ji, 1981).

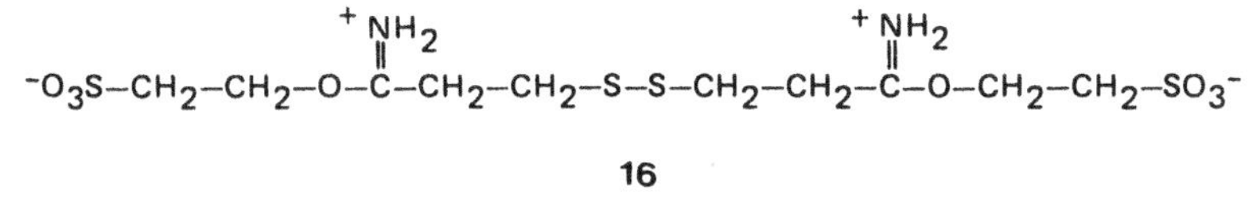

16

In addition to imidates, *N*-hydroxy succinimide esters are known to react primarily with amino groups, yet interaction of the label with histidine imidazole or cysteine

SH may occur (Anderson *et al.*, 1964; Cuatrecasas and Parikh, 1972). *N*-Hydroxysuccinimide esters have been applied in membrane modification and crosslinking studies (Capaldi *et al.*, 1979; Zisapel and Littauer, 1978; Ji, 1979; Ji and Ji, 1982). The study of Henderson *et al.* (1978), using biotinyl-*N*-hydroxy succinimide ester as a selective surface label, may serve as an example. The reagent has been applied to determine the sided orientation of the purple membrane from *Halobacterium halobium*. Covalently bound biotinyl residues were recovered on the extracellular surface of the purple membrane, visualized by electron microscopy upon formation of the biotin–avidin–ferritin complex.

2. Chemical Reagents for Cysteine Thiols

In membrane modification studies, chemical reagents designed to react with cysteine thiol functions are specific for membrane proteins unless side reactions with nucleophiles other than R—SH compete for covalent binding. The polar reagent 5,5′-dithiobis-(2-nitrobenzoic acid) (17) interacts with membrane proteins by a sulfhydryl–disulfide exchange mechanism (Habeeb, 1972). The reagent is applicable for titration of surface-accessible cysteine residues (Andersen and Møller, 1977). In contrast to the sulfhydryl–disulfide exchange, thiol addition to activated double bonds, as available in *N*-ethylmaleimide, leads to an irreversible covalent derivative.

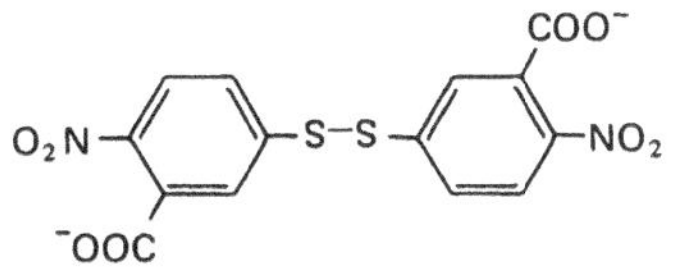

17

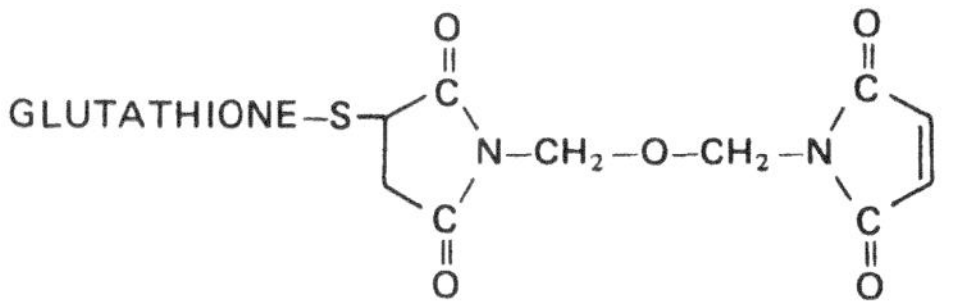

18

The ambient polarity of *N*-ethylmaleimide does not allow a selective modification of only hydrophobically located or externally oriented membrane cysteine thiol functions. Specificity can, however, be gained by sequential application of the membrane impermeant glutathione-maleimide (18) and *N*-ethylmaleimide (Abbott and Schachter, 1976). Absolute specificity for thiol group modification has been proposed for the fluorescent reagent 4-(*N*-2-aminoethyl-2′-pyridyldisulfide)-7-nitrobenz-2-oxa-1,3-diazole which encompasses the attractive properties of a site-specific, environmentally sensitive spectroscopic reporter group (Stuchbury *et al.*, 1975).

3. Chemical Modification of Tyrosine Residues

Aryl diazonium cations are electrophilic reagents and appear to react predominantly with four amino acids: lysine, tyrosine, histidine, and cysteine. Lysine, histidine, and tyrosine residues can be mono- and disubstituted by the polar reagents (Glazer, 1976). Diazotized sulfanilic acid (19) has commonly been utilized in surface-labeling studies (Hubbard and Cohn, 1976; Ludwig *et al.*, 1979). The use of a radiolabeled analog, diazotized 2,6-diiodo-sulfanilic acid, provides a most sensitive method to determine the surface exposure of membrane proteins (Edwards *et al.*, 1979).

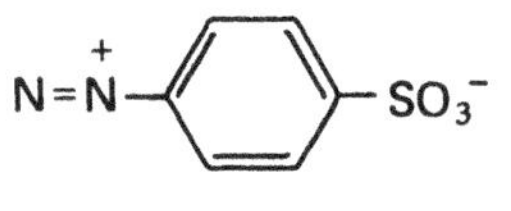

19

Selective modification of a single tyrosine residue (tyrosine 64) has been reported for bacteriorhodopsin. In the study of Lemke *et al.*(1982), lysine residues were protected with methylacetimidate prior to the treatment of purple membranes with diazotized sulfanilic acid.

4. *Selective Modification of Arginine Residues*

Reagents that selectively modify arginyl residues in proteins have played a decisive role in the identification of arginyl residues as essential binding sites for negatively charged substrates and cofactors in numerous enzymes (Riordan, 1979). Arginine residues have been proposed to be essential for the anion translocation process in human erythrocytes (Knauf, 1979). In this system, chloride exchange is irreversibly inhibited with phenylglyoxal, a reagent specific for the modification of arginyl residues (Wieth *et al.*, 1982; Zaki, 1983). In the future, selective arginine modification studies may be extended in that the environment of the primary (arginyl) binding site can be explored by *p*-azidophenylglyoxal, a heterobifunctional photoactivatable crosslinking reagent (Ngo *et al.*, 1981).

5. *Membrane Surface Photolabeling*

Specificity can be ascribed to *N*-(4-azido-2-nitrophenyl)-2-aminoethyl sulfonate: NAP-taurine (20) in that the reagent serves as general label for surface proteins. The reactivity of the photogenerated nitrene is nonselective (Staros and Richards, 1974).

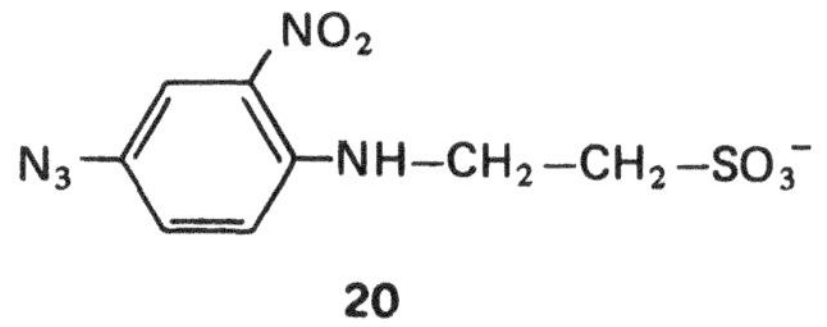

20

The topographical arrangement of the mitochondrial cytochrome oxidase subunits has been inferred from data obtained by surface labeling techniques including the NAP-taurine photolabel (Prochaska *et al.*, 1980). In erythrocytes, the reagent seems to interact preferentially with anion binding sites. Its transport through the membrane, mediated by the anion transport protein, is inhibited by DIDS (Knauf and Rothstein, 1980). The anionic probe NAP-taurine thus bears the characteristics of a photoaffinity label. Similarly, nucleotide binding sites (Hoyer *et al.*, 1980; Czarnecki *et al.*, 1982), membrane transport systems (Fannin *et al.*, 1981; Boulay *et al.*, 1983), and membrane receptors (Das and Fox, 1979; Raftery *et al.*, 1980; Yip *et al.*, 1982) have been specifically tagged by photolabeled substrate or inhibitor analogs.

6. *Enzyme-Mediated Protein Modification*

Selective radioactive labeling of cell surface sialoglycoproteins has been reported by Gahmberg and Andersson (1977). Low concentrations of sodium metaperiodate induce specific oxidative cleavage of sialic acids between carbon 7 and 8 or carbon 8 and 9. The aldehydes formed can easily be reduced with tritiated borohydride. As described for glycolipids, glycoproteins are equally accessible to the side-specific action of galactose oxidase (Steck and Dawson, 1974; Schweizer *et al.*, 1982). Enzyme-mediated modifications of surface-exposed segments of membrane proteins are inherent specific processes. Iodination of tyrosine residues by lactoperoxidase has widely been applied in membrane research. Advantages and limits of the method have been described in detail by Hubbard and Cohn (1976). The lactoperoxidase–glucose oxidase enzyme system is, at present, available in the form of immobilized enzymes coupled to hydrophilic polyacrylamide beads. By the latter approach, selective surface labeling of accessible tyrosine residues has been attained in bacteriorhodopsin (Katre and Stroud, 1981).

7. *Proteases*

Selective proteolytic fragmentation of membrane proteins has been utilized in both topographical studies and in establishing membrane protein sequences. Due to membrane integration, many proteolytic cleavage sites are not accessible to the enzymes. The site of action, involving the enzyme and the polypeptide chain, is restricted to the aqueous phase. Side-specific enzymatic cleavage sites are known for bacteriorhodopsin, where chymotrypsin cleaves at Phe 72/Gly 73 (Gerber and Khorana, 1982). In retina rhodopsin, the chymotryptic action is limited to Phe 146/Arg 147 and Gln 244/Lys 255 yielding three major peptides from the membrane-bound protein (Ovchinnikov, 1982). A 55,000-dalton segment and a 38,000-dalton segment of band 3 are obtained by chymotryptic treatment of erythrocytes from the outer membrane surface (Steck *et al.*, 1976; Rothstein, 1982). With regard to other membrane-integrated proteins, selective cleavage sites by trypsin have been described for the Na^+/K^+-ATPase (Jørgenson *et al.*, 1982), the sarcoplasmic Ca^{2+}-ATPase (MacLennan and Campbell, 1979), and glycophorin (Furthmayr *et al.*, 1978). Papain has preferably been chosen for the release of brush border membrane hydrolases which are anchored via a low-molecular-weight membrane-integrated peptide (Spiess *et al.*, 1982; Frielle *et al.*, 1982).

B. *Hydrophobic Labeling of Membrane Protein Components*

1. *Hydrophobic Photolabeling*

Crucial steps in membrane processes occur, at least in part, within the apolar membrane phase. Improved understanding of the structural and functional characteristics of the intramembraneous protein domains is thus mandatory for a general insight into transmembrane phenomena. Among the various methodological approaches available for exploration of the intrinsic membrane domain (Seelig and Seelig, 1980;

Andersen, 1978; Chapman *et al.*, 1982), the technique of chemical modification may advantageously be applied to obtain structural information. Hydrophobic photogenerated reagents serve to chemically modify the domain of membrane proteins which are in proximity with the hydrocarbon core of the lipid bilayer. The broad reactivity of photogenerated probes, however, implies nonselective modification of various amino acid side chains and lipid acyl chains (Section II-B).

Among the arylazides used in hydrophobic protein modification studies are the tritiated or ^{125}I-labeled phenylazide (8) (Wells and Findlay, 1979, 1980), [^{125}I]hexanoyldiiodo-*N*-(4-azido-2-nitrophenyl)tyramine (Owen *et al.*, 1980; Bijlenga *et al.*, 1982), and 5-iodonaphthyl-1-azide (21) (Gitler and Bercovici, 1980, and references therein; Cerletti and Schatz, 1979; Tarrab-Hazdai *et al.*, 1980; Jørgensen *et al.*, 1982). Photogenerated nitrenes inserted into defined protein segments. Alternatively, covalently-bound photoreagents were recovered within distinct subunits of membrane–protein complexes. Hydrophobic labeling in combination with controlled surface proteolysis of the labeled membrane systems provided evidence for label binding selectivity for the apolar membrane domain.

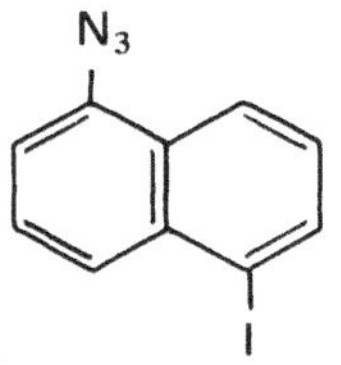

21

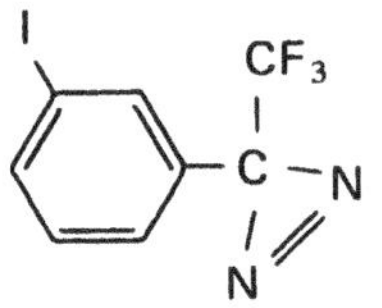

22

Photogenerated carbenes are preferably used for detection of membrane-embedded protein segments. Only the most reactive reagents will derivatize the predominantly hydrophobic and chemically inert amino acid side chains. Adamantane diazirine (10) has been characterized as an effective apolar modifier in the erythrocyte membrane system (Bayley and Knowles, 1980) and the microsomal Na^+/K^+-ATPase (Farley *et al.*, 1980). A significant improvement in the hydrophobic photolabeling technique was the synthesis and application of 3-trifluoromethyl-3-(*m*-iodophenyl)-diazirine (22) (Brunner and Semenza, 1981). Upon photolysis of this reagent, the highly reactive carbene precursors do not detectably rearrange. Utilizing this reagent in biological membrane systems, protein modification occurring specifically within the apolar membrane domain was accomplished (Spiess *et al.*, 1982; Frielle *et al.*, 1982).

Photoactivatable carbene and nitrene precursors were incorporated synthetically at defined positions of the fatty acyl chains which were then used to synthesize phospholipids (Brunner and Richards, 1980; Bisson and Montecucco, 1981; Radhakrishnan *et al.*, 1981). Incorporation of photoreactive fatty acid derivatives into phospholipids was also achieved biosynthetically using microorganisms (Quay *et al.*, 1981) or cell culture techniques (Leblanc *et al.*, 1982). The prospective approach to measure the depth of membrane protein insertion into the lipid bilayer is supported by the studies of Bisson *et al.*, 1979; cytochrome *c* oxidase) but questioned by the data of Brunner and Richards (1980; reconstituted gramicidin A) and Ross *et al.* (1982; erythrocyte

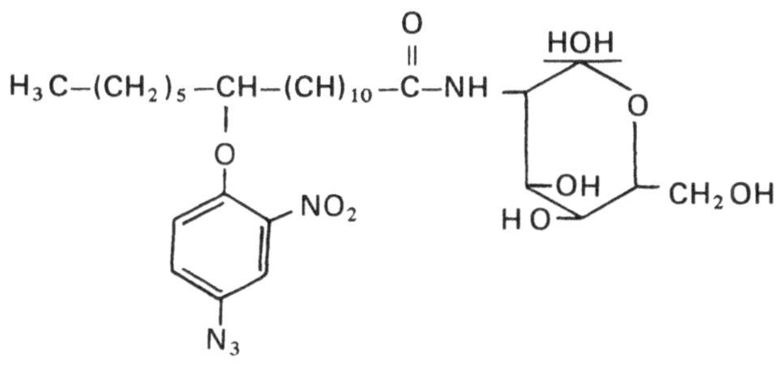

23

glycophorin). At present, it cannot be excluded that the photoactivated fatty acid chain mobility allows the covalent reaction to occur close to the membrane surface. In a most elegant approach, Wisnieski *et al.* (1979) and Wisnieski and Bramhall (1981) investigated the dynamic aspects of a transmembrane process by photolabeling cholera toxin subunits during membrane penetration. The photoreactive glycolipid probe 12-(4-azido-2-nitrophenoxy)stearoylglucosamine (23) effected preferential modification of the membrane-traversing A_1 subunit of cholera toxin.

2. *Group-Directed Hydrophobic Modification*

In recent years, increasing effort has been put into the selective modification of membrane-integrated parts of membrane proteins. In these studies, group-specific reagents have been utilized to achieve covalent modification of defined amino acid side chain functional groups (Lys-$^{\varepsilon}NH_2$, Asp/Glu-$^{\beta/\gamma}COOH$, Tyr-OH, Cys-SH) localized within apolar membrane domains. The presence of these groups within apolar phases is obviously required for group-directed modifications. Several independent investigations in various membrane systems have indicated the availability of hydrophobically located functional groups for selective modification.

Membrane protein labeling by hydrophobic *arylisothiocyanates* combines the advantages of topological selectivity and group specificity, allowing facile binding site identification by standard analytical procedures. Arylisothiocyanates are known to form covalent bonds with nucleophilic groups in their nonprotonated form (Drobnica *et al.*, 1977). For modification of membrane proteins, the pH-dependent reactivity of proteinaceous nucleophilic groups can therefore be used to obtain selective labeling. At neutral pH, the ε-amino functions of lysine residues exposed to the aqueous phase are protonated and accordingly nonreactive with arylisothiocyanates. In contrast, the buried bulk pH-independent amino groups may be in a reactive (deprotonated) state, making modification with hydrophobic arylisothiocyanates feasible. Hydrophobically located ε-amino groups are favored for apolar arylisothiocyanate modification (Sigrist and Zahler, 1982a). Group-directed hydrophobic modification of membrane proteins by phenylisothiocyanate (1) or 4-*N*,*N*-dimethylamino-azobenzene-4′-isothiocyanate (24) results in covalent thiocarbamylation of a unique lysine ε-amino group of bacteriorhodopsin. Lysine 41 has been identified as the reagent binding site by established sequencing techniques (Sigrist *et al.*, 1981, 1984; Stauffer *et al.*, 1982).

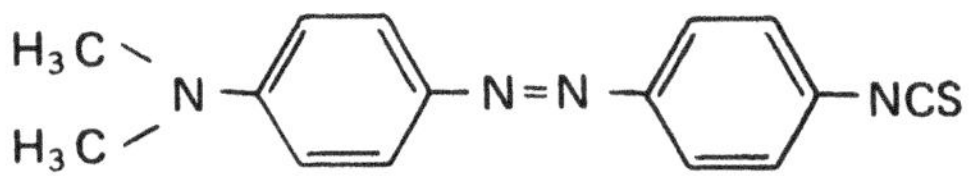

24

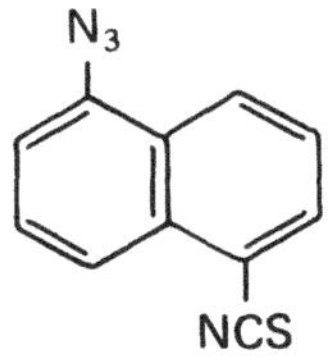

25

The apolar disposition of Lys 41 within the purple membrane has been ascertained by the inaccessibility of the nucleophilic ε-amino group to the polar analog *p*-sulfophenylisothiocyanate and by the inaccessibility of the introduced *N*,*N*-dimethylazobenzene group to protonation. In the latter case, protonation of the azo label is achieved only upon protein denaturation by detergents (Sigrist *et al.*, 1984). Covalent binding of phenylisothiocyanate to the erythrocyte anion transport protein, band 3, has been reported (Sigrist *et al.*, 1980; Kempf *et al.*, 1981). The heterobifunctional azidoarylisothiocyanates, *p*-azidophenylisothiocyanate and 5-isothiocyanato-1-naphthalene azide (25), compete for the monofunctional phenylisothiocyanate binding sites in both the purple membrane (bacteriorhodopsin) and the erythrocyte membrane (band 3) systems (Sigrist and Zahler, 1982b; Sigrist *et al.*, 1982). Homopolymers of the modified proteins were recovered upon photolysis of the heterofunction (arylazide) indicating that the selective thiocarbamylation (Figure 1) and the nitrene insertion produced a covalent link between neighboring protein molecules.

The *arylhalide* 7-chloro-4-nitrobenz-2-oxa-1,3-diazole (26) is known to react with nucleophilic groups (Fager *et al.*, 1973). The fluorescence produced upon binding is specific for the nucleophilic group (Ferguson *et al.*, 1975; Deters *et al.*, 1975; Birkett *et al.*, 1970). Site-directed modification of bacteriorhodopsin has been achieved by the reaction of 7-chloro-4-nitrobenz-2-oxa-1,3-diazole at neutral pH with purple membranes. The reaction mechanism leading to a covalent derivative of Lys 41 implies the availability of the nucleophilic ε-amino group in the reactive (deprotonated) R-NH_2 state. Furthermore, the introduced fluorophore is not accessible to polar quencher molecules (Allegrini *et al.*, 1983) providing further evidence for the apolar disposition of Lys-41 of bacteriorhodopsin.

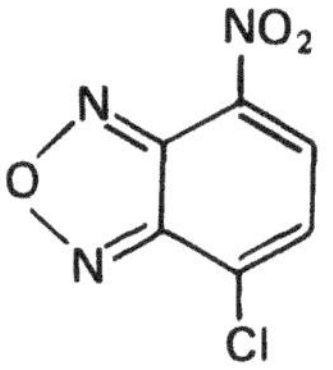

26

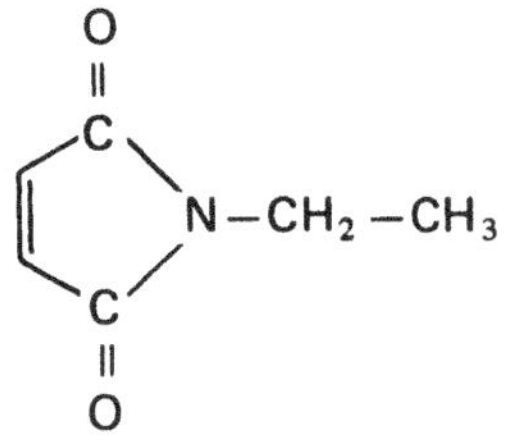

27

Selective covalent modification of lysine residues has further been achieved by *borohydride reduction of the Schiff base* formed between the ε-amino group of a lysine residue and an (aldehyde) carbonyl group. The Schiff base salt or tertiary iminium salt formed by the reactants under mildly acidic conditions (pH 4.5–6.5) is actually reduced by BH_4^- (Schellenberg, 1963). The procedure has been used to covalently bind pyridoxal phosphate within the transport site of the erythrocyte anion transport protein (Cabantchik *et al.*, 1978). Borohydride reduction was also applied to stabilize the Schiff base in retinal binding proteins allowing then identification of the chromophore binding sites. Chemical reduction by BH_4^- proceeds best either under illumination (pumping conditions) of bacteriorhodopsin (Huang *et al.*, 1982) or, as applied for halorhodopsin, at pH 5 using cyanoborohydride (Lanyi and Oesterhelt, 1982). Upon Schiff base reduction, the retinal binding site in bacteriorhodopsin was alloted to Lys 216 (Bayley *et al.*, 1981; Katre *et al.*, 1981; Lemke and Oesterhelt, 1981b; Mullen *et al.*, 1981) and in retina rhodopsin, to Lys 296 (Ovchinnikov, 1982).

The modification of membrane sulfhydryl groups by *N-ethylmaleimide* (27) occurs by nucleophile addition to the activated C═C double bond. The reaction leads to a covalent irreversible derivative of the thiol function (Peach, 1974). *N*-Ethylmaleimide-reactive thiol functions may be available in membranes exposed to the aqueous phase and/or buried to various extent within apolar membrane domains. Selectivity for the hydrophobic domain may therefore be attained only by protecting externally accessible thiols with a polar reagent prior to modification of the embedded cysteine residues with *N*-ethylmaleimide (Abbott and Schachter, 1976; Haest *et al.*, 1979). Maleimides containing apolar alkyl chains are expected to be stabilized by hydrophobic interactions leading to preferential modification of buried cysteine residues (Fonda and Anderson, 1969). Reversible heterobifunctional crosslinking directed towards membrane protein thiol functions has been achieved by various reagents (for a review, see Ji, 1979). The reagent *N*-(4-azidophenylthio)phthalimide follows a reaction mechanism which does not involve thiol–disulfide interchange. Phthalimide is the leaving group in the thiol modification reaction. The crosslink is formed by light-activation of the arylazide heterofunction (Moreland *et al.*, 1982).

Selective modification of tyrosine residues may be achieved with *tetranitromethane* (28). The mechanism of this reaction has been examined by Bruice *et al.* (1968). Nitration of phenols with tetranitromethane was shown to be a first order reaction in both the reagent and the phenoxide anion. The major product of the tetranitromethane reaction with nonsulfhydryl proteins has been identified as 3-nitrotyrosine. In bacteriorhodopsin, which does not contain cysteine residues, modification of less than 3–4 equivalents of tyrosine residues did not cause inhibition of the protein's function (Campos-Cavieres *et al.*, 1979). Tyrosine 26 and 64 were fully nitrated, whereas tyrosine 131 and 133 were partially modified in the experiments of Lemke and Oesterhelt (1981a). Reduction of the nitrated purple membrane with dithionite left only Tyr-26 nitrated, indicating that the residues 64, 131, and 133 are located at the outside-facing membrane surface.

Carbodiimides are probably the most widely used reagents for forming amide bonds. The reaction of a carboxyl group with a carbodiimide yields first an *O*-acylisourea that, if an amine is present, reacts to form an amide. Competing with this reaction is the rearrangement to an *N*-acylurea, which is stable and no longer reacts with nucleophiles. The latter reaction, first described by Hoare and Koshland (1967), preferentially occurs in the absence of nucleophiles, e.g., amino groups, and water (Previero *et al.*, 1973). It is therefore not surprising that apolar carbodiimides such as *N,N'*-

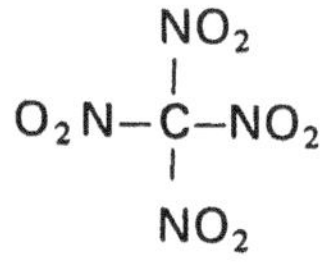

28

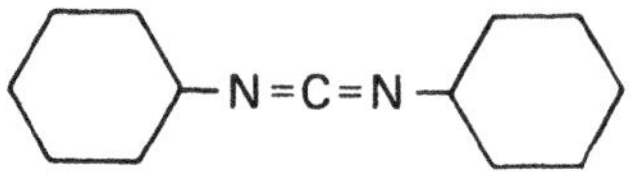

29

dicyclohexyl-carbodiimide (29) have frequently been used for hydrophobic modification of membrane-buried carboxyl groups.

The most outstanding example for dicyclohexylcarbodiimide modification are the dicyclohexylcarbodiimide-binding proteins which are membrane-integrated subunits of proton-translocating ATPase complexes. The sequences of at least nine of these proteins from various sources have been established. Dicyclohexylcarbodiimide binds covalently to γ(β)-carboxyl groups of glutamic (aspartic) acid (Hoppe and Sebald, 1981). The apolar disposition of the carbodiimide binding site has been documented by inaccessibility either to the polar analog 1-ethyl-3-(3-dimethylaminopropyl)carbodiimide (Sigrist-Nelson and Azzi, 1980; Pick and Racker, 1979) or to ascorbic acid when spin-labeled proteins were investigated (Sigrist-Nelson and Azzi, 1979). Modification of glutamic acid residues of F_1-ATPase complexes derived from various membrane systems has equally been attained by dicyclohexylcarbodiimide. Environmentally promoted label binding paralleled by inhibition of the enzymatic function is restricted to the β-subunit of the F_1-complexes (Yoshida *et al.*, 1982; Esch *et al.*, 1981). In beef heart cytochrome oxidase, the major site for dicyclohexylcarbodiimide binding has been found to be glutamic acid 90 located in subunit III (Prochaska *et al.*, 1981). Finally, selective covalent modification of the hydrophobically located glutamic acid 70 was determined upon labeling of reconstituted glycophorin A with dicyclohexylcarbodiimide (Ross *et al.*, 1982).

IV. INFORMATION ACQUIRED THROUGH SELECTIVE MODIFICATION

A. Membrane Structure: Sidedness, Asymmetry, and Protein Topography

Selective modification of biological membranes offers the possibility of efficient probing of hydrophilic and hydrophobic membrane domains. Reagents which interact with membrane constituents selectively from the aqueous phase are applicable for the study of membrane sidedness and asymmetry (Op den Kamp, 1979; Lodish and Rothman, 1979). Techniques used to investigate ***membrane sidedness and asymmetry*** exploit the specific labeling of one half of the bilayer in sealed membrane vesicles. The components labeled under such conditions are then compared with the labeling pattern obtained in preparations of broken membranes and, if possible, inverted sealed membranes. To conclusively interpret the data, several criteria need to be fulfilled: (1) all the vesicles or cells must have a unique sidedness and be impermeable to the reagent; (2) the labeling reagent must be able to react with the exposed membrane components; therefore, the use of surface labels differing in the target function is advisable; and (3) membrane permeability should not be affected by the modification procedures. Some of the reagents listed in Sections II-A and III-A are appropriate for membrane surface labeling. Further descriptions of reaction conditions and membrane permeabilities of individual probes are quoted in the review of Hubbard and Cohn (1976).

Investigations delineating the assembly of membrane components may advantageously include side-specific surface labels. Defined target functions in growing cells or subcellular membrane systems may be tagged by directed modifications (Sabban *et al.*, 1981).

Selective membrane labeling by polar and apolar reagents will further yield information on membrane organization and *membrane protein topography*. Chemical crosslinking reagents, providing defined distances between the reactive functions, are being employed in membrane protein investigations (Peters and Richards, 1977; Dellweg and Sumper, 1978). Delineation of subtle interrelationships among membrane components may be assessed by heterobifunctional crosslinking studies (Ji, 1979). Improved knowledge is acquired by utilizing bifunctional reagents which possess binding selectivity for one of the functions. A specifically allocated crosslinking reagent enables structural exploration of the membrane environment within linking distance and facilitates identification of crosslinked components.

The data obtained through selective modification of bacteriorhodopsin may illustrate the potency and versatility of the methods used for investigation of the protein's topography.

1. Selective Surface Labeling of Bacteriorhodopsin in Purple Membranes

Lemke *et al.* (1982) reported amidination of five lysine residues by methylacetimidate. Exposed lysine residues were assumed to participate in selective trimer formation upon crosslinking of purple membranes with bisimidoesters (Dellweg and Sumper, 1978). Biotinyl-*N*-hydroxy succinimide ester was found to label (a) lysine residue(s) accessible only on the extracellular surface of the membrane (Henderson *et al.*, 1978). It is inferred from the topographical information presently available that Lys 129 is selectively modified by the procedure mentioned. Tyrosine 64 is exposed to the aqueous phase as indicated by the modification with diazotized sulfanilic acid (Lemke *et al.*, 1982). Together with the tyrosine residues 131 and 133, Tyr 64 is accessible to dithionite as indicated by reduction of the preformed nitrotyrosine derivatives (Lemke and Oesterhelt, 1981a). Allocation of Tyr 131 and Tyr 133 at the protein–water interphase has been derived from surface iodination studies utilizing immobilized lactoperoxidase (Katre and Stroud, 1981). Dansylhydrazine was coupled by a soluble carbodiimide to glutamic and/or aspartic acid residues allocated in the carboxyl terminal end (residues 232–248) of bacteriorhodopsin (Renthal *et al.*, 1983). Proteolytic digestion of the membrane-integrated protein by papain produced three membrane-spanning segments, and chymotrypsin is known to cleave the protein into two homogeneous segments (Ovchinnikov *et al.*, 1979; Gerber and Khorana, 1982).

2. Selective Hydrophobic Modification of Bacteriorhodopsin

The reagents used for selective modification of hydrophobic domains are directed towards lysine and tyrosine side chain functional groups. Arylisothiocyanates and the arylhalide 7-chloro-4-nitrobenz-2-oxa-1,3-diazole were found to selectively bind to Lys 41 whose reactive ε-amino group is situated within an apolar environment (Sigrist *et al.*, 1984; Allegrini *et al.*, 1983). The retinal binding site, allocated on Lys 216 upon BH_4^- reduction of the Schiff base, is close to the midpoint of a putative membrane-spanning α-helix (Bayley *et al.*, 1981; Huang *et al.*, 1982). As a product of the tetranitromethane treatment of purple membranes the nitrated Tyr 26 resisted reduction by dithionite (Lemke and Oesterhelt, 1981a) indicating apolar disposition of the modified tyrosine residue.

The described chemical modifications provide the experimental data required for alignment of the polypeptide chain within the purple membrane. Experimental evidence in combination with valid theoretical considerations lead to a refined structural model of the light-driven proton pump (Engelman *et al.*, 1980; Huang *et al.*, 1982; Ovchinnikov, 1982).

B. Membrane Protein Function and Mechanism

Significant functional information may be gained by selective labeling of membrane proteins involved in membrane catalysis and transmembrane processes. Participation of a selectively modified residue in the protein's function may be considered probable if (1) the extent of label binding is strictly correlated with the loss of enzymatic activity, and (2) a single binding site is identified within the protein structure. The same arguments apply to chemical reversal of an introduced inhibitory modification. However, unequivocal involvement of the modified amino acid in the catalytic process is ascertained only if modification-induced conformational changes can be excluded.

Specific inhibition of H^+-translocation in H^+-ATPase systems is effected by selective modification of the dicyclohexylcarbodiimide-binding protein. Due to the concomitant selective modification of a single carboxyl function (Glu/Asp) by the carbodiimide, the involvement of the carboxyl groups in the H^+-transport process has been implied. In bacteriorhodopsin modification of Tyr 64 by diazotized sulfanilic acid abolished light-induced proton pumping. Reversal of the inhibition, attained by reduction of the mono (*p*-azobenzene sulfonic acid) tyrosine with dithionite to the corresponding amino derivative, strongly indicates participation of the tyrosine residue in the H^+-translocation process (Lemke *et al.*, 1982).

Understanding the mechanism by which membrane proteins catalyze reactions requires detailed knowledge of the events occurring within the tertiary structure of proteins. Obtaining such information is difficult experimentally since it requires the selective visualization of a minute region of a macromolecule in a short time interval. Chemical probes used in absorption, fluorescence, or circular dichroism are sensitive to the probes' immediate environment. Vibrational transitions of Raman resonance probes respond to conformational changes (Carey and Schneider, 1978). In all cases, signal transitions between the resting and acting protein can be interpreted functionally, if selective modification is achieved without significant alteration of protein function. Selectively introduced "reporter" groups may thus provide the necessary physiochemical properties for structural and mechanistic probing of the functionally active protein. Mandatory specificity must be combined with minimal perturbation of the protein, leaving the functional biological system in an intact native state.

In conclusion, it is the authors' wish that the small percentage of studies cited here effectively illustrates the utility of selective covalent modification and may thus serve to stimulate the reader in creative exploration of biological membranes in further work.

ACKNOWLEDGMENTS

Stimulating discussions and critical reading of the manuscript by Dr. K. Sigrist-Nelson are thankfully acknowledged. The authors wish to thank E. Kislig and I. Sutter for their assistance in the preparation of this manuscript. Experimental work carried out in the authors' laboratory was supported by the Swiss National Science Foundation (Grant No. 3.674-0.80) and the Central Laboratories of the Swiss Blood Transfusion of the Swiss Red Cross.

REFERENCES

Abbott, R. E., and Schachter, D., 1976, Impermeant maleimides, *J. Biol. Chem.* **251**:7176–7183.

Abu-Salah, K. M., and Findlay, J. B. L., 1977, Labeling of egg phosphatidylcholine vesicles and myeline membrane with a photoreactive lipophilic reagent, *Biochem. J.* **161**:223–228.

Allegrini, P. R., Sigrist, H., Schaller, J., and Zahler, P., 1983, Site-directed fluorogenic modification of bacteriorhodopsin by 7-chloro-4-nitrobenz-2oxa-1,3-diazole, *Eur. J. Biochem.* **132**:603–608.

Andersen, H. C., 1978, Probes of membrane structure, *Annu. Rev. Biochem.* **47**:359–383.

Andersen, J. P., and Møller, J. V., 1977, Reaction of sarcoplasmic reticulum Ca^{2+}-ATPase in different functional states with 5,5'-dithiobis(2-nitrobenzoate), *Biochim. Biophys. Acta* **485**:188–202.

Anderson, G. W., Zimmermann, J. E., and Callahan, F. M., 1964, The use of esters of *N*-hydroxysuccinimide in peptide synthesis, *J. Am. Chem. Soc.* **86**:1839–1842.

Barzilay, M., Ship, S., and Cabantchik, Z. I., 1979, Anion transport in red blood cells. I. Chemical properties of anion recognition sites as revealed by structure–activity relationships of aromatic sulfonic acids, *Membr. Biochem.* **2**:227–254.

Bayley, H., 1982, Photoactivated hydrophobic reagents for integral membrane proteins, in: *Membranes and Transport*, Vol. 1 (A. N. Martonosi, ed.), Plenum Press, New York, pp. 185–194.

Bayley, H., and Knowles, J. R., 1978a, Photogenerated reagents for membrane labeling. 1. Phenylnitrene formed within the lipid bilayer, *Biochemistry* **17**:2414–2419.

Bayley, H., and Knowles, J. R., 1978b, Photogenerated reagents for membrane labeling. 2. Phenylcarbene and adamantylidene formed within the lipid bilayer, *Biochemistry* **17**:2420–2423.

Bayley, H., and Knowles, J. R., 1980, Photogenerated reagents for membranes: Selective labeling of intrinsic membrane proteins in the human erythrocyte membrane, *Biochemistry* **19**:3883–3892.

Bayley, H., Huang, K. S., Radhakrishnan, R., Ross, A. H., Takagaki, J., and Khorana, H. G., 1981, Site of attachment of retinal in bacteriorhodopsin, *Proc. Natl. Acad. Sci. USA* **78**:2225–2229.

Bercovici, T., and Gitler, C., 1978, 5-[^{125}I]Iodonaphtyl azide, a reagent to determine the penetration of proteins into the lipid bilayer of biological membranes, *Biochemistry* **17**:1484–1489.

Bijlenga, R. K. L., Briottet, C., and Jaton, J. C., 1982, Structural differences between heavy chains of secreted and membrane-bound IGM of a human lymphoblastoid cell line, *Mol. Immunol.* **19**:45–49.

Birkett, D. J., Price, N. C., Radda, G. K., and Salmon, A. G., 1970, The reactivity of SH groups with a fluorogenic reagent, *FEBS Lett.* **6**:346–348.
Bishop, D. G., Op den Kamp, J. A. F., and Van Deenen, L. L. M., 1977, The distribution of lipids in the protoplast membranes of Bacillus subtilis, *Eur. J. Biochem.* **80**:381–391.
Bishop, D. G., Bevers, E. M., van Meer, G., Op den Kamp, J. A. F., and Van Deenen, L. L. M., 1979, A monolayer study of the reaction of trinitrobenzene sulfonic acid with amino phospholipids, *Biochim. Biophys. Acta* **551**:122–128.
Bisson, R., and Montecucco, C., 1981, Photolabelling of membrane proteins with photoactive phospholipids, *Biochem. J.* **193**:757–763.
Bisson, R., Montecucco, C., Gutweniger, H., and Azzi, A., 1979, Cytochrome c oxidase subunits in contact with phospholipids, *J. Biol. Chem.* **254**:9962–9965.
Boulay, F., Lauquin, G. J. M., Tsugita, A., and Vignais, P. V., 1983, Photolabeling approach to the study of the topography of the atractyloside binding site in mitochondrial adenosine 5′-diphosphate/adenosine 5′-triphosphate carrier protein, *Biochemistry* **22**:277–484.
Bretscher, M. S., 1972, Phosphatidyl-ethanolamine: Differential labelling in intact cells and cell ghosts of human erythrocytes by a membrane-impermeable reagent, *J. Mol. Biol.* **71**:523–528.
Bruice, T. C., Gregory, J. J., and Walters, S. L., 1968, Reactions of tetranitromethane. I. Kinetics and mechanism of nitration of phenols by tetranitromethane, *J. Am. Chem. Soc.* **90**:1612–1619.
Brunner, J., 1981, Labelling the hydrophobic core of membranes, *TIBS* **6**:44–46.
Brunner, J., and Richards, F. M., 1980, Analysis of membranes photolabeled with lipid analogues, *J. Biol. Chem.* **255**:3319–3329.
Brunner, J., and Semenza, G., 1981, Selective labeling of the hydrophobic core of membranes with 3-(trifluoromethyl)-3-(*m*-[^{125}I]iodophenyl)diazirine, a carbene-generating reagent, *Biochemistry* **20**: 7174–7182.
Bunnett, J. F., 1963, Nucleophilic reactivity, *Annu. Rev. Phys. Chem.* **14**:271–290.
Cabantchik, Z. I., and Rothstein, A., 1974, Membrane proteins related to anion permeability of human red cells. I. Localization of disulfonic stilbene binding sites in proteins involved in permeation, *J. Membr. Biol.* **15**:207–266.
Cabantchik, Z. I., Knauf, P. A., and Rothstein, A., 1978, The anion transport system of the red blood cell: The role of membrane protein evaluated by the use of "probes," *Biochim. Biophys. Acta* **515**:239–302.
Campos-Cavieres, M., Moore, T. A., and Perham, R. N., 1979, Effects of modification of the tyrosine residues of bacteriorhodopsin with tetranitromethane, *Biochem. J.* **179**:233–238.
Capaldi, R. A., Briggs, M. M., and Smith, R. J., 1979, Cleavable bifunctional reagents for studying near neighbour relationships among mitochondrial inner membrane complexes, *Meth. Enzymol.* **56**:630–642.
Carey, P. R., and Schneider, H., 1978, Resonance Raman labels: A submolecular probe for interactions in biochemical and biological systems, *Acc. Chem. Res.* **11**:122–128.
Carraway, K., 1975, Covalent labeling of membranes, *Biochim. Biophys. Acta* **415**:379–410.
Cerletti, N., and Schatz, G., 1979, Cytochrome *c* oxidase from bakers' yeast: Photolabeling of subunits exposed to the lipid bilayer, *J. Biol. Chem.* **254**:7746–7751.
Chakrabarti, P., and Khorana, H. G., 1975, A new approach to the study of phospholipid–protein interactions in biological membranes. Synthesis of fatty acids and phospholipids containing photosensitive groups, *Biochemistry* **14**:5021–5033.
Chapman, D., Gomez-Fernandez, J. C., and Goñi, F. M., 1982, The interaction of intrinsic proteins and lipids in biomembranes, *TIBS* **7**:67–70.
Chowdhry, V., and Westheimer, F. H., 1979, Photoaffinity labeling of biological systems, *Annu. Rev. Biochem.* **48**:293–325.
Cuatrecasas, P., and Parikh, I., 1972, Adsorbents for affinity chromatography. Use of *N*-hydroxysuccinimide esters of agarose, *Biochemistry* **11**:2291–2299.
Czarnecki, J. J., Abbott, M. S., and Selman, B. R., 1982, Photoaffinity labeling with 2-azidoadenosine diphosphate of a tight nucleotide binding site on chloroplast coupling factor 1, *Proc. Natl. Acad. Sci. USA* **79**:7744–7748.
Das, M., amd Fox, C. F., 1979, Chemical cross-linking in biology, *Annu. Rev. Biophys. Bioeng.* **8**:165–193.
Dellweg, H. G., and Sumper, M., 1978, Selective formation of bacterio-opsin trimers by chemical cross-linking of purple membrane, *FEBS Lett.* **90**:123–126.

Deters, D. W., Racker, E., Nelson, N., and Nelson, H., 1975, Partial resolution of the enzymes catalyzing photophosphorylation. XV. Approaches to the active site of coupling factor 1, *J. Biol. Chem.* **250:**1041–1047.

Drobnica, L., Kristian, P., and Augustin, J., 1977, The Chemistry of the –NCS group, in: *The Chemistry of Cyanates and Their Thioderivatives,* Part 2 (S. Patai, ed.), Wiley, New York, pp. 1002–1222.

Edwards, R. M., Kempson, S. A., Carlson, G. L., and Dousa, T. P., 1979, Diazotized [^{125}I]diiodosulfanilic acid as a label for cell surface membranes, *Biochim. Biophys. Acta* **553:**54–65.

Engelman, D. M., Henderson, R., McLachan, A. D., and Wallace, B. A., 1980, Path of the polypeptide in bacteriorhodopsin, *Proc. Natl. Acad. Sci. USA* **77:**2023–2027.

Esch, F. S., Böhlen, P., Otsuka, A. S., Yoshida, M., and Allison, W. S., 1981, Inactivation of the bovine mitochondrial F_1-ATPase with dicyclohexyl[^{14}C]carbodiimide leads to the modification of a specific glutamic acid residue in the β-subunit, *J. Biol. Chem.* **256:**9084–9089.

Fager, R. S., Kutina, C. B., and Abrahamson, E. W., 1973, The use of NBD chloride (7-chloro-4-nitrobenzo-2-oxa-1,3-diazole) in detecting amino acids and as *N*-terminal reagent, *Anal. Biochem.* **53:**290–294.

Fannin, E. F., Evans, J. O., Gibbs, E. M., and Diedrich, D. F., 1981, Phloretinyl-3′-benzylazide: A high affinity probe for the sugar transporter in human erythrocytes. 1. Hexose transport inhibition and photolabeling of mutarotase, *Biochim. Biophys. Acta* **649:**189–201.

Farley, R. A., Goldman, D. W., and Bayley, H., 1980, Identification of regions of the catalytic subunit of (Na – K)-ATPase embedded within the cell membrane, *J. Biol. Chem.* **38:**234–236.

Ferguson, S. J., Lloyd, W. J., Lyons, M. H., and Radda, G. K., 1975, The mitochondrial ATPase. Evidence for a single essential tyrosine residue, *Eur. J. Biochem.* **54:**117–126.

Fonda, M. L., and Anderson, B. M., 1969, D-Amino acid oxidase. IV. Inactivation by maleimides, *J. Biol. Chem.* **244:**666–674.

Freedman, R. B., 1979, Cross-linking reagents and membrane organization, *TIBS* **4:**193–197.

Freedman, R. B., and Radda, G. K., 1968, The reaction of 2,4,6,-trinitrobenzenesulfonic acid with amino acids, peptides and proteins, *Biochem. J.* **108:**383–391.

Frei, E., and Zahler, P., 1979, Phospholipase A_2 from sheep erythrocyte membranes: Ca^{2+} dependence and localization, *Biochim. Biophys. Acta* **550:**450–463.

Frielle, T., Brunner, J., and Curthoys, N. P., 1982, Isolation of the hydrophobic membrane binding domain of rat renal γ-glutamyl transpeptidase selectively labeled with 3-trifluoromethyl-3-(m[^{125}I]-iodophenyl)diazirine, *J. Biol. Chem.* **257:**14979–14982.

Furthmayr, H., Galardy, R. E., Tomita, M., and Marchesi, V. T., 1978, The intramembranous segment of human erythrocyte glycophorin A, *Arch. Biochem. Biophys.* **185:**21–29.

Gahmberg, C. G., 1981, Membrane glycoproteins and glycolipids: Structure, localization and function of the carbohydrate, in: *Membrane Structure,* Vol. 1 (J. B. Finean and R. H. Michell, eds.), Elsevier/North-Holland Biomedical Press, Amsterdam, pp. 127–160.

Gahmberg, C. G., and Andersson, L. C., 1977, Selective radioactive labeling of cell surface sialoglycoproteins by periodate-tritiated borohydride, *J. Biol. Chem.* **252:**5888–5894.

Gahmberg, C. G., and Hakomori, S., 1973, External labeling of cell surface galactose and galactosamine in glycolipid and glycoprotein of human erythrocytes, *J. Biol. Chem.* **248:**4311–4317.

Gerber, G. E., and Khorana, H. G., 1982, Primary structure of bacteriorhodopsin: Sequencing methods for membrane proteins, *Meth. Enzymol.* **88:**56–74.

Gitler, C., and Bercovici, T., 1980, Use of lipophilic photoactivatable reagents to identify the lipid embedded domains of membrane proteins, *Ann. N.Y. Acad. Sci.* **346:**199–211.

Glazer, A. N., 1976, The chemical modification of proteins by group-specific and site-specific reagents, in: *The Proteins,* Vol. 1 (N. Neurath and R. L. Hill, eds.), Academic Press, London, pp. 1–103.

Gordesky, S. E., Marinetti, G. V., and Love, R., 1975, The reaction of chemical probes with the erythrocyte membrane, *J. Membr. Biol.* **20:**111–132.

Gupta, C. M., Radhakrishnan, R., Gerber, G. E., Olsen, W. L., Quay, S. C., and Khorana, H. G., 1979, Intermolecular crosslinking of fatty acyl chains in phospholipids: Use of photoactivable carbene precursors, *Proc. Natl. Acad. Sci. USA* **76:**2595–2599.

Habeeb, A. F. S. A., 1972, Reaction of protein sulfhydryl groups with Ellman's reagent, *Meth. Enzymol.* **25:**457–464.

Haest, C. W. M., Kamp, D., and Deutike, B., 1979, Topology of membrane SH-groups in the human erythrocyte. Demonstration of a nonreactive population in intrinsic proteins, *Biochim. Biophys. Acta* **557**:363–371.

Hand, E. S., and Jenks, W. P., 1962, Mechanism of the reaction of imido esters with amines, *J. Am. Chem. Soc.* **84**:3505–3514.

Henderson, R., Jubb, J. S., and Whytock, S., 1978, Specific labelling of the protein and lipid on the extracellular surface of purple membrane, *J. Mol. Biol.* **123**:259–274.

Hoare, D. G., and Koshland, D. E., 1967, A method for quantitative modification and estimation of carboxylic acid groups in proteins, *J. Biol. Chem.* **242**:2447–2453.

Hoppe, J., and Sebald, W., 1981, An essential carboxyl group for H^+ conduction in the proteolipid subunit of the ATP synthase, in: *Chemiosmotic Proton Circuits in Biological Membranes* (V. P. Skulachev and P. C. Hinkle, eds.), Addison-Wesley Publishing Company, London, pp. 449–458.

Hoyer, P. B., Owens, J. R., and Haley, B. E., 1980, The use of photoaffinity probes to elucidate molecular mechanisms of nucleotide-regulated phenomena, *Ann. N.Y. Acad. Sci.* **346**:280–301.

Hu, V. W., and Wisnieski, B. J., 1979, Photoreactive labeling of M 13 coat protein in model membranes by use of a glycolipid probe, *Proc. Natl. Acad. Sci. USA* **76**:5460–5464.

Huang, K. S., Radhakrishnan, R., Bayley, H., and Khorana, H. G., 1982, Orientation of retinal in bacteriorhodopsin as studied by cross-linking using a photosensitive analog of retinal, *J. Biol. Chem.* **257**:13616–13623.

Hubbard, A. L., and Cohn, Z. A., 1976, Specific labels of cell surfaces, in: *Biochemical Analysis of Membranes* (A. H. Maddy, ed.), Wiley, New York, pp. 427–501.

Iddon, V. B., Meth-Cohn, O., Criven, E. F. V., Suschitzky, H., and Gallagher, P. T., 1979, Entwicklungen in der Arylnitren-Chemie: Synthesen und Mechanismen, *Angew. Chem.* **91**:965–982.

Ji, T. H., 1977, A novel approach to the identification of surface receptors. The use of photosensitive hetero-bifunctional cross-linking reagent, *J. Biol. Chem.* **252**:1566–1570.

Ji, T. H., 1979, The application of chemical crosslinking for studies on cell membranes and the identification of surface reporters, *Biochim. Biophys. Acta* **559**:39–69.

Ji, T. H., and Ji, I., 1982, Macromolecular photoaffinity labeling with radioactive photoactivable hetero-bifunctional reagents, *Anal. Biochem.* **121**:286–289.

Johnstone, A. P., and Crumpton, M. J., 1979, Comparison of diiodosulphophenylisothiocyanate with other reagents as surface labels for lymphocytes, *FEBS Lett.* **108**:119–123.

Jørgensen, P. L., Karlish, S. J. D., and Gitler, C., 1982, Evidence for the organization of the transmembrane segments of (Na, K)-ATPase based on labeling lipid-embedded and surface domains of the α-subunit, *J. Biol. Chem.* **257**:7435–7442.

Kampmann, L., Lepke, S., Fasold, H., Fritzsch, G., and Passow, H., 1982, The kinetics of intramolecular cross-linking of the band 3 protein in the red blood cell membrane by 4,4′-diisothiocyano dihydrostilbene-2,2′-disulfonic acid (H_2DIDS), *J. Membr. Biol.* **70**:199–216.

Katre, N. V., and Stroud, R. M., 1981, A probable linking sequence between two transmembrane components of bacteriorhodopsin, *FEBS Lett.* **136**:170–174.

Katre, N. V., Wolber, P. K., Stoeckenius, W., and Stroud, R. M., 1981, Attachment site(s) of retinal in bacteriorhodopsin, *Proc. Natl. Acad. Sci. USA* **78**:4068–4072.

Kempf, Ch., Brock, C., Sigrist, H., Tanner, M. J. A., and Zahler, P., 1981, Interaction of phenylisothiocyanate with human erythrocyte band 3 protein. II. Topology of phenylisothiocyanate binding sites and influence of *p*-sulfophenylisothiocyanate on phenylisothiocyanate modification, *Biochim. Biophys. Acta* **641**:88–98.

Kirmse, E., (ed.), 1971, *Carbene Chemistry*, 2nd ed., Academic Press, New York.

Klip, A., and Gitler, C., 1974, Photoactive covalent labeling of membrane components from within the lipid core, *Biochem. Biophys. Res. Commun.* **60**:1155–1162.

Knauf, P. A., 1979, Erythrocyte anion exchange and the band 3 protein: Transport kinetics and molecular structure, *Curr. Top. Membr. Trans.* **12**:251–263.

Knauf, P. A., and Rothstein, A., 1980, Use of Nap-taurine as a photoaffinity probe for the human erythrocyte anion exchange system, *Ann. N.Y. Acad. Sci.* **346**:212–231.

Lanyi, J. K., and Oesterhelt, D., 1982, Identification of the retinal-binding protein in halorhodopsin, *J. Biol. Chem.* **257**:2674–2677.

Leblanc, P., Capone, J., and Gerber, G. E., 1982, Synthesis and biosynthetic utilization of radioactive photoreactive fatty acids, *J. Biol. Chem.* **257**:14586–14589.

Lemke, H. D., and Oesterhelt, D., 1981a, The role of tyrosine residues in the function of bacteriorhodopsin. Specific nitration of tyrosine 26, *Eur. J. Biochem.* **115**:595–604.

Lemke, H. D., and Oesterhelt, D., 1981b, Lysine 216 is a binding site of the retinyl moiety in bacteriorhodopsin, *FEBS Lett.* **128**:255–260.

Lemke, H. D., Bergmeyer, J., Straub, J., and Oesterhelt, D., 1982, Reversible inhibition of the proton pump bacteriorhodopsin by modification of tyrosine 64, *J. Biol. Chem.* **257**:9384–9388.

Lodish, H. F., and Rothman, J. E., 1979, The assembly of cell membranes, *Sci. Am.* **240**:38–53.

Ludwig, B., Downer, N. W., and Capaldi, R. A., 1979, Labeling of cytochrome *c* oxidase with [^{35}S]diazobenzenesulfonate. Orientation of this electron transfer complex in the inner mitochondrial membrane, *Biochemistry* **18**:1401–1407.

MacLennan, D. H., and Campbell, K. P., 1979, Structure, function and biosynthesis of sarcoplasmic reticulum proteins, *TIBS* **4**:148–151.

Marfey, S. P., and Tsai, K. H., 1975, Cross-linking of phospholipids in human erythrocyte membrane, *Biochem. Biophys. Res. Commun.* **65**:31–38.

Marinetti, G. V., and Love, R., 1976, Differential reaction of cell membrane phospholipids and proteins with chemical probes, *Chem. Phys. Lipids* **16**:239–254.

Mawby, W. J., and Findlay, J. B. C., 1982, Characterization and partial sequence of di-iodosulphophenyl isothiocyanate-binding peptide from human erythrocyte anion-transport protein, *Biochem. J.* **205**:465–475.

Means, G. E., and Feeney, R. E., 1971, *Chemical Modification of Proteins,* Holden-Day, San Francisco.

Mitchinson, C., Wilderspin, A. F., Trinnaman, B. J., and Green, N. M., 1982, Identification of a labelled peptide after stoichiometric reaction of fluorescein isothiocyanate with the Ca^{2+}-dependent adenosine triphosphatase of sarcoplasmic reticulum, *FEBS Lett.* **146**:87–92.

Miyakawa, T., Takemoto, L. J., and Fox, C. F., 1978, Membrane permeability of bifunctional, amino site-specific, cross-linking reagents, *J. Supramol. Struct.* **8**:303–310.

Moreland, R. B., Smith, P. K., Fujimoto, E. K., and Dockter, M. E., 1982, Synthesis and characterization of *N*-(4-azidophenylthio)-phthalimide: A cleavable, photoactivable crosslinking reagent that reacts with sulfhydryl groups, *Anal. Biochem.* **121**:321–326.

Mullen, E., Johnson, A. H., and Akhtar, M., 1981, The identification of lysine 216 as the retinal binding residue in bacteriorhodopsin, *FEBS Lett.* **130**:187–193.

Ngo, T. T., Yam, C. F., Lenhoff, H. M., and Ivy, J., 1981, *p*-Azidophenylglyoxal. A heterobifunctional photoactivable cross-linking reagent selective for arginyl residues, *J. Biol. Chem.* **256**:11313–11318.

Nicolson, G. L., and Singer, S. J., 1974, The distribution and asymmetry of mammalian cell surface saccharides utilizing ferritin-conjugated plant agglutinins as specific saccharide stains, *J. Cell. Biol.* **60**:236–248.

Nielsen, P. E., and Buchardt, O., 1982, Aryl azides as photoaffinity labels. A photochemical study of some 4-substituted aryl azides, *Photochem. Photobiol.* **35**:317–323.

Op den Kamp, J. A. F., 1979, Lipid asymmetry in membranes, *Annu. Rev. Biochem.* **48**:47–71.

Ovchinnikov, Yu. A., 1982, Rhodopsin and bacteriorhodopsin: Structure–function relationships, *FEBS Lett.* **148**:179–191.

Ovchinnikov, Yu. A., Abdulaev, N. G., Feigina, M. Yu., Kiselev, A. V., and Lobanov, N. A., 1979, The structural basis of the functioning of bacteriorhodopsin: An overview, *FEBS Lett.* **100**:219–224.

Owen, M. J., Knott, J. C. A., and Crumpton, M. J., 1980, Labeling of lymphocyte surface antigens by the lipophilic, photoactivatable reagent hexanoyldiiodo-*N*-(4-azido-2-nitrophenyl)tyramine, *Biochemistry* **19**:3092–3099.

Passow, H., Fasold, H., Jennings, M. L., and Lepke, S., 1982, The study of the anion transport protein (band 3 protein) in the red cell membrane by means of tritiated 4,4′-diisothiocyano-dihydrostilbene-2,2′-disulfonic acid (H_2DIDS), in: *Chloride Transport in Biological Membranes* (A. Zadunaisky, ed.), Academic Press, New York, pp. 1–31.

Peach, M. E., 1974, Thiols as nucleophiles, in: *The Chemistry of the Thiol Group,* Part 2 (S. Patai, ed.), John Wiley and Sons, New York, pp. 721–784.

Peters, K., and Richards, F. M., 1977, Chemical cross-linking: Reagents and problems in studies of membrane structure, *Annu. Rev. Biochem.* **46**:523–551.

Phillips, A. T., 1977, Differential labeling: A general technique for selective modification of binding sites, *Meth. Enzymol.* **46**:59–69.

Pick, U., and Racker, E., 1979, Inhibition of the (Ca^{2+})ATPase from sarcoplasmic reticulum by dicyclohexylcarbodiimide: Evidence for location of the Ca^{2+} binding site in a hydrophobic region, *Biochemistry* **18**:*108–113*.

Previero, A., Devancourt, J., Coletti-Previero, M. A., and Laursen, R. A., 1973, Solid phase sequential analysis: Specific linking of acidic peptides by their carboxyl ends to insoluble resins, *FEBS Lett.* **33**:135–138.

Prochaska, L. J., Bisson, R., and Capaldi, R. A., 1980, Structure of the cytochrome *c* oxidase complex: Labeling by hydrophilic and hydrophobic protein modifying reagents, *Biochemistry* **19**:3174–3179.

Prochaska, L. J., Bisson, R., Capaldi, R. A., Steffens, G. C. M., and Buse, G., 1981, Inhibition of cytochrome *c* oxidase function by dicyclohexylcarbodiimide, *Biochim. Biophys. Acta* **637**:360–373.

Quaroni, A., and Semenza, G., 1976, Partial amino acid sequences around the essential caroboxylate in the active sites of the intestinal sucrase isomaltase complex, *J. Biol. Chem.* **251**:3250–3253.

Quay, S. C., Radhakrishnan, R., and Khorana, H. G., 1981, Incorporation of photosensitive fatty acids into phospholipids of *Escherichia coli* and irradiation-dependent cross-linking of phospholipids to membrane proteins, *J. Biol. Chem.* **256**:4444–4449.

Radhakrishnan, R., Robson, R. J., Takagaki, Y., and Khorana, H. G., 1981, Synthesis of modified fatty acids and glycerophospholipid analogs, *Meth. Enzymol.* **72**:408–433.

Raftery, M. A., Witzemann, V., and Blanchoud, S. G., 1980, The use of photochemical probes for studies of structure and function of purified acetylcholine receptor preparations, *Ann. N.Y. Acad. Sci.* **346**:458–474.

Ramjeesingh, M., Gaarn, A., and Rothstein, A., 1980, The location of a disulfonic stilbene binding site in band 3, the anion transport protein of the red blood cell membrane, *Biochim. Biophys. Acta* **599**:127–139.

Ramjeesingh, M., Gaarn, A., and Rothstein, A., 1981, The amino acid conjugate formed by the interaction of the anion transport inhibitor 4,4′-diisothiocyano-2,2′-stilbenedisulfonic acid (DIDS) with band 3 protein from human red blood cell membranes, *Biochim. Biophys. Acta* **641**:173–182.

Renthal, R., Dawson, N., Tuley, J., and Horowitz, P., 1983, Constraints on the flexibility of bacteriorhodopsin's carboxyl-terminal tail at the purple membrane surface, *Biochemistry* **22**:5–12.

Riordan, J. F., 1979, Arginyl residues and anion binding sites in proteins, *Mol. Cell. Biochem.* **26**:71–92.

Ross, A. H., Radhakrishnan, R., Robson, R. J., and Khorana, H. G., 1982, The transmembrane domain of glycophorin A as studied by crosslinking using photoactivatable phospholipids, *J. Biol. Chem.* **257**:4152–4161.

Rothstein, A., 1982, Functional structure of band 3, the anion transport protein of the red blood cell, as determined by proteolytic and chemical cleavages, in: *Membranes and Transport*, Vol. 2 (A. Martonosi, ed.), Plenum Press, New York, pp. 435–440.

Sabban, E., Marchesi, V., Adesnik, M., and Sabatini, D. D., 1981, Erythrocyte membrane protein band 3: Its biosynthesis and incorporation into membranes, *J. Cell. Biol.* **91**:637–646.

Schäfer, R., Hinnen, R., and Franklin, R. M., 1974, Structure and synthesis of a lipid-containing bacteriophage. Properties of the structural proteins and distribution of the phospholipid, *Eur. J. Biochem.* **50**:15–27.

Schellenberg, K., 1963, The synthesis of secondary and tertiary amines by borohydride reduction, *J. Org. Chem.* **28**:3259–3261.

Schweizer, E., Angst, W., and Lutz, H. U., 1982, Glycoprotein topology on intact human red blood cells reevaluated by cross-linking following amino group supplementation, *Biochemistry* **21**:6807–6818.

Seelig, J., and Seelig, A., 1980, Lipid conformation in model membranes and biological membranes, *Q. Rev. Biophys.* **13**:19–61.

Shaw, A. B., and Marinetti, G. V., 1980, Cross-linking of erythrocyte membrane proteins and phospholipids by chemical probes, *Membr. Biochem.* **3**:1–19.

Ship, S., Shami, Y., Breuer, W., and Rothstein, A., 1977, Synthesis of tritiated 4,4′-diisothiocyano-2,2′-stilbene disulfonic acid (H_2DIDS) and its covalent reaction with sites related to anion transport in red blood cells, *J. Membr. Biol.* **33**:311–324.

Sigman, D. S., and Mooser, G., 1975, Chemical studies of enzyme active sites, *Annu. Rev. Biochem.* **44**:889–931.

Sigrist, H., and Zahler, P., 1982a, Hydrophobic labeling and cross-linking of membrane proteins, in: *Membranes and Transport*, Vol. 1 (A. N. Martonosi, ed.), Plenum Press, New York, pp. 173–184.

Sigrist, H., and Zahler, P., 1982b, Heterobifunctional crosslinking of bacteriorhodopsin by hydrophobic azidoarylisothiocyanates, *Meth. Enzymol.* **88:**207–212.

Sigrist, H., Kempf, Ch., and Zahler, P., 1980, Interaction of phenylisothiocyanate with human erythrocyte band 3: I. Covalent modification and inhibition of phosphate transport, *Biochim. Biophys. Acta* **597:**137–144.

Sigrist, H., Allegrini, P. R., Strasser, R. J., and Zahler, P., 1981, Chemical modification of bacteriorhodopsin by phenylisothiocyanate: Effect on the photocycle, in: *The Blue Light Syndrome* (H. Senger, ed.), Springer Verlag, Berlin, pp. 30–37.

Sigrist, H., Allegrini, P. R., Kempf, Ch., Schnippering, Ch., and Zahler, P., 1982, 5-Isothiocyanato-1-naphthalene azide and *p*-azidophenylisothiocyanate: Synthesis and application in hydrophobic heterobifunctional photoactive cross-linking of membrane proteins, *Eur. J. Biochem.* **125:**197–201.

Sigrist, H., Allegrini, P. R., Stauffer, K., Schaller, J., Abdulaev, N. G., Rickli, E. E., and Zahler, P., 1984, Group-directed modification of bacteriorhodopsin by arylisothiocyanates. *J. Mol. Biol.* **173:**93–108.

Sigrist-Nelson, K., and Azzi, A., 1979, The proteolipid subunit of the chloroplast adenosine triphosphatase complex: Mobility, accessibility and interaction studied by a spin label technique, *J. Biol. Chem.* **254:**4470–4474.

Sigrist-Nelson, K., and Azzi, A., 1980, The proteolipid subunit of the chloroplast adenosine triphosphatase complex, *J. Biol. Chem.* **255:**10638–10643.

Spiess, M., Brunner, J., and Semenza, G., 1982, Hydrophobic labeling isolation and partial characterization of the NH_2-terminal membraneous segment of sucrase–isomaltase complex, *J. Biol. Chem.* **257:**2370–2377.

Staros, J. V., and Richards, F. M., 1974, Photochemical labeling of the surface proteins of human erythrocytes, *Biochemistry* **13:**2720–2727.

Staros, J. V., Morgan, D. G., and Appling, D. R., 1981, A membrane-impermeant, cleavable cross-linker. Dimers of human erythrocyte band 3 subunits cross-linked at the extracytoplasmic membrane face, *J. Biol. Chem.* **256:**5890–5893.

Stauffer, K., Sigrist, H., and Zahler, P., 1982, Azo dye labeling of bacteriorhodopsin, *Experientia* **38:**733.

Steck, T. L., and Dawson, G., 1974, Topographical distribution of complex carbohydrates in the erythrocyte membrane, *J. Biol. Chem.* **249:**2135–2142.

Steck, T. L., Ramos, B., and Strapazon, E., 1976, Proteolytic dissection of band 3, the predominant transmembrane polypeptide of the human erythrocyte membrane, *Biochemistry* **15:**1154–1161.

Stoffel, W., Salm, K. P., and Müller, M., 1982, Syntheses of phosphatidylcholines, sphingomyelines and cholesterol substituted with azido fatty acids, *Hoppe-Seyler's Z. Physiol. Chem.* **363:**1–18.

Stuchbury, T., Shipton, M., Norris, R., Malthouse, J. P. G., Brocklehurst, K., Herbert, J. A. L., and Suschitzky, H., 1975, A reporter group delivery system with both absolute and selective specificity for thiol groups and an improved fluorescent probe containing the 7-nitrobenzo-2-oxa-1,3-diazole moiety, *Biochem. J.* **151:**417–432.

Tarrab-Hazdai, R., Bercovici, T., Goldfarb, V., and Gitler, C., 1980, Identification of the acetylcholine receptor subunit in the lipid bilayer of *Torpedo* electric organ excitable membranes, *J. Biol. Chem.* **255:**1204–1209.

Vallee, B. L., and Riordan, J. F., 1969, Chemical approaches to the properties of active sites of enzymes, *Annu. Rev. Biochem.* **38:**733–794.

Van Deenen, L. L. M., 1981, Topology and dynamics of phospholipids in membranes, *FEBS Lett.* **123:**3–15.

Van den Bosch, H., 1980, Intracellular phospholipases A, *Biochim. Biophys. Acta* **604:**191–246.

Vanin, E. F., and Ji, T. H., 1981, Synthesis and application of cleavable photoactivable heterobifunctional reagents, *Biochemistry* **20:**6754–6760.

Verkleij, A. J., Zwaal, R. F. A., Roelofsen, B., Comfurius, P., Kastelijn, D., and Van Deenen, L. L. M., 1973, The asymmetric distribution of phospholipids in the human red cell membrane, *Biochim. Biophys. Acta* **323:**178–193.

Wang, K., and Richards, F. M., 1975, Reaction of dimethyl-3,3′-dithiobispropionimidate with intact human erythrocytes. Cross-linking of membrane proteins and hemoglobin, *J. Biol. Chem.* **250:**6622–6626.

Wells, E., and Findlay, J. B. C., 1979, Labelling of the intramembraneous region of the major sialoglycoprotein of human erythrocytes with a photosensitive hydrophobic probe, *Biochem. J.* **179:**265–272.

Wells, E., and Findlay, J. B. C., 1980, The isolation of human-erythrocyte band-3 polypeptide labelled with a photosensitive hydrophobic probe, *Biochem. J.* **187:**719–725.

Wieth, J. O., Bjerrum, P. J., and Borders, C. L., 1982, Irreversible inactivation of red cell chloride exchange with phenylglyoxal, an arginine-specific reagent, *J. Gen. Physiol.* **79:**283–312.

Wisnieski, B. J., and Bramhall, J. S., 1981, Photolabeling of cholera toxin subunits during membrane penetration, *Nature* **289:**319–321.

Wisnieski, B. J., Shiflett, M. A., Mekalanos, J., and Bramhall, J. S., 1979, Analysis of transmembrane dynamics of cholera toxin using photoreactive probes, *J. Supramol. Struct.* **10:**191–197.

Wold, F., 1972, Bifunctional reagents, *Meth. Enzymol.* **25:**623–651.

Yip, C. C., Moule, M. L., and Yeung, C. W. T., 1982, Subunit structure of insulin receptor of rat adipocytes as demonstrated by photoaffinity labeling, *Biochemistry* **21:**2940–2945.

Yoshida, M., Allison, W. S., Esch, F. S., and Futai, M., 1982, The specificity of carboxyl group modification during the inactivation of the *Escherichia coli* F_1-ATPase with dicyclohexyl[^{14}C]carbodiimide, *J. Biol. Chem.* **257:**10033–10037.

Zahler, P., and Wolf, M., 1982, Side-specific transfer of membrane-bound phospholipase A_2 to other membranes by fusion, in: *Protides of the Biological Fluids,* Vol. 29 (H. Peeters, ed.), Pergamon Press, Oxford, pp. 267–270.

Zaki, L., 1983, Anion transport in red blood cells and arginine specific reagents. (1) Effect of chloride and sulfate ions on phenylglyoxal sensitive sites in the red blood cell membrane, *Biochem. Biophys. Res. Commun.* **110:**616–624.

Zisapel, N., and Littauer, U. Z., 1978, A cationic hydroxysuccinimide ester. A reagent for labeling exterior membrane proteins, *Biochim. Biophys. Acta* **512:**156–162.

Zwaal, R. F. A., and Roelofsen, B., 1976, Applications of pure phospholipases in membrane studies, in: *Biochemical Analysis of Membranes* (A. H. Maddy, ed.), Chapman and Hall, London, pp. 352–377.

11

Calcium Ions, Enzymes, and Cell Fusion

Jack A. Lucy

I. INTRODUCTION

Membrane fusion is a phenomenon of major importance in cell biology. It embraces events at the subcellular level, e.g., endocytosis and exocytosis, and it is an essential dynamic component of secretion, membrane flow, and lysosomal activity. Cell fusion is a special case of membrane fusion that occurs naturally in a comparatively small number of situations. Of these, fertilization and the fusion of myoblasts in the development of skeletal muscle are probably the most extensively investigated. There are also, however, a number of experimental ways of inducing cells to fuse artificially by, for example, treating them with viruses, e.g., Sendai virus, with chemicals, e.g., poly(ethylene glycol), and most recently by the application of electrical forces (Zimmermann, 1982). In addition to its physiological significance, cell fusion has therefore recently become valuable as a laboratory tool, particularly in the production of hybridoma cells that secrete monoclonal antibodies.

Ca^{2+} ions play an important role both in membrane fusion in general and in cell fusion in particular. This is illustrated by the fact that the multi-author book on membrane fusion edited by Poste and Nicolson (1978) has considerably more than 100 references to Ca^{2+} in its subject index. The present article is primarily concerned with three aspects of Ca^{2+} in membrane fusion. First, the role of Ca^{2+} in the fusion of myoblasts will be considered in view of the fact that recent experimental work indicates that Ca^{2+} has intracellular functions in the fusion of myoblasts. This differs from the idea, which has been held for several years, that the Ca^{2+}-sensitive site in myoblast fusion is on the outside surface of the cells.

Jack A. Lucy • Department of Biochemistry and Chemistry, Royal Free Hospital School of Medicine, University of London, London NW3 2PF, England.

Second, some general hypotheses regarding Ca^{2+} and membrane fusion will be discussed. Experimental work in my own laboratory on the chemically induced fusion of erythrocytes will then be considered, which has demonstrated that cell fusion can occur at high incidence in the presence of EGTA. This contrasts with the view of some investigators that Ca^{2+} is needed to trigger phase separations in membrane phospholipids that, in turn, lead to membrane fusion. The attention of the reader is also drawn to a number of reports on the fusion of lipid bilayers in model systems in which Ca^{2+} is equally not required.

Finally, a new hypothesis is considered in which the writer has proposed that hydrophobic proteins or peptides, which are formed by the endogenous proteolysis of membrane proteins and which act like viral fusogenic proteins, may mediate biomembrane fusion in general.

II. THE FUSION OF MYOBLASTS

A. Dependence on Ca^{2+}

It has been known for some 15 years that the fusion of myoblasts is dependent on Ca^{2+} ions. This was apparently first reported by Shainberg *et al.* (1969) who observed that primary cultures of rat skeletal muscle cells failed to fuse but continued to proliferate when the concentration of Ca^{2+} in the culture medium was decreased to approximately 270 μM. Fusion nevertheless occurred within 3–4 hr of the concentration of Ca^{2+} being restored to normal (1400 μM). Subsequently, it was found by others that myoblasts from a variety of species showed a similar dependence on the presence of Ca^{2+} for fusion, and also that the addition of chelating agents for Ca^{2+}, e.g., EGTA, can be used to inhibit their fusion. The ability to control and partially synchronize the fusion of myoblasts in cell cultures by varying the concentration of Ca^{2+}, or by using chelating agents, has in fact been used as an experimental device by numerous investigators, as in the ultrastructural studies of Kalderon and Gilula (1979) and of Fumigalli *et al.* (1981). The latter workers noted, incidentally, that rat myoblasts that had been inhibited for more than three days partially lost the ability to fuse when Ca^{2+} was replaced. In early investigations, it was shown by van der Bosch *et al.* (1972) that the fusion of myoblasts proceeded from 10 to 90% when the concentration of Ca^{2+} was increased by only a factor of three. Also, Schudt *et al.* (1973) demonstrated that Mg^{2+}, Zn^{2+}, Mn^{2+}, Cu^{2+}, La^{2+}, and Li^{+} cannot substitute for Ca^{2+}, although limited fusion occurred with Sr^{2+}.

In studies on the mechanism of action of Ca^{2+} in myoblast fusion, Dahl *et al.* (1978) investigated the fusion of plasma membranes that were isolated from myoblasts grown in tissue culture. In this way, it was hoped to avoid problems of interpretation arising from the aggregation, motility, and metabolism of intact cells during the fusion process. In these experiments, the fusion of isolated plasma membrane vesicles was characterized by the appearance of twinned vesicles in electron micrographs. The percentage of fusion in 24-hr myoblast membranes that occurred after the addition of 1.4 mM Ca^{2+} was no different from that without Ca^{2+}. However, in membrane preparations derived from myoblasts after 30, 50, or 60 hr in culture, Ca^{2+}-specific fusion progressively increased. As with intact myoblasts, Mg^{2+}, Sr^{2+}, and Ba^{2+} were either unable or only weakly able to support fusion by comparison with Ca^{2+}. Lysolecithin (10^{-6} M) decreased the fusion of the vesicles in a similar manner to its effect on the fusion of myoblasts (Schudt *et al.*, 1976) but, by contrast with intact cells, fusion of isolated myoblast membranes was enhanced by concentrations of Ca^{2+} in excess of 5 mM.

Important early experiments undertaken by Schudt and Pette (1975) employed the divalent cation ionophore A23187 to determine whether Ca^{2+} acts on the outside surfaces of myoblasts or in their interior during the fusion process. Because of the difficulty of directly measuring an increase in the intracellular concentration of free Ca^{2+}, these investigators tried to show indirectly that A23187 influences the various pools of Ca^{2+} in chick myoblasts. They found that efflux of $^{45}Ca^{2+}$ from prelabeled cells was stimulated by A23187, thereby indicating that the ionophore could also increase the intracellular concentration of Ca^{2+} by enhancing the permeability of the membrane to the exogenous ion. It was also observed, however, that the Ca^{2+} dependence of myoblast fusion was not affected by the ionophore at any concentration of Ca^{2+}, apart from an inhibition by toxic concentrations of A23187. Schudt and Pette therefore concluded that the dependence of myoblast fusion on the concentration of Ca^{2+} reflects the concentration dependence of binding sites of Ca^{2+} that are expressed at the outer surface of the plasma membrane.

Comparable observations have been made by other workers. For example, Horwitz and his colleagues found that, at levels of ionophore that catalyze a net transfer of Ca^{2+} across cellular membranes, myoblasts did not form myotubes in a medium containing slightly suboptimal concentrations of Ca^{2+}. In addition, the presence of the ionophore did not inhibit the formation of myotubes in media containing concentrations of Ca^{2+} that support fusion. Other related experiments showed that ruthenium red, an impermeable reagent that competes for Ca^{2+}-binding sites, inhibits myotube formation (unpublished observations referred to in Knudsen and Horwitz, 1978). In their work on the fusion of membrane vesicles isolated from myotubes that is referred to above, Dahl *et al.* (1978) also observed that vesicle fusion was not increased at suboptimal concentrations of Ca^{2+} in the presence of A23187.

These various findings on myoblasts, Ca^{2+}, and A23187 stand in direct contrast to observations made in this laboratory with a quite different fusion system, namely, hen erythrocytes treated with A23187 and Ca^{2+}. Erythrocytes that were aggregated by incubation with dextran (or better by treatment with neuraminidase and dextran) were readily fused into large multinucleated cells in media containing A23187 and 2 mM Ca^{2+} (Ahkong *et al.*, 1975a). On the basis of this observation, it was proposed that Ca^{2+} may mediate cell fusion, occurring either naturally or in response to fusogenic agents, by interacting with the cytoplasmic surface of the plasma membrane following an initial increase in membrane permeability to exogenous Ca^{2+}. In subsequent work, Blow *et al.* (1979) investigated the effects of a number of different chemical fusogens, including benzyl alcohol and poly(ethylene glycol) on the permeability of erythrocytes to Ca^{2+} and to other ions. From this study, it was concluded that both lipid-soluble and water-soluble fusogens increase the permeability of erythrocyte membranes to cations, including Ca^{2+}, and that the entry of Ca^{2+} into the cells initiates the fusion of cells treated in this way.

B. Some Recent Developments

1. The Concentration of Intracellular Ca^{2+}

An important recent paper by David *et al.* (1981) appears to have clarified partially the complexities of the interactions of Ca^{2+} and myoblasts in the fusion reaction. First, these investigators found that, although EGTA blocked the alignment (and hence the fusion) of embryonic chick myoblasts when it was added to cell cultures at 24 hr, the addition of EGTA at 48 and 64 hr still inhibited cell fusion even though much of the alignment process was already complete at the later times. This indicated that Ca^{2+} functions in more than one way, and that an initial function of the ion is to facilitate cellular alignment, possibly involving extracellular Ca^{2+}-sensitive sites. Second, they found that there is a measurable increase in the net cellular influx of Ca^{2+} just prior to observable cell fusion. In addition, the drug D600, which blocks Ca^{2+} channels in membranes, inhibited cell fusion while myoblast alignment apparently proceeded normally. Finally, ionophore A23187 was observed to induce precocious fusion, bringing the time course of fusion forward by some 6–8 hr, without either affecting the general shape of the curve for the time course of fusion or increasing the final percentage of nuclei in myotubes. David and co-workers have therefore concluded that an increased rate of net Ca^{2+} influx into myoblasts is a normal and essential step in the myoblast fusion reaction.

These observations constitute an interesting parallel with those recently made by Cuthbertson *et al.* (1981) on the activation of mouse oocytes. They have investigated the concentration of free Ca^{2+} in the cytoplasm of single mouse oocytes during artificial activation by benzyl alcohol and during fertilization, and they have reported an increase from a resting concentration of less than 0.1 M to more than 5 M in 10–30 min.

In the light of these recent developments, it would now seem that, although the three systems have many individual distinguishing features, the fusion of myoblasts, fertilization, and the chemically induced fusion of erythrocytes (by, for example, benzyl alcohol) each involves an entry into the cytoplasm of extracellular Ca^{2+}.

2. Events Occurring before the Entry of Ca^{2+}

David and Higginbotham (1981) have further explored the influx of Ca^{2+} ions into myoblasts prior to fusion in relation to Zalin's earlier findings that a transient rise in cyclic AMP, dependent on prostaglandin PGE_1, occurs 5–6 hr prior to the fusion of myoblasts (Zalin, 1977). In this connection, they have shown that (1) the increase in cyclic AMP precedes and/or is independent of the influx of Ca^{2+}, (2) the Ca^{2+} influx is either directly or indirectly dependent on PGE_1 activity and perhaps on the synthesis of PGE_1, and (3) although the increase in cyclic AMP may be essential for fusion, it is not sufficient in the absence of an influx of Ca^{2+} ions. In an interesting, very recent paper, Wakelam (1983) has reported that radioactively labeled phosphatidylinositol 4-phosphate and phosphatidylinositol 4,5-bisphosphate are degraded, and diacylglycerol and phosphatidic acid are formed, when fusion-competent chick embryonic myoblasts are stimulated to fuse by increasing the concentration of Ca^{2+} in the medium to 1.4 mM. Similar findings were made with phosphatidylinositol itself (Wakelam and Pette, 1982), but the other common phospholipids were unaffected. Sr^{2+} which, as mentioned above, can replace Ca^{2+} in stimulating myoblast fusion, also stimulated the breakdown of all three inositol phospholipids, whereas Mg^{2+} inhibited both the Ca^{2+}-stimulated fusion and breakdown of the phospholipids. In view of the possibility that the breakdown of phosphatidylinositol 4,5-bisphosphate is a primary event in Ca^{2+}-gating in membranes (Michell, 1982), and of the fact that Ca^{2+} enters myoblasts before fusion, Wakelam has suggested that the changes observed in the myoblast phospholipids are part of a gating process for the entry into these cells of extracellular Ca^{2+}, and that the diacylglycerol that is formed (a weak chemical fusogen; Ahkong *et al.*, 1973) may stimulate fusion. Since it has also been observed that the fusion of myoblasts does not occur in the absence of embryo extract, even when 1.4 mM Ca^{2+} is present, Wakelam and Pette (1984) have further proposed that myoblast fusion is initiated by a factor, present in the extract, that binds to a cell surface receptor to initiate the breakdown of inositol phospholipids.

3. *Events after the Entry of Ca^{2+}*

To return to a comparison of the fusion of myoblasts with the chemically induced fusion of erythrocytes, it is relevant to consider the events that are associated with the entry of Ca^{2+} into red blood cells that result from the increase in membrane permeability caused by chemical fusogens. In rat erythrocytes treated with benzyl alcohol, cell fusion was inhibited by EGTA, and also by thiol reagents (Ahkong *et al.*, 1980). In the treated cells, membrane proteins that correspond to bands 2 and 3 in human erythrocytes were decreased, and a polypeptide with a slightly greater mobility than band 3 was produced. These changes were also inhibited by EGTA and by thiol reagents, indicating that a Ca^{2+}-activated thiol proteinase is important in the fusion reaction. Since several recent investigations have indicated that other cells have structures that are comparable to the spectrin–actin skeleton of the erythrocyte membrane (Haest 1982; Baines 1983), these findings may well be relevant to the initiation of other biomembrane fusion reactions. Muscle tissue, for example, contains a well-documented Ca^{2+}-dependent thiol proteinase. Furthermore, Ca^{2+}-activated neutral proteinase activity has been reported to appear concomitantly with myotube formation during the differentiation of a cell line (L_6) of rat skeletal myoblasts, while other proteases such as cathepsin D and plasminogen activator did not show any change in their activities (Kaur and Sanwal, 1981). The appearance of the protease activity was apparently not fusion dependent, as judged by the high protease activity of a fusion-defective mutant line of myoblasts. On the other hand, preliminary experiments indicated that treatment with 5-bromodeoxyuridine, which can inhibit myoblast fusion, decreased the protease activity, and also that prefusion myoblasts may contain an inhibitor of the enzyme. It should be noted with regard to the experiments with 5-bromodeoxyuridine, however, that, as pointed out by Bischoff (1978), agents that act at the nuclear level probably affect the generation of fusion-competent cells from precursors and are not likely to provide information about the events directly involved in fusion.

Another investigation on rat myoblasts, but with primary cultures, has shown that the Ca^{2+}-dependent fusion of these cells appears to require the activity of a neutral metalloendoprotease at the time of fusion (Couch and Strittmatter, 1983). The metalloproteinase inhibitor 1,10-phenanthroline and the synthetic dipeptide substrate CBZ-Gly-Phe-amide prevented fusion, but inhibitors of serine, thiol, and carboxy proteases had no effect. The role of the metalloendoprotease in the fusion process is at present unknown. It would seem, however, that the enzyme is probably concerned with a Ca^{2+}-dependent function occurring at the outside surface of the cell, since 1,10-phenanthroline and the synthetic substrate inhibitor may not be able to enter the cells.

Transglutaminase is a Ca^{2+}-dependent enzyme that is widely distributed in tissues and, in this connection, it is interesting to make a further comparison between the chemically induced fusion of erythrocytes and the myoblast fusion reaction. The transglutaminase of human erythrocytes is essentially inactive until, for example, the cells are incubated in the presence of extracellular Ca^{2+} and ionophore A23187. After activation of the intracellular enzyme in this way, crosslinked protein polymers derived from spectrin and bands 4.1 and 3 are found in preparations of erythrocyte membrane proteins (Lorand *et al.*, 1980). In view of the increase in membrane permeability to Ca^{2+} that is associated with the fusion of hen erythrocytes induced by chemical fusogens discussed above, it is to be expected, and has in fact been observed, that the membrane proteins of fused erythrocytes would show changes consistent with the activation of transglutaminase activity (Quirk *et al.*, 1978; Ahkong *et al.*, 1980). As recent work has indicated that an entry of exogenous Ca^{2+} into myoblasts also occurs prior to the fusion reaction, and indeed seems to be necessary for fusion to occur, it is relevant to ask if a transglutaminase is similarly activated in myoblasts during their fusion and, if so, whether the enzyme plays a functional role. Very recently, it has been shown that lysates of embryonic chick myoblasts contain transglutaminase activity that is inhibited by EGTA, histamine, and glycine methyl ester (Berstein *et al.*, 1983). It was also observed that histamine and dansyl cadaverine each reversibly inhibits myotube formation by freshly isolated chick myoblasts. Glycine methyl ester inhibited myotube formation but sarcosine methyl ester, which is not a competitive inhibitor of transglutaminase, had little if any inhibitory action. In addition, the formation of myotubes was not inhibited by cultivation in normal media adjusted to pH 8.0–8.1, indicating that the observed inhibitory effects were not mediated by a lysosomotropic effect. It was apparent from these experiments that myoblast fusion was not, in fact, inhibited in the treated cells. Instead, the fused cells failed to elongate into myotubes, the products of cell fusion thus resembling the myosacs observed with colchicine-treated myoblast cultures by Bischoff and Holtzer (1968). Histamine may perhaps have inhibited myotube formation by preventing myoblasts from adhering to the substrate. It seemed more likely, however, that crosslinking by a transglutaminase of proteins involved in the development of the structure of the myotube, such as the 58,000 mol. wt. protein described by Fulton *et al.* (1981), may be an essential feature of the formation of the normal structure of myotubes following the fusion of myoblasts.

III. GENERAL HYPOTHESES: Ca^{2+}, PHOSPHOLIPIDS, AND MEMBRANE FUSION

A. Ca^{2+} and ATPase Activity

Ten years ago, Poste and Allison (1973) surveyed the large number of membrane fusion reactions in exocytosis that are Ca^{2+} dependent: in the adrenal medulla, neurohypophysis, polymorphonuclear leukocytes, mast cells, platelets, synapses, endocrine pancreas, and salivary gland. They stressed that these membrane fusion reactions are inhibited by an excess of Ca^{2+}. As mentioned above, it is also known (van der Bosch *et al.*, 1972; Schudt and Pette, 1975) that high concentrations of Ca^{2+} inhibit the fusion of myoblasts. Poste and Allison drew attention in relation to the inhibition of fusion by Ca^{2+} to previous observations made by Woodin and Wieneke (1963, 1964) on the release by exocytosis of lysosomal enzymes from polymorphonuclear leukocytes treated with staphylococcal leucocidin, vitamin A, and streptolysin O, and to a proposal put forward in this connection by the earlier workers. Woodin and Wieneke proposed that, since high concentrations of Ca^{2+} and ATP inhibited the release reaction, the displacement of membrane-associated Ca^{2+} and ATP might produce areas in the membrane having a decreased structural rigidity and an increased susceptibility to fusion. Woodin and Wieneke further suggested that the displacement of Ca^{2+} and ATP from the membrane was initiated by the hydrolysis of membrane-bound ATP involving an ATPase present in the membrane granules. In their review article, Poste and Allison (1973) developed a general theory for membrane fusion based on the earlier proposals of Woodin and Wieneke, in which the Ca^{2+} dependence displayed by membrane fusion reactions in different situations and the inhibitory effects of high concentrations of Ca^{2+} on fusion were attributed to involvement of an ATPase.

Particular emphasis was placed by Poste and Allison on the ability of Ca^{2+} to associate with and dissociate from membrane sites in order to allow the membrane to exist in two states: (1) the "normal" Ca^{2+}-associated state, and (2) the "fusion-susceptible" Ca^{2+}-dissociated state. Thus, it was considered that the actual process of membrane fusion could occur only after the removal of Ca^{2+} from membranes, which was thought to (1) increase the motional freedom of the membrane lipids and possibly membrane proteins, (2) increase the number of potential interaction sites in the membrane by increasing the number of free ionogenic sites, and (c) induce a general "disorder" perturbation that would favor contact and a linkage between macromolecular components in the two membranes undergoing fusion.

When putting forward their hypothesis, Poste and Allison noted that it has been reported that avian erythrocytes can be induced to fuse by the addition of very high concentrations of Ca^{2+} (40 mM; Toister and Loyter, 1971), but that attempts to repeat this observation in Poste's laboratory had been unsuccessful. Subsequently, Ahkong *et al.* (1975a) showed that ionophore A23187 promotes the fusion of erythrocytes in the presence of Ca^{2+}, indicating that Ca^{2+} mediates the fusion of these cells by interacting with the cytoplasmic side of the plasma membrane following an increase in membrane permeability to exogenous Ca^{2+}. It has also been demonstrated that the fusion of isolated secretory vesicles from pancreatic islets, liver, adrenal medulla, and neurohypophysis is triggered by Ca^{2+} (reviewed in Gratzl *et al.*, 1980), and innumerable papers have reported in recent years that Ca^{2+} is involved in the induction of membrane fusion. It therefore seems most unlikely that membranes are "fusion-susceptible" when they are in a Ca^{2+}-dissociated state as proposed by Poste and Allison. Even so, there is still no clear understanding of the problem that was originally addressed in their review, namely, why membrane fusion reactions are inhibited by high concentrations of Ca^{2+}.

B. Phase Separations of Membrane Lipids

Papahadjopoulos and his colleagues have developed a general theory for the role of Ca^{2+} in membrane fusion reactions on the basis of an extensive series of investigations on the behavior of phospholipid vesicles. The first report on the role of Ca^{2+} in the fusion of such vesicles presented evidence that incubation of phosphatidylserine vesicles with Ca^{2+} induced an increase in size that could not be reversed by excess EDTA (Papahadjopoulos *et al.*, 1974). Later, it was shown by freeze-fracture electron microscopy that the addition of Ca^{2+} (1–10 mM) to sonicated vesicles of this phospholipid resulted in the formation of large spiral (cochleated) lipid cylinders that became large, closed, spherical vesicles on the addition of excess EDTA (Papahadjopoulos *et al.*, 1975). An interesting feature of the work of this group with model membrane systems is the specificity for divalent cations that was observed. For example, Mg^{2+} did not induce the fusion of vesicles of phosphatidylserine or of phosphatidylglycerol under the conditions used for Ca^{2+}-induced fusion, but Ca^{2+} and Mg^{2+} were equally effective in inducing the fusion of vesicles of phosphatidic acid. It was considered that the effects of Ca^{2+} on the fusion of vesicles of phosphatidylserine and phosphatidylglycerol are due to specific binding to the head-group charges (Papahadjopoulos *et al.*, 1977). More recently, it has been proposed that this involves a polydentate chelation of Ca^{2+} with the ionic head groups of phosphatidylserine from apposed membranes in an anhydrous intermembrane arrangement ("trans" complex; Portis *et al.*, 1979; Ekerdt and Papahadjopoulos, 1982).

Observations made in further studies by these investigators on the mechanism of the Ca^{2+}-induced fusion of phosphatidylserine vesicles were taken to indicate that a phase change from a fluid to a solid state is a key event during fusion induced by divalent cations (Papahadjopoulos *et al.*, 1977). Ca^{2+} ions were therefore suggested to have a dual role in membrane fusion. First, they promote the close apposition of adjacent membranes by enhancing electrostatic interactions between them. The actual fusion of apposed bilayers was proposed to involve a second step regarded as *the crucial event in fusion* (their italics) responsible for triggering most, and perhaps all membrane fusion phenomena; namely, a Ca^{2+}-induced separation of acidic phospholipids such as phosphatidylserine into rigid crystalline domains, with fusion occurring at domain boundaries between adjacent membranes (Papahadjopoulos, 1978; Papahadjopoulos *et al.*, 1979). This was put forward as a single mechanism to explain both extracellular fusions (cell-to-cell, enveloped virus-to-cell, lipid vesicle-to-cell) as well as intracellular fusions (endocytosis, exocytosis). It was additionally envisaged that, since acidic phospholipids vary in their ability to undergo Ca^{2+}-induced phase separations, the susceptibility to fusion of any given membrane may be regulated not only by the absolute content of acidic phospholipids but also by their specific distribution within the bilayer. In the case of exocytosis induced by physiological agents, however, it was considered that fusion requires only that the intracellular concentration of Ca^{2+} be raised to the level necessary to induce a phase separation of acidic phospholipids since the inner face of the plasma membrane and the membranes of secretory granules are characterized by the presence of acidic phospholipids such as phosphatidylserine.

Papahadjopoulos and his co-workers have themselves subsequently commented that, as the percentage of phosphatidylserine in most biological membranes is less than 50% of the total phospholipids, the use of pure phosphatidylserine model membranes for studying the mechanisms of membrane fusion can be questioned (Duzgunes *et al.*, 1981). They have nevertheless shown that vesicles containing only 10–25% phosphatidylserine are able to fuse extensively in the presence of Ca^{2+}, and that Mg^{2+} reduces the concentration of Ca^{2+} required to induce fusion. In addition, it has been observed that synexin, a soluble protein obtainable from the adrenal medulla and from liver which causes Ca^{2+}-dependent aggregation of isolated chromaffin granules, decreases the concentration of Ca^{2+} required for the fusion of large unilamellar vesicles of phosphatidylserine and of a mixture of phosphatidylserine with phosphatidylethanolamine. Synexin also increases the initial rate of fusion, and it has been proposed that this protein may play a role in exocytosis by promoting the fusion in the presence of Ca^{2+} of membranes containing specific phospholipids (Hong *et al.*, 1981). Despite these interesting observations, Papahadjopoulos and his colleagues would, however, appear to have overstated their case in regarding the Ca^{2+}-induced phase separation of acidic phospholipids as the crucial event that triggers most, or perhaps all membrane fusion phenomena since, as is discussed below, the fusion of cells and of phospholipid vesicles can occur in the absence of Ca^{2+}.

In another publication from this group it was reported that, in the presence of physiological concentrations of phosphate (10 mM), the concentration of Ca^{2+} required for the fusion of vesicles of phosphatidylserine is lowered by 3- to 4-fold, and the rate of vesicle fusion is increased by up to 1000-fold (Fraley *et al.*, 1980). Other experiments indicated that a significant enhancement in vesicle fusion occurred only when the precipitation of calcium phosphate was initiated in the presence of the vesicles, suggesting that crystal nucleation on the surface of the vesicles is a prerequisite for fusion. This is relevant to a number of reports on the chemically induced fusion of cells that involve the precipitation of a calcium salt. In early work, it was noted that hen erythrocytes are fused by the addition of fluoride ions in the presence of Ca^{2+} (Lucy *et al.*, 1971). Zakai *et al.* (1976) have since described in detail the fusion of human erythrocyte ghosts that is promoted by the combined action of calcium and phosphate ions. All of the divalent metal ions that formed precipitates with Ca^{2+} were observed to cause agglutination of erythrocyte ghosts. However, fusion was distinct from agglutination since Zn^{2+} and Mn^{2+} in phosphate buffer did not promote fusion although these ions were active in agglutination. More recently, it has been reported that intact human erythrocytes (Baker and Kalra, 1979) and hen erythrocytes (Majumdar and Baker, 1980) are fused by calcium and phosphate ions. Also, incubation of human erythrocytes with either uranyl ions or the ions of rare earth metals at 37°C for 30–40 min results in cell fusion (Majumdar *et al.*, 1980).

C. Nonbilayer Structures

One of the first hypotheses on possible changes in the organization of the phospholids of membranes that may occur during membrane fusion was that proposed in relation to lysophosphatidylcholine, which was predicted (Lucy, 1969, 1970) and subsequently shown (Howell and Lucy, 1969; Poole *et al.*, 1970) to be capable of inducing cell fusion. It was suggested that a primary requisite for the fusion of two membranes might be that both of the membranes should have a relatively high proportion of their phospholipid molecules in the micellar configuration that is favored by the presence of lysophosphatidylcholine (Haydon and Taylor, 1963; Bangham and Horne, 1964). In the context of the present review, it is interesting to note that phospholipase A enzymes, which generate lysophospholipids, are Ca^{2+} dependent for their activity. However, as noted by Poste and Allison (1973) and by others, there is little evidence to support the role of lysophospholipids as a mediator of biomembrane fusion other than under experimental conditions.

When the interactions of simple fusogenic lipids, e.g., oleoylglycerol and oleic acid, with phospholipids were later investigated using the negative-staining technique of electron microscopy, mixtures of phosphatidylcholine with fusogenic lipids exhibited lamellar and vesicular structures and a hexagonal phase with a repeat structure of 8 nm (Howell *et al.*, 1973). These structures were observed both in the presence and in the absence of added Ca^{2+}, but none of the new macromolecular assemblies were observed when phosphatidylcholine was treated with glyceryl di- and tri-oleate, palmitic, or stearic acid, which have little or no ability to cause cell fusion. It was therefore suggested that a primary effect of low-melting lipids like oleoylgylcerol and oleic acid in causing cell fusion is on the organization of the phospholipids of the plasma membranes. Cullis and Hope (1978) have confirmed these early findings, at least implicitly, in the interpretation of changes in the ^{31}P-NMR spectra of erythrocyte ghosts treated with oleic acid in terms of a progressive conversion of the membrane phospholipids from the bilayer phase to the hexagonal (H_{II}) phase. They have proposed a model for cell fusion induced by oleic acid in which the two outer monolayers of the fusing cells combine to make an intermediate hexagonal (H_{II}) phase that is subsequently restabilized to a bilayer structure to complete the fusion process. They also noted that the presence of Ca^{2+} would seem to be particularly important in the formation of the H_{II} phase, and further, that the restabilization event may be envisaged to occur by way of an active (ATP-dependent) removal of Ca^{2+} from the cytoplasm of the partially fused cells that would encourage a reversion to the bilayer phase.

Hope and Cullis (1981) have further compared the behavior of several lipid fusogens with nonfusogens (including glyceryl tri-oleate and stearic acid) and they similarly found that only the fusogens can induce a conformational change in membrane phospholipids, from the bilayer to the hexagonal (H_{II}) phase. They have therefore suggested a common mechanism of action for fusogenic lipids whereby lipid structures such as long cylinders (H_{II} configuration) or inverted micelles are formed, which serve as intermediates in the fusion event (see also Cullis and Verkleij, 1979). It seems to me, however, that this extrapolation from experimental studies on lipids in bulk to the fusion of biomembranes may not be justified, at least as far as the hexagonal (H_{II}) phase is concerned, since this may not be a stable structure in a paucimolecular membrane. In this connection, it is noteworthy that Hope and Cullis have themselves drawn attention to the fact that examination with the light microscope of erythrocytes treated with oleoylglycerol (oleoylglycerol:phospholipid ratio of approximately 4 : 1) reveals that, despite considerable cell damage, intact membranes are still present indicating that a significant proportion of the membranes must be in a bilayer configuration. They comment that this is potentially inconsistent with the observations that similar membrane concentrations of oleoylglycerol in ghost membranes induce the hexagonal (H_{II}) configuration for the vast majority of the endogenous phospholipids. This inconsistency was attributed to the centrifugation step employed in the preparations of the ghost membranes and, in an experiment bearing on this point, data were obtained indicating that membranes rich in fusogen *are able to enter the (H_{II}) phase more easily when in close proximity* (my italics). The possibility therefore remains that the hexagonal phase is present in the centrifuged, packed preparations of ghosts and in bulk artificial mixtures of phospholipids and fusogenic lipids dispersed in an aqueous environment, but not in the membranes of individual cells.

A further criticism of this work is that X-ray diffraction analysis has not been employed to confirm that the interpretation of the ^{31}P-NMR spectra in terms of the hexagonal (H_{II}) phase is correct. It is thus conceivable that the narrow spectra with a high field shoulder attributed to the hexagonal phase may correspond to a minor change in the organization of bilayer membrane phospholipids. In this connection, Thayer and Kohler (1981) have undertaken a pertinent theoretical examination of the relationship between the conformation of phosphatidylethanolamine and the shape of the predicted ^{31}P-NMR spectrum, using a simple model for reorientation of the head group accompanied by fast rotation of the phospholipid about its long axis in the bilayer. They have shown that it is possible to generate spectra considered to be typical of the hexagonal (H_{II}) phase, simply by changing the conformation of the phospholipid head group, while *retaining the molecules in the bilayer phase* (their italics). They comment that their results show that the NMR line shape will be highly dependent on the orientation of the head group, and that unambiguous interpretations of ^{31}P-NMR spectra for such systems is possible only if the orientation can be determined.

In freeze-fracture studies of lipid systems that exhibit the ^{31}P-NMR spectra attributed to the hexagonal phase, numerous small (60–120 Å diameter) particles have been observed on the fracture faces, and it has been suggested that these particles represent inverted micelles lying within phospholipid bilayers (de Kruijff *et al.*, 1980). The spherical particles may be regarded as being related to the hexagonal (H_{II}) phase in that progressive shortening of the inverted micellar cylinders that comprise the H_{II} phase will, at the limit, yield an inverted spherical micelle. Thus, the lipidic particles or inverted micelles may fuse to form the cylinders of the H_{II} phase. Ca^{2+} has been found, for example, to induce cardiolipin to exhibit the narrow line shape NMR spectra (de Kruijff *et al.*, 1979), and the fusion of vesicles of phosphatidylcholine–cardiolipin induced by Ca^{2+} is accompanied by the appearance at the fusion interface of lipidic particles that are thought to be intermediates between bilayer and hexagonal (H_{II}) structures (Verkleij *et al.*, 1979, 1980). In the light of these several observations, it has been concluded that factors which promote the H_{II} phase in various multilamellar systems cause fusion of corresponding unilamellar, sonicated vesicle systems of the same lipid composition, and that fusion proceeds via the intermediary formation of nonlamellar, inverted micellar structures (van Venetie and Verkleij, 1981; Hope *et al.*, 1983).

Two criticisms have been made of the above interpretations of the nature of the lipid particles. First, evidence has been presented that the observed particles represent conical deformations of membranes at sites of intermembrane attachments (Miller, 1980; Hui and Stewart, 1981; but see also Verkleij and de Kruijff, 1981). Second, the validity of lipidic particles as intermediates in membrane fusion has been questioned because of recent findings made with quick-freezing, freeze-fracture electron microscopy. Using this technique, which eliminates the use of glycerol as a cryoprotectant, Bearer *et al.* (1982) have examined very early events in the Ca^{2+}-induced fusion of large (0.2 μm diameter) unilamellar vesicles of cardiolipin, phosphatidylcholine, and of phosphatidylserine–phosphatidylethanolamine. Freeze-fracture replicas of the suspensions of vesicles frozen after 1–2 sec of stimulation revealed that the majority of the vesicles had already undergone membrane fusion, as evidenced by the presence of dumbbell-shaped structures and large vesicles. In the absence of glycerol, lipidic particles were not observed at the sites of fusion. By contrast, when cardiolipin–phosphatidylcholine liposomes were treated for 1 min with Ca^{2+} (5 mM) and then incubated for 1–2 hr in glycerol, many large vesicles emerged, 10–20% of which were embossed with 9- to 10-nm particles. Bearer *et al.*, (1982) have concluded that these studies provide no evidence that lipidic particles are necessary intermediates for membrane fusion. On the contrary, they indicate that lipidic particles are structures that are produced at equilibrium long after membrane fusion has occurred and, furthermore, that the particles become particularly prevalent in the presence of glycerol. As these workers also point out, however, this conclusion does not exclude a transitory intermediate at the site of membrane fusion, which involves a lipid conformation that differs from the unmodified stable bilayer configuration. This "elusive" intermediate might be an inverted micellar or some other nonbilayer structure, or a small domain of a more condensed or crystalline lipid bilayer. Bearer and co-workers further comment that for lack of concrete evidence, the "intermediate" could be characterized simply as a local perturbation of the lipid bilayer structure which allows mixing of lipid molecules between the two closely apposed membranes. It is interesting to note that this conclusion is not too dissimilar from the proposal made earlier by Ahkong *et al.* (1975b) that cell fusion induced by exogenous chemical agents seems to involve, *inter alia*, the following events: first, a perturbation of the bilayer structure of membrane lipids may occur which increases the fluidity of the lipid region, or in the extreme case results in micelle formation; second, the interaction and intermixing of the disturbed lipid molecules of closely apposed membranes in regions denuded of intramembranous proteins and glycoproteins may allow adjacent cells to fuse.

It should be added here, in relation to changes in membrane fluidity, that the number of molecules of a fusogenic lipid that penetrate into the phospholipid bilayer of treated cells is probably quite high. For example, it appears that oleoylglycerol induces cell fusion when it is present in an approximately equimolar ratio to the phospholipids (Hope and Cullis, 1981). Unfortunately, however, the concept that an increase in the overall fluidity of the lipid region of membranes occurs when they contain a high proportion of exogenous lipids that have fluid hydrocarbon chains (Ahkong *et al.*, 1973) has been misinterpreted by some workers as implying that exogenous membrane-active substances increase the intrinsic fluidity of the endogenous phospholipid molecules of membranes. Model experiments with dimethylsulfoxide and poly(ethylene glycol) on phospholipid vesicles (Lyman and Preisler, 1976; Tilcock and Fisher, 1979) and with phospholipid monolayers (Maggio and Lucy, 1978) have, in fact, indicated that whether or not the transition temperature of a phospholipid is increased or decreased by the presence of fusogenic molecules depends, at least with water-soluble fusogens, on the parameters of each individual system.

IV. CELL FUSION AND VESICLE FUSION WITHOUT Ca^{2+}

A. Cell Fusion

Data discussed in the preceding sections of this chapter have illustrated, on one hand, the undoubted importance of Ca^{2+} in a major example of a biomembrane fusion reaction (the fusion of myoblasts) and shown, on the other hand, that Ca^{2+} plays an important role in the fusion of phospholipid vesicles, particularly those containing acidic phospholipids. It does not necessarily follow, however, that Ca^2 is important in biomembrane fusion because of its role(s) in the fusion of phospholipid vesicles in model systems. As has already been mentioned, the behavior of vesicles of phosphatidylserine as a model for biomembrane fusion reactions is open to question from the outset since relatively small quantities of this phospholipid are present in the majority of biomembranes. Additional doubts on the overriding importance of the interactions of Ca^{2+} with phospholipids in the fusion of biomembranes are cast by two further sets of observations: (1) that cells are observed to fuse in the absence of Ca^{2+}, and (2) that phospholipid vesicles also fuse in the absence of Ca^{2+}.

Experiments undertaken in this laboratory on the chemically induced fusion of rat erythrocytes have indicated that, independently of any requirement for the ion in the fusion of their phospholipid bilayers, Ca^{2+} is involved in activating an endogenous proteinase that participates in the fusion process. Rat erythrocytes were found to fuse readily when treated with benzyl alcohol (100 mM) in the presence of Ca^{2+} (4 mM; Ahkong *et al.*, 1980). Fusion was very effectively prevented by 4 mM EGTA and by 1 mM *N*-ethylmaleimide. Other thiol reagents (iodoacetamide, tetrathionate, cystamine) were also inhibitory. In addition to inhibiting fusion, EGTA and *N*-ethylmaleimide also completely prevented both decreases in certain membrane proteins (bands 2 and 3) and production of a polypeptide moving slightly faster than band 3 that were associated with the fusion process. By contrast, histamine (a competitive inhibitor of the Ca^{2+}-dependent intracellular transglutaminase) had no effect either on fusion or on the changes in membrane proteins, indicating that alterations in membrane proteins associated with the fusion reaction were due to proteolysis and not to the crosslinking of membrane proteins. Since the susceptibility to cold-induced aggregation of the intramembranous protein particles in the benzyl alcohol-treated cells was also markedly inhibited by EGTA and partially inhibited by *N*-ethylmaleimide, it was proposed that a Ca^{2+}-activated cysteine proteinase, which acts to degrade membrane proteins and to give freedom of lateral movement to intramembranous particles, may be an essential feature of membrane fusion in rat erythrocytes treated with benzyl alcohol. It was also suggested that degradation of spectrin-binding proteins in the band 2 region by such an enzyme may be specifically involved in the fusion reaction.

By contrast with the complete inhibition by EGTA of the fusion of rat erythrocytes treated with benzyl alcohol, it was found in a related study that human erythrocytes treated with the fusogen oleoylglycerol in the presence of EGTA were fused to a level of 25% of that occurring in the presence of Ca^{2+} (Quirk *et al.*, 1978). In recent work on the fusion of human erythrocytes by chlorpromazine hydrochloride (2 mM) at pH 7.4, fusion was only weakly inhibited by EGTA (10 mM) and by *N*-ethylmaleimide

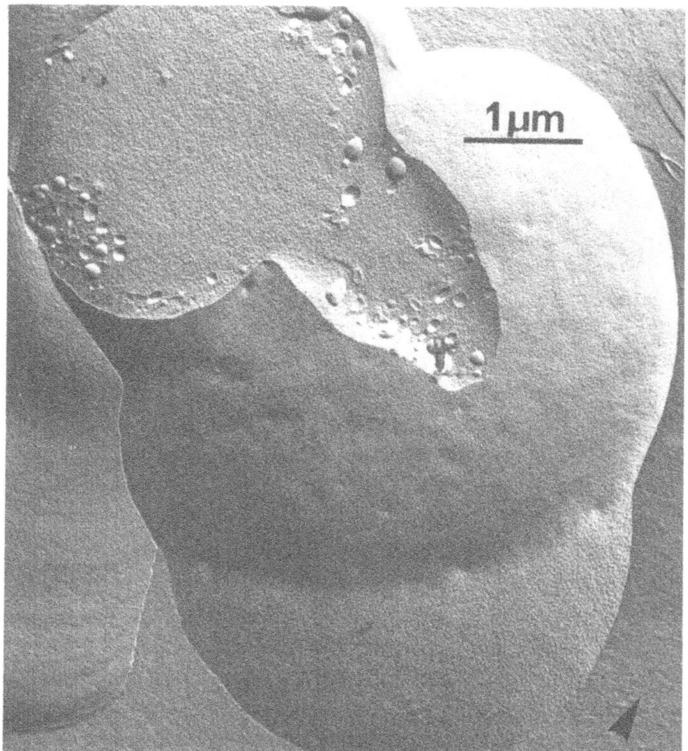

Figure 1. An electron micrograph of a freeze-fractured preparation of human erythrocytes showing a multicellular body formed after 30 min of incubation at 37°C, pH 7.4, in the presence of 2 mM chlorpromazine and 10 mM EGTA. Reproduced with permission from Lucy (1984).

(50 mM). Thus, in the presence of 10 mM EGTA, fusion indices of 25% and 31% were observed after 30 min and 60 min, respectively, with 2 mM chlorpromazine, compared with 35% and 65% in the presence of 10 mM $CaCl_2$ (Lang *et al.*, 1984). An electron micrograph of human erythrocytes fused by chlorpromazine in the presence of EGTA is shown in Figure 1. By contrast with the ineffectiveness of EGTA, however, pretreatment of the cells with Tos-Lys-CH_2Cl (7.5 mM) markedly inhibited fusion (fusion index 9% after 30 min).

Changes in the membrane proteins of human erythrocytes fused by chlorpromazine were investigated in this work. In addition, fusion experiments were undertaken with cells in which band 3 had been removed by pretreatment with chymotrypsin in order to monitor the production during fusion of protein(s), which are known to be produced by the endogenous proteolysis of ankyrin and which migrate in the band 3 region. From data on the fusion indices of human erythrocytes treated with chlorpromazine under various conditions (Table 1), and from observations on the degradation of membrane proteins and on the mobility of the intramembranous particles, it was concluded that the Ca^{2+}-insensitive component of this biomembrane fusion reaction is associated with degradation of band 2.1 to bands 2.3–2.6 by a serine proteinase. Conversely, the component of fusion inhibited by EGTA and by *N*-ethylmaleimide is apparently associated with degradation of band 3 to band 4.5 by a Ca^{2+}-activated cysteine proteinase. Proteolysis of ankyrin (and the consequent release of molecules of band 3 from their attachment to spectrin) appeared to be sufficient to permit the chlorpromazine-induced fusion of erythrocytes, but fusion occurred more rapidly when band 3 was also degraded in the presence of Ca^{2+}.

In the light of this work, it appears that whether or not Ca^{2+} is required for biomembrane fusion reactions depends on the characteristics of individual membrane

Table 1. Fusion Indices of Human Erythrocytes Treated with Chlorpromazine[a]

Pretreatment	Addition on incubation	Incubation time (minutes) after addition of chlorpromazine	Fusion
None	10 mM $CaCl_2$	30	34.7 ± 6 [8]
None	10 mM EGTA	30	24.6 ± 3.9 [8]
None	10 mM $CaCl_2$	60	64.7 ± 10.1 [10]
None	10 mM EGTA	60	30.6 ± 9.6 [9]
1 mM Tos-Lys-$CHCl_2$	10 mM $CaCl_2$ + 1 mM Tos-Lys-$CHCl_2$	30	24.1 ± 2.7 [4]
1 mM Tos-Lys-$CHCl_2$	10 mM EGTA + 1 mM Tos-Lys-$CHCl_2$	30	23.4 ± 2.3 [4]
7.5 mM Tos-Lys-$CHCl_2$	10 mM $CaCl_2$	30	9.2 ± 1.3 [4]
7.5 mM Tos-Lys-$CHCl_2$	10 mM EGTA	30	8.8 ± 1.4 [4]
50 mM *N*-ethylmaleimide	10 mM $CaCl_2$	30	30.8 ± 4.3 [5]
50 mM *N*-ethylmaleimide	10 mM EGTA	30	24.9 ± 3.4 [5]
α-Chymotrypsin	10 mM $CaCl_2$	30	75.0 ± 5.9 [5]
α-Chymotrypsin	10 mM EGTA	30	73.6 ± 6.3 [5]

[a] The concentration of chlorpromazine hydrochloride used was 2 mM in all experiments, except for those with α-chymotrypsin-treated cells when 1 mM chlorpromazine was used. The fusion indices shown are mean values, ± *SD*, for the number of independent experiments given in brackets. (Reproduced with permission from Lang *et al.*, 1984.)

systems and that, in some instances, the fusion reaction can proceed without Ca^{2+}. When Ca^{2+} does participate, it seems reasonable to suppose, as has been suggested elsewhere (Lucy, 1984), that it may directly or indirectly affect the membrane lipids or the membrane proteins, or both. In the above work, chlorpromazine was, incidentally, unlikely to have mobilized free ionic Ca^{2+} (that remained within the erythrocyte cytoplasm despite the EGTA outside), since the presence of polymerized proteins in the treated cells indicated that chlorpromazine acts like other chemical fusogens (Blow *et al.*, 1979; Aldwinckle *et al.*, 1982) to increase membrane permeability to Ca^{2+}. It is also apparent that the Ca^{2+}-activated cysteine proteinase activity is much more important in the fusion of rat erythrocytes by benzyl alcohol than it is with human erythrocytes fused by chlorpromazine. This may partly reflect different effects on the membranes of the two fusogens. It is nevertheless noteworthy that the content of calpastatin, an endogenous inhibitor of the Ca^{2+}-activated cysteine proteinase known as calpain I, is extremely small in rat erythrocytes, while in human erythrocytes there is much more inhibitor than enzyme (Murachi *et al.*, 1981).

Cell fusion without Ca^{2+} is not a special feature of human erythrocytes treated with oleoylglycerol and chlorpromazine. For example, although we observed that Lesch–Nyhan human skin fibroblasts treated with poly(ethylene glycol) 6000 in the presence of 8 mM EGTA gave no greater fusion than that observed in control untreated cultures (Blow *et al.*, 1978), it has subsequently been reported that toxic effects of poly(ethylene glycol) are decreased, and the yield of hybrid cells is increased, when this polymer is used in a Ca^{2+}-medium and the cells then maintained in a Ca^{2+}-free medium for at least 15 min (Schneiderman *et al.*, 1979). Also, hen erythrocytes are fused by lysophophatidylcholine without added Ca^{2+} (Howell and Lucy, 1969), and it has recently been reported that the fusion of human erythrocytes by tetracaine occurs in the absence of added Ca^{2+} and in the presence of EDTA (Coakley *et al.*, 1983).

There are also a number of reports on the induction of cell fusion by viruses in the absence of Ca^{2+}. Peretz *et al.* (1974) reported that human erythrocytes treated with Sendai virus will fuse in the absence of bivalent metal cations or in the presence of EDTA provided that low concentrations of virus are used to minimize hemolysis. In similar work with hen erythrocytes, we found that maximum cell fusion by Sendai occurred in the presence of EGTA, and that cell fusion was significantly decreased by Ca^{2+}, even at a concentration of 0.2 mM, possibly because it diminished interactions between virus particles and the erythrocyte membranes (Hart *et al.*, 1976). It is also relevant to note that Knutton and Pasternak (1979), from studies on the fusion of erythrocytes by "1 day" Sendai virus, have concluded that the entry of Ca^{2+} into the treated cells is a consequence, not a cause, of the fusion event.

B. Vesicle Fusion

It seems that even phospholipid liposomes, free from protein, can fuse at least without added Ca^{2+} and in some instances in the presence of chelating agents for Ca^{2+}.

Boni *et al.* (1981) have reported that fusion between unilamellar vesicles of egg phosphatidylcholine, and of bovine phosphatidylserine, is induced in 1 min by 50%

poly(ethylene glycol). In their experiments, the lipids were dispersed in an aqueous buffer containing 0.2 M EDTA, and fusion was monitored by electron microscopy and by turbidity measurements. Electron microscopy was also used by Aldwinckle *et al.* (1982) to monitor the fusion of vesicles composed of egg phosphatidylcholine, egg phosphatidic acid, and cholesterol (7 : 2 : 1, molar ratios). Large vesicles were observed after 3 hr with 30% poly(ethylene glycol) in the absence of added Ca^{2+}. Tilcock and Fisher (1982), in studies on the calorimetric behavior of mixed phospholipids in the presence of poly(ethylene glycol) and 5 mM EDTA, noted changes that were consistent with vesicle fusion. Furthermore, it has also been reported that alkyl bromide-induced fusion of both egg phosphatidylcholine and phosphatidylcholine–phosphatidic acid vesicles shows no dependence on the concentration of Ca^{2+}, over a range of 0.5–10 mM, assayed either by light-scattering or by direct microscopical observations (Mason *et al.*, 1980).

Finally, there are a number of reports in the literature on the fusion of phospholipid vesicles occurring on their interaction with various peptides and proteins. These include alamethicin (Lau and Chan, 1975), concanavalin A (van der Bosch and McConnell, 1975), serum albumin (Shenkman *et al.*, 1981), tobacco mosaic virus coat protein (Banerjee *et al.*, 1981), and melittin (Morgan *et al.*, 1982; Eytan and Almary, 1983). It is noteworthy that Ca^{2+} was not present in any of these systems, and that fusion with melittin and with concanavalin A occurred in the presence of EGTA.

V. CONCLUDING COMMENTS

In conclusion, attention is drawn to a new proposal (Lucy, 1984) which has arisen from the findings described above on the endogenous proteolysis of erythrocyte membrane proteins that accompanies cell fusion induced by chlorpromazine. It has been suggested that endogenous proteinases, activated by Ca^{2+} or otherwise, may release hydrophobic fragments (derived from integral proteins at the cytoplasmic surface or from membrane skeleton proteins) that penetrate and perturb the lipid bilayers of cells and thus induce fusion. This process would be complementary to the induction of fusion by penetration into the lipid bilayer, from outside cells, of the fusogenic glycoprotein of Sendai virus. The production *in situ* of fusogenic peptides or proteins by endogenous proteinases that are activated *in vivo* by appropriate stimuli might, indeed, be a general mechanism for the fusion of lipid bilayers in biomembrane fusion reactions in a wide variety of situations, including endocytosis and exocytosis, some aspects of which are imitated by the fusogenic viruses.

The potential plausibility of this proposal is perhaps indicated by a remark made by Helenius (see Helenius and Marsh, 1982) that fusogenic viruses may have acquired the mechanism by which they induce fusion from a preexisting cellular mechanism. In this regard, it is interesting to reflect that when Sendai virus is grown in chick embryos, it contains the fusogenic F_1 glycoprotein, but that when the virus is grown in bovine kidney (MDBK) cells, it contains the precursor F_0 which is inactive in fusion. Activation is dependent on cleavage of the F_0 protein by a *cellular* enzyme, which is present in the chick embryo but not in MDBK cells (Scheid and Choppin, 1974).

REFERENCES

Ahkong, Q. F., Fisher, D., Tampion, W., and Lucy, J. A., 1973, The fusion of erythrocytes by fatty acids, esters, retinol and α-tocopherol, *Biochem. J.* **136:**147–155.

Ahkong, Q. F., Tampion, W., and Lucy, J. A., 1975a, Promotion of cell fusion by divalent cation ionophores, *Nature* **256:**208–209.

Ahkong, Q. F., Fisher, D., Tampion, W., and Lucy, J. A., 1975b, Mechanisms of cell fusion, *Nature* **253:**194–195.

Ahkong, Q. F., Botham, G. M., Woodward, A. W., and Lucy, J. A., 1980, Calcium-activated thiol-proteinase activity in the fusion of rat erythrocytes induced by benzyl alcohol, *Biochem. J.* **192:**829–836.

Aldwinckle, T. J., Ahkong, Q. F., Bangham, A. D., Fisher, D., and Lucy, J. A., 1982, Effects of poly(ethylene glycol) on liposomes and erythrocytes. Permeability changes and membrane fusion, *Biochim. Biophys. Acta* **689:**548–560.

Baines, A. J., 1983, Cell membrane skeleton. The spread of spectrin, *Nature* **301:**377–378.

Baker, R. F., and Kalra, V. K., 1979, Chemically induced fusion of fresh human erythrocytes, *Biochem. Biophys. Res. Commun.* **86:**920–928.

Banerjee, S., Vandenbranden, M., and Ruysschaert, J-M., 1981, Tobacco mosaic virus protein induces fusion of liposome membranes, *Biochim. Biophys. Acta* **646:**360–364.

Bangham, A. D., and Horne, R. W., 1964, Negative staining of phospholipids and their structural modification by surface-active agents as observed in the electron microscope, *J. Mol. Biol.* **8:**660–668.

Bearer, E. L., Duzgunes, N., Friend, D. S., and Papahadjopoulos, D., 1982, Fusion of phospholipid vesicles arrested by quick-freezing. The question of lipidic particles as intermediates in membrane fusion, *Biochim. Biophys. Acta* **693:**93–98.

Bersten, A. M., Ahkong, Q. F., Hallinan, T., Nelson, S. J., and Lucy, J. A., 1983, Inhibition of the formation of myotubes *in vitro* by inhibitors of transglutaminase, *Biochim. Biophys. Acta* **762:** 429–436.

Bischoff, R., 1978, Myoblast fusion, in: *Cell Surface Reviews*, Vol. 5 (G. Poste and G. L. Nicholson, eds.), Elsevier/North-Holland Biomedical Press, Amsterdam, pp. 127–179.

Bischoff, R., and Holtzer, H., 1968, The effect of mitotic inhibitors on myogenesis *in vitro, J. Cell Biol.* **36:**111–127.

Blow, A. M. J., Botham, G. M., Fisher, D., Goodall, A. M., Tilcock, C. P. S., and Lucy, J. A., 1978, Water and calcium ions in cell fusion induced by poly(ethylene glycol), *FEBS Lett.* **94:**305–310.

Blow, A. M. J., Botham, G. M., and Lucy, J. A., 1979, Calcium ions and cell fusion. Effects of chemical fusogens on the permeability of erythrocytes to calcium and other ions, *Biochem. J.* **182:**555–563.

Boni, L. T., Stewart, T. P., Alderfer, J. L., and Hui, S. W., 1981, Lipid-polyethylene glycol interactions: I. Induction of fusion between liposomes, *J. Membr. Biol.* **62:**65–70.

Coakley, W. T., Nwafor, A., and Deeley, J. O. T., 1983, Tetracaine modifies the fragmentation mode of heated human erythrocytes and can induce heated cell fusion, *Biochim. Biophys. Acta* **727:**303–312.

Couch, C. B., and Strittmatter, W. J., 1983, Rat myoblast fusion requires metalloendoprotease activity, *Cell* **32:**257–265.

Cullis, P. R., and Hope, M. J., 1978, Effects of fusogenic agent on membrane structure of erythrocyte ghosts and the mechanism of membrane fusion, *Nature* **271:**672–674.

Cullis, P. R., and Verkleij, A. J., 1979, Modulation of membrane structure by Ca^{2+} and dibucaine as detected by 31P NMR, *Biochim. Biophys. Acta* **552:**546–551.

Cuthbertson, K. S. R., Whittingham, D. G., and Cobbold, P. H., 1981, Free Ca^{2+} increases in exponential phases during mouse oocyte activation, *Nature* **294:**754–757.

Dahl, G., Schudt, C., and Gratzl, M., 1978, Fusion of isolated myoblast plasma membranes. An approach to the mechanism, *Biochim. Biophys. Acta* **514:**105–116.

David, J. D., and Higginbotham, C-A., 1981, Fusion of chick embryo skeletal myoblasts: Interactions of prostaglandin E_1, adenosine 3′ : 5′ monophosphate, and calcium influx, *Dev. Biol.* **82:**308–316.

David, J. D., See, W. M., and Higginbotham, C-A., 1981, Fusion of chick embryo skeletal myoblasts: Role of calcium influx preceding membrane union, *Dev. Biol.* **82:**297–307.

de Kruijff, B., Verkley, A. J., van Echteld, C. J. A., Gerritsen, W. J., Mombers, C., Noordam, P. C., and de Gier, J., 1979, The occurrence of lipidic particles in lipid bilayers as seen by 31 P NMR and freeze-fracture electron microscopy, *Biochim. Biophys. Acta* **555:**200–209.

de Kruijff, B., Cullis, P. R., and Verkleij, A. J., 1980, Nonbilayer lipid structures in model and biological membranes, *Trends Biochem. Sci.* **5**:79–81.

Duzgunes, N., Wilschut, J., Fraley, R., and Papahadjopoulos, D., 1981, Studies on the mechanism of membrane fusion. Role of head-group composition in calcium- and magnesium-induced fusion of mixed phospholipid vesicles, *Biochim. Biophys. Acta* **642**:182–195.

Ekerdt, R., and Papahadjopoulos, D., 1982, Intermembrane contact affects calcium binding to phospholipid vesicles, *Proc. Natl. Acad. Sci. USA* **79**:2273–2277.

Eytan, G. D., and Almary, T., 1983, Melittin-induced fusion of acidic liposomes, *FEBS Lett.* **156**:29–32.

Fraley, R., Wilschut, J., Duzgunes, N., Smith, C., and Papahadjopoulos, D., 1980, Studies on the mechanism of membrane fusion: Role of phosphate in promoting calcium ion induced fusion of phospholipid vesicles, *Biochemistry* **19**:6021–6029.

Fulton, A. B., Prives, J., Farmer, S. R., and Penman, S., 1981, Developmental reorganization of the skeletal framework and its surface lamina in fusing muscle cells, *J. Cell Biol.* **91**:103–112.

Fumigalli, G. M., Brigonzi, A., Tachikawa, T., and Clementi, F., 1981, Rat myoblast fusion: Morphological study of membrane apposition, fusion, and fission during controlled myogenesis *in vitro, J. Ultrastruct. Res.* **75**:112–125.

Gratzl, M., Schudt, C., Ekerdt, C., and Dahl, G., 1980, Fusion of isolated biological membranes, in: *Membrane Structure and Function*, Vol. 3 (E. E. Bittar, ed.), John Wiley, Chichester, pp. 59–92.

Haest, C. W. M., 1982, Interactions between membrane skeleton proteins and the intrinsic domain of the erythrocyte membrane, *Biochim. Biophys. Acta* **694**:331–352.

Hart, C. A., Fisher, D., Hallinan, T., and Lucy, J. A., 1976, Effects of calcium ions and the bivalent cation ionophore A23187 on the agglutination and fusion of chicken erythrocytes by Sendai virus, *Biochem. J.* **158**:141–145.

Haydon, D. A., and Taylor, J., 1963, The stability and properties of bimolecular lipid leaflets in aqueous solutions, *J. Theoret. Biol.* **4**:281–296.

Helenius, A., and Marsh, M., 1982, *Membrane Recycling*, CIBA Foundation Symposium No. 92 (D. Evered and G. M. Collins, eds.) Pitman, London, pp. 59–74.

Hong, K., Duzgunes, N., and Papahadjopoulos, D., 1981, Role of synexin in membrane fusion, *J. Biol. Chem.* **256**:3641–3644.

Hope, M. J., and Cullis, P. R., 1981, The role of nonbilayer lipid structures in the fusion of human erythrocytes induced by lipid fusogens, *Biochim. Biophys. Acta* **640**:82–90.

Hope, M. J., Walker, D. C., and Cullis, P. R., 1983, Ca^{2+} and pH induced fusion of small unilamellar vesicles consisting of phosphatidylethanolamine and negatively charged phospholipids: A freeze fracture study, *Biochem. Biophys. Res. Commun.* **110**:15–22.

Howell, J. I., and Lucy, J. A., 1969, Cell fusion induced by lysolecithin, *FEBS Lett.* **4**:147–150.

Howell, J. I., Fisher, D., Goodall, A. H., Verrinder, M., and Lucy, J. A., 1973, Interactions of membrane phospholipids with fusogenic lipids, *Biochim. Biophys. Acta* **332**:1–10.

Hui, S. W., and Stewart, T. P., 1981, "Lipidic particles" are intermembrane attachment sites, *Nature* **290**:427–428.

Kalderon, N., and Gilula, N. B., 1979, Membrane events involved in myoblast fusion, *J. Cell Biol.* **81**:411–425.

Kaur, H., and Sanwal, B. D., 1981, Regulation of the activity of a calcium-activated neutral protease during differentiation of skeletal myoblasts, *Can. J. Biochem.* **59**:743–747.

Knudsen, K. A., and Horwitz, A. F., 1978, Toward a mechanism of myoblast fusion, *Prog. Clin. Biol. Res.* **23**:563–568.

Knutton, S., and Pasternak, C. A., 1979, The mechanism of cell–cell fusion, *Trends Biochem. Sci.* **4**:220–223.

Lang, R. D., Wickenden, C., Wynne, J., and Lucy, J. A., 1984, Proteolysis of ankyrin and of band 3 in chemically-induced cell fusion: Ca^{2+} is not mandatory for fusion, *Biochem. J.*, **218**:295–305.

Lau, A. L. Y., and Chan, S. I., 1975, Alamethicin-mediated fusion of lecithin vesicles, *Proc. Natl. Acad. Sci. USA* **72**:2170–2174.

Lorand, L., Siefring, G. E., and Lowe-Krentz, L., 1980, Ca^{2+}-triggered and enzyme-mediated cross-linking of membrane proteins in intact human erythrocytes, in: *Membrane Transport in Erythrocytes*, Alfred Benzon Symposium 14 (U. V. Lassen, H. H. Ussing, and J. O. Wieth, eds.), Munksgaard, Copenhagen, p. 285–299.

Lucy, J. A., 1969, Lysosomal membranes, in: *Lysosomes in Biology and Pathology*, Vol. 2 (J. T. Dingle and H. B. Fell, eds.), North Holland, Amsterdam, pp. 313–341.

Lucy, J. A., 1970, The fusion of biological membranes, *Nature* **227**:814–817.

Lucy, J. A., 1984, Fusogenic mechanisms, in: *Cell Fusion*, CIBA Symposium No 103 (J. Evered and J. Whelan, eds.), Pitman, London, pp. 28–44.

Lucy, J. A., Ahkong, Q. F., Cramp, F. C., Fisher, D., and Howell, J. I., 1971, Cell fusion without viruses, *Biochem. J.* **124**:46–47P.

Lyman, G. H., and Preisler, H. D., 1976, Membrane action of DMSO and other chemical inducers of Friend leukaemic cell differentiation, *Nature* **262**:360–363.

Maggio, B., and Lucy, J. A., 1978, Interactions of water-soluble fusogens with phospholipids in monolayers, *FEBS Lett.* **94**:301–304.

Majumdar, S., and Baker, R. F., 1980, Phosphate-calcium induced fusion of chicken erythrocytes, *Exp. Cell Res.* **126**:175–182.

Majumdar, S., Baker, R. F., and Kalra, V. K., 1980, Fusion of human erythrocytes induced by uranyl acetate and rare earth metals, *Biochim. Biophys. Acta* **598**:411–416.

Mason, W. T., Lane, N. J., Miller, N. G. A., and Bangham, A. D., 1980, Fusion of liposome membranes by the *n*-alkyl bromides, *J. Membr. Biol.* **55**:69–79.

Michell, R. H., 1982, Is phosphatidylinositol really out of the calcium gate? *Nature* **296**:492–493.

Miller, R. G., 1980, Do "lipidic particles" represent intermembrane attachment sites? *Nature* **87**:166–167.

Morgan, C. G., Thomas, E. W., Moras, T. S., and Yianni, Y. P., 1982, The use of a phospholipid analogue of diphenyl-1,3,5-hexatriene to study melittin-induced fusion of small unilamellar phospholipid vesicles, *Biochim. Biophys. Acta* **692**:196–201.

Murachi, T., Tanaka, K., Hatanaka, M., and Murakami, T., 1981, Intracellular Ca^{2+}-dependent protease (calpain) and its high-molecular-weight endogenous inhibitor (calpastatin), in: *Advances in Enzyme Regulation*, Vol. 19 (G. Weber, ed.), Pergamon Press, New York and Oxford, pp. 407–424.

Papahadjopoulos, D., 1978, Calcium-induced phase changes and model membranes, in: *Cell Surface Reviews*, Vol. 5 (G. Poste and G. L. Nicolson, eds.), Elsevier/North-Holland, Amsterdam, pp. 765–790.

Papahadjopoulos, D., Poste, G., Schaeffer, B. E., and Vail, W. J., 1974, Membrane fusion and molecular segregation in phospholipid vesicles, *Biochim. Biophys. Acta* **352**:10–28.

Papahadjopoulos, D., Vail, W. J., Jacobson, K., and Poste, G., 1975, Cochleate lipid cylinders: Formation by fusion of unilamellar lipid vesicles, *Biochim. Biophys. Acta* **394**:483–491.

Papahadjopoulos, D., Vail, W. J., Newton, C., Nir, S., Jacobson, K., Poste, G., and Lazo, R., 1977, Studies on membrane fusion. III. The role of calcium-induced phase changes, *Biochim. Biophys. Acta* **465**:579–598.

Papahadjopoulos, D., Poste, G., and Vail, W. J., 1979, Studies on membrane fusion with natural and model membranes, in: *Methods in Membrane Biology*, Vol. 10 (E. D. Korn, ed.), Plenum Press, New York, pp. 1–121.

Peretz, H., Toister, Z., Laster, Y., and Loyter, A., 1974, Fusion of intact human erythrocytes and erythrocyte ghosts, *J. Cell Biol.* **63**:1–11.

Poole, A. R., Howell, J. I., and Lucy, J. A., 1970, Lysolecithin and cell fusion, *Nature* **227**:810–813.

Portis, A., Newton, C., Pangborn, W., and Papahadjopoulos, D., 1979, Studies on the mechanism of membrane fusion: Evidence for an intermembrane Ca^{2+}-phospholipid complex, synergism with Mg^{2+}, and inhibition by spectrin, *Biochemistry* **18**:780–790.

Poste, G., and Allison, A. C., 1973, Membrane fusion, *Biochim. Biophys. Acta* **300**:421–465.

Poste, G., and Nicolson, G. L., 1978, Membrane fusion, *Cell Surface Reviews*, Vol. 5, Elsevier/North-Holland, Amsterdam.

Quirk, S. J., Ahkong, Q. F., Botham, G. M., Vos, J., and Lucy, J. A., 1978, Membrane proteins in human erythrocytes during cell fusion induced by oleoylglycerol, *Biochem. J.* **176**:159–167.

Scheid, A., and Choppin, P. W., 1974, Identification of biological activities of paramyxovirus glycoproteins. Activation of cell fusion, hemolysis, and infectivity by proteolytic cleavage of an inactive precursor protein of Sendai virus, *Virology* **57**:475–490.

Schenkman, S., de Araujo, P. S., Sesso, A., Quina, F. H., and Chaimovich, H., 1981, A kinetic and structural study of two-step aggregation and fusion of neutral phospholipid vesicles promoted by serum albumin at low pH, *Chem. Phys. Lipids* **28**:165–180.

Schneiderman, S., Farber, J. L., and Baserga, R., 1979, A simple method for decreasing the toxicity of polyethylene glycol in mammalian cell hybridization, *Somat. Cell Genet.* **5**:263–269.

Schudt, C., and Pette, D., 1975, Influence of the ionophore A23187 on myogenic cell fusion, *FEBS Lett.* **59**:36–38.

Schudt, C., van der Bosch, J., and Pette, D., 1973, Inhibition of muscle cell fusion *in vitro* by Mg^{2+} and K^+ ions, *FEBS Lett.* **32**:296–298.

Schudt, C., Dahl, G., and Gratzl, M., 1976, Calcium-induced fusion of plasma membranes isolated from myoblasts grown in culture, *Cytobiologie* **13**:211–223.

Shainberg, A., Yagil, G., and Yaffe, D., 1969, Control of myogenesis *in vitro* by Ca^{2+} concentration in nutritional medium, *Exp. Cell Res.* **58**:163–167.

Thayer, A. M., and Kohler, S. J., 1981, Phosphorous-31 nuclear magnetic resonance spectra characteristic of hexagonal and isotropic phospholipid phases generated from phosphatidylethanolamine in the bilayer phase, *Biochemistry* **20**:6831–6834.

Tilcock, C. P. S., and Fisher, D., 1979, Interaction of phospholipid membranes with poly(ethylene glycol)s, *Biochim. Biophys. Acta* **577**:53–61.

Tilcock, C. P. S., and Fisher, D., 1982, The interaction of phospholipid membranes with poly(ethylene glycol) vesicle aggregation and lipid exchange, *Biochim. Biophys. Acta* **688**:645–652.

Toister, Z., and Loyter, A., 1971, Ca^{2+}-induced fusion of avian erythrocytes, *Biochim. Biophys. Acta* **241**:719–724.

van der Bosch, J., and McConnell, H. M., 1975, Fusion of dipalmitoylphosphatidylcholine vesicle membranes induced by concanvalin A, *Proc. Natl. Acad. Sci USA* **72**:4409–4413.

van der Bosch, J., Schudt, C., and Pette, D., 1972, Quantitative investigation on Ca^{++} and pH dependence of muscle cell fusion *in vitro*, *Biochem. Biophys. Res. Commun.* **48**:326–332.

van Venetie, R., and Verkleij, A. J., 1981, Analysis of the hexagonal II phase and its relations to lipidic particles and the lamellar phase. A freeze fracture study, *Biochim. Biophys. Acta* **645**:262–269.

Verkleij, A. J., and de Kruijff, B., 1981, "Lipidic particles" are intermembrane attachment sites. (A reply), *Nature* **290**:427–428.

Verkleij, A. J., Mombers, C., Gerritsen, W. J., Leunissen-Bijvelt, L., and Cullis, P. R., 1979, Fusion of phospholipid vesicles in association with the appearance of lipidic particles as visualized by freeze fracturing, *Biochim. Biophys. Acta* **555**:358–361.

Verkleij, A. J., van Echteld, C. J. A., Gerritsen, W. J., Cullis, P. R., and de Kruijff, B., 1980, The lipidic particle as an intermediate structure in membrane fusion processes and bilayer to hexagonal H_{II} transitions, *Biochim. Biophys. Acta* **600**:620–624.

Wakelam, M. J. O., 1983, Inositol phospholipid metabolism and myoblast fusion, *Biochem. J.* **214**:77–82.

Wakelam, M. J. O., and Pette, D., 1982, The breakdown of phosphatidylinositol in myoblasts stimulated to fuse by the addition of Ca^{2+}, *Biochem. J.* **202**:723–729.

Wakelam, M. J. O., and Pette, D., 1984, Myoblast fusion–inositol phospholipid breakdown—Relevant or coincidence? in: *Cell Fusion*, CIBA Symposium No. 103 (J. Evered and J. Whelan, eds.), Pitman, London, pp. 100–118.

Woodin, A. M., and Wieneke, A. A., 1963, The accumulation of calcium by the polymorphonuclear leucocyte treated with staphylococcal leucidin and its significance in the extrusion of protein, *Biochem. J.* **87**:487–495.

Woodin, A. M., and Wieneke, A. A., 1964, The participation of calcium, adenosine triphosphate and adenosine triphosphatase in the extrusion of the granule proteins from the polymorphonuclear leucocyte, *Biochem. J.* **90**:498–509.

Zakai, N., Kulka, R. G., and Loyter, A., 1976, Fusion of human erythrocyte ghosts promoted by the combined action of calcium and phosphate ions, *Nature* **263**:696–699.

Zalin, R. J., 1977, Prostaglandins and myoblast fusion, *Dev. Biol.* **59**:241–248.

Zimmermann, U., 1982, Electric field-mediated fusion and related electrical phenomena, *Biochim. Biophys. Acta* **694**:227–277.

12

Role of Membrane Fluidity in the Expression of Biological Functions

Juan Yguerabide and Evangelina E. Yguerabide

I. INTRODUCTION

Biological membranes are composed of proteins, lipids, and carbohydrates. It is generally agreed that proteins are the components most directly responsible for the great diversity of functions displayed by natural membranes while the lipids, arranged in a bimolecular leaflet, provide a highly impermeable and supportive matrix for the proteins. Some of the membrane proteins, the so-called integral membrane proteins, extend into the hydrophobic regions of the lipid bilayer and are exposed on at least one of the membrane surfaces or may span the lipid bilayer and be exposed on both surfaces. Other membrane proteins, the peripheral proteins, are noncovalently attached to the membrane surface and do not extend significantly into hydrophobic regions of the membrane. The carbohydrates reside on the membrane surface, covalently attached to proteins or lipids. The lipid bilayer is not a static structure but can exist in different dynamic states in which the lipid molecules exhibit different degrees of rotational, segmental, and lateral mobilities. When mobility is high, the membrane is said to be in a high fluid state and free membrane proteins can readily translate and rotate in the lipid bilayer.

The idea that biological membranes are fluid structures became generally accepted around 1972 (Singer and Nicolson, 1972), and its acceptance was immediately followed by the suggestion that changes in membrane fluidity could play an important role in the triggering or modulation of membrane functions (Singer, 1974; Nicolson, 1976)

Juan Yguerabide and Evangelina E. Yguerabide • Department of Biology, University of California at San Diego, La Jolla, California 92093.

through the modulation of the molecular conformations and mobilities of proteins. Since then, hundreds of articles have appeared in which changes in membrane fluidity are claimed to be indeed involved in the modulation or generation of a wide range of normal and diseased functions such as cell growth and differentiation, hormone responses, immunological processes, electron transport, temperature acclimation, sensory reception, recovery from cell injury, drug effects, aging, a variety of cancers and red cell diseases, atherosclerosis, and liver diseases due to diet.

In spite of the impressive literature, it is not yet possible, in most instances, to assign a precise role to fluidity in the expression of membrane functions. Disagreements between laboratories and uncertainties which have only recently been recognized in the techniques used to measure membrane fluidity have questioned the validity of some of the original conclusions. In addition, there is the problem of cause and effect. That is, are changes in plasma-membrane fluidity responsible for changes in function or are they merely side effects of functional changes brought about by other mechanisms. The object of this article is to present a critical review of the present state of knowledge concerning relations between membrane fluidity and function.

II. MEANING AND MEASUREMENT OF MEMBRANE FLUIDITY

A. Definition of Fluidity and Viscosity of Ordinary Liquids

The term fluidity was originally introduced to describe the resistance which an ordinary liquid offers to its own flow or to the movement of a solid object through it. At the molecular level, an increase in fluidity is characterized by an increase in the mobility (lateral, rotational, and segmental) of the molecules of the liquid. Fluidity of an ordinary liquid can be defined quantitatively in terms of the rate at which the liquid flows under gravitational pull, as in an Oswald viscometer, or the rate at which an object falls through the liquid under gravitational pull. Since the two methods yield the same value for a given liquid, fluidity has an absolute quantitative meaning. Viscosity is the reciprocal of fluidity and is usually measured in units of poise.

B. Measurement of Membrane Fluidity through Fluorescence Anisotropy

The method most commonly used to measure membrane fluidity is the steady-state fluorescence anisotropy technique with 1,6-diphenylhexatriene (DPH) as the fluorescent probe (Shinitzky *et al.*, 1971; Shinitzky and Barenholz, 1978). This technique measures fluidity in terms of the resistance which the lipid bilayer offers to the rotational motions of DPH molecules dissolved in it. To apply this technique, the membrane of interest is first labeled by injection of a solution of DPH in an organic solvent into the bathing electrolyte and the emission or fluorescence anisotropy of the DPH which spontaneously incorporates into the membrane, presumably dissolved in the lipid bilayer, is then measured with a standard spectrofluorimeter equipped with emission and excitation polarizers. Emission anisotropy A is defined by the expression

$$A = (I_{VV} - I_{VH})/(I_{VV} + 2I_{VH}) \quad (1)$$

where I_{VV} and I_{VH} are the fluorescence intensities measured with excitation polarizer oriented in the vertical direction and emission polarizer oriented first in a vertical direction I_{VV}, and then in a horizontal direction I_{VH}. Fluorescence anisotropy has a maximum value of 0.4 in a highly viscous membrane and decreases toward zero with decrease in viscosity or increase in fluidity. Theory shows that the value of A actually depends on the ratio $T\,\tau/\eta$ where T is temperature, τ is lifetime, and η is microviscosity (viscosity measured with a molecular probe). The value of fluorescence anisotropy is essentially determined by the average angle over which the excited DPH molecules rotate during their lifetime (about 8 nsec for DPH). The rate of rotation, in turn, is determined by viscosity and temperature.

In practice, the exact quantitative relation between fluorescence anisotropy A and the ratio $T\,\tau/\eta$ is obtained experimentally by measuring the fluorescence anisotropy and lifetime of DPH dissolved in an oil with known viscosities at different temperatures. The microviscosity of a DPH-labeled membrane is then obtained from the resulting calibration graph, and values of A and lifetime measured for the membrane-bound DPH (Shinitzky *et al.*, 1971). This procedure assumes that (1) the DPH molecules reside in the hydrocarbon region of the lipid bilayer, (2) the DPH molecules in this membrane region can rotate freely in all directions as in an isotropic oil, (3) the Einstein theory of the Brownian motion of large particles applies to the rotational motions of DPH molecules in an isotropic environment, and (4) macroscopic and microscopic viscosities have the same value for a given liquid. The validity of these assumptions is discussed in a later section.

Microviscosities that have been measured for the plasma membranes of a variety of cells range from 1 to 15 poise. Examples in the order of increasing microviscosity are (Schachter and Shinitzky, 1977) lymphoma cells (1.2P), mouse 3T3d cells (1.5P), rat liver cells (1.7P), beef heart mitochondria (2.1P), human chronic lymphatic leukemic lymphocytes (3.4P), normal human lymphocytes (4.4P), human erythrocyte ghost membranes (6.3P), rat duodenal microvillus membrane (12.4P), rat jejunal microvillus membrane (14.7P), and rat ileal microvillus membrane (14.7P).

C. Measurement of the Conformation (Order) and Dynamics (Fluidity) of Membranes by NMR and ESR

Nuclear magnetic resonance (NMR) and electron spin resonance spectroscopy (ESR) have also been extensively used to measure the fluidity of natural membranes, and synthetic lipid bilayers (Berliner, 1976; Horwitz, 1972; Yguerabide and Foster, 1981). In its simplest form, NMR evaluates fluidity from the line width of the NMR signals produced by protons of the lipid molecules in the membrane. The proton magnetic spectrum of a lipid bilayer actually consists of two absorption bands. One comes from protons of the lipid hydrocarbon chains while the other comes from protons in the lipid polar heads. The width of each band is related to the rotational motions and orientations of the corresponding protons. Line width decreases with increase in rotational rate. The rotational motion of each proton is described in terms of the rotational motion of a vector which lies along the C–H bond. Since lipid molecules have a net orientation in the membrane, the C–H vector cannot rotate freely in all directions. Such restricted motion is usually described in terms of the half angle θ_H of a cone within which the C–H vector can rotate freely with a rotational correlation time ϕ. The half angle gives structural or conformational information since it is related to the degree of order in the bilayer while ϕ gives dynamic information related to fluidity. These two parameters are said to give information on the conformational dynamics of a lipid bilayer. The parameter ϕ is usually too short, a few nanoseconds, to be actually evaluated by NMR which has a time resolution around 10^{-5} sec. Instead, the line widths of the NMR spectrum are most often used to measure the structural parameter S, defined by

$$S = (3 \cos^2\theta_H - 1)/2 \tag{2}$$

S has a value of 1 for perfect alignment and a value of zero for a completely random distribution of orientations. NMR, thus, usually yields information on the degree of orientation of the hydrocarbon chains and polar heads but in its simplest form, as described above, gives no direct information on membrane fluidity which is contained in the rotational relaxation time ϕ. It is often assumed that fluidity increases with decrease in S and that the value of S is thus an indicator of fluidity. This assumption, however, is incorrect since experiments have shown that S and fluidity have no systematic relationship (Seelig and Seelig, 1974; Ligeti and Horvath, 1980). It is therefore incorrect to assume that a membrane with a higher value of S necessarily has a lower fluidity or vice versa.

The proton magnetic resonance signal from the methylene protons gives information on the average value of S for the entire hydrocarbon chain. Information on the structural parameter for individual segments of the hydrocarbon chains, however, can be obtained using lipids in which deuterium (Seelig and Seelig, 1980) or ^{13}C (Levine *et al.*, 1972; Metcalfe *et al.*, 1973) has been introduced into specific segments of these chains. In its more complex forms, NMR can also give information on ϕ or fluidity.

ESR also gives information on the value of S for different segments of the hydrocarbon chains or polar heads (Jost *et al.*, 1971; McConnell and McFarland, 1970). To obtain this information, it is necessary to attach a spin label to the lipid segment of interest. In addition, ESR can also give information on the dynamic or fluidity-related parameter ϕ.

D. Limitations of the DPH Technique for Measuring Fluidity

Although the fluorescence anisotropy technique has been used in a great number of membrane studies to establish a relationship between membrane fluidity and function, it has recently been recognized that the technique suffers from a variety of problems that seriously question previous conclusions on fluidity–function relations obtained by this technique. The problems in outline are discussed below.

1. Microviscosity and Membrane Fluidity Have No Simple Quantitative Meaning

Recent measurements of the nanosecond fluorescence anisotropy (decay of fluorescence anisotropy after excitation by a short pulse of light) of DPH in lipid bilayers indicate that the DPH molecule is not randomly oriented in the bilayer and does not rotate freely in all directions as assumed in the evaluation of fluidity with this probe (Yguerabide and Foster, 1981; Sherwood and Yguerabide, 1974; Kawato *et al.*, 1977; Dale *et al.*, 1977; Chen *et al.*, 1977; Lakowicz and Prendergast, 1978; Lakowicz and Knutson, 1980; Kinosita *et al.*, 1977). Instead, the long axis of the molecule is preferentially aligned along the hydrocarbon chains and displays restricted rotational motion. The rotational motions of DPH in the membrane, therefore, cannot be simply described in terms of fluidity or the related dynamic parameter ϕ, but require the introduction of a structural parameter S as well (Yguerabide and Foster, 1981; Kinosita *et al.*, 1977).

Likewise, the steady-state emission anisotropy also depends on the structural parameter in addition to fluidity so that previous interpretations based purely on fluidity considerations must be modified. As described above, nanosecond fluorescence anisotropy measurements separately evaluate S and ϕ. However, these parameters cannot be separately evaluated from steady-state emission anisotropy measurements alone, and it has, therefore, been proposed that results obtained by the steady-state technique should be presented for theoretical reasons simply as values of $\langle (A_0/A) - 1\rangle^{-1}$ (Schachter and Shinitzky, 1977), instead of erroneously converting them to fluidity, and that these values should be taken as measures of conformational dynamics instead of fluidity alone as is commonly done. An increase in the value of this parameter can result from a decrease in orientation, increase in fluidity, or both, but the exact nature of the change cannot be unraveled from the steady-state data alone.

Although S and ϕ can be evaluated separately for DPH in any membrane by nanosecond fluorescence anisotropy measurements, the nanosecond technique is difficult to apply and is presently available in only a few laboratories. The majority of studies of fluidity with DPH have therefore been done and continue to be done by the steady-state method.

In addition to the problem mentioned above, it has now been recognized that microviscosity (viscosity measured with a molecular probe) has no absolute quantitative meaning even in the case of isotropic liquids. In fact, Perrin (1936) indicated that the theory of Brownian motion, initially developed for relatively large particles, might not apply to the rotations of small particles with dimensions comparable to the dimension of the solvent molecules. For these small particles, the resistance to rotation is not simply related to the macroscopic viscosity, as measured, for example, with an Oswald viscometer, but also depends on the detailed molecular interactions between particle and solvent molecules. In addition, local order of solvent molecules at the microscopic level restricts their rotational motions, i.e., the rotational motions of the small probe particles are not random over the short time interval of one fluorescence lifetime. The practical significance of these considerations has recently been shown by experiments which demonstrate that the calibration graph of A vs. $T\ \tau/\eta$ for DPH depends on the oil used for the calibration (Hare and Laussan, 1977). This means that an absolute value of microviscosity cannot be assigned to any liquid and that microviscosity has no absolute quantitative meaning in contrast to macroscopic viscosity.

In spite of the difficulties encountered in defining membrane fluidity, it is difficult to abandon the use of this word since it is so ingrained in our description of membranes. In this article, we will continue to use the term membrane fluidity with the understanding that it has no quantitative meaning and that in a loose sense it can refer to the conformation, dynamics, or conformational dynamics of a membrane. High fluidity in this sense can mean high conformational disorder, high molecular mobility, or both.

2. *DPH Is Not Exclusively Located in the Plasma Membrane of Intact Cells*

Most fluidity experiments with intact cells are aimed at measuring fluidity of the plasma membrane, and it has often been assumed that in intact labeled cells the DPH molecules are chiefly located in the plasma membrane. However, recent studies by fluorescence microscopy, autoradiography, and isolation of internal cellular structures have shown that DPH does not exclusively label the plasma membrane but migrates into the cell interior where it also labels internal membranes, lipid granules, fat droplets, and lipoproteins (Berlin and Fera, 1977; Pagano *et al.*, 1977; McVey *et al.*, 1981; Esko *et al.*, 1977; Pessin *et al.*, 1978). This probe internalization is a serious problem in the use of DPH to measure plasma-membrane fluidity and questions the validity of conclusions previously drawn about the role of plasma-membrane fluidity in the expression of normal and abnormal membrane functions. Indeed, experiments which have compared the fluidity (as determined with DPH) of intact cells with the fluidity of plasma membranes isolated from these cells indicate that probe internalization in the intact cell may (1) obscure fluidity changes in the plasma membrane (Roozemond and Urli, 1979), (2) report changes when none actually occur in the plasma membranes (McVey *et al.*, 1981), or (3) report changes that are opposite to those actually occurring in the plasma membrane (Berlin and Fera, 1977).

In an effort to resolve the problem of DPH internalization, we have tested several probes for their ability to selectively label the plasma membrane and detect fluidity changes with high sensitivity (Yguerabide and Yguerabide, unpublished results). All of the nonpolar probes which we have tested, e.g., pyrene, DPH, perylene, and anthracene, readily penetrate the plasma membrane and label internal membranes. The behavior of amphipathic probes is complex and their ability to penetrate the plasma membrane depends on the nature of the polar head, hydrocarbon chains, and cell type. Phosphatidylyethanolamine (PE) labeled with rhodamine, through covalent linkage with the amino group of PE, readily penetrates the outer plasma membrane of many cells. A net charge on the polar head does not guarantee that the probe remains exclusively in the plasma membrane. The positively-charged carbocyanine dyes with two long hydrocarbon chains covalently attached to the carbocyanine moiety readily penetrate the membrane if the chain length is less than 12 carbon atoms per chain. On the other hand, these dyes do not penetrate or even label the plasma membrane if the chain length is greater than 16 carbon atoms. Carbocyanine dyes with 14 or 16 carbon atoms per chain, however, seem to label the plasma membrane without labeling internal membranes during the time of our experiments as determined by observations under the fluorescence microscope. It should be noted, however, that all the fluorophores which we have tested eventually become internalized and label internal membranes. The criterion for internalization is therefore a relative one, i.e., a probe is considered nonpenetrating if it does not internalize during the time required to do an experiment which may be anywhere from 15 min to 1 hr.

Although some carbocyanine dyes are good probes in the sense that they selectively label the plasma membrane, their fluorescence anisotropy is unfortunately insensitive to changes in fluidity due to their very short lifetime. We have, therefore, not yet found a fluorescent probe that exclusively labels the plasma membrane and also has a high sensitivity for changes in fluidity through its polarized emission.

To avoid the problems of probe internalization in intact cells, many investigators recently have started using isolated plasma-membrane preparations obtained by cell disruption followed by purification through density gradient centrifugation (Schachter and Shinitzky, 1977; Berlin and Fera, 1977; Esko *et al.*, 1977; Pessin *et al.*, 1978; McVey *et al.*, 1981; Van Hoeven *et al.*, 1979; Van Blitterswijk *et al.*, 1977; Roozemond and Urli, 1979). The use of isolated membrane preparations, however, carries the risk of working with membranes whose natural properties have been altered by the isolation procedure, for example, by disruption of cytoskeletal structures and rearrangement of membrane proteins. In addition, plasma membranes can fragment into membrane fragments of different protein and lipid composition, and one of these membrane fragment subsets can be inadvertently selected by the isolation or purification procedure. Nevertheless, in many instances, values of fluidity obtained with isolated membrane preparations correlate much better with other types of data, for example, lipid composition, than values obtained with intact cells.

An alternate method for measuring plasma-membrane fluidity in intact cells is the fluorescence recovery after photobleaching technique (FRAP; Yguerabide, 1972a; Peters *et al.*, 1974; Koppel *et al.*, 1976). This technique measures the lateral mobility of a fluorescent lipid probe in the membrane and is independent of the lifetime of the fluorophore. Thus, fluorescent probes such as the carbocyanine dyes which selectively label the plasma membrane (but whose emission anisotropy is insensitive to fluidity) can be used in this technique. The FRAP technique has the additional advantage of being able to give information on the fluidity of different regions of a cell as discussed below.

3. *Heterogeneity in the Lipid Bilayer of a Natural Membrane*

The lipid bilayer of a natural membrane is heterogeneous with respect to the distribution, conformation, and dynamics of lipids in its plane and in each of its lipid monolayers. It may, for example, consist of contiguous lipid bilayer patches some of which are in a fluid state and others in the gel state, a condition called lateral phase separation. In addition, lipid molecules in the vicinity of membrane proteins, so-called boundary lipids, have a different orientation, mobility, and composition than the bulk lipids. Finally, the two lipid monolayers (cytoplasmic and external) usually have different compositions.

The interpretation of data obtained with a probe such as DPH requires consideration of the lipid bilayer heterogeneity. It is convenient in these considerations to explicitly view the bilayer in terms of regions consisting of (1) the two lipid monolayers, (2) the different liquid and gel phases present in each monolayer, (3) the boundary lipid, and (4) the water–lipid interface at each lipid monolayer surface. In addition, it is also convenient to consider the lipid polar heads (polar region) and hydrocarbon chains (hydrophobic region) as subregions of the various liquid and gel phases and boundary lipid present in each lipid monolayer. Each region has its own lipid composition and each subregion has its own conformation and dynamics. The regions are coupled to each other through lipid exchange and physical interactions. A chemical or physical change in one region or subregion can affect protein functions either directly or indirectly through its coupling with other regions.

The interpretation of the data obtained with a fluorescent probe requires knowledge of which bilayer regions and subregions are being sensed by the probe. A variety of experiments have shown that DPH and perylene partition equally well between fluid and gel phases (Lentz *et al.*, 1976; Foster and Yguerabide, 1979) and might also partition into the hydrophobic region of boundary lipid. Because of these partitioning properties, some investigators have considered DPH as a general fluidity probe. However, one must keep in mind that DPH seems to be excluded from polar regions, at least in the fluid state and, therefore, can sense changes in these regions only indirectly through coupling of polar regions with hydrophobic lipid tails. That such coupling does exist is supported by the fact that the fluidity of the hydrophobic region of a lipid bilayer depends on the nature of the polar heads. This is due to polar head attractions or repulsions which respectively increase or decrease packing of hydrocarbon tails. However, there are a variety of experiments which indicate that at least in some instances the coupling is not sufficiently high for DPH to sense changes in the polar regions. Thus, Ca^{2+} which interacts with charged polar heads has been found to have little effect (Schachter and Shinitzky, 1977) on fluidity when measured with DPH, but significant effects are seen when fluidity was measured with 12-(9-anthroyl) stearate or retinol. Additional examples are: sea urchin eggs where fluidity changes accompanying fertilization are found to be confined to the polar region (Campis and Scandella, 1980), and pressure and anesthetics have fluidizing effects only in the hydrocarbon region of the lipid bilayer (Finch and Kiesow, 1979).

Caution must, therefore, be exercised in the interpretation of results when using a single probe such as DPH. The experimental evidence indicates that several probes should be used for proper assessment of fluidity changes. Examples of probes which have been suggested for the various bilayer regions are parinaric acid for polar and hydrophobic regions of gel phases (Sklar *et al.*, 1975, 1977a,b), *N*-phenyl-1-naphthylamine (Overath and Trauble, 1973; Shroeder, 1978) and retinol (Schachter and Shinitzky, 1977) for the region intermediate between the polar and hydrophobic regions, and the family of probes 2-anthroyl stearate (2AS), 6AS, 9AS, and 12AS (where AS stands for anthroyl stearate) to sample the bilayer at different depths (Thulborn *et al.*, 1979).

Charged probes can also be used to selectively label the inner or outer lipid monolayers of a membrane. Because of the lipid bilayer asymmetry of natural membranes, negatively charged probes tend to selectively label one monolayer while positively charged probes label the other monolayer. Thus, in the case of red blood cells, positively and negatively charged amphipathic molecules bind to the inner (cytoplasmic) and outer lipid monolayers, respectively (Sheetz and Singer, 1974). As another example, lidocaine has been found to exert its effects on liver plasma membranes by selectively binding to the inner lipid monolayer where it interacts with acidic phospholipids (Houslay *et al.*, 1980). Parinaric acid has been used to probe bilayer asymmetry (Sklar and Doody, 1980).

Selective labeling of the boundary lipid at the interface between lipids and integral proteins is also of interest. A variety of experiments indicate that the fluidity or conformational dynamics of this region can have more pronounced effects on protein functions than the fluidity of the bulk lipid (for details, see Section III-B). However, it is difficult to measure the fluidity of boundary lipid with a fluorescent probe because this lipid accounts for only about 20% of the total lipid in a natural membrane. It could, however, be studied with a probe that partitions preferentially into the boundary lipid region. The importance of the boundary lipid in the expression of function and the difficulties involved in detecting its conformational dynamics are demonstrated by studies with (Ca^{2+}-Mg^{2+})-ATPase in a synthetic bilayer of dipalmitoylphosphatidylcholine (DPPC; Hesketh *et al.*, 1976). The results show two discontinuities in an Arrhenius plot of enzymatic activity, one associated with the phase transition of bulk lipid at 39°C and a larger one associated with the phase transition of boundary lipid at 30°C. However, the Arrhenius plot of fluorescence anisotropy of DPH in the same ATPase-lipid bilayer shows only the bulk lipid transition unless the protein (ATPase) to lipid ratio is made high enough to enhance the contribution of fluorescent labeled boundary lipid to the fluorescence signal. These results indicate that the presence of a transition in an Arrhenius plot of functional activity without a corresponding transition in the Arrhenius plot of emission anisotropy cannot be used as evidence that a lipid transition is not involved.

On the other hand, some discontinuities on an Arrhenius plot of DPH fluorescence anisotropy can be the result of protein instead of lipid transitions. Some investigators maintain that, at least in some membranes, the DPH molecules bind preferentially to membrane proteins (Mely-Goubert and Freedman, 1980; Miller and Raison, 1980; Mely-Goubert *et al.*, 1979). In this case, protein conformational transitions would dominate the Arrhenius plot of fluorescence anisotropy of DPH. In addition, the apparent fluidity determined from the emission anisotropy of DPH would correspond to protein immobilization of probe rather than to lipid bilayer fluidity. An approach which has been used by some investigators to determine the extent to which protein–probe interactions contribute to the DPH emission anisotropy from a given membrane has been to compare Arrhenius profiles of emission anisotropy of the probe in the membrane with profiles obtained with probe in liposomes formed from the membrane lipids and no proteins (Schachter and Shinitzky, 1977; Miller and Raison, 1980; Mely-Goubert *et al.*, 1979). In principle, discontinuities and apparent fluidity effects produced by protein–probe interactions can be identified from the absence in the liposome Arrhenius profile of transitions which are present in the natural membrane profile. This approach, of course, severely suffers from uncertainties due to differences in the lipid arrangement in a liposome and in a natural membrane with the same lipid composition, e.g., liposomes lack the lipid bilayer asymmetry of natural membranes.

Protein–probe interactions have also been evaluated from studies of the interaction of probe with isolated integral membrane proteins solubilized in detergent. For example, the fluidity of intestinal microvillus membranes as measured with the probes DPH, retinol, and 12AS is lower than that of the liposomes formed from the same membrane lipids (Schachter and Shinitzky, 1977). This difference could be due to boundary lipid, compression of bulk lipid by proteins in the membrane, or direct binding of the probe to sites on the protein. Fluorescence anisotropy measurements of probe added to detergent-solubilized integral membrane proteins from the microvillus membranes indicate that retinol binds directly to membrane proteins (Schachter and Shinitzky, 1977). Care, however, must be exercised in reaching conclusions from this type of measurement since protein–probe interactions are dependent on the type of environment around the protein.

The discussions presented above indicate that the interpretation of emission anisotropy data in terms of membrane fluidity is not trivial because of membrane heterogeneity. Questions which must be considered in order to properly assess with a given probe the effects of lipid conformation and dynamics on protein functions are: which regions of the membrane are sensed by the probe, does the probe fail to detect changes in some of the lipid regions, to what extent is membrane fluidity determined by protein, are Arrhenius plot discontinuities due to conformational transitions of proteins or lipids, and can the probe bind directly to proteins?

4. *Anatomically or Functionally Polarized Cells*

It is well known that the plasma membrane of a tissue or tissue culture cell is anatomically and functionally polarized. That is, different areas of the membrane along the sides and length of the cell have different structures and functions. Epithelial cells, for example, have lateral surfaces that are specialized for cell–cell contact and formation of tight junctions, while the lumenal and basal surfaces differ from each other in transport, enzymatic, and receptor functions. Furthermore, anchored cells such as macrophages and tissue culture cells such as developing myoblasts are differentiated structurally and functionally along their length.

It is customary in the study of the fluidity of tissue or tissue culture cells to perform the fluidity measurements in cell suspensions prepared by releasing the cells through enzymatic digestion (trypsin or collagenase) or removal of Ca^{2+} with EDTA. These procedures at best give an average of the fluidity of the various differentiated areas of the cell. However, it is more likely that the measured fluidity does not even represent a true average since the breakage of contacts between cells is now known to obliterate cell polarity by allowing the mixing of lipids and proteins from the different specialized areas (Pisam and Ripoche, 1976; Ziomek *et al.*, 1980). Moreover, it has been demonstrated that trypsinization of tissue culture cells changes membrane fluidity (De Laat *et al.*, 1978). The validity of the many studies which have been made with polarized cells in cell suspension are therefore subject to question.

Reliable information on the fluidity of polarized cells, however, can be obtained by using pieces of intact tissue or whole tissue cultures (with minimum disruption of cell–cell and cell–substrate contacts) in conjunction with a fluorescence microscope and a microbeam to elicit fluorescence from a defined small area (2 diameter) on the membrane of a single cell. With this arrangement, the plasma membrane of an individual cell can be studied under conditions of minimum disruption and the spatial variation of fluidity over the plasma membrane can be evaluated by moving the microbeam over the different areas. Fluidity can be determined by either fluorescence anisotropy measurements (De Laat *et al.*, 1978; Elson and Yguerabide, 1979) or by measurement of lateral mobility using the technique of fluorescence recovery after photobleaching (FRAP; Yguerabide, 1972a; Peters *et al.*, 1974; Koppel *et al.*, 1976; Elson and Yguerabide, 1979). The latter technique has the advantage that nonpenetrating fluorescent probes are readily available for measurement of lateral mobility, whereas sensitive, nonpenetrating probes are not available for the fluorescence anisotropy measurements as previously discussed. Moreover, the FRAP technique is capable of measuring the lateral mobility (diffusion coefficient) of proteins as well as of lipids. A limitation of the FRAP techinque is its dependence on the size (diameter) of the illuminated area which makes the technique dependent on the presence of microvilli (Elson and Yguerabide, 1979). Since the area contributed by the microvilli cannot be determined by fluorescence microscopy, it is not possible without additional information to state unequivocally that differences in rate of fluorescence recovery (between two different plasma membranes or between areas on a given plasma membrane or the same membrane area under different physiological conditions) are not due to differences in the number or size of microvilli.

In FRAP studies, however, interest is usually focused on the mobility of proteins and whether differences in protein mobilities are due to differences in lipid fluidity or in cytoskeletal interactions. This can be answered from the ratio R of the lateral diffusion coefficient of the protein to that of the lipid probe (Elson and Yguerabide, 1979). If a change in protein diffusion coefficient as measured by FRAP is due to a change in fluidity or number of microvilli, then the value of R should not be greatly affected. If the change is due to cytoskeletal interactions with the protein, then the value of R should change significantly. Emission anisotropy, on the other hand, is not sensitive to area and can, therefore, be used to unravel contributions of microvilli to FRAP data if a sensitive nonpenetrating fluorescent probe can be found for the anisotropy measurement.

The ideas presented above concerning R assume that the mobilities of components in the microvillar membrane are the same as in the contiguous planar membrane between the microvilli. Several considerations, however, indicate that this assumption may not always be correct. Cytoskeletal structures are responsible for producing some types of microvilli and could cause specific proteins or lipids to enter or be excluded from the microvillar membrane. Moreover, theoretical considerations indicate that curvature favors the presence of unsaturated lipids in the microvillar membrane and excludes cholesterol and large protein molecules from these structures (Israelachvili *et al.*, 1980). Finally, some microvilli could owe their existence to asymmetric insertion of proteins or amphipathic molecules into the lipid bilayer (Sheetz and Singer, 1974;

Papahadjopoulos, 1977) as in the case of discocyte to echinocyte transformation in red blood cells. In any case, the microvillar membranes and the planar membrane contiguous with them are expected to differ in composition and cytoskeletal interactions and possibly in the lateral and rotational mobility of proteins and lipids.

The number of studies on cell polarity and microvilli by fluorescence microscopy are still few, but the application so far indicates high potential for obtaining new insights by these studies. Thus, in the case of anchored cells, FRAP measurements have shown that proteins move more readily in a direction parallel to the stress fibers than in a perpendicular direction (Smith *et al.*, 1979). In a study of anchored myoblasts (Elson and Yguerabide, 1979), it was shown that the fluidity of the membrane varied along the length of the myoblast cell, the fluidity being lowest at the edges and extended arms of the cell. It has also been established that the mobility of a lipid probe in the region between spicules of echinocyte red cells is the same as in a discoid cell.

III. FACTORS THAT INFLUENCE MEMBRANE FLUIDITY

Elucidation of the mechanisms by which membrane fluidity affects membrane functions requires some understanding of the factors which influence fluidity. Here we present a brief discussion of the most pertinent of these factors.

A. Lipid Composition

Lipid composition is one of the main determinants of membrane fluidity and changes in lipid composition can be triggered by a wide range of physiological stimuli. Some of these changes in composition can occur relatively fast. Phosphatidylinositol turnover rate, for example, changes significantly in some cells within a few minutes after presentation of an appropriate physiological stimulus (Smith *et al.*, 1979). In this section, we discuss how lipid composition influences membrane fluidity and mechanisms by which lipid composition can be altered in a cell.

Lipid bilayer fluidity depends on the nature of the lipid hydrocarbon tails and polar heads (Shinitzky and Barenholz, 1978; Yguerabide, 1980). For a bilayer composed of lipid with a given polar head, fluidity increases with decrease in chain length and with increase in the degree of unsaturation of hydrocarbon tails. If the lipid chains are kept constant and the polar head is varied, it is found that phosphatidylethanolamine (PE) yields a lipid bilayer with a lower fluidity than phosphatidylcholine (PC), while sphingomyelin (SM) produces a bilayer with a much lower fluidity than one made from PE. The dependence of fluidity on polar heads is due to differences in attractive or repulsive forces (hydrogen bonding and electrostatic) between head groups. Increase in attractive or decrease in repulsive interactions increases the packing density of the hydrocarbon tails which consequently decreases bilayer fluidity.

Cholesterol decreases fluidity of phospholipid bilayers by immobilizing the segmental motions of the hydrocarbon tails through its rigid steroid structure. Much interest has recently focused on the roles of cholesterol and sphingomyelin on membrane functions. Both of these lipids are present in eukaryotic cell membranes and both decrease membrane fluidity of lipid bilayers in the fluid state (Shinitzky and Inbar, 1976; Shinitzky and Barenholz, 1974). Cholesterol seems to be involved in a wide range of diseases associated with increased cholesterol levels in the blood serum including atherosclerosis, red cell abnormalities, and decreased responses in lymphocytes (Cooper and Shattil, 1980). Sphingomyelin is of special interest because it produces lipid bilayers with very low fluidity and because it seems to be the only natural lipid whose bilayers display lateral phase separations in the range 20–40°C (Shinitzky and Barenholz, 1974; Schmidt *et al.*, 1977). The role of these lipids in membrane function is discussed in Section V.

The extent to which changes in membrane lipid composition modify lipid bilayer fluidity depends on the value of fluidity prior to the change in composition. Thus, little change in fluidity was observed (Herring *et al.*, 1980; by ESR and DPH) when large amounts of polyunsaturated lipids were added to highly unsaturated *Dictyostelium discoideum* membranes which indicates that an increase in the lipid unsaturation of a membrane has little effect on bilayer fluidity, at least as detected by ESR and DPH, if the membrane is already highly unsaturated.

Living cells are constantly renewing their lipids which accounts for their ability to respond quickly to specific stimuli through changes in lipid compositions. Mechanisms used for lipid renewal include (Thompson, 1973; McMurray, 1973) (1) *de novo* synthesis in the endoplasmic reticulum of PC, PE, and PI from phosphatidic acid and corresponding polar head moieties, (2) conversions such as PS to PE to PC, (3) direct base exchange by which PE, PC, and PS are interconverted, and (4) exchange of fatty acid chains by phospholipase deacylation and acyltransferase reacylation. Much of the lipid metabolic activity of nongrowing cells involves fatty acid exchange which permits rapid changes of fatty acid composition in response to changes in dietary fats or physiological stimuli.

In addition to these chemical mechanisms, whole lipid molecules can also be removed and inserted into membranes by lipid-exchange proteins with a wide range of specificities. Exchange of lipids between ER and other cellular membranes is of great importance because the ER seems to be the only site capable of *de novo* synthesis in a cell (McMurray, 1973). The plasma membrane however contains lipases, acyltransferases, and base exchange proteins which allow this membrane to alter lipids received from the ER and, thus, assume a lipid composition that is different from other membranes in the cell.

There is also much interest in processes which exchange whole lipid molecules between plasma membranes and blood lipoproteins because of the role which these processes play in cases of abnormal lipid composition of the blood due to diseases or diet (Cooper, 1977). Plasma membranes can rapidly exchange lipids with serum lipoproteins (LDH, HSL, and VHDL; Cooper and Shattil, 1980; Cooper, 1977; Rivnay *et al.*, 1980; King and Quinn, 1980; Gottlieb, 1980; Kuehl *et al.*, 1980; Berlin *et al.*, 1980; Jackson and Gotto, 1976; Mistry *et al.*, 1981; Cooper *et al.*, 1980), and in the case of cells with limited metabolic activity such as red cells and platelets, the lipid composition of the plasma membrane approaches the lipid composition of the serum lipoproteins. In the case of cells with greater metabolic capacity, the lipid composition of the cell plasma membrane is determined by both serum exchange and internal metabolic activity. Internal membranes are also affected by serum lipid composition (McMurchie and Raison, 1979; Morre *et al.*, 1979), probably through lipid exchange between plasma and internal membranes. It should, however, be noted that each internal membrane has its own distinct lipid composition.

Fusion between internal granule and plasma membranes provides another path for altering lipid composition of the cell membrane. Such fusion, for example, is triggered in toad bladder epithelial cells when exposed to the antidiuretic hormone vasopressin (Masur *et al.*, 1972; Gronowicz *et al.*, 1980).

B. Proteins and Boundary Lipid

We have already noted in Section II-D that proteins can influence lipid bilayer fluidity. Here, we present additional details.

Integral as well as peripheral proteins can influence membrane fluidity. Our attention will first be directed to the integral proteins. Many of the effects which integral proteins have on lipid bilayers are due to immobilization of an annulus of boundary lipid molecules in their immediate vicinity (Jost *et al.*, 1973; Stier and Sackmann, 1973). Experimental results indicate that there are about 30 boundary lipid molecules per integral protein molecule (Jost *et al.*, 1973) or about 0.2 mg of boundary lipid per mg of protein (on the average, 20% of the total lipid in a natural membrane). This number corresponds to a single layer of lipid molecules on the circumference of each integral protein molecule. However, theoretical calculations (Owicki *et al.*, 1978; Marcelja, 1976) suggest that the lipid perturbation of an integral protein may extend beyond this strongly immobilized layer and reach as far as three or four molecular layers into the bulk lipid. If this is the case, the perturbation of the integral membrane proteins may affect as much as 80% of the lipid in a natural membrane. Twenty percent of the lipid would be strongly perturbed (boundary lipid) while 60% would be weakly perturbed.

Boundary lipid is not static. Boundary and bulk lipid molecules, for example, can exchange in times greater than 10^{-4} sec (Seelig and Seelig, 1978). Moreover, boundary lipid can undergo thermal phase transitions or separations with transition and separation temperatures that differ from those of the bulk lipid. Thus, for a bilayer composed of dipalmitoylphosphatidylcholine containing (Ca^{2} + Mg^{2+})-ATPase, the bulk transition temperature is found at 39°C while the boundary lipid transition is at 30°C (Hesketh *et al.*, 1976). In general, experiments indicate that the phase transition temperature of bulk lipid compared to boundary lipid will be lower for saturated lipids and higher for unsaturated lipids.

Integral proteins can also modify the fluidity and phase separation temperature T_c of bulk lipid by preferential binding of lipids with saturated or unsaturated hydrocarbon chains or specific polar heads, e.g., sphingomyelin. Both electrostatic and hydrophobic forces seemed to be involved in the formation of boundary lipid (Brotherus *et al.*, 1980), and cholesterol seems to be excluded from the boundary area (Warren *et al.*, 1975). Bulk bilayer fluidity and T_c can be decreased or increased depending on the type of bulk lipid which is preferentially included or excluded from the boundary lipid. It should, however, be noted that the extent to which an integral protein affects membrane fluidity also depends on the inherent fluidity of the lipid bilayer, i.e., the fluidity in the absence of proteins. For example, a bilayer of low fluidity due to a high cholesterol content would be less affected by an integral protein than one with low cholesterol. It should also be noted that some investigators have presented evidence that glycophorin and acetylcholine receptor protein increase instead of decrease the fluidity of the lipid bilayer (Brulet and McConnell, 1976; Bienvenuë *et al.*, 1977).

Extrinsic proteins can also affect lipid bilayer properties. Basic proteins can bind electrostatically to acidic lipids and cluster them into domains which separate from the bulk lipid and have a high molecular packing density (Birrell and Griffith, 1976). Fluidity would be decreased by the increased packing density. However, some basic proteins such as cytochrome *c* can also partially penetrate a lipid bilayer and, in contrast, increase fluidity by disruption of the fatty acid tails.

The discussion presented above indicates that care must be exercised in relating changes in membrane fluidity to changes in membrane function since changes in fluidity could be the result and not the cause of changes in protein conformations and function.

C. pH

Changes in pH have their most significant effects on lipid bilayers through changes in the degree of ionization of the lipid polar heads (Yguerabide, 1980; Papahadjopoulos, 1968; Trauble and Eibl, 1974). The lipids most likely to be affected are phosphatidylserine (PS) and phosphatidic acid (PA) since they have ionizable groups with p*K* values around physiological pH. A decrease in pH decreases the electrical charge on PA and PS, and consequently decreases repulsion between polar heads and increases packing density. The increased packing density is expected to decrease membrane fluidity. A decrease in the fluidity of the polar region of lipid bilayers with a decrease in electrical charge has indeed been observed experimentally by ESR techniques (Flanagan and Hesketh, 1973). However, in a study in which membrane electrical charge was increased with dicetylphosphate (Shinitzky and Barenholz, 1974), no change in membrane fluidity could be detected with DPH which indicates that DPH may not always be capable of detecting fluidity changes induced by changes in electrical charge. Such changes are expected to be greatest in the polar regions of the membrane.

A change in pH can also indirectly affect membrane fluidity through its effect on proteins. Indeed, pH-induced fluidity changes in intestinal microvillus membranes have been attributed to effects on proteins possibly involving the side chains of glutamic acid and histidine (Schachter and Shinitzky, 1977).

D. Calcium

It is well known that Ca^{2+} modulates many membrane functions and that it decreases the fluidity of a wide range of membranes. The most direct effects of Ca^{2+} on bilayer fluidity seem to involve interactions between calcium ions and negatively charged (acidic) phospholipids (Yguerabide, 1980; Trauble and Eibl, 1974; Lussan and Faucon, 1974; Verkleij *et al.*, 1974; Georgescauld *et al.*, 1979). Binding of Ca^{2+} to an anionic lipid can have two effects. First, neutralization of the negative charge by Ca^{2+} reduces repulsion between polar heads which facilitates packing of lipid molecules and decreases fluidity. Second, Ca^{2+} can induce aggregation of anionic lipids into clusters consisting of anionic lipid and bound Ca^{2+}.

Calcium can also have indirect effects on fluidity by inducing changes in metabolic activity that alter the lipid composition. For example, in a study of the fluidity of isolated rat hepatocyte membranes with DPH (Livingstone and Schachter, 1980), it was found that Ca^{2+} changed fluidity in two steps. First, upon addition of Ca^{2+}, there was an immediate decrease in membrane fluidity with a half-life of about 15 min. This effect could be reversed with EDTA and seemed to involve an interaction of Ca^{2+} with lipids. The second step occurred more slowly, with a half-life of about 75 min, and produced a larger effect on fluidity; this step seemed to involve stimulation of membrane-bound enzymes which altered lipid composition.

E. Salt Concentration

Changes in salt concentration have their most significant effects on bilayer properties through effects on lipids with a net electrical charge. An increase in salt concentration increases the electrical shielding of electrically charged groups on the membrane surface. This increased shielding diminishes the repulsive forces between lipids with similar charge and increases their packing density with a consequent decrease in membrane fluidity (Yguerabide, 1980; Trauble and Eibl, 1974).

IV. MECHANISMS BY WHICH MEMBRANE FLUIDITY INFLUENCES MEMBRANE FUNCTIONS

It is intuitively clear that processes which depend on collision rates are sensitive to fluidity. Indeed, the effects of fluidity on a number of membrane processes can be largely explained in terms of effects on collision rates. Examples are the reaction between NADH–cytochrome b_5 reductase and cytochrome b_5 (Rogers and Strittmatter, 1974), coupling between adenylate cyclase and adrenergic receptors in the presence of hormone (Rimon *et al.*, 1978), and the rate of concanavalin-A-mediated agglutination of PyBHA cells (Rule *et al.*, 1979). There are, however, many processes which do not depend on collision rates but are nevertheless sensitive to membrane fluidity. Such processes are presumably affected by fluidity through changes in the conformation and disposition of corresponding proteins in the membrane. In this section, attention will be focused on the latter types of effects.

We first consider fluidity changes involving lipid composition. Lipids can affect protein molecular conformations and functions through either specific protein–lipid interactions or through changes in bulk lipid fluidity (Lands, 1980). Specific interactions may range from very strong protein–lipid attractions which lead to strongly bound lipid to weaker interactions which determine the fluidity and lipid composition of boundary lipid. The relative effects of bulk and boundary lipid have been studied by the lipid removal technique. Thus, removal of bulk lipid is found to have some effect on enzymatic functions but activity remains high, at least for some of the enzymes which have so far been studied, as long as boundary lipid is not removed (Warren *et al.*, 1975). However, removal of boundary lipid significantly diminishes enzymatic activity which becomes vanishingly small when boundary lipid is reduced much below 30 lipid molecules per protein molecule. Finally, removal of tightly bound lipid leads to protein denaturation (Gazzotti and Peterson, 1977). These results suggest that boundary lipid is essential for membrane protein function while tightly bound lipid is necessary to maintain a stable protein conformation.

Membrane protein functions, however, depend not only on the mere presence of boundary lipid but more specifically on the fluidity or hydrophobic environment provided by these lipids (Gazzotti and Peterson, 1977). This is demonstrated by experiments in which a cytochrome oxidase preparation was partially delipidated to remove bulk lipid and the effect of boundary lipid composition on enzymatic activity was then evaluated by addition of specific lipids (Warren *et al.*, 1975). The results showed that activity is highest with dioleoyllecithin (DOL) and decreases when DOL is partially replaced with a saturated lipid such as dimyristoyl- (DML) or dipalmitoyl- lecithin (DPL). Activity decreases linearly with the mole fraction of saturated lipid. Since DML and DPL are less fluid than DOL, the results suggest a dependence on boundary lipid fluidity.

The fluidity of bulk lipid can also independently affect protein function as demonstrated by an experiment in which (Ca^{2+}-Mg^{2+})-ATPase was incorporated into a lipid bilayer of dipalmitoyllecithin (PC). The Arrhenius plot of enzymatic activity shows two discontinuities, one at 39°C corresponding to the bulk phase transition temperature of PC and another at 30°C corresponding to the phase transition of boundary lipid. These experiments suggest that protein functions depend on the fluidity of bulk as well as boundary lipid.

Protein functions depend not only on fluidity but also on the polar environment provided by different lipids. This is suggested by experiments that, for a variety of protein functions, imply a need for small amounts of lipids with specific polar heads. Experiments of this type, however, are difficult to interpret and it has been stated that a requirement for specific lipids has never been clearly demonstrated. Nevertheless, examples of proteins which have been presented as requiring specific lipids for expression of function are (Gazzotti and Peterson, 1977) (Na^{+}-K^{+})-ATPase (acidic phospholipid, especially phosphatidylserine), erythrocyte acetylcholinesterase (cardiolipin), PEP-phosphotransferase of enzyme II (anionic lipid, phosphatidylglycerol), mitochondrial-hydroxybutyrate dehydrogenase (phosphatidylcholine), and (Ca^{2+}-Mg^{2+})-ATPase of sarcoplasmic reticulum (phosphatidylethanolamine). Specific lipids exert their effects at low concentration and, therefore, presumably act through interactions with specific sites on membrane proteins. However, since specifically required lipids can exchange with bulk lipids, they presumably reside in the boundary lipid area or some weakly binding site on the protein, in contrast to tightly bound lipid.

Taken together, the results presented above indicate that the lipid bilayer can influence protein functions through the fluidity and polar environment of boundary lipid and fluidity of bulk lipid. Changes in lipid composition are likely to influence both boundary and bulk lipid since these two types of lipid readily exchange with each other.

The exact mechanism by which the fluidity of bulk or boundary lipid influences protein function is unknown (except for collisional processes), but it presumably involves restrictions on protein conformation and arrangement in the membrane. Decrease in membrane fluidity, in a loose sense, is due to an increase in the packing of hydrocarbon chains and is equivalent to an increased lateral pressure. Such a pressure would be expected to restrict protein conformations and to "squeeze" proteins out of the bilayer. Alternatively, one can view a decrease of fluidity as an increase in lipid packing density due to increase in lipid–lipid interaction and a relative decrease in lipid–protein interaction. All of these effects could result in protein aggregation and displacement of proteins in a vertical direction (Schachter and Shinitzky, 1977).

Recent experiments have shown that a decrease in membrane fluidity exposes reactive groups on integral proteins, and this effect has been interpreted as supportive of a vertical displacement of membrane protein molecules produced by a decrease in membrane fluidity (Borochov and Shinitzky, 1976; Borochov *et al.*, 1975; Shinitzky and Souroujon, 1979). On the other hand, Amar *et al.* (1979) find that varying growth conditions vary the exposure of reactive protein groups, but no consistant correlation could be found between the degree of exposure and membrane fluidity.

Changes in lipid fluidity could also influence protein functions by inducing aggregation of membrane proteins as mentioned above. The idea that membrane functions can be triggered through aggregation of membrane proteins has, in fact, gained much popularity recently and has been postulated to be involved in a wide range of modulated functions. A variety of experimental observations and theoretical calculations have given some insights into various aspects of membrane protein aggregation. Thus, many membrane receptors appear to exist in clusters in natural membranes, integral proteins are found to aggregate (or be segregated from the lipids) below the membrane phase transition temperature in many natural membranes (Grant and McConnell, 1974; Kleemann and McConnell, 1976), and aggregation is inhibited by branched lipids and cholesterol. Theoretical considerations indicate that, for certain lipid compositions, boundary lipid can induce protein aggregation at physiological temperatures (Marcelja, 1976; Owicki *et al.*, 1978; Israelachivili *et al.*, 1977; Israelachivili, 1978; Vanderkooi *et al.*, 1974). These theoretical ideas, however, have not yet been tested experimentally.

Finally, as discussed in a previous section, lipid fluidity can be influenced by Ca^{2+}, pH, and salt concentration. Changes in fluidity induced by changes in any of these parameters can influence protein functions through the mechanism discussed above involving bulk and boundary lipid.

V. ROLE OF MEMBRANE FLUIDITY IN SOME MEMBRANE FUNCTIONS

A. Effects of Cholesterol

Some interest has recently focused on cholesterol because of its possible involvement in a wide range of diseases. Cholesterol is present in the membrane of all eukaryotic cells and seems to play an important role in the regulation of cell fluidity. Each of the different membranes in a cell, for example, plasma membrane, endoplasmic reticuluum, and nuclear membrane, maintains normal functions only within certain narrow limits of cholesterol to phospholipid (C/PL) ratio.

Much of the present interest in cholesterol stems from the fact that the C/PL ratio in many cells depends on the C/PL ratio in the blood serum which can become abnormally high in a variety of diseases and types of lipid diets. Cells with limited metabolic activity, such as red cells and platelets, are particularly sensitive to the serum C/PL ratio as discussed previously, although other types of cells such as lymphocytes and liver cells are also affected by the serum C/PL ratio. Examples of normal C/PL ratios (moles of cholesterol to moles of phopholipid) for different membranes are (Cooper, 1977) platelets (0.5), mitochondria (0.1), and red cells (0.95). It should be noted that, although the C/PL ratio varies from one membrane to another, the ratio of cholesterol to phospholipid plus glycolipid, C/(PL + C), is fairly constant between different membranes under normal conditions (Cooper, 1977). This observation indicates that the latter ratio is the controlling factor under normal conditions.

In the case of red cells, a decrease in the C/PL ratio from its normal value of 0.95 increases membrane fragility while an increase in this ratio leads to spur cell anemia which is common in patients with severe liver diseases (Cooper, 1977). The spur cells have a spiculed appearance, a C/PL ratio as high as 1.6, and are prematurely destroyed by the liver. Normal red cells become spurred when placed in serum from patients with spur cell anemia. The C/PL ratio of red cells, as well as other types of cells, has been found to increase linearly with the C/PL ratio of the serum. It should be noted that cholesterol is not soluble in blood and is actually carried by the serum lipoproteins. Fluidity measurements indicate that spur cells have a higher microviscosity than normal cells and it has been suggested that decreased fluidity is responsible for the spur cell appearance (Vanderkooi *et al.*, 1974). However, spur cells are produced by a wide range of abnormal C/PL ratios and, while it is true that microviscosity decreases in abnormal cells at very high C/PL ratios, it is not affected in the lower range where spur cell formation begins (Sherwood and Yguerabide, 1974; Sherwood *et al.*, unpublished results). Therefore, microviscosity does not seem to be the controlling factor in the production of spur cells.

The effects of fluidity and cholesterol on platelet functions have been the subject of several studies. Platelets normally participate in the control of bleeding by aggregating and plugging damaged blood vessels. However, under abnormal conditions, e.g., high serum cholesterol, platelets also participate in the formation of atherosclerotic plaques which are the major cause of cardiovascular disease. Although the mechanism for the formation of these plaques is not completely understood, the plaque seems to begin by abnormal growth of endothelial cells and accumulation of cholesterol on the wall of a blood vessel. The forming plaque then triggers the aggregation of platelets which, in turn, accumulate on the plaque and participate in its growth. Experiments indicate that an increase in the C/PL ratio of platelets increases microviscosity. Microviscosity, for example, changes from 2.48 to 6.0P for an 118% increase in the C/PL ratio from its normal value (Shattil and Cooper, 1976). In addition, the platelets become more sensitive to the aggregating agents epinephrine and ADP and less sensitive to the aggregation inhibitor prostaglandin E_1 (Shattil and Cooper, 1976). Other experiments indicate that platelets do not adhere to fluid bilayers made, for example, from egg lecithin but do adhere to less fluid bilayers made from saturated lecithins. These results are consistant with the observation that atherosclerotic plaques are stimulated by diets of high cholesterol and saturated fatty acids. Such diets increase platelet cholesterol which facilitates platelet aggregation and, presumably, also decrease the fluidity of the endothelial cells which facilitates adhesion of the platelets to the blood vessel surface during plaque formation. However, the effects of high cholesterol and saturated fatty acid diets are subject to large variations between individuals due to differences in ability to stabilize serum cholesterol levels by excretion and decreased production of cholesterol by the liver.

Changes in cholesterol and fluidity of plasma membranes have also been implied in various aspects of cancer. Initial studies (Shinitzky and Inbar, 1974; Inbar and Ben-Bassat, 1976; Ben-Bassat *et al.*, 1977) using DPH to measure fluidity of intact cells indicated that the fluidities of intact malignant lymphoma cells from leukemic mice and humans were higher than normal cells. The increased fluidity was attributed to an experimentally observed decreased C/PL ratio in the plasma membranes of cancerous cells. Later experiments by other investigators, however, have questioned the validity of these initial observations and interpretations. Speigel *et al.* (1981), for example, find that there are no differences in the fluidities of membranes isolated from malignant and normal lymphocytes, although differences are indeed observed between intact cells with DPH as probe. The differences between intact cells, however, can be simply explained by the additional observation that malignant lymphocytes are characterized by a high content of lipid droplets which have high fluidities, readily take up DPH, and are highly fluorescent. Differences in fluidity between intact normal and malignant cells as measured by DPH, therefore, appear to be due to differences in lipid droplet content and not differences in plasma membrane fluidity. Similar results have been reported by Johnson and Robinson (1979) who, in addition, find very small differences in the lipid composition of leukemic and normal cell membranes. The fluidities of the plasma cell membranes of normal and SV40-transformed cells have also been reported to be very similar (Eldridge *et al.*, 1980). Plasma membranes are undoubtedly involved in the abnormal behavior of cancerous cells but the role, if any, which plasma membrane fluidity plays in the abnormal properties displayed by cancerous cells is at present uncertain.

B. Anesthetics

Anesthetics are substances which induce sleep, attenuate reflexes, and produce muscular relaxation. They produce these effects by reversibly blocking nerve conduction through inhibition of sodium channels which prevents nerve depolarization. Since anesthesia is produced by a wide range of structurally unrelated substances, anesthetics are usually considered to be nonspecific drugs which implies that they do not produce their effects by interacting with specific receptors but rather by interacting with some common membrane structure. A common feature of all anesthetics is their lipid solubility, i.e., there is a strong correlation between anesthetic potency and solubility of anesthetic in a reference oil (Roth, 1980) or partition coefficient between water and reference oil. These observations, first made by Overton (1901) and Meyer

(1901), led to the hypothesis that anesthetics exert their effects by simply partitioning into and interacting with hydrophobic regions of the lipid bilayer of nerve plasma membranes. In recent times, this hypothesis has been extended by the additional postulate that anesthetics exert their effects by fluidizing the hydrophobic regions of lipid bilayers. Experimentally, it has been found that general and local anesthetics indeed fluidize lipid bilayers at high concentrations (Feinstein *et al.*, 1975; Paphadjopoulos *et al.*, 1975; Trudell, 1977). In fact, a fair correlation has been found between the potency of an anesthetic and the fluidity change which it induces in a lipid bilayer.

In spite of these correlations, there are indications that fluidity changes may not be the only mechanism through which anesthetics exert their effects. It is well known that different anesthetics produce subtle but significantly different effects on the respiratory, cardiovascular, and nervous systems contrary to the nonspecific effects expected for a fluidity mechanism (Roth, 1980). Several alternative mechanisms have been proposed to account for these nonspecific effects including interactions of anesthetics with proteins and boundary lipid and effects on phase transitions. An interesting mechanism which retains fluidity as its chief parameter but at the same time has an element of specificity has been proposed by Miller and Pang (1976). They have found experimentally that the fluidizing effect of an anesthetic depends on the lipid composition of the lipid bilayer. They, therefore, propose that the subtle specificities shown by different anesthetics in whole animals are due to differences in lipid composition of neurons innervating different organs and tissues.

The net electrical charge carried by some anesthetics introduces additional complications in attempts to explain anesthetic effects in terms of one common mechanism. Thus, the acid barbiturates have a negative electrical charge while the local anesthetics such as lidocaine have a positive charge. When these charged anesthetics bind to a membrane, they modify the electrostatic surface potential which, in turn, modifies several membrane properties such as ionic permeabilities which are essential for the production of action potentials and nerve impulse propagation (Yguerabide, 1972b). A positive surface charge enhances transport of negative ions and inhibits positive ion transport. It also displaces Ca^{2+} bound to the membrane in association with anionic groups of acidic phospholipids. Such displacement of Ca^{2+} has been shown to affect lipid fluidity and also induce disassembly of cytoskeletal structures by increasing cytoplasmic Ca^{2+} concentration (Poste *et al.*, 1975; Papahadjopoulos *et al.*, 1975). Any or all of these effects could be responsible for the state of anesthesia. It should be noted that the total surface charge produced by an electrically charged anesthetic is proportional to the anesthetic concentration in the membrane which, in turn, is proportional to the lipid solubility of the anesthetic for a given concentration in the bathing medium. A correlation would then indeed be expected to exist between anesthetic lipid solubility and anesthetic potency in the case of a surface charge mechanism. Therefore, a correlation between anesthetic potency and lipid solubility cannot be used to argue that the effect of the anesthetic is produced through changes induced in the hydrophobic region of the membrane.

In summary, anesthetics may operate through several mechanisms involving fluidity, surface electrostatic charge, boundary lipid, direct effects on protein, and effects on cytoskeletal structure. The exact contribution of each of these mechanisms has not yet been established. The well-known correlation between anesthetic potency and partition coefficient into an oil or hydrophobic environment simply reflects the mode by which the anesthetics, which are usually amphipathic or hydrophobic, enter and become concentrated in the membrane. Once concentrated in the membrane, the anesthetic can cause anesthesia by any of the mechanisms enumerated above.

C. Aging

Aging is accompanied by many different cellular changes including (1) the morphology of the cell and internal organelles, (2) lipid composition of membranes, (3) enzymatic activities, (4) biosynthesis, (5) sensitivity and density of surface receptors, and (6) ionic composition of cytoplasm. There are indications that at least some of the functional changes are due to alterations of the lipid bilayer. For example, some membrane enzymatic activities which are found to be altered in aged cells display normal activity when placed in a standard lipid bilayer. Here, we present a brief review of the few studies which have so far been done on the role of fluidity in aging.

Cells which have been studied include erythrocytes, lymphocytes, adipocytes, and rat liver microsomes. In these cells, membrane fluidity decreases with age but the cause of the decrease seems to differ from cell to cell. In the case of lymphocytes (Rivnay *et al.*, 1979, 1980), the decrease in fluidity is correlated with the ratio of cholesterol to phospholipid of the membrane and serum, both of which increase with age. In the case of red cells, the decrease in fluidity with age seems to be due to a decrease in the lipid/protein ratio (Shiga *et al.*, 1979), while for an adipocyte (Hubbard and Garratt, 1980), the decrease seems to be due to an increase in saturated fatty acids.

D. Cell Growth and Differentiation

Several studies have reported fluidity changes associated with cell growth and differentiation. In a study by ESR of Chinese hamster ovary cells, fluidity was found to change during the cell cycle, being highest in mitosis and lowest in G_1 and early S phases (Lai *et al.*, 1980). In contrast, during the cell cycle of neuroblastoma cells, the fluidity, as inferred from diffusion coefficient measurements by FRAP, was a minimum in mitosis, increased during G_1, remained constant at maximal value during S, and decreased shortly before mitosis (De Laat, 1980). In the case of Yoshida sarcoma cells, it has been found that an increase in membrane fluidity produced by supplementation of the growth medium with unsaturated fatty acids stimulates growth but saturated fatty acids have no effect (Yamane and Tomioka, 1979). In a study of developing myoblasts in which FRAP was used to measure fluidity, the fluidity was found to increase shortly before cell fusion but decreased after fusion (Elson and Yguerabide, 1979). Finally, in the case of rat liver microsomes, a dramatic increase in fluidity is seen to occur at birth, increasing by about a factor of two between the last day of fetal life and the first day after birth. This increase in fluidity has been correlated with an increase in the molar ratio of phospholipid to cholesterol (Kapitulnik *et al.*, 1979).

REFERENCES

Amar, A., Rottem, S., and Razin, S., 1979, Is the vertical disposition of mycoplasma membrane proteins affected by membrane fluidity? *Biochim. Biophys. Acta* **552**:457–467.

Ben-Bassat, H., Polliak, A., Rosenbaum, S. M., Naparstek, E., Shouval, D., and Inbar, M., 1977, Fluidity of membrane lipids and lateral mobility of concanavalin A receptors in the cell surface of normal lymphocytes and lymphocytes from patients with malignant lymphomas and leukemias, *Cancer Res.* **37**:1307–1312.

Berlin, R. D., and Fera, J. P., 1977, Changes in membrane microviscosity associated with phagocytosis: Effects of colchicine, *Proc. Natl. Acad. Sci. USA* **74**:1072–1076.

Berlin, E., Matusik, E. J., and Young, C., Jr., 1980, Effect of dietary fat on the fluidity of platelet membranes, *Lipids* **15**:604–608.

Berliner, J. L. (ed.), 1976, *Spin Labelling, Theory and Application*, Academic Press, New York.

Bienvenue, A., Rousselet, A., Kato, G., and Devaux, P. F., 1977, Fluidity of the lipids next to the acetylcholine receptor protein of *Torpedo* membrane fragments. Use of amphiphilic reversible spin-labels, *Biochemistry* **16**:841–848.

Birrell, G. B., and Griffith, D. H., 1976, Cytochrome *c* induced lateral phase separation in a diphosphatidylglycerol–steroid spin-label model membrane, *Biochem. N.Y.* **15**:2925–2929.

Borochov, H., and Shinitzky, M., 1976, Vertical displacement of membrane proteins mediated by changes in microviscosity, *Proc. Natl. Acad. Sci. USA* **73**:4526–4530.

Borochov, H., Abbott, R. E., Schachter, D., and Shinitzky, M., 1975, Modulation of erythrocyte membrane proteins by membrane cholesterol and lipid fluidity, *Biochemistry* **18**:251–255.

Brotherus, J. R., Jost, P. C., Griffith, O. H., Keana, J. F. W., and Hokin, L. E., 1980, Charge selectivity at the lipid–protein interface of membranous Na, K-ATPase, *Proc. Natl. Acad. Sci. USA* **77**:272–276.

Brulet, P., and McConnell, H. M., 1976, Protein–lipid interactions: Glycophorin and dipalmitoylphosphatidylcholine, *BBRC* **68**:363–368.

Campis, J., and Scandella, C. J., 1980, Bulk membrane fluidity increases after fertilization of partial activation of sea urchin eggs, *J. Biol. Chem.* **255**:5411–5419.

Chen, L. A., Dale, R. E., Roth, S., and Brand, L., 1977, Nanosecond time-dependent fluorescence depolarization of diphenylhexatriene in dimyristoyllecithin vesicles and the determination of "microviscosity," *J. Biol. Chem.* **252**:2163–2169.

Cooper, R. A., 1977, Abnormalities of cell membrane fluidity in the pathogenesis of disease, *New Engl. J. Med.* **297**:371–377.

Cooper, R. A., and Shattil, S. J., 1980, Membrane cholesterol—Is enough too much? *New Engl. J. Med.* **302**:49–50.

Cooper, R. A., Leslie, M. H., Knight, D., and Detweiler, D. K., 1980, Red cell cholesterol enrichment and spur cell anemia in dogs fed a cholesterol-enriched atherogenic diet, *J. Lipid Res.* **21**:1082–1089.

Dale, R. E., Chen, L. A., and Brand, L., 1977, Rotational relaxation of the "microviscosity" probe diphenylhexatriene in paraffin oil and egg lecithin vesicles, *J. Biol. Chem.* **252**:7500–7510.

De Laat, S. W., Van Der Saag, P. T., and Shinitzky, M., 1978, Microviscosity modulation during the cell cycle of neuroblastoma cells, *Proc. Natl. Acad. Sci. USA* **74**:4458–4461.

De Laat, S. W., Van Der Saag, P. T., Elson, E. L., and Schlessinger, J., 1980, Lateral diffusion of membrane lipids and proteins during the cell cycle of neuroblastoma cells, *Proc. Natl. Acad. Sci. USA* **77**:1526–1528.

Eldridge, C. A., Elson, E. L., and Webb, W. W., 1980, Fluorescence photobleaching recovery measurements of surface lateral mobilities on normal and SV40-transformed mouse fibroblasts, *Biochemistry* **19**:2075–2079.

Elson, H. F., and Yguerabide, J., 1979, Membrane dynamics of differentiating cultured embryonic chick skeletal muscle cells by fluorescence microscopy techniques, *J. Supramol. Struct.* **12**:47–61.

Esko, J., Gilmore, J., and Glasser, M., 1977, Use of a fluorescent probe to determine the viscosity of LM cell membranes with altered phospholipid compositions, *Biochemistry* **16**:1881–1890.

Feinstein, M. B., Fernandez, S. M., and Sha'afi, R. I., 1975, Fluidity of natural membranes and phosphatidylserine and ganglioside dispersions: Effects of local anesthetics, cholesterol and protein, *Biochim. Biophys. Acta* **413**:354–370.

Finch, E. D., and Kiesow, L. A., 1979, Pressure, anaesthetics and membrane structure: A spin-probe study, *Undersea Biomed. Res.* **6**:41–45.

Flanagan, M. T., and Hesketh, T. R., 1973, Electrostatic interactions in the binding of fluorescent probes to lipid membranes, *Biochim. Biophys. Acta* **298**:535–545.

Foster, M. C., and Yguerabide, J., 1979, Partition of a fluorescent molecule between liquid–crystalline and crystalline regions of membranes, *J. Membr. Biol.* **45**:125–146.

Gazzotti, P., and Peterson, S. W., 1977, Lipid requirement of membrane-bound enzymes, *J. Bioenerg. Biomembr.* **9**:373–386.

Georgescauld, D., Desmazes, J. P., and Duclohier, H., 1979, Temperature dependence of the fluorescence of pyrene labelled crab nerve membranes, *Mol. Cell. Biochem.* **27**:147–153.

Gottlieb, M. H., 1980, Rates of cholesterol exchange between human erythrocytes and plasma lipoproteins, *Biochim. Biophys. Acta* **600**:530–541.

Grant, C. W. M., and McConnell, H. M., 1974, Glycophorin in lipid bilayers, *Proc. Natl. Acad. Sci. USA* **71**:4653–4657.

Gronowicz, G., Masur, S. K., and Holtzman, E., 1980, Quantitative analysis of exocytosis and endocytosis in the hydroosmotic response of toad bladder, *J. Membr. Biol.* **52**:221–235.

Hare, F., and Laussan, C., 1977, Variations in microviscosity values induced by different rotational behavior of fluorescent probes in some aliphatic environments, *Biochim. Biophys. Acta* **467**:267–272.

Herring, F. G., Tatischeff, I., and Weeks, G., 1980, The fluidity of plasma membranes of Dictyostelium discoideum: The effects of incorporation assessed by fluorescence depolarization and electron paramagnetic resonance, *Biochim. Biophys. Acta* **602**:1–9.

Hesketh, T. R., Smith, G. A., Houslay, M. D., McGill, K. A., Birdsall, N. J. M., Metcalfe, J. C., and Warren, G. B., 1976, Annular lipids determine the ATPase activity of a calcium transport protein complexed with dipalmitoyllecithin, *Biochemistry* **15**:4145–4151.

Horwitz, A. F., 1972, Nuclear magnetic resonance studies on phospholipids and membranes, in: *Membrane Molecular Biology* (C. F. Fox and A. D. Keith, eds.), Sinauer Associates, Sunderland, Massachusetts, pp. 164–191.

Houslay, M. D., Dipple, I., Rawal, S., Sauerheber, R. D., Esgate, J. A., and Gordon, L. M., 1980, Glucagon-stimulated adenylate cyclase detects a selective perturbation of the inner half of the liver plasma-membrane bilayer achieved by the local anaesthetic prilocaine, *Biochem. J.* **190**:131–137.

Hubbard, R. E., and Garratt, C. J., 1980, The composition and fluidity of adipocyte membranes prepared from young and adult rats, *Biochim. Biophys. Acta* **600**:701–704.

Inbar, M., and Ben-Bassat, H., 1976, Fluidity difference in the surface membrane lipid core of human lymphoblastoid and lymphoma cell lines, *Int. J. Cancer* **18**:293–297.

Israelachivili, J. N., 1978, *Light Transducing Membranes: Structure, Function and Evolution* (D. W. Deamer, ed.), Academic Press, New York, p. 91.

Israelachivili, J. N., Mitchell, D. J., and Ninham, B. W., 1977, Theory of self-assembly of lipid bilayers and vesicles, *Biochim. Biophys. Acta* **470**:185–201.

Israelachivili, J. N., Marcelja, S., and Horn, R. G., 1980, Physical principles of membrane organization, *Quart. Rev. Biophys.* **13**:121–200.

Jackson, R. L., and Gotto, A. M., Jr., 1976, Hypothesis concerning membrane structure, cholesterol, and atherosclerosis, in: *Atherosclerosis Reviews*, Vol. 1, Raven Press, New York, pp. 1–21.

Johnson, S. M., and Robinson, R., 1979, The composition and fluidity of normal and leukaemic or lymphomatous lymphocyte plasma membranes in mouse and man, *Biochim. Biophys. Acta* **558**:282–295.

Jost, P., Waggoner, A. S., and Griffith, O. H., 1971, Spin labeling and membrane structure, in: *Structure and Function of Biological Membranes*, Academic Press, New York and London.

Jost, P. C., Griffith, O. H., Capaldi, R. A., and Vanderkooi, G., 1973, Evidence for boundary lipid in membranes, *Proc. Natl. Acad. Sci. USA* **70**:480–484.

Kapitulnik, J., Tshershedsky, M., and Barenholz, Y., 1979, Fluidity of the rat liver microsomal membrane: Increase at birth, *Science* **206**:843–844.

Kawato, S., Kinosita, K., Jr., and Ikegami, A., 1977, Dynamic structure of lipid bilayers studied by nanosecond fluorescence techniques, *Biochemistry* **16**:2319–2324.

King, M. D., and Quinn, P. J., 1980, The use of phospholipid exchange processes to modulate the fluidity of biological membranes, *Biochem. Soc. Trans.* **8**:322–323.

Kinosita, K., Kawato, S., and Ikegami, A., 1977, A theory of fluorescence polarization decay in membranes, *Biophys. J.* **20**:289–305.

Kleemann, W., and McConnell, H.M., 1976, Interactions of proteins and cholesterol with lipids in bilayer membranes, *Biochim. Biophys. Acta* **419**:206–222.

Koppel, D. E., and Sheetz, M. P., 1981, Fluorescence photobleaching does not alter the lateral mobility of the erythrocyte membrane glycoproteins, *Nature* **293**:159–161.

Koppel, D. E., Axelrod, D., Schlessinger, J., Elson, E. L., and Webb, W. W., 1976, Dynamics of fluorescence marker concentration as a probe of mobility, *Biophys. J.* **16**: (2, Part 2):216a (Abst.).

Kuehl, K. S., Yeroushalmy, S., and Holloway, P. W., 1980, Modulation of membrane composition of swine vascular smooth muscle cells by homologous lipoproteins in culture, *Biochim. Biophys. Acta* **600**:689–700.

Lai, C. S., Hopwood, L. E., and Swartz, H. M., 1980, Electron spin resonance studies of changes in membrane fluidity of Chinese hamster ovary cells during the cell cycle, *Biochim. Biophys. Acta* **602**:117–126.

Lakowicz, J. R., and Knutson, J. R., 1980, Hindered depolarization rotations of perylene in lipid bilayers. Detection by lifetime-resolved fluorescence anisotrophy measurements, *Biochemistry* **19**:905–911.

Lakowicz, J. R., and Prendergast, F. G., 1978, Quantitation of hindered rotations of diphenylhexatriene in lipid bilayers by differential polarized phase fluorometry, *Science* **200**:1399–1401.

Lands, W. E. M., 1980, Membrane enzymes in their lipid environment, *Biochem. Soc. Trans.* **8**:25–27.

Lentz, B. R., Barenholz, Y., and Thompson, T. E., 1976, Fluorescence depolarization studies of phase transitions and fluidity in phospholipid bilayers. 2. Two-component phosphatidylcholine liposomes, *Biochemistry* **15**:4529–4537.

Levine, Y. K., Partington, P., Roberts, G. C. K., Birdsall, N. J. M., Lee, A. G., and Metcalf, J. C., 1972, ^{13}C nuclear magnetic relaxation times and models for chain motion in lecithin vesicles, *FEBS Lett.* **23**:203–207.

Ligeti, E., and Horvrath, L. I., 1980, Effect of Mg^{2+} on membrane fluidity and K^+ transport in rat liver mitochondria, *Biochim. Biophys. Acta* **600**:150–156.

Livingstone, C. J., and Schachter, D., 1980, Calcium modulates the lipid dynamics of rat hepatocyte plasma membranes by direct and indirect mechanisms, *Biochemistry* **19**:4823–4827.

Lussan, C., and Faucon, J., 1974, Effects of ions on vesicles and phospholipid dispersions studied by polarization of fluorescence, *Biochim. Biophys. Acta* **345**:83–101.

Marcelja, S., 1976, Lipid-mediated protein interaction in membranes, *Biochim. Biophys. Acta* **455**:1–7.

Masur, S. K., Holtzman, E., and Walter, R., 1972, Hormone-stimulated exocytosis in the toad urinary bladder, *J. Cell Biol.* **52**:211–219.

McConnell, H. M., and McFarland, E. G., 1970, Physics and chemistry of spin labels, *Quart. Rev. Biophys.* **3**:91–136.

McMurchie, E. J., and Raison, J. K., 1979, Membrane lipid fluidity and its effect on the activation energy of membrane associated enzymes, *Biochim. Biophys. Acta* **554**:364–374.

McMurray, W. C., 1973, Phospholipids in subcellular organelles and membranes, in: *Form and Functions of Phospholipids* (G. B. Ansell, J. N. Hawthorne, and R. M. C. Dawson, eds.), Elsevier, New York, pp. 205–251.

McVey, E., Yguerabide, J., Hanson, D. C., and Clark, W. R., 1981, The relationship between plasma membrane lipid composition and physical–chemical properties I. Fluorescence polarization studies of fatty acid-altered EL4 tumor cell membranes, *Biochim. Biophys. Acta* **642**:106–118.

Mely-Goubert, B., and Freedman, M. H., 1980, Lipid fluidity and membrane protein monitoring using 1,6-diphenyl-1,3,5-hexatriene, *Biochim. Biophys. Acta* **601**:315–327.

Mely-Goubert, B., Calvo, F., and Rosenfeld, C., 1979, Study of platelet membrane protein through fluorescence polarization of diphenyl hexatriene, *Biomedicine* **31**:*155–156*.

Metcalfe, J. C., Birdsall, N. J. M., and Lee, A. G., 1973, NMR studies of lipids in bilayers and membranes, *Ann. N.Y. Acad. Sci. USA* **222**:460–467.

Meyer, H. H., 1901, Zur theory der alkoholnarkose. III.Mit. der einfluss wechselnder temperatur auf wirkungstarke und teilungskoefficient der narkotika, *Arch. Exp. Pathol. Pharmakol* **46**:338.

Miller, K. W., and Pang, K. Y. Y., 1976, General anaesthetics can selectively perturb lipid nature bilayer membranes, *Nature* **263**:253–255.

Miller, R. G., and Raison, J. K., 1980, The fluidity of chloroplast thylakoid membranes and their constituent lipids, *Biochim. Biophys. Acta* **599**:63–72.

Mistry, P., Miller, N. E., Laker, M., Hazzard, W. R., and Lewis, B., 1981, Individual variation in the effects of dietary cholesterol on plasma lipoproteins and cellular cholesterol homeostasis in man, *J. Clin. Invest.* **67**:493–502.

Morre, D. J., Kartenbeck, J., and Franke, W. W., 1979, Membrane flow and interconversions among endomembranes, *Biochim. Biophys. Acta* **559**:71.

Nicolson, G. L., 1976, Transmembrane control of the receptors on normal and tumor cells. I. Cytoplasmic influence over cell components, *Biochim. Biophys. Acta* **457**:57–108.

Overath, P., and Trauble, H., 1973, Phase transitions in cells, membranes, and lipids of *Escherichia coli.* Detection by fluorescent probes, light scattering, and dilatometry, *Biochemistry* **12**:2625–2634.

Overton, E., 1901, *Studien über die Narkose zugleich ein Beitrag zur allgemeinen Pharmakologie,* G. Fischer, Jena.

Owicki, J. C., Springgate, M. W., and McConnell, H. M., 1978, Theoretical study of protein–lipid interactions in bilayer membranes, *Proc. Natl. Acad. Sci. USA* **75**:1616–1619.

Pagano, R. E., Ozato, K., and Ruysschaert, J. M., 1977, Intracellular distribution of lipophilic fluorescent probes in mammalian cells, *Biochim. Biophys. Acta* **465**:661–671.

Papahadjopoulos, D., 1968, Phospholipids model membranes I. Structural characteristics of hydrated liquid crystals, *Biochim. Biophys. Acta* **135**:624–638.

Papahadjopoulos, D., 1977, Effects of bivalent cations and proteins on thermotropic properties of phospholipid membranes, *J. Colloid Interface Sci.* **58**:459.

Papahadjopoulos, D., Jacobson, K., Poste, G., and Shepherd, G., 1975, Effects of local anesthetics on membrane properties I. Changes in the fluidity of phospholipid bilayers, *Biochim. Biophys. Acta* **394**:504–519.

Perrin, F., 1936, Movement Brownien d'un ellipsoide (II). Rotation libre et depolarisation des fluorescences. Translation et diffusion de molecules ellipsoidales, *J. Phis. Rad.* **7**:1.

Pessin, J., Salter, D., and Glaser, M., 1978, Use of a fluorescent probe to compare the plasma membrane properties in normal and transparent cells. Evaluation of the interference by triacylglycerols and alkylacylglycerols, *Biochemistry* **17**:1997–2004.

Peters, R. J., Peters, J., Tews, K. H., and Bahr, W., 1974, A microfluorometric study of translational diffusion in erythrocyte membranes, *Biochim. Biophys. Acta* **367**:282–294.

Pisam, M., and Ripoche, P., 1976, Redistribution of surface macromolecules in dissociated epithelial cells, *J. Cell Biol.* **71**:907–920.

Poo, M. M., and Cone, R. A., 1974, Lateral diffusion of rhodopsin in the photoreceptor membrane, *Nature* **247**:438–441.

Poste, G., Papahadjopoulos, D., Jacobson, K., and Vail, W. J., 1975, Effects of local anaesthetics of membrane properties II. Enhancement of the susceptibility of mammalian cells to agglutination by plant lectins, *Biochim. Biophys. Acta* **394**:520–539.

Rimon, G., Hanski, E., Braun, S., and Levitzki, A., 1978, Mode of coupling between hormone receptors and adenylate cyclase elucidated by modulation of membrane fluidity, *Nature* **276**:394–396.

Rivnay, B., Globersor, A., and Shinitzky, M., 1979, Viscosity of lymphocyte plasma membrane in aging mice and its possible relation to serum cholesterol, *Mech. Aging Dev.* **10**:71–79.

Rivnay, B., Bergman, S., Shinitzky, M., and Globerson, A., 1980, Correlations between membrane viscosity serum cholesterol, lymphocyte activation and aging in man, *Mech. Aging Dev.* **12**: 119–126.

Rogers, M. J., and Strittmatter, P., 1974, Evidence for random distribution and translational movement of cytochrome b_5 in endoplasmic reticulum, *J. Biol. Chem.* **249**:895–900.

Roozemond, R. C., and Urli, D. C., 1979, Lipid composition and microviscosity of subcellular fraction from rabbit thymocytes: Differences in the microviscosity of plasma membranes from subclasses of thymocytes, *Biochim. Biophys. Acta* **556**:17–37.

Roth, S. H., 1980, Mechanisms of anaesthesia: A review, *Can. Anaesth. Soc. J.* **27**:433–439.

Rule, G. S., Kruuv, J., and Lepock, J. R., 1979, Membrane lipid fluidity as rate limiting in the concanavalin A-mediated agglutination of PyBHK cells, *Biochem. Biophys. Acta* **556**:399–407.

Schachter, D., and Shinitzky, M., 1977, Fluorescence polarization studies of rat intestinal microvillus membranes, *J. Clin. Invest.* **59**:536.

Schmidt, C. F., Barenholz, Y., and Thompson, T. E., 1977, A nuclear magnetic resonance study of sphingomyelin in bilayer systems, *Biochemistry* **16:**2649–2656.

Seelig, A., and Seelig, J., 1974, The dynamic structure of fatty acyl chains in a phospholipid bilayer measured by deuterium magnetic resonance, *Biochemistry* **13:**4839–4845.

Seelig, A., and Seelig, J., 1978, Lipid–protein interaction in reconstituted cytochrome *c* oxidase/phospholipid membranes, *Hoppe-Seyler's Z. Physiol. Chem.* **359:**1747.

Seelig, A., and Seelig, J., 1980, Lipid conformation in model membranes and biological membranes, *Quart. Rev. Biophys.* **13:**19–61.

Shattil, S. J., and Cooper, R. A., 1976, Membrane microviscosity and human platelet function, *Biochemistry* **15:**4832–4837.

Sheetz, M. P., and Singer, S. J., 1974, Biological membranes as bilayer couples. A molecular mechanism of drug–erythrocyte interactions, *Proc. Natl. Acad. Sci. USA* **71:**4457–4461.

Sherwood, D., and Yguerabide, J., 1974, Ph.D. thesis, University of California, San Diego.

Shiga, T., Maeda, N., Suda, T., Kon, K., and Schiya, M., 1979, The decreased membrane fluidity of *in vivo* aged, human erythrocytes: A spin label study, *Biochim. Biophys. Acta* **553:**84–95.

Shinitzky, M., and Barenholz, Y., 1974, Dynamics of the hydrocarbon layer in liposomes of lecithin and sphingomyelin containing dicetylphosphate, *J. Biol. Chem.* **249:**2652–2657.

Shinitzky, M., and Barenholz, Y., 1978, Fluidity parameters of lipid regions determined by fluorescence polarization, *Biochim. Biophys. Acta* **515:**367–394.

Shinitzky, M., and Inbar, M., 1974, Difference in microviscosity induced by different cholesterol levels in the surface membrane lipid layer of normal lymphocytes and malignant lymphoma cells, *J. Mol. Biol.* **85:**603–615.

Shinitzky, M., and Inbar, M., 1976, Microviscosity parameters and protein mobility in biological membranes, *Biochim. Biophys. Acta* **433:**133–149.

Shinitzky, M., and Souroujon, M., 1979, Passive modulation of blood-group antigens, *Proc. Natl. Acad. Sci. USA* **76:**4438–4440.

Shinitzky, M., Dianoux, A. C., Gitler, C., and Weber, G., 1971, Microviscosity and order in the hydrocarbon region of micells and membranes determined with fluorescent probes. I. Synthetic micells, *Biochemistry* **10:**2106–2113.

Shroeder, F., 1978, Isothermal regulation of membrane fluidity in marine fibroblasts with altered phospholipid polar head groups, *Biochim. Biophys. Acta* **511:**356–376.

Singer, S. J., 1974, The molecular organization of membranes, *Annu. Rev. Biochem.* **43:**805–833.

Singer, S. J., and Nicolson, G. L., 1972, The fluid mosaic model of the structure of cell membrane, *Science* **175:**720–730.

Sklar, L. A., and Doody, M. C., 1980, Differences in fluidity between bilayer halves of plasma cell membranes, *Nature* **287:**255–256.

Sklar, L. A., Hudson, B. S., and Simoni, R. D., 1975, Conjugated polyene fatty acids as membrane probes: Preliminary characterization, *Proc. Natl. Acad. Sci. USA* **72:**1649–1653.

Sklar, L. A., Hudson, B. S., Peterson, M., and Diamond, J., 1977a, Conjugated polyene fatty acids on fluorescent probes: Spectroscopic characterization, *Biochemistry* **16:**813–819.

Sklar, L. A., Hudson, B. S., and Simoni, R. D., 1977b, Conjugated polyene fatty acids as fluorescent probes: Synthetic phospholipid membrane studies, *Biochemistry* **16:**819–828.

Smith, B. A., Clark, W. R., and McConnell, H. M., 1979, Anisotropic molecular motion on cell surfaces, *Proc. Natl. Acad. Sci. USA* **76:**5641–5644.

Speigel, R. J., Magrath, I. T., and Shutte, J. A., 1981, Role of cytoplasmic lipids in altering diphenylhexatriene fluorescence polarization in malignant cells, *Cancer Res.* **41:**452–458.

Stier, A., and Sackmann, E., 1973, Spin labels as enzyme substrates heterogeneous lipid distribution in liver microsomal membranes, *Biochim. Biophys. Acta* **311:**400–408.

Thompson, G. A., Jr., 1973, Phospholipid metabolism in animal tissues, in: *Form and Functions of Phospholipids*, Elsevier, New York, pp. 67–116.

Thulborn, K. R., Tilley, L. M., Sawyer, W. H., and Treloar, F. E, 1979, The use of *n*(9-anthroyloxy) fatty acids to determine fluidity and polarity gradients in phospholipid bilayers, *Biochim. Biophys. Acta* **558:**166–178.

Traüble, A., and Eibl, H., 1974, Electrostatic effects on lipid phase transitions: Membrane structure and ionic environment, *Proc. Natl. Acad. Sci. USA* **71:**214–219.

Trudell, J. R., 1977, A unitary theory of anaesthesia based on lateral phase separations in nerve membranes, *Anaesthesiology* **46**:5–10.

Van Blitterswijk, W. J., Emmelot, P., Hillkmann, H. A. M., Obmen-Meulemans, E. P. M., and Inbar, M., 1977, Differences in lipid fluidity among isolated plasma membranes of normal and leukemic lymphocytes and membranes exfoliated from their cell surface, *Biochim. Biophys. Acta* **467**:309–320.

Vanderkooi, J., Fischott, S., Chance, B., and Cooper, R. A., 1974, Fluorescent probe analysis of the lipid architecture of natural and experimental-rich membranes, *Biochemistry* **13**:1589–1595.

Van Hoeven, R. P., Von Blitterswijk, W. J., and Emmelot, P., 1979, Fluorescence polarization measurements on normal and tumor cells and their corresponding plasma membranes, *Biochim. Biophys. Acta* **551**:44–54.

Verkleij, A. J., DeKruyff, B., Ververgaert, P. H. J. Th., Tocanne, J. F., and Van Deenen, L. L. M., 1974, The influence of pH, Ca^{2+} and protein on the thermotropic behaviour of the negatively charged phospholipid, phosphatidylglycerol, *Biochim. Biophys. Acta* **339**:432–437.

Warren, G. B., Houslay, M. D., Metcalfe, J. C., and Birdsall, N. J. M., 1975, Cholesterol as excluded from the phospholipid annulus surrounding an active calcium transport protein, *Nature* **255**:684–687.

Yamane, I., and Tomioka, F., 1979, The concomitant effect of unsaturated fatty acid supplemented to medium on cellular growth and membrane fluidity of cultured cells, *Cell Biol. Int. Rep.* **3**:515–523.

Yguerabide, J., 1972a, Lecture presented at the 16th Annual Meeting, Biophysical Society, Toronto, Canada.

Yguerabide, J., 1972b, Nanosecond fluorescence spectroscopy of biological macromolecule and membrane, in: Fluorescence Techniques in Cell Biology (A. A. Thaer and M. Serentz, eds.), Springer-Verlag, New York, pp. 311–331.

Yguerabide, J., 1980, Spectroscopic and calorimetric probe methods, in: *Cell Membranes and Viral Envelopes*, Academic Press, New York, pp. 71–140.

Yguerabide, J., and Foster, M. C., 1981, Fluorescence spectroscopy of biological membranes, in: *Membrane Spectroscopy* (E. Grell, ed.), Springer-Verlag, New York, pp. 199–269.

Ziomek, C. A., Schulman, S., and Edidin, M., 1980, Redistribution of membrane proteins in isolated mouse intestinal epithelial cells, *J. Cell Biol.* **86**:849–857.

13

Rotational Diffusion of Membrane Proteins

Optical Methods

Peter B. Garland and Pauline Johnson

I. HISTORICAL BACKGROUND

Optical methods for measurement of Brownian rotational diffusion depend upon the use of linearly polarized light. Furthermore, the molecule under study must be capable of being photoexcited to some state which can be detected separately from the non-excited state. The measurements of rotation depend upon photoselection, which briefly works as follows. An isotropic (random) array of molecules is partially converted to the excited state by a flash of polarized light of appropriate wavelength. Because of the relationship between molecular orientation and the probability of absorbing linearly polarized light, the population of excited molecules is anisotropic. This anisotropy can be detected optically in various ways: by polarization of light emitted from the excited state, e.g., prompt and delayed fluorescence and phosphorescence, by polarized absorption measurements of the excited state absorption bands, i.e., linear dichroism, or of the remaining and also anisotropic ground-state absorption bands. Rotational diffusion will abolish the flash-established anisotropy in a time-dependent fashion. Conversely, measurements of the decay of flash-induced anisotropy enable rotational diffusion coefficients to be calculated. These concepts were established more than half a century ago (Perrin, 1926, 1929) and have been extensively reviewed (Albrecht, 1961, 1970).

Studies of the physical properties of lipids in model and natural membranes over a number of years, reviewed by Chapman and Restall (1982), led to the notion,

Peter B. Garland and Pauline Johnson • Department of Biochemistry, University of Dundee, Dundee DD1 4HN, Scotland, United Kingdom.

popularized by Singer and Nicholson (1972), that the hydrophobic core of a lipid bilayer provides a fluid phase. Consequently, proteins plugged into this phase would exhibit Brownian diffusion, both rotational and translational. Two-dimensional translation of a membrane protein was first shown by Frye and Edidin (1970) with a now classical experiment in which the mixing of differently colored immunofluorescently labeled cell surface antigens of two differing cell types was observed by fluorescence microscopy following cell fusion. Fluorescence depolarization methods (see below) for measuring the rotational diffusion of macromolecules in solution had been available for many years (Weber, 1953). Their application to membrane proteins by Wahl *et al.* (1971), Inbar *et al.*, (1973), and Shinitzky *et al.* (1973) failed to detect rotation, at least on the time scale of 0–100 nsec over which this technique was sensitive to rotational movement. However, in 1972, Cone showed that rhodopsin of invertebrate retinal rod outer segment disks exhibited Brownian rotation, but on a time scale of a few microseconds, not nanoseconds.

Mathematical models for Brownian diffusion of proteins in membranes were presented in 1975 by Saffman and Delbrück. For any realistic values of the viscosity of the membrane hydrophobic core (in the liquid crystalline state) and of the size of the protein component spanning this core, it was predicted that rotational diffusion would occur on a time scale of a few microseconds or more. Accordingly, the failure of fluorescence depolarization methods to detect rotation of membrane proteins is due to the relative briefness of fluorescence lifetimes, generally less than 100 nsec, when compared to the time scale of the rotations. It was precisely this problem that Cone (1972) circumvented by exploiting the long-lived photoexcited state of rhodopsin.

Long-lived (several milliseconds) spectral changes occur when the carbon monoxide complexes of CO-binding cytochromes are reversibly photolyzed by a flash of intense light. Junge (1972) exploited such changes to study the rotation of cytochrome oxidase in the mitochondrial inner membrane. However, the number of membrane-bound proteins that contain light-activated chromophores with long lifetimes (a few microseconds upwards) is few, being limited to rhodopsin (bacteriorhodopsin and visual rhodopsin), photodissociable CO-binding cytochromes, components of the photosynthetic system (Mar *et al.*, 1981), and possibly some flavoproteins. Such light-sensitive probes may be called intrinsic. As an alternative strategy, Naqvi *et al.* (1973) proposed the use of triplet-forming molecules as extrinsic probes. The light-activated triplet state of a molecule can be as long as many milliseconds in aqueous solution at room temperature provided that triplet quenchers such as oxygen are excluded. Experimental realization of this idea was made by Cherry and colleagues who synthesized eosin isothiocyanate and successfully used it in a number of interesting applications to membrane biology (Cherry *et al.*, 1976a; Cherry and Schneider, 1976; Cherry, 1978, 1979). The triplet probe technique was also developed by Lavalette *et al.* (1977) with pyrene maleimide and dibromofluorescein isothiocyanate.

In all of the instances mentioned so far, the light-activated spectroscopic state used to monitor membrane protein rotation was measured spectrophotometrically as a change of visible wavelength light extinction. In the case of rhodopsin (Cone, 1972), sensitivity was achieved by using a microspectrophotometer that could exploit the stacking of rod outer segment disks to maximize the amount of pigment in the experimental light path. In all other cases, such favorable geometry did not exist, and sensitivity was limited to that which could be achieved with a conventional spectrophotometric cuvette, namely, about 1 nmole per mg of membrane protein. This sensitivity is sufficient for studying the rotation of proteins present in the membrane at relatively high occupancy, for instance, cytochrome oxidase of mitochondria, cytochrome P-450 of liver microsomes, and band 3 protein of erythrocytes. These proteins are present in their membranes at an occupancy some 10^2–10^4 times greater than that of hormone and other receptors of great interest in the plasma membranes of cells.

In 1979, three laboratories introduced delayed emission methods for the detection of the triplet state. Greinhert *et al.* (1979) described the use of delayed fluorescence from eosin extrinsic probes. Austin *et al.* (1979) described the use of phosphorescence from eosin and diiodofluorescein, and our own laboratory developed the synthesis of erythrosin isothiocyanate as an extrinsic phosphorescent probe for measuring rotational motion of membrane proteins (Moore and Garland, 1979; Garland and Moore, 1979; Moore *et al.*, 1979). When used with a conventional suspension of membranes in a cuvette, e.g., 1 ml of suspension containing 1 mg of membrane protein, the sensitivity of the phosphorescence method is improved roughly ten-fold using eosin and fifty-fold using erythrosin. These light emission methods of triplet-state detection can be adapted for use with a fluorescence microscope and a modulated argon ion laser as the excitation source. With such an arrangement, the sensitivity for erythrosin is about 10^7–10^8 molecules (Johnson, 1983).

A further improvement in sensitivity is provided by the so-called fluorescence depletion method for detection of the triplet state (Johnson and Garland, 1981, 1982a,b). This method combines the sensitivity of fluorescence detection with the long lifetime of the triplet state and can detect as few as 10^4 molecules of probe on the surface of a single cell or large liposome. Whereas the choice of probe for the phosphorescence method is restricted to just a few with adequate quantum yields for phosphorescence, the choice for the fluorescence depletion method is much wider (Johnson and Garland, 1982a,b). Overall, these advances in method enable a move away from the high power lasers and cuvette suspension samples of the flash photolysis chemist toward the low power laser and microscope sample of the cell biologist.

The rotation of membrane proteins is anisotropic; it occurs about a single axis normal to the plane of the membrane. Extraction of rotational diffusion coefficients from the experimental data is therefore not straightforward. Cone (1972) provided a solution for the special case where the membrane was perpendicular to the direction of the photoselecting light beam. The solution to the more common case where the membranes are randomly orientated was provided by several laboratories (Rigler and Ehrenberg, 1973; Belford *et al.*, 1972; Chuang and Eisenthal, 1972).

The other methods of measuring slow molecular rotations should be mentioned. The first uses the phosphorescence of protein tryptophan residues, and may be of most value in reconstituted systems where its low specificity is less problematical (Strambini and Galley, 1976, 1980). The other uses a fluorescence polarization correlation method in which random fluctuations of orientation in a population of fluorescent molecules cause "noise" in the degree of polarization of continuously excited fluorescence (Ehrenberg and Rigler, 1976). In principle, the latter method could have many advantages, including high sensitivity, wide range of probes and microscopic application, but practical difficulties such as probe bleaching have hindered its development and application to membranes.

II. PHYSICAL MODEL FOR ROTATIONAL DIFFUSION OF A MEMBRANE PROTEIN

Diffusion of a molecule due to Brownian motion is described by diffusion coefficients D_L and D_R for translational and rotational diffusion, respectively. Translational motion describes displacement of the molecule in space, whereas rotational motion describes angular rotation of the molecule. If $\bar{x}^2$ is the mean square displacement in time t in one dimension, then

$$\bar{x}^2 = 2\, D_L t \tag{1}$$

Similarly, if $\bar{\theta}^2$ is the mean square angular rotation about an axis in time t, then

$$\bar{\theta}^2 = 2\, D_R t \tag{2}$$

The simplest model for a membrane protein represents the protein as a cylinder of radius r spanning a hydrophobic bilayer of height h and viscosity μ (Saffman and Delbrück, 1975). If K is Boltzmann's constant and T the absolute temperature, then

$$D = \frac{K\,T}{4\,\pi\,\mu\,r^2 h} \tag{3}$$

It is important to note that D_R is highly dependent on the radius of that part of the protein spanning the bilayer. By contrast, the lateral diffusion coefficient D_L is relatively insensitive to size:

$$D_L = \frac{K\,t}{4\,\pi\,\mu\,h}\left(\log\frac{\mu\,h}{\mu' r} - \gamma\right) \tag{4}$$

where μ' is the viscosity of the aqueous phase and γ is Euler's constant (0.5772).

Although this model is a considerable oversimplification, its predictions for values of D_L and D_R are reasonably well met in the case of bacteriorhodopsin reconstituted into dimyristoylphosphatidylcholine bilayers (Peters and Cherry, 1982). The intramembrane part of bacteriorhodopsin has seven spans of polypeptide, and a cross-sectional diameter of 3.5 nm (Henderson and Unwin, 1975). By contrast, if the acyl chains of a phospholipid molecule are modeled as a cylinder, then its cross-sectional diameter is about 0.9 nm. The Saffman–Delbrück model is a hydrodynamic model, and treats the lipid bilayer as a fluid continuum. This approximation can be met if the diameter of the transmembranous part of the protein is relatively large compared to that of the phospholipid molecules. Different models may be needed for the diffusion of phospholipids themselves (Trauble and Sackmann, 1973; Vaz *et al.*, 1982) or for proteins that are attached to membranes by a long chain fatty acyl group rather than by themselves being plugged into the bilayer.

III. PHYSICAL PRINCIPLES OF PHOTOSELECTION

When a molecule such as fluorescein absorbs a photon of light of an energy corresponding to an electronic transition, then the electronic energy level of the molecule can change according to the scheme of Figure 1. The absorption bands of the molecule arise from singlet–singlet transitions, S_0–S_1, S_0–S_2, and so on. The higher singlet states usually relax rapidly and nonradiatively to S_1, the lowest excited singlet state. Return of S_1 to the ground state S_0 may occur either directly or indirectly via the triplet state T_1. The S_1–T_1 transition is favored by heavy atom substituents. The S_1–S_0 transition may be radiative, in which case the emission is known as prompt fluorescence. The lifetime of the S_1 state is typically a few nanoseconds or less, corresponding to the lifetime of prompt fluorescence. Return of T_1 to the ground state can occur by return to S_1 and thence by an S_1–S_0 transition that yields fluorescence spectrally identical with prompt fluorescence but with a lifetime governed by that of the T_1 state which, in the absence of quenching, can last for several microseconds or milliseconds. Fluorescence arising from the T_1–S_1–S_0 transition is therefore called delayed fluorescence. The T_1 state can also return to the ground state nonradiatively, by collisional quenching with oxygen, or radiatively, giving rise to phosphorescence. Due to energy losses in the S_1–T_1 transition and also in dropping to the lowest T_1 level, phosphorescence is red shifted compared to fluorescence. The lifetime of the triplet state T_1 is appropriate for measuring rotational motion in the microsecond–millisecond time range. The triplet state can be detected optically by ground-state

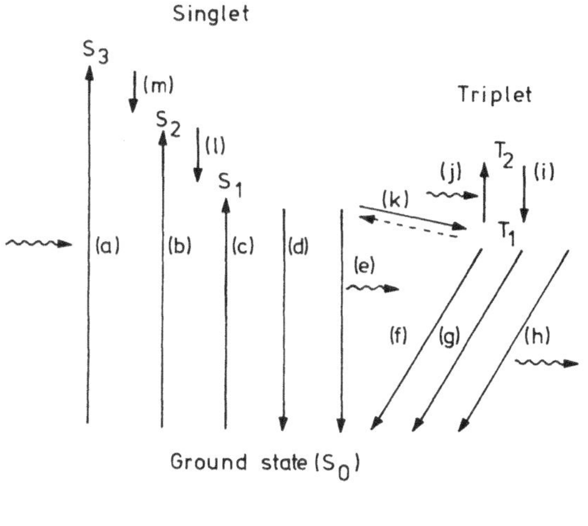

Figure 1. Electronic energy levels and transitions for fluorescence and phosphorescence. The transitions are (a), (b), and (c), light-driven transitions from S_0 to S_3, S_2 and S_1, respectively, (d) nonradiative S_1–S_0 transition, (e) radiative (fluorescence) S_1–S_0 transition, (f) nonradiative quencher, (h) radiative (phosphorescence) T_1–S_0 transition, (i) T_2–T_1 transition, (j) light-driven T_1–T_2 transition (triplet absorption bands), (k) S_1–T_1 transition (intersystem crossing. Thermally driven reversal is shown with an interrupted line), (l) and (m) S_2–S_1 and S_3–S_2 transitions. Vibrational states at the various levels are omitted in the interests of clarity.

depletion, i.e, the absorption of S_0 is diminished because S_0 is depleted to populate the triplet state, by measurements of delayed fluorescence or phosphorescence or by triplet–triplet absorption. The quantum yield for triplet formation (ϕ_T) of an organic molecule can be increased by heavy atom substituents. For example, ϕ_r for fluorescein is about 0.05, rising to 0.7 for tetrabromofluorescein (eosin) and 1.0 for tetraiodofluorescein (erythrosin). Quantum yields for phosphorescence (ϕ_p) also increase, but the triplet lifetime falls (Garland and Moore, 1979).

The absorption of linearly polarized light by an organic molecule depends not only on the wavelength of the light but also on the orientation of the molecule. If we define an angle α between a particular geometric axis within the molecule (the transition dipole moment for absorption) and the plane of polarization of the incident light, then the probability of light absorption is proportional to $\cos^2 \alpha$.

Thus, a molecule with its dipole moment lying in the polarization plane of incident light has a greater probability of absorbing a photon than does a molecule with its absorption dipole moment lying out of plane. This, therefore, is the basis of photoselection. The anisotropy set up by photoselection is maximal if the molecule has but a single absorption dipole moment orientated along a unique molecular axis. If the absorption dipole moments lie in a plane rather than along a unique axis, as with cytochromes, then photoselection is less effective in establishing an anisotropic distribution of molecules.

The photoselected population of molecules differs from the parent population with respect to the orientational distribution of their absorption dipole moments. Photoselection will, therefore, alter the abilities of the population to absorb light polarized either parallel ($\|$) or perpendicular ($\perp$) to the plane of polarization of the photoselecting flash. Differential absorption by a set of molecules of light beams polarized orthogonally to each other is known as linear dichroism. Decay of linear dichroism following photoselection is caused by rotational diffusion.

The emission dipole moment of a fluorescent or phosphorescent molecule is an axis within the molecule, and the emitted light is preferentially polarized in a plane containing that axis. If a population of molecules in the S_1 or T_1 state is anisotropic, then the fluorescent or phosphorescent emission will also be anisotropic, i.e., polarized. Rotational motion of molecules in the S_1 or T_1 state will cause decay of polarization of prompt fluorescence or delayed emission (phosphorescence or delayed fluorescence), respectively. Depolarization will not occur if rotational motion is insignificant during the lifetime of the excited state.

Measurements of linear dichroism or polarized emission involve measurements of light signals viewed through plane polarizers orientated either parallel or perpendicular to the plane of the photoselection flash. If we call these signals $S_{\|}$ and $S_{\perp}$, how are they related immediately after photoselection, i.e., at zero time, and how do they decay? The zero-time values of $S_{\|}$ and $S_{\perp}$ are independent of rotation, but depend on (1) the optical geometries for photoselection and measurements, and (2) the angles between the dipole moments used for photoselection and measurement. Reference to Figure 2 shows two common optical geometries used for photoselection experiments involving linear dichroism or emission. Table 1 gives relative values of $S_{\|}$ and $S_{\perp}$ for

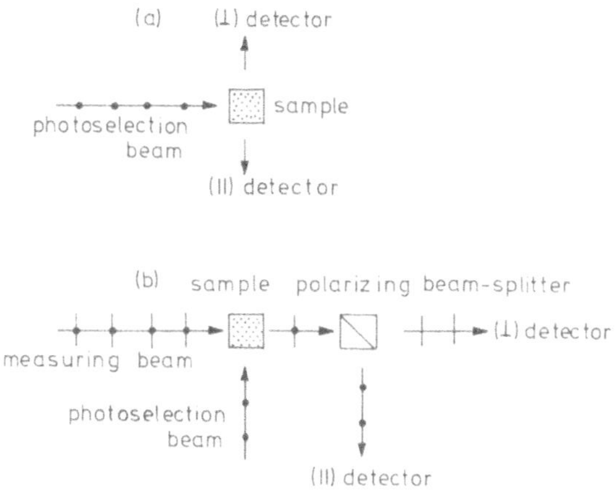

Figure 2. Optical geometries for the measurement of (a) polarized emission and (b) linear dichroism. In each case, the plane of polarization of the photoselecting beam is perpendicular to the plane of the paper. Filters specific for wavelength or polarization plane are not shown, but would normally be placed in front of the light detectors. Light polarized in the plane of the paper is shown with crosslines, whereas, light polarized perpendicular to the plane of the paper has solid round symbols.

some idealized cases. Also given in Table 1 are values of r_0, the zero-time anisotropy parameter where (Jablonski, 1961):

$$r(t) = \frac{S_{\parallel}(t) - S_{\perp}(t)}{S_{\parallel}(t) + 2S_{\perp}(t)} \tag{5}$$

The importance of the anisotropy parameter is as follows. The difference between $S_{\parallel}$ and $S_{\perp}$ reflects the anisotropy of the photoselected population. This difference will decay by rotation, and also if the absolute values of $S_{\parallel}$ and $S_{\perp}$ themselves decay, which they do unless the photoactivated state is exceedingly long relative to the

Table 1. Ideal Ratios of Signals $S_{(\parallel)}$ and $S_{(\perp)}$ and Anisotropy Parameter r_0 at Zero Time in Photoselection Experiments on Non-Orientated Membranes

Polarization of photoselection beam	Dipole moments for photoselection and measurement	$S_{(\parallel)}/S_{(\perp)}$	$r_0 = \frac{S_{(\parallel)} - S_{(\perp)}}{S_{(\parallel)} + 2S_{(\perp)}}$
As in Figure 2	Parallel	3	0.4
As in Figure 2	Perpendicular	1/2	−0.2
As in Figure 2	Circularly degenerate in plane	4/3	0.1
Rotate through 90°	Any	1	0

observation time. However, the sum ($S_{\|} + 2\ S_{\perp}$) is independent of rotation (for an initially random set of molecules prior to photoselection) and measures in effect the size of the photoselected population. Consequently decay of $(S_{\|} - S_{\perp})/(S_{\|} + 2\ S_{\perp})$ is due to rotation free of changes in the number of photoselected molecules (Jablonski, 1961).

For the special case of isotropic rotation, as with rotation of a spherical molecule in aqueous solution

$$r(t) = r(0) \exp(-6\ D_R t) \tag{6}$$

The time-dependent constant characterizing this decay of $r(t)$ is $1/6\ D_R$; it is known as the relaxation time. For the case of anisotropic rotation about a single axis (Rigler and Ehrenberg, 1973),

$$r(t) = A_1 \exp(-D_R t) + A_2 \exp(-4\ D_R t) + A_3 \tag{7}$$

where

$$A_1 = 1.2 \sin \theta_p \sin \theta_m \cos \theta_p \sin \theta_m \cos \phi \tag{8}$$
$$A_2 = 0.3 \sin^2 \theta_p \sin^2 \theta_m (\cos^2 \phi - \sin^2 \phi) \tag{9}$$
$$A_3 = 0.1 (3 \cos^2 \theta_p - 1)(3 \cos^2 \theta_m - 1) \tag{10}$$

where θ_p and θ_m are the angles between the transition dipole moments for either photoselection (θ_p) or measurement (θ_m) and the membrane normal, and ϕ is the angle between the projections of these dipole moments onto the membrane plane (see Figure 3).

If the dipole moments used for photoselection and measurement are parallel, then Eqs. (8–10) simplify to

$$A_1 = 1.2 \sin^2 \theta \cos^2 \theta \tag{11}$$
$$A_2 = 0.3 \sin^4 \theta \tag{12}$$
$$A_3 = 0.1(3 \cos^2 \theta - 1)^2 \tag{13}$$

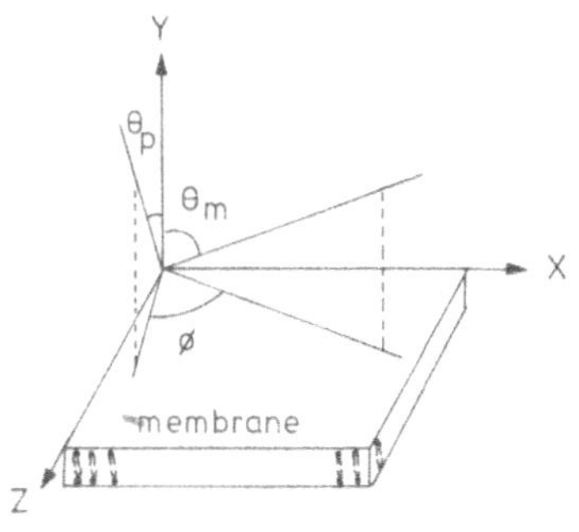

Figure 3. Orientation of transition dipole moments in photoselection experiments. Of the three orthogonal axes shown, the X and Z axes lie in the plane of the membrane, whereas the Y axis is normal to that plane. The transition dipole moments for photoselection and measurement make angles θ_p and θ_m relative to the Y axis. The angle ϕ is that between the projection of the dipole moments onto the membrane plane. θ_p and θ_m are interchangeable.

In the simplest case ($\theta_p = \theta_m = \theta$), the r_0 value has a maximum of 0.4, and with uniaxial rotation, decays towards a time-independent value given by A_3 (Eqs. 7 and 13). Because A_1, A_2, and A_3 depend on θ which is not usually known, the interpretation of anisotropy decay curves for uniaxial rotation is complex. Moreover, the situation is even worse for phosphorescence where θ_p and θ_m are not necessarily equal. Consequently, the r_0 value for phosphorescence can lie anywhere between -0.2 and 0.4, and may either grow or decay within these limits. A further difficulty is provided by the possibility that the direction of the dipole moment for phosphorescence emission can change with environment.

IV. INTRINSIC AND EXTRINSIC PROBES

The retinal chromophore of rhodopsin and bacteriorhodopsin and the heme chromophore of photodissociable CO-binding are examples of intrinsic probes for measuring slow rotation. They have a single and relatively fixed position within their host molecules, and the angles that their dipole moments make relative to the membrane plane can be obtained by independent means (Heyn *et al.*, 1981; Erecinska *et al.*, 1978). Extrinsic triplet probes such as eosin isothiocyanate are capable of attaching to multiple sites on a protein, unless by good fortune they show specificity for some structural feature such as cleft or substrate site as probably happens with band 3 protein (Cherry *et al.*, 1976b). Site specific labeling by an extrinsic probe can be arranged by first attaching the probe specifically to some ligand, e.g., insulin, epidermal growth factor, or β-bungarotoxin, and then allowing the ligand to bind to its receptor. Extrinsic probes, unless they replace an intrinsic probe, are unlikely to be held fully immobilized by the protein framework to which they are attached. As a result, they may exhibit fast (nanosecond) movement that is limited in extent by surrounding structures. This leads to diminution of the experimentally detectable r_0 value due to unresolved rotation

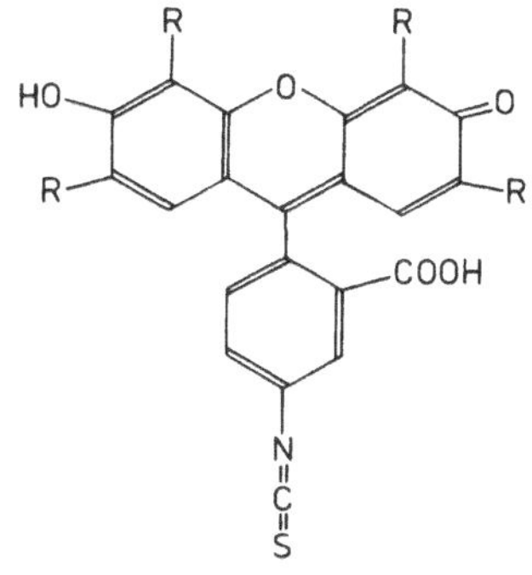

If R = H, dye = Fluorescein isothiocyanate
R = Br, dye = Eosin "
R = I, dye = Erythrosin "

Figure 4. Structures of some xanthene dyes.

during the photoselection flash (nanosecond–microsecond) or immediately thereafter. The site of attachment of the extrinsic probe may also move independently of the whole protein, for example, movement of a polypeptide segment or domain. Such segmental motion presents an interesting topic in its own right, particularly in the little explored region from about 100 nsec upwards where fluorescence depolarization studies are inapplicable. Nevertheless, segmental flexion like rapid motion of the probe itself can degrade r_0 values to the detriment of attempts to measure D_R for the whole protein. (Kawato and Kinosita, 1981; Burkli and Cherry, 1981; Spiers *et al.*, 1983).

Figure 4 gives structures for some commonly used extrinsic triplet probes.

V. TIME-RESOLVED AND STEADY-STATE METHODS

Rotational relaxation times for proteins in aqueous solution have been determined from steady-state measurements of extrinsic probe fluorescence (Weber, 1953; Brand and Witholt, 1967). Following an earlier suggestion (Garland and Moore, 1979), Murray *et al.* (1983) have explored time-averaged measurements of phosphorescence polarization of erythrosin-labeled sarcoplasmic reticulum Ca^{2+}-activated ATPase. Because of the high intensity of the long-wavelength tail of prompt fluorescence in the spectral region of phosphorescence, time steady-state measurements could not be made. Instead, the exciting source (arc-lamp) was mechanically shuttered to be repetitively on for 30 μsec and off for 30 msec, and the phosphorescence was averaged over the "off" period. It is known from time-resolved studies, both by linear dichroism and by phosphorescence polarization, that the rotational relaxation time of the ATPase decreases as temperature increases from 5 to 40°C, and this general trend was confirmed by the time-averaged method (Murray *et al.*, 1983). More subtle features revealed by time-resolved methods were, not surprisingly, unobserved, but a continuous readout of anisotropy may well have value in the empirical detection of conditions where interesting conformational or rotational transitions occur. A good approximation to steady-state conditions of phosphorescence could be achieved by using an acousto-optically modulated argon ion laser as the excitation source (514.5 nm) for exciting erythrosin phosphorescence in conjunction with gated amplifier electronics and a repetitive cycle of 2 μsec excitation followed by 2 μsec measurement.

VI. LINEAR DICHROISM

The majority of measurements made so far of membrane protein rotational diffusion have used linear dichroism methods. A microspectrophotometer was used for rhodopsin of rod outer segment disks. In all other cases, a more conventional flash spectrophotometer has been used with a sample cuvette containing 1 ml or more of a membrane suspension that was 1–10 μM with respect to the probe. Either pulsed dye lasers or frequency-doubled Nd-YAG lasers have provided the photoselection flash. Signal-to-noise ratio is problematical, and worsens if faster time resolution is sought. Improvements in signal-to-noise ratio are obtained with signal averaging of repetitive cycles of photoselection and relaxation of the photoactivated state to the ground state. Further improvements are provided by the use of high-intensity measuring beams such as argon ion or He–Ne lasers, or pulsed flash lamps. Descriptions of suitable apparatus are to be found in Cherry (1978), Austin *et al.* (1979), Moore and Garland (1979), Hoffman *et al.* (1980), and Jovin *et al.* (1981).

The main advantage of the linear dichroism method is that the photoselection and measuring wavelength can be sufficiently close to ensure that the same transition dipole moment is used for each. This enables the simpler form of Eq. (7) to be used

$$r(t) = 1.2 \sin^1 \theta \cos^2 \theta \exp(-Dt) + 0.3 \sin^4 \theta \exp(-4\,Dt) + 0.1(3 \cos^2 \theta - 1)^2 \quad (14)$$

If the probe is circularly degenerate, as with a cytochrome where the orthogonally related x and y transition dipole moments lie in the heme plane (Gouterman and Stryer, 1962), then Eq. (14) becomes

$$r(t) = 0.3 \sin^2 \theta' \cos^2 \theta' \exp(-Dt) + 0.075 \sin^4 \theta' \exp(-4\,Dt) + 0.025(3 \cos^2 \theta' - 1)^2 \quad (15)$$

where θ is the tilt of the heme plane from the membrane plane (Kawato *et al.*, 1981).

The theoretical maximum r_0 value is rarely, if ever, observed. Reasons for this include multiplicity of dipole moments, excessive ground-state depletion (>20%), rapid probe movement, depolarization by light scattering by cells and membrane vesicles (Teale, 1969), and instrumental factors such as the solid angle of measuring light collection (Lindmo and Steen, 1977). Typically, the experimental data for triplet decay and the computed anisotropy parameter decay curve might resemble those of Figure 5. Because of the time-independent term A_3 of Eq. (7), the anisotropy parameter does not decay to zero except when θ is about 55°C. In principle then, θ could be

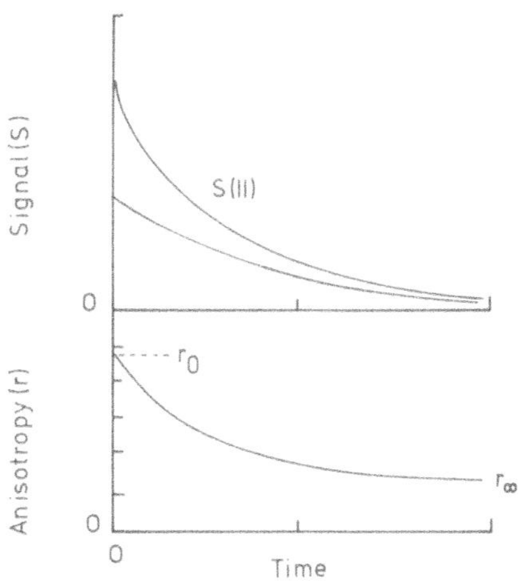

Figure 5. Idealized experimental curves for $S(\parallel)$, $S(\perp)$, and $r(t)$ for anisotropic rotation of a membrane protein.

calculated from the observed value of r_∞, the time-independent anisotropy. In practice, this is rendered difficult by uncertainty as to whether or not the preparation contains aggregated species of protein that did not rotate significantly on the experimental time scale.

Calculation of D_R from anisotropy decay curves is complicated by the presence of two exponential terms each of usually unknown amplitude. It is usual, therefore, to fit the decay curve to a single diffusion coefficient. This apparent value will lie between D_R and $D_R/4$, depending on the values of A_1 and A_2 in Eq. (7). The reciprocal of the apparent diffusion coefficient has the dimension of time, and can be referred to as the rotational relaxation time.

Membrane vesicles, liposomes, and cells may themselves rotate and cause decay of the anisotropy parameter. For instance, vesicles of less than 100 nm diameter will rotate in water with a rotational relaxation time of about 100 μsec or less. To distinguish protein movement from vesicle tumbling, it may be necessary to increase the medium viscosity with glycerol or sucrose, or to aggregate the vesicles. However, these treatments themselves may induce changes in structure and protein rotation. In very small vesicles, lateral diffusion around the vesicle surface could also mimic the effect of rotation in causing the decay of anisotropy.

Applications of linear dichroism techniques are described below.

VII. DELAYED FLUORESCENCE

The possible use of delayed fluorescence for measuring slow rotations of membrane proteins were first prepared by Naqvi and Wild (1975), and realized experimentally by Greinhert *et al.* (1979). As with emission methods, in general, the sensitivity is potentially high. It is also likely that the dipole moments for photoselection (absorption) and measurement (delayed fluorescence) will be parallel. A severe practical drawback arises from the intense prompt fluorescence flash excited by the photoselection flash. This is spectrally identical to delayed fluorescence, and cannot usefully be blocked by cut-off filters. Mechanical shuttering can be used to protect the photomultiplier tube, but it is likely to give an unacceptable dead time, e.g., 70 μsec in the instrument of Greinhert *et al.* (1979). Photomultiplier gating can be achieved electronically (Austin *et al.*, 1979; Jovin *et al.*, 1981), but it is still difficult to suppress postgating artifacts to levels acceptable for high-sensitivity measurements or fast-time resolution. For these reasons, and because of the higher sensitivity available from phosphorescence measurements (Moore and Garland, 1979), the delayed fluorescence technique has not been widely adopted.

VIII. PHOSPHORESCENCE

Phosphorescence depolarization methods are the long lifetime equivalent of fluorescence depolarization methods for measuring rotational motion. Instrument requirements are simpler than for linear dichroism. For many purposes where the fastest time resolution is not needed, e.g., 1–2 μsec resolution, photomultiplier gating can be omitted, a long-pass filter being sufficient to protect the photomultiplier tube from prompt fluorescence excited by the photoselection flash. Relatively low-power laser flashes can be used for photoselection, as there is no requirement (as in linear dichroism) to obtain other than low population of the ground state. Accordingly, a simple N_2-pumped dye laser operating at about 0.05–1.0 mJ per flash and 20–40 Hz provides an excellent photoselection source, used in conjunction with signal averaging. The two major drawbacks of the method are, first, the limited range of probes with adequate phosphorescence (erythrosin and eosin) and, second, the unpredictable angles of the emission dipole moment (Pilipovich, 1961; Spiers *et al.*, 1983). The main advantage of the method is its sensitivity, about 10^{-11}–10^{-12} mole in cuvette samples and 10^8–10^9 molecules in a microscopic version (Johnson, 1983). Some applications of phosphorescence depolarization are discussed below.

Certain hemoproteins with rhodium replacing Fe exhibit phosphorescence. Vaz *et al.* (1979) exploited such luminescence to study cytochrome b_5 rotation, and Dixit *et al.* (1982) made a phosphorescent cytochrome *c* derivative to use as a probe for mitochondrial cytochrome oxidase to which it binds.

IX. FLUORESCENCE DEPLETION

Linear dichroism measurements are usually made by measuring the changes of the light absorption caused by ground-state depletion of photoexcited triplet probe molecules. Most triplet probes are also fluorescent. We reasoned that changes of absorption of light due to changes in the ground-state population could well be measured by detection of the changes in fluorescence that such light absorption excites. This would, in effect, convert the linear dichroism method into an emission method, combining the sensitivity of prompt fluorescence with the long lifetime of the triplet state. A full description of the method has been published elsewhere (Johnson and Garland, 1981, 1982a,b) but, briefly, we use a photon-counting fluorescence microscope equipped with an acousto-optically modulated argon ion laser to enable microscope samples to be studied. We have also developed a convenient anaerobic microscope slide for removal of triplet-quenching oxygen, and extended the method to include measurements of the slow rotation of lipid probes (Johnson and Garland, 1983). The combination of a fluorescence microscope with an argon ion laser can also be used to measure lateral diffusion coefficients by the fluorescence recovery after photobleaching method (Peters *et al.*, 1974; Axelrod *et al.*, 1976); indeed, our instrument was first constructed for that purpose (Garland, 1981).

The main advantages of the fluorescence depletion method are, first, range of actual potential probes and, second, high sensitivity. However, the equipment is not simple. We are currently constructing a second instrument to be used solely for measurements of D_R with the intent of avoiding design and operating compromises if one and the same machine is used for measuring D_R or D_L in a variety of different projects.

Table 2. Summary of Studies Made on the Rotational Diffusion of Membrane Proteins

Protein and membrane	Probe	Question(s) posed (other methods used)	References[b]
Rhodopsin of frog retinal rods	Intrinsic	How rapidly can a membrane protein rotate?	1
Cytochrome a_3 of mitochondria	Intrinsic (CO-complex), extrinsic (cytochrome c)[a]	Free to rotate or in complexes with other mitochondrial proteins? (reconstitution with bc_1)	2–4, 5
Cytochrome o and a_1 of *E. coli*	Intrinsic (CO-complex)	Free to rotate or in respiratory assemblies? (varied growth conditions)	6
Cytochrome P-450 of microsomes	Intrinsic (CO-complex)	Association with its NADPH-dependent reductase? (antisera, reconstitution, crosslinking)	7–9
Bacteriorhodopsin in liposomes	Intrinsic	Is the Saffman–Delbrück model correct? (D_L measurements)	10, 11
Ca^{2+}-dependent ATPase of sarcoplasmic reticulum	Eosin, erythrosin[a]	What is the oligomeric state of the enzyme? (reconstitution, temperature dependence)	12–15
Band 3 of erythrocytes	Eosin	Why is its lateral diffusion so slow? What are its contacts with other proteins? (antisera, genetic deficiences, D_L measurements)	16–18
Acetylcholine receptor of electric organ	Erythrosin,[a] eosin[a]	What is the role of peripheral membrane proteins in immobilizing the receptor? (peripheral protein extraction, reconstitution)	19, 20
Epidermal growth factor receptor of cell surface	Erythrosin[a]	Does growth factor binding cause microaggregation of receptors? (temperature dependence, D_L measurements, fluorescence microscopy)	21
H^+-translocating ATPase of chloroplasts	Eosin	What are the effects of "membrane energization" on ATPase conformation? (photosynthetic)	22, 23
ADP/ATP antiporter of mitochondria	Eosin	Fixed to respiratory complexes?	24

[a] Studies made with phosphorescence. All other studies used linear dichroism.

[b] (1) Cone, 1972, (2) Junge and Devault, 1975, (3) Kunze and Junge, 1977, (4) Kawato *et al.*, 1980, 1981, 1982b, (5) Dixit *et al.*, 1982, (6) Garland *et al.*, 1979, (7) Richter *et al.*, 1979, (8) Kawato *et al.*, 1982a, (9) Gut *et al.*, 1982, 1983, (10) Cherry and Godfrey, 1981, (11) Cherry *et al.*, 1977, (12) Spiers *et al.*, 1983, (13) Hoffman *et al.*, 1979, (14) Hoffman *et al.*, 1980, (15) Bürkli and Cherry, 1981, (16) Cherry *et al.*, 1976a,b, (17) Nigg and Cherry, 1979, 1980, (18) Matayoshi *et al.*, 1982, (19) Lo *et al.*, 1980, (20) Bartholdi *et al.*, 1981, (21) Zidovetzki *et al.*, 1981, (22) Wagner and Junge, 1982, (23) Wagner *et al.*, 1981, 1982, 1983, (24) Muller *et al.*, 1982.

Nevertheless, the number and scope of applications are likely to be large. An analogous method using irreversible photobleaching was described by Smith *et al.* (1981), and is suited for very slow rotations (20 msec upwards).

X. APPLICATIONS

Measurements of D_R are, in themselves, not of greatly compelling interest. Nor, for reasons discussed above, is it usually possible to obtain an accurate or meaningful absolute value for the diameter of the transmembranous domain of a membrane protein. However, intelligent and imaginative use of D_R measurements in association with other biochemical and biological techniques can yield important new information. Examples of such approaches are listed in Table 2 where we emphasize the use of other techniques in conjunction with D_R measurements to address specific questions.

XI. PROSPECTS

Over the past decade, the rate of technological advance in methods for measuring D_R of membrane proteins has been fast. Contributory factors have been numerous: the increased availability of lasers, fast signal averagers and on-line computers of higher performance but lower cost, the widespread introduction of fluorescence emission microscopy for the visualization of cell-surface membrane receptor (Willingham and Pastan, 1978), the stimulus provided by developments in the measurement of D_L (Peters *et al.*, 1974; Axelrod *et al.*, 1976; Koppel *et al.*, 1976), and a growing awareness that the mobility of membrane proteins is no less important than the biological mechanisms, such as cytoskeletal interactions with the membrane, that cause immobility.

We can identify several areas where there is still scope for considerable technological improvements for the measurement of D_R. They are (1) exploration of a wider range of fluorescent triplet probes and matching laser wavelengths for the fluorescence depolarization method, (2) selection of fluorescent triplet probes that are more resistant to photochemical damage (Johnson and Garland, 1982a,b), (3) identification of conditions such as addition of free-radical scavengers that may diminish photochemical damage to fluorescent triplet probes, (4) development of phase and amplitude modulation methods appropriate for triplet lifetimes and anisotropic rotations (Valeur and Weber, 1977; Weber, 1977; Lakowicz, *et al.*, 1979), (5) development of pseudo-steady-state methods for fluorescence depletion and phosphorescence, and (6) a scaled-up two-laser fluorescence depletion method for fast-time resolution studies in the 100 nsec to 10 μsec time domain of segmental protein movements.

ACKNOWLEDGMENTS

Work reported from our own laboratory was supported by the Medical Research Council, the Science and Engineering Research Council, and the Royal Society.

REFERENCES

Albrecht, A. C., 1961, Polarisations and assignments of transitions: The method of photoselection, *J. Mol. Spectrosc.* **6**:301–334.

Albrecht, A. C., 1970, The method of photoselection and some recent applications, *Prog. React. Kinet.* **5**:301–334.

Austin, R. H., Chan, S. S., and Jovin, T. M., 1979, Rotational diffusion of cell surface components by time-resolved phosphorescence anisotropy, *Proc. Natl. Acad. Sci. USA* **76**:5650–5654.

Axelrod, D., Koppel, D. E., Schlessinger, J., Elson, E., and Webb, W. W., 1976, Mobility measurement by analysis of fluorescence photobleaching recovery kinetics, *Biophys. J.* **16**:1055–1069.

Bartholdi, M., Barrantes, F. J., and Jovin, T. M., 1981, Rotational molecular dynamics of the membrane-bound acetylcholine receptor revealed by phosphorescence spectroscopy, *Eur. J. Biochem.* **120**:389–396.

Belford, G. G., Belford, R. L., and Weber, G., 1972, Dynamics of fluorescence polarizations in macromolecules, *Proc. Natl. Acad. Sci. USA* **69**:1392–1393.

Brand, L., and Witholt, B., 1967, Fluorescence measurements, *Meth. Enzymol.* **11**:776–857.

Burkli, A., and Cherry, R. J., 1981, Rotational motion and flexibility of Ca^{2+} and Mg^{2+} dependent adenosine 5′-triphosphatase in sarcoplasmic reticulum membranes, *Biochemistry* **20**:138–145.

Chapman, D., and Restall, C. J., 1982, Rotational and lateral movements in biomembranes: The dynamics of biomembrane components, *Biochem. Soc. Symp.* **46**:139–154.

Cherry, R. J., 1978, Measurement of protein rotational diffusion in membranes by flash photolysis, *Meth. Enzymol.* **54**:47–61.

Cherry, R. J., 1979, Rotational and lateral diffusion of membrane proteins, *Biochim. Biophys. Acta* **559**:289–327.

Cherry, R. J., and Godfrey, R. E., 1981, Anisotropic rotation of bacteriorhodopsin in lipid membranes. Comparison of theory with experiment, *Biophys. J.* **36**:257–276.

Cherry, R. J., and Schneider, G., 1976, A spectroscopic technique for measuring slow rotational diffusion of macromolecules. 2: Determination of rotational correlation times of proteins in solution, *Biochemistry* **15**:3657–3661.

Cherry, R. J., Cogoli, A., Oppliger, M., Schneider, G., and Semenza, G., 1976a, A spectroscopic technique for measuring slow rotational diffusion of macromolecules. I: Preparation and properties of a triplet probe, *Biochemistry* **15**:3653–3656.

Cherry, R. J., Burkli, A., Busslinger, M., Schneider, G., and Parish, G. R., 1976b, Rotational diffusion of band 3 proteins in the human erythrocyte membrane, *Nature (London)* **263**:389–393.

Cherry, R. J., Muller, U., and Schneider, G., 1977, Rotational diffusion of bacteriorhodopsin in lipid membranes, *FEBS Lett.* **80**:465–469.

Chuang, T. J., and Eisenthal, K. B., 1972, Theory of fluorescence depolarization by anisotropic rotational diffusion, *J. Chem. Phys.* **57**:5094–5097.

Cone, R. A., 1972, Rotational diffusion of rhodopsin in the visual receptor membrane, *Nature (London) New Biol.* **236**:39–43.

Dixit, R. P. S., Waring, A. J., Wells, K. O., Wong, P. S., Woodrow, G. V., and Vanderkooi, J. M., 1982, Rotational motion of cytochrome *c* derivatives bound to membranes measured by fluorescence and phosphorescence anisotropy, *Eur. J. Biochem.* **126**:1–9.

Ehrenberg, M., and Rigler, R., 1976, Fluorescence correlation spectroscopy applied to rotational diffusion of macromolecules, *Quart. Rev. Biophys.* **9**:69–81.

Erecinska, M., Wilson, D. F., and Blasie, J. K., 1978, Studies on the orientation of the mitochondrial redox carriers. I. Orientation of the hemes of cytochrome *c* oxidase with respect to the plane of a cytochrome oxidase lipid model membrane, *Biochim. Biophys. Acta* **501**:53–62.

Frye, L. D., and Edidin, M., 1970, The rapid intermixing of cell surface antigens after formation, *J. Cell Sci.* **7**:319–335.

Garland, P., 1981, Letter to the editor. Fluorescence photobleaching recovery: Control of laser intensities with an acousto-optic modulator, *Biophys. J.* **33**:481–482.

Garland, P. B., and Johnson, P., 1983, Rotational diffusion measured by depolarization of fluorescence depletion, in: *Spectroscopy and the Dynamics of Biological Systems* (P. Bailey and R. E. Dale), in press.

Garland, P. B., and Moore, C. H., 1979, Phosphorescence of protein-bound eosin and erythrosin. A possible probe for measurements of slow rotational mobility, *Biochem. J.* **183**:561–572.

Garland, P. B., Davison, M. T., and Moore, C. H., 1979, Rotational mobility of membrane-bound cytochrome *o* of *Escherichia coli* and cytochrome *a*1 of *Thiobacillus ferro-oxidans*, *Biochem. Soc. Trans.* **7**:1112–1113.

Gouterman, M., and Stryer, L., 1962, Fluorescence polarization of some porphyrins, *J. Chem. Phys.* **37**:2260–2266.

Greinhert, R., Staerk, H., Stier, A., and Weller, A., 1979, E-type delayed fluorescence depolarization, a technique to probe rotational motion in the microsecond range, *J. Biochem. Biophys. Meth.* **1**:77–83.

Gut, J., Richter, C., Cherry, R. J., Winterhalter, K. H., and Kawato, S., 1982, Rotation of cytochrome P-450. II. Specific interactions of cytochrome P-450 with NADPH-cytochrome P-450 reductase in phospholipid vesicles, *J. Biol. Chem.* **257**:7030–7036.

Gut, J., Richter, C., Cherry, R. J., Winterhalter, K. H., and Kawato, S., 1983, Rotation of cytochrome P-450. Complex formation of cytochrome P-450 with NADPH-cytochrome P-450 reductase in liposomes, *J. Biol. Chem.* **258**:8588–8594.

Henderson, R., and Unwin, P. N. T., (1975), Three-dimensional model of purple membrane obtained by electron microscopy, *Nature (London)* **257**:28–32.

Heyn, M. P., Cherry, R. J., and Dencher, N. A., 1981, Lipid-protein interactions in bacteriorhodopsin-dimyristoylphosphatidylcholine vesicles, *Biochemistry* **20**:840–849.

Hoffman, W., Sarzala, M. G., and Chapman, D., 1979, Rotational motion and evidence for oligomeric structures of sarcoplasmic reticulum Ca^{2+}-activated ATPase, *Proc. Natl. Acad. Sci. USA* **76**:3860–3864.

Hoffman, W., Sarzala, M. G., Gomezfernandez, J., Goni, F. M., Restall, C. J., Chapman, D., 1980, Protein rotational diffusion and lipid structure of reconstituted systems of Ca^{2+}-activated adenosine triphosphatase, *J. Mol. Biol.* **141**:119–132.

Inbar, M., Shinitzky, M., and Sachs, L., 1973, Rotational relaxation time of concanavalin A bound to the surface membrane of normal and malignant transformed cells, *J. Mol. Biol.* **81**:245–253.

Jablonski, V. A., 1961, Uber die abklingungsvorgange polarisierter photolunineszenz, *Aeitschrift Naturforsch.* **16**:1–4.

Johnson, P., 1983, The lateral and rotational mobility of membrane components measured by fluorescence recovery after photobleaching and fluorescence depletion recovery, Ph.D. thesis, University of Dundee.

Johnson, P., and Garland, P. B., 1981, Depolarization of fluorescence depletion: A microscopic method for measuring rotational diffusion of membrane proteins on the surface of a single cell, *FEBS Lett.* **132**:252–256.

Johnson, P., and Garland, P. B., 1982a, Fluorescent triplet probes for measuring the rotational diffusion of membrane proteins, *Biochem. J.* **203**:313–321.

Johnson, P., and Garland, P. B., 1982b, Carbocyanine dyes used as fluorescent triplet probes for measuring slow rotational diffusion of lipids in membranes, *Biochem. J.* **203**:313–394.

Jovin, T. M., Bartholdi, M., Vaz, W. L. C., and Austin, R. H., 1981, Rotational diffusion of biological macromolecules by time-resolved delayed luminescence (phosphorescence, fluorescence) anisotropy, *Ann. N.Y. Acad. Sci. USA* **399**:176–196.

Junge, W., 1972, Brownian rotation of the cytochrome oxidase in the mitochondrial membrane, *FEBS Lett.* **25**:109–112.

Junge, W., and Devault, D., 1975, Symmetry, orientation and rotational mobility in the a_3 heme of cytochrome *c* oxidase in the inner membrane of mitochondria, *Biochem. Biophys. Acta* **408**:200–214.

Kawato, S., and Kinosita, K., 1981, Time dependent absorption anisotropy and rotational diffusion of proteins in membranes, *Biochem. J.* **36**:227–296.

Kawato, S., Sigel, E., Carafoli, E., and Cherry, R. J., 1980, Cytochrome oxidase rotates in the inner membrane of intact mitochondria and submitochondrial particles, *J. Biol. Chem.* **255**:5508–5510.

Kawato, S., Sigel, E., Carafoli, E., and Cherry, R. J., 1981, Rotation of cytochrome oxidase in phospholipid vesicles. Investigations of interactions between cytochrome oxidases and between cytochrome oxidase and cytochrome bc_1 complex, *J. Biol. Chem.* **256**:7518–7527.

Kawato, S., Gut, J., Cherry, R. J., Winterhalter, K. H., and Richter, C., 1982a, Rotation of cytochrome P-450. I. Investigation of protein–protein interactions of cytochrome P-450 in phospholipid vesicles and liver microsomes, *J. Biol. Chem.* **257**:7023–7029.

Kawato, S., Lehner, C., Muller, M., and Cherry, R. J., 1982b, Protein-protein interactions of cytochrome oxidase in inner mitochondrial membranes. The effect of liposome fusion on protein rotational mobility, *J. Biol. Chem.* **257**:6470–6476.

Koppel, D. E., Axelrod, D., Schlessinger, J., Elson, E. L., and Webb, W. W., 1976, Dynamics of fluorescence marker concentration as a probe of mobility, *Biophys. J.* **16:**1315–1329.

Kunze, U., and Junge, W., 1977, Ellipticity of cytochrome a_3 and rotational mobility of cytochrome *c*-oxidase in the cristae membrane of mitochondria, *FEBS Lett.* **80:**429–434.

Lakowicz, J. R., Prendergast, F. G., and Hogan, D., 1979, Differential polarized phase fluorometric investigations of diphenylhexatriene in lipid bilayers. Quantitation of hindered depolarizing rotations, *Biochemistry* **18:**508–519.

Lavalette, D., and Amand, B., and Pochon, F., 1977, Rotational relaxation of 70S ribosome by a depolarization method using triplet probes, *Proc. Natl. Acad. Sci. USA* **74:**1407–1411.

Lindmo, T., and Steen, H. B., 1977, The effect of numerical aperture of detector optics on polarization values, *Biophys. J.* **18:**173–187.

Lo, M. M. S., Garland, P. B., Lamprecht, J., and Barnard, E. A., 1980, Rotational mobility of the membrane-bound acetylcholine receptor of *Torpedo* electric organ measured by phosphorescence depolarization, *FEBS Lett.* **111:**407–412.

Mar, T., Picorel, R., and Gingras, G., 1981, Rotational mobility of the photoreaction center in chromophore membranes of *Rhodospirillum rubrum, Biochim. Biophys. Acta* **637:**546–550.

Matayoshi, E. D., Corin, A. F., Zidovetzki, R., Sawyer, W. H., and Jovin, T. M., 1982, Rotational dynamics of cell surface proteins, in: *Proceedings of the FEBS Symposium Konstanz. Mobility and Recognition in Cell Biology.* (H. Sund and C. Veeger, eds.), Walter de Gruyter & Co., Berlin.

Moore, C. H., and Garland, P. B., 1979, Synthesis of erythrosin isothiocyanate and its use as a phosphorescence depolarization probe for slow rotational mobility of membrane proteins, *Biochem. Soc. Trans.* **7:**945–946.

Moore, C., Boxer, D., and Garland, P., 1979, Phosphorescence depolarization and the measurement of rotational motion of proteins in membranes, *FEBS Lett.* **108:**161–166.

Muller, M., Krebs, J. J. R., Cherry, R. J., and Kawato, S., 1982, Selective labeling and rotational diffusion of the ADP/ATP translocator in the inner mitochondrial membrane, *J. Biol. Chem.* **257:**1117–1120.

Murray, E. K., Restall, C. J., and Chapman, D., 1983, Monitoring membrane protein rotational diffusion using time-averaged phosphorescence, *Biochim. Biophys. Acta* **732:**347–351.

Naqvi, K. R., and Wild, U. P., 1975, The use of E-type delayed fluorescence for probing rotational relaxation, *Chem. Phys. Lett.* **36:**222–224.

Naqvi, K. R., Gonzales, R. J., Cherry, R. J., and Chapman, D., 1973, Spectroscopic technique for studying protein rotation in membranes, *Nature (London) New Biol.* **245:**249–251.

Nigg, E. A., and Cherry, R. J., 1979, Influence of temperature and cholesterol on the rotational diffusion of band 3 in the human erythrocyte membrane, *Biochemistry* **18:**3457–3465.

Nigg, E. A., and Cherry, R. J., 1980, Anchorage of a band 3 population at the erythrocyte cytoplasmic membrane surface: Protein rotational diffusion measurements, *Proc. Natl. Acad. Sci. USA* **77:**4702–4706.

Perrin, P. F., 1926, Polarisation de la lumiere de fluorescence. Vie moyenne des molecules dans l'etat excite, *J. Phys. Rad.* **7:**390.

Perrin, P. F., 1929, La fluorescence des solutions. Induction moleculaire.—polarization et duree d'emission-photochimie, *Annde Phys.* **12:**169–275.

Peters, R., and Cherry, R. J., 1982, Lateral and rotational diffusion of bacteriorhodopsin in lipid bilayers: An experimental test of the Saffmann–Delbrück equations, *Proc. Natl. Acad. Sci. USA* **79:**4317–4321.

Peters, R., Peters, J., Tews, K. H., and Bahr, W., 1974, A microfluorometric study of translational diffusion in erythrocyte membranes, *Biochim. Biophys. Acta* **367:**282–294.

Pilipovich, V. A., 1961, Polarization of phosphorescence in organophors, *Opt. Spectrosc.* **10:**104–107.

Richter, C., Winterhalter, K. H., and Cherry, R. J., 1979, Rotational diffusion of cytochrome P-450 in rat liver microsomes, *FEBS Lett.* **102:**151–154.

Rigler, R., and Ehrenberg, M., 1973, Molecular interactions and structure as analysed by fluorescence relaxation spectroscopy, *Quart. Rev. Biophys.* **6:**139–199.

Saffman, R. G., and Delbrück, M., 1975, Brownian motion in biological membranes, *Proc. Natl. Acad. Sci. USA* **72:**3111–3113.

Shinitzky, M., Inbar, M., and Sachs, L., 1973, Rotational diffusion of lectins bound to the surface membrane of normal lymphocytes, *FEBS Lett.* **34:**247–250.

Singer, S., and Nicholson, G. L., 1972, The final mosaic model of the structure of cell membranes, *Science* **175:**720–731.

Smith, L., Weis, R. M., and McConnell, H. M., 1981, Measurement of rotational motion in membranes using fluorescence recovery after photobleaching, *Biophys. J.* **36**:73–91.

Spiers, A., Moore, C. H., Boxer, D. H., and Garland, P. B., 1983, Segmental motion and rotational diffusion of the Ca^{2+}-translocating ATPase of sarcoplasmic reticulum, measured by time-resolved phosphorescence depolarization, *Biochem. J.* **213**:67–74.

Strambini, G. B., and Galley, W. C., 1976, Detection of slow rotational motions of proteins by steady-state phosphorescence anisotropy, *Nature (London)* **260**:554–556.

Strambini, G. B., and Galley, W. C., 1980, Time-dependent phosphorescence anisotropy measurements of the slow rotational motions of proteins in viscous solution, *Biopolymers* **19**:383–394.

Teale, F. W. J., 1969, Fluorescence depolarization by light-scattering in turbid solutions, *Photobiochem. Photobiophys.* **10**:363–374.

Trauble, H., and Sackmann, E., 1973, Lipid motion and rhodopsin rotation, *Nature (London)* **245**:210–211.

Valeur, B., and Weber, G., 1977, Anisotropic rotations in 1-naphthylamine. Existence of a red-edge transition moment normal to the ring plane, *Chem. Phys. Lett.* **45**:140–144.

Vaz, W. L. C., Austin, R. H., and Vogel, H., 1979, The rotational diffusion of cytochrome b_5 in lipid bilayer membranes. Influence of the lipid physical state, *Biophys. J.* **26**:415–426.

Vaz, W. L. C., Criado, M., Madeira, V. M. C., Schoellmann, G., and Jovin, T. M., 1982, Size dependence of the translational diffusion of large integral membrane proteins in liquid–crystalline phase lipid bilayers. A study using FRAP, *Biochemistry* **21**:5608–5612.

Wagner, R., and Junge, W., 1982, Coupling factor for photophosphorylation labelled with eosin isothiocyanate: Activity, size and shape in solution, *Biochemistry* **21**:1890–1899.

Wagner, R., Carrillo, N., Junge, W., and Vallejos, R. H., 1981, Heat-activated conformational changes of isolated coupling factor of photophosphorylation CF1, *FEBS Lett.* **136**:208–212.

Wagner, R., Carillo, N., Junge, W., and Vallejos, R. H., 1982, On the conformation of reconstituted ferredoxin: NADP oxidoreductase in the thylakoid membrane. Studies via triplet lifetime and rotational diffusion with eosin isothiocyanate as label, *Biochim. Biophys. Acta* **680**:317–330.

Wagner, R., Andreo, C., and Junge, W., 1983, Evidence for a sequestered solvent space in the chloroplast ATPase, *Biochim. Biophys. Acta* **723**:123–127.

Wahl, P., Kasai, M., and Changeux, J. P., 1971, A study on the motion of proteins in excitable membrane fragments by nanosecond fluorescence polarization spectroscopy, *Eur. J. Biochem.* **36**:257–276.

Weber, G., 1953, Rotation, Brownian motion and polarization of the fluorescence of solutions, *Adv. Protein Chem.* **8**:415–459.

Weber, G., 1977, Theory of differential phase fluorometry: Detection of anisotropic molecular rotations, *J. Chem. Phys.* **66**:4081–4091.

Willingham, M. C., and Pastan, A., 1978, The visualization of fluorescent proteins in living cells by video intensification microscopy, *Clin. Endocrinol.* **13**:501–507.

Zidovetzki, R., Yarden, Y., Schlessinger, J., and Jovin, T. M., 1981, Rotational diffusion of epidermal growth factor complexed to cell surface receptors reflects rapid microaggregation and endocytosis of occupied receptors, *Proc. Natl. Acad. Sci. USA* **78**:6981–6985.

Index

GPSR Compliance

The European Union's (EU) General Product Safety Regulation (GPSR) is a set of rules that requires consumer products to be safe and our obligations to ensure this.

If you have any concerns about our products, you can contact us on ProductSafety@springernature.com

In case Publisher is established outside the EU, the EU authorized representative is:

Springer Nature Customer Service Center GmbH
Europaplatz 3
69115 Heidelberg, Germany

Batch number: 10838292

Printed by Printforce, the Netherlands